D0944934

DIGITAL COMPUTER TREATMENT OF PARTIAL DIFFERENTIAL EQUATIONS

V. VEMURI
State University of New York at Binghamton

WALTER J. KARPLUS
University of California at Los Angeles

Prentice-Hall, Inc., Englewood Cliffs, New Jersey 07632

Library of Congress Cataloging in Publication Data

VEMURI, V.
Digital computer treatment of partial differential equations.

(Prentice-Hall series in computational mathematics)
Includes bibliographies and index.
1. Differential equations, Partial—Numerical solutions—Data processing. 2. Distributed parameter systems—Data processing. I. Karplus, Walter J. II. Title. III. Series.
QA374.V45 515.3′53′02854 81-5223
ISBN 0-13-212407-6 AACR2

Prentice-Hall Series in Computational Mathematics
Cleve Moler, Advisor

Editorial/Production Supervision: *Lynn S. Frankel*
Interior Design: *Karen J. Clemments*
Cover Design: *Edsal Enterprises*
Manufacturing Buyer: *Gordon Osbourne*

Printed in the United States of America

10 9 8 7 6 5 4 3 2 1

Prentice-Hall International, Inc., *London*
Prentice-Hall of Australia Pty. Limited, *Sydney*
Prentice-Hall of Canada, Ltd., *Toronto*
Prentice-Hall of India Private Limited, *New Delhi*
Prentice-Hall of Japan, Inc., *Tokyo*
Prentice-Hall of Southeast Asia Pte. Ltd., *Singapore*
Whitehall Books Limited, *Wellington, New Zealand*

To
Uma and Takako

Contents

||5|| BOUNDARY VALUE PROBLEMS *109*

||6|| INITIAL VALUE PROBLEMS *139*

||7|| SPECIAL TOPICS *175*

||8|| INTERPOLATION AND APPROXIMATION *193*

Preface

Field problems involving the analysis, design, and control of distributed parameter systems constitute the heart of virtually every physical science and engineering discipline. Such important contemporary problems as the control of air and water pollution and the prediction of earthquakes fall into this category. More "classical" field problems include the analysis of heat transfer, the flow of fluids, and many more. All these problems are studied with the aid of mathematical models that take the form of partial differential equations (often referred to in this book as PDEs).

Because most fields of interest to scientists and engineers manifest irregular boundaries, nonuniform parameter distributions, and nonlinearities, it is generally impossible to obtain analytical solutions to the characterizing PDEs. Numerical methods are therefore used almost exclusively to treat such problems. Prior to the 1960s, analog simulation methods were widely used. In recent years, however, these have been almost completely supplanted by the digital computer, which is now considered virtually the only viable tool for solving field problems.

It is the purpose of this text to provide the reader with a comprehensive view of alternative digital computer methods to the treatment of field problems characterized by PDEs as well as an insight into the underlying mathematical issues. Because of the inherent practical difficulties involved in obtaining meaningful solutions to these equations, the reader is also exposed to the physical concepts that enter into the derivation of these equations or, more generally, the mathematical modeling of a physical system.

This book is directed at first-year graduate students in engineering or in the physical sciences, as well as individuals in government or industry with specific field problems to solve. The reader is assumed to have at least an elementary background in the principles of linear algebra, particularly matrix notation, and of calculus. Familiarity with the elements of numerical analysis, including such topics as interpolation and numerical quadrature, while useful, is definitely not essential. It is also assumed that the reader has an acquaintance with the utilization of digital computers to solve scientific problems. In particular, some understanding of the FORTRAN language is essential to understand the sample programs that have been provided.

The book is organized into three major parts. Part I, "Formulations," is a discussion of the origin of mathematical models and more particularly the manner in which PDEs characterize physical systems. In Part II, "Transformations," the accent is on changing these PDEs into systems of algebraic equations that can actually be handled by a digital computer. The solution of these algebraic equations using a digital computer is considered in detail in Part III, "Computations."

More specifically, Chapter 1 is intended to provide the reader with a perspective on the subject of modeling and simulation. In Chapter 2 it is shown how the major partial differential equations, as well as various modified forms of these equations, are derived from basic physical principles. It is demonstrated that the general class of PDEs characterizing a physical system, whether these equations be elliptic, parabolic, hyperbolic, or biharmonic partial differential equations, is determined entirely by the physical properties of the medium being modeled, regardless of the specific application area into which the field falls. Such a unified approach provides the student with an insight into the dynamic characteristics of the computer solution. Chapter 3 is devoted to the formulation of the mathematical model in more formal mathematical terms, which in turn provides a starting point for the discussions of the succeeding chapters.

The major objective of Part II is to demonstrate how the PDEs formulated in Part I can be translated into sets of algebraic equations suitable for digital computer solution. In Chapter 4, the classical Taylor series method for deriving finite difference approximations is presented. The importance of symmetry and positive definiteness in the characterizing matrix is discussed in some detail. In Chapter 5, boundary value problems are treated using the self-adjoint elliptic partial differential equation as a vehicle. It is pointed out that even a self-adjoint operator fails to yield a symmetric difference operator if either the boundaries are irregular or the grid is nonuniform. Two methods of arriving at symmetric difference equations are presented. These methods are in fact very similar to those used in the finite element method. Thus Chapter 5 paves the way for the subsequent treatment of finite element methods without an extensive recourse to concepts from structural mechanics. Initial value problems associated with parabolic and hyperbolic partial differential equations are collectively treated in Chapter 6. Here the major emphasis is on a variety of practical finite difference schemes. In Chapter 7 several important special situations are treated, including particularly nonlinearities, singularities and moving boundaries. The basic ideas of interpolation and approximation in one and two dimensions are presented in Chapter 8 while the finite element method is presented from a number of viewpoints in Chapter 9.

Material in Part III, "Computations," has been gathered and arranged to give the student a flavor of what can be expected in a realistic problem solving environment. Toward this end, Chapter 10 contains a review of direct and iterative methods for solving simultaneous linear algebraic equations. Chapters 11 and 12 constitute an attempt to introduce the reader to the impact upon the solution of field problems to be expected from major hardware and software advances occurring since the late 1970s. In this respect, Chapter 11 is an attempt to summarize the state of the art in software packages intended for solving PDEs while Chapter 12 focuses on the use of parallelism in the architecture of digital computer systems, a major departure from classical or Von Neumann architecture. Finally, in the book's last two chapters, the process of generating a numerical solution to a typical problem from beginning to end is demonstrated. It is useful to note that Chapters 11 and 12 represent the state of the art and could therefore appeal to the research-oriented student while the tutorial nature of Chapters 13 and 14 would appeal more to the beginner.

A comment on the use of this book as a text is in order. All material can be covered at a comfortable speed in two semesters. If the interest is primarily in finite difference methods, Chapters 1 to 7 and Chapters 10 to 13 would form a coherent unit. If the interest is primarily in finite element methods, then Chapters 1 to 3, Chapters 8 to 10, and Chapters 12 and 14 would be a logical unit.

Finally, it is time for an expression of our appreciation to a number of individuals. We are indebted to Granino Korn and James M. Hyman who read initial drafts of the manuscript and made a number of useful suggestions; to D. Greenspan for his kind permission to let us use the example problem discussed in Chapter 13; and to a number of authors who readily gave permission to use material from their works. Some of the programs in Chapter 13 were written by George Spisak, and Michael Adler helped in the preparation of the index. A note of thanks is also due to Karl Karlstrom, Lynn Frankel, and all others at Prentice-Hall, Inc. who contributed to the production of this book. Finally, we would like to express our appreciation to our families for all their patience and understanding.

V. VEMURI
W. J. KARPLUS

To the Reader

WHERE TO FIND ADDITIONAL INFORMATION

The novice embarking on a serious study of computational methods to solve PDEs should know where to look for up-to-date information. The subject matter of interest may be found under any of the following broad categories.

1. Finite difference methods
2. Finite element methods
3. Matrix methods in structural mechanics
4. Analog and hybrid computer methods
5. Supercomputers and networks of microcomputers

Most of the material in these categories is widely scattered in textbooks, technical journals, proceedings of conferences and symposia, and reports available from NTIS (National Technical Information Service, Springfield, Virginia, 22151).

As the finite element method started off with structural engineers, most of the early work on this subject can be found in structural engineering literature. For example, reports of government agencies and other research groups are indexed and abstracted in STAR (*Scientific and Technical Aerospace Reports*) and GRA/GRI (*Government Reports Announcements* and *Government Reports Index*).

Most of the early finite difference work was done in a nonstructural mechanics context with emphasis on applications pertaining to nuclear reactor designs, geological explorations, and the like, and the related literature is scattered throughout the mathematical and engineering literature.

A good source to look at for the latest developments is *Dissertation Abstracts International* in which doctoral dissertations are indexed. For pursuit of ongoing developments in a particular specialty, the *Science Citation Index* may be helpful; for a specific paper it cites subsequent publications that have used that paper as a reference. Computer-assisted literature searches are being offered by a number of agencies such as NTIS, the National Aeronautics and Space Administration, and Defense Documentation Center. The best source to get information on what is available at your institution is your librarian. Today, many university libraries are offering literature search services.

Listings of computer programs are frequently contained in reports. The National Aeronautics and Space Administration has begun a quarterly journal, *Computer Program Abstracts*, to announce documented computer programs offered for public sale. This journal is available from Superintendent of Documents, U.S. Government Printing Office, Washington, D.C. 20402. Program decks, tapes, and documents are also sold by COSMIC (Barrow Hall, University of Georgia, Athens, Georgia 30601).

Finally, the student should remember that there are tremendous time lags involved in the publication of results. It may take more than a year from the time a researcher finished writing a paper to the time it appears in a journal. It may take a like amount of time from the time it gets cited by one of the abstracting services. The serious student should always be on the lookout. The following are a representative set of journals where one is likely to find relevant and timely information.

S. I. A. M. Journal of Numerical Analysis

Numerische Mathematik

Mathematics of Computation

Journal of Computational and Applied Mathematics

Computers and Mathematics

Computer Journal

Computational Physics

Computers and Structures

International Journal for Numerical Methods in Engineering

ACM Transactions on Mathematical Software

Mathematics and Computers in Simulation

Advances in Computer Methods for Partial Differential Equations
(Proceedings of International Symposia on Computer Methods for Partial Differential Equations, R. Vichnevetsky, editor, Rutgers University, New Brunswick, New Jersey, 08903)

Of course, there is nothing comparable to maintaining personal contacts with colleagues in the field.

I

FORMULATIONS

1

Mathematical Models and the Computer

1.1. MODELS, SIMULATION, AND PROBLEM SOLVING

This book is concerned with digital computer methods for solving partial differential equations, but not any and every partial differential equation that may come to the mind of a mathematician. Rather our attention is limited to certain classes of partial differential equations: those which are of interest to scientists and engineers because they characterize or serve as mathematical models for natural phenomena. These phenomena occur in a vacuum or a solid, liquid, or gaseous medium, a region of space termed a ***field***. For this reason, scientific or engineering problems involving partial differential equations are often called ***field problems***. The physical areas in which field problems arise include heat transfer, fluid mechanics, electrostatics, electromagnetics, acoustics, particle diffusion, to mention just a few. In each of these areas the space variables, and sometimes time, serve as independent variables, while the dependent problem variables can usually be described either as flux or potential.

A scientist or engineer turns to the computer to solve field problems with some specific objectives in mind. He or she may wish to design a device to determine how best to control the environment, to predict events within the field in the future, and so on. The digital computer solution of field problems, which constitutes the principal subject of this book, is in fact but one link in a chain of activities. At one end of this chain lies the scientific or engineering problem in need of solution, while the ultimate utilization or application of the solution lies at the other end. The following are some of the important elements of the problem-solving methodology:

1. ***Problem specification:*** to create a clear and concise, though not necessarily quantitative, specification of what is given and what is to be found.
2. ***Mathematical modeling:*** to translate the problem into formal mathematical terms, usually as a system of equations.
3. ***Mathematical transformations and approximations:*** to make the problem more amenable to solution.
4. ***Solution generation:*** to get analytical, numerical, or graphical outputs that constitute an acceptable compromise between the desired accuracy and the cost of the computational effort.
5. ***Verification:*** to determine that the computational tool is correctly solving the equations comprising the mathematical model.
6. ***Validation:*** to determine that the mathematical model as implemented adequately represents the problem to be solved.
7. ***Simulation:*** a succession of solution runs to provide the answer to problems as formulated.
8. Final interpretation and utilization.

The effective use of a computer as a problem-solving tool requires that the analyst have an insight into all the above stages of the problem-solving process. Only in this way can there be a degree of assurance that the computer outputs constitute meaningful and useful solutions to a specific problem. Toward this end, this book is organized into three parts. Part I, "Formulations," is a discussion of the origin of mathematical models and more particularly the manner in which partial differential equations characterize physical systems. In Part II, "Transformations," the accent is on changing these partial differential equations into systems of algebraic equations that can actually be handled by a digital computer. The solution of these algebraic equations using a digital computer is considered in detail in Part III, "Computations." The rest of this chapter is intended to provide the reader with a general perspective on the subject of modeling and simulation on the one hand and on the role of computers as working tools in the history of modeling and simulation on the other.

1.2. DIRECT AND INVERSE SYSTEMS PROBLEMS

A physical system is defined as a physical entity with clearly definable boundaries, capable of being studied by means of physical measurements, and that reacts to external stimuli in a known or predictable manner. The term physical system includes electric circuits, linkages of mechanical elements, thermal conductors, the fields surrounding antennas, and so on. Fundamental to all physical systems theory is the cause-and-effect relationship between excitation (inputs) and response (outputs). An excitation generally takes the form of matter or energy applied to the system, while the response describes the resulting steady-state or transient mass or energy transfers within the system and at its boundaries. This is illustrated in general form in Figure 1.1a, where the excitations are represented by E, the physical system by S, and the responses by R. Using an electrical circuit as an example, the system S might be a network of resistors, inductors, and capacitors. The excitation E might be expressions of the type $e = f(t)$, characterizing transient voltage sources acting upon the network. The response R would then be functions of time characterizing the voltages or currents as measured at specified locations in the system.

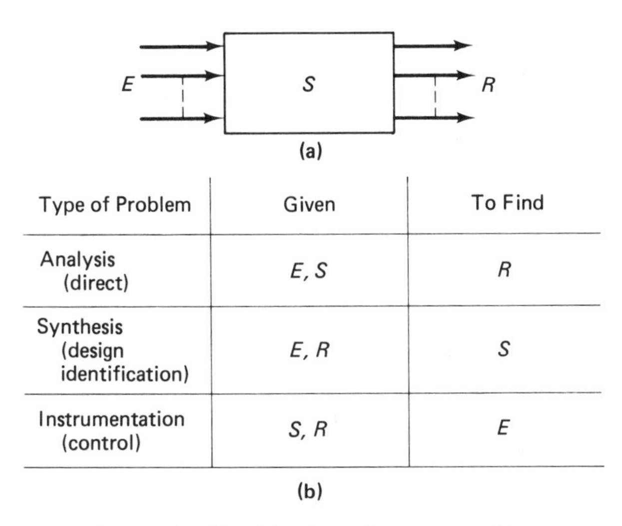

Type of Problem	Given	To Find
Analysis (direct)	E, S	R
Synthesis (design identification)	E, R	S
Instrumentation (control)	S, R	E

(b)

Figure 1-1 Classification of systems problems

The problems confronted by system scientists or engineers can be classified broadly according to which two of these three items, E, S, and R, are given and which one is to be found. As shown in Figure 1.1b, in analysis problems the system and the excitation are specified, and the response is to be determined. In a synthesis problem the excitation and response are given, and the system having this excitation-response relationship is to be designed or realized. The third type of problem, sometimes called the instrumentation or control problem, involves the determination of the excitation given the system and the response. The analysis problem is termed a ***direct*** problem, whereas the synthesis and instrumentation problems are termed ***inverse*** problems. While a direct problem generally has a unique solution, that is, for a given excitation and system configuration there can exist only one set of responses, the same is not true of an inverse problem. In fact, there always exists an infinite number of correct solutions to inverse problems. For example, an infinite variety of combinations of resistors will draw a current of 1 ampere from a 1-volt source: a 1-ohm resistor, two 2-ohm resistors in parallel, two $\frac{1}{2}$-ohm resistors in series, and so on. The solution of such a synthesis problem therefore involves not only the determination of a system manifesting the specified excitation-response relationship, but also the selection of one "optimum" S from the infinitude of possible solutions. To this end, the solution of an inverse problem requires the imposition of separately specified constraints such as minimum cost, minimum weight, etc.

As discussed in some detail next, mathematical modeling involves a combination of system analysis and system synthesis. The differences in the nature of the mathematical models as they arise and are used in various application areas can be related directly to the relative amount of analysis and synthesis involved.

1.3. MODELING METHODOLOGY

The construction of a mathematical model is based upon a number of fundamental assumptions. The basic tenets of mathematical modeling have been so deeply inculcated upon its practitioners that they are rarely if ever questioned in practice. It is useful to review briefly the basic assumptions or paradigms that are the foundations of all mathemetical models. Models can be considered credible or valid only to the extent that the system being modeled satisfy these successively more, restrictive conditions.

Of fundamental importance in all systems studies is the notion of ***separability***. To some extent all objects and phenomena in the universe are interrelated. In defining a system for modeling and analysis, however, it is assumed that most of these interactions can be ignored, so that the system can be studied as a separate entity. This separation frequently involves the definition of a "boundary" of the system or an enumeration of all the elements or components comprising the system. Thus in making a mathematical representation of the behavior of an electric circuit, it is assumed that the circuit is not linked to the outside world by leakage resistance or electromagnetic coupling or other phenomena other than those explicitly specified.

Once a system has been defined as a separate entity, it is necessary to describe the interaction of the system with the external world. To arrive at a model of manageable proportions, it is necessary to invoke a ***selectivity*** condition. That is, it is necessary to assume that of all possible interactions, only a relatively small subset is relevant to a specific inquiry or purpose. For example, in modeling an electrical circuit, we ignore thermal, acoustic, optical, and mechanical interactions and consider only electrical variables. The selectivity condition therefore permits the description of system behavior in terms of a limited number of inputs and outputs.

Having invoked selectivity, it is now necessary in mathematical modeling to assume ***causality***. That is, it is assumed that the inputs and outputs of the system are functionally related, that the outputs are "caused" by the inputs. The subject of causality has been widely discussed by philosophers and all types of scientists. It continues to be a challenging and perplexing problem. In general, causality can only be assumed if it is possible to identify a complete chain of causally related events linking inputs and outputs. It is not enough to observe that a given output invariably follows a given input, since both may be responses to a common cause.

A mathematical model then is a set of equations that characterizes a real-life system, the prototype system, in the sense that some of the excitation-response relationships of the prototype system are correctly represented. A selected subset of all the prototype system inputs is expressed mathematically and serves as the excitation of the mathematical model; the solutions of the model equations then constitute mathematical representations of the corresponding (causally related) subset of system responses.

The construction of the mathematical model of a system entails the utilization of two kinds of information: (1) knowledge and insight about the system being modeled, and (2) experimental data constituting observations of system inputs and outputs. The utilization of the former class of information involves ***deduction***, while modeling using empiric observations involves ***induction***. The process of constructing a mathematical model purely deductively constitutes the solution of a system analysis problem, and hence provides a unique solution of the modeling problem. By contrast, the construction of a mathematical model by inductive methods constitutes a system synthesis problem, so that there exist always an infinite number of models satisfying the observed input-output relationship. In inductive modeling it is therefore necessary to introduce additional assumptions or constraints to help select the optimum model from the infinitude of possible models. For this reason, in constructing a mathematical model an attempt is always made to carry deduction as far as possible, and to rely upon system observations and experiments only when absolutely necessary.

Deduction is reasoning from known principles to deduce an unknown; it proceeds from the general to the specific. In modeling deductively, one derives the mathematical model analytically and uses experimental observations only to fill in certain gaps. This analytical process makes use of a series of progressively more specific concepts, which may be broadly categorized as laws, structure, and parameters.

Laws. Laws are the basic principles that determine the general nature of the equations characterizing the system. In physical systems, these laws are usually expressions of the principles of conservation and continuity. The application of these laws permits the derivation of elliptic, parabolic, or hyperbolic partial differential equations in distributed systems (fields); it leads to systems of first- and second-order ordinary differential equations in circuits, and it leads to systems of algebraic equations for systems in steady state. These basic laws are usually formulated for specific application areas and include such well-known principles as Kirchhoff's laws, Newton's laws, Fourier's law, Maxwell's equations, and the Navier-Stokes equations, to mention just a few. The application of a law to a system usually involves the focusing of attention upon a single physical area. For example, when utilizing Kirchhoff's laws or Maxwell's equations in analyzing an electrical system, one ignores chemical and thermal processes that may be going on simultaneously within the system.

Structure. The system being modeled is usually regarded as consisting of a large number of interconnected components or elements. This view of a system frequently involves the making of simplifying assumptions or approximations. However, because this approach lends itself so well to eventual computer implementations, even systems that are continuous in space (as, for example, a water reservoir or the atmosphere) are often viewed as being comprised of an array of closely spaced elements. In physical systems these elements can be broadly classified as being either dissipators of energy, reservoirs of potential, or reservoirs of flux. In nonphysical systems there is a wider assortment of possible elements. In any event, the construction of a valid mathematical model demands a knowledge of the types of elements that are present in the system and how these elements are interconnected. The interconnections specify the paths over which matter or energy flows within the system; the types of elements determine what happens to this matter or energy as it flows through the elements. Mathematically, this in turn determines the number of simultaneous equations in the mathematical model, as well as the types of terms in each equation (first derivatives, second derivatives, and so on). Whereas deductive knowledge of the laws governing a system are obtained from the study of a scientific discipline, knowledge about the structure of a system can only come from insight into the specific system being modeled.

Parameters. The parameters in a mathematical model are the numerical values assigned to the various coefficients appearing in the equations. These are related to the specific magnitudes of all the elements comprising the system, as well as to the boundary and initial conditions, which together with the governing equation constitute a completely specified model.

If fortunate enough to have available to him all the information regarding the laws, structure, and parameters relevant to a specific problem, the system analyst is able to construct the mathematical model entirely deductively. The model then represents the solution of a direct rather than an inverse system problem. On most occasions, however, some essential ingredients required to construct the mathematical model are not explicitly provided. Some components of the mathematical model must therefore be inferred indirectly either by observing the system in operation and measuring certain excitations and responses or by performing specific experiments upon the system. Whenever inductively obtained information must be used in the modeling effort, the resulting mathematical model constitutes the solution of an inverse problem and is therefore not unique. That is, it is always possible to find an infinite number of mathematical models that match perfectly any set of observed excitations and responses. The "optimum" solution of the inverse problem is the model that bears the closest similarity to the prototype system being modeled. Only a model that is sufficiently close to that optimum can be expected to be useful in predicting responses to excitations other than those employed as part of the inductive process in constructing the model.

1.4. CLASSES OF MATHEMATICAL MODELS

All systems exist in a time-space continuum in the sense that inputs and outputs can generally be measured at an infinite number of points in space and at an infinitude of instants of time. Mathematically, this means that time and the three space variables should actually be considered to constitute continuous independent variables in all system studies. Within the closed region defined by the system boundary, all dependent variables could therefore be expressed as functions of time and three space variables. Since this usually leads to unnecessarily detailed models and to unmanageably complicated equations the time-space continuum is most often represented by an array of discretely spaced points in one or more of the four principal coordinates. Attention is then focused upon the magnitudes of the dependent variables at those points rather than at all intermediate points. For example, if the time variable is discretized with a uniform discretization interval of 1 hour, the system variables are only measured or computed at hourly intervals. Solution values may, for example, be computed only for 1 P.M., 2 P.M., 3 P.M., and so on, and special interpolation techniques must be employed if it is desired to predict the magnitudes of a problem variable at other instants.

To classify mathematical models, it is expedient to recognize three broad classes of discretization:

1. ***Distributed parameter models:*** All relevant independent variables are maintained in continuous form.
2. ***Lumped parameter models:*** All space variables are discretized, but the time variable appears in continuous form.
3. ***Discrete time models:*** All space and time variables are discretized.

It should be recognized that this classification refers to the approximations made in deriving the sets of equations characterizing the system. If a digital computer is to be employed to simulate the system or solve the equations comprising the mathematical model, additional discretizations may be required, since digital computers are unable to perform other than arithmetic operations. The three classes described result in mathematical models comprised, respectively, of partial differential equations, ordinary differential equations, and algebraic equations.

Discretization of space variables involves essentially the representation of the system as an interconnection of two-terminal elements or subsystems. Although each element actually occupies a substantial amount of space, all activities internal to the element are ignored, and attention is focused only at the two terminals of the element. Such a system is usually represented as a circuit, a collection of elements interconnected in a specified manner. The lines or channels constituting these interconnections are assumed to be "ideal," having themselves no effect upon the system variables other than to act as conduits between elements.

To construct a mathematical model of a system assumed to be comprised entirely of lumped elements, a conservation principle (such as Newton's or Kirchhoff's laws) is employed to provide a separate equation at each of the system nodes, the junction points of two or more elements. For the types of elements occurring in nature, these equations are always either algebraic equations or first- or second-order differential equations with time as the independent variable. The mathematical model is then a system of equations, one equation per network node. By straightforward techniques, this system of equations can then be converted into a larger system of first-order differential equations, and the dependent variables of these equations are designated as the ***state variables***. The mathematical model of the system is then expressed conveniently in vector form as

$$\dot{\mathbf{y}} = \mathbf{f}(\mathbf{y}, \mathbf{a}, \mathbf{u}, t), \qquad \mathbf{y}(0) = \mathbf{y}_0 \tag{1.1}$$

where $\dot{\mathbf{y}}$ are the first derivatives of the state variable, $\mathbf{a}$ are the parameters and $\mathbf{u}$ the inputs. Where the system contains a large number of elements, the state variable vector may become very large, so that the simultaneous solution of all the state equations becomes excessively time consuming even where large computers are available. In principle, however, a state variable representation constitutes a convenient and powerful mathematical model.

In many application areas, it is generally not possible to regard the system as an interconnection of lumped elements. For example, in modeling the atmosphere in an air-pollution problem or an underground water reservoir in an aquifer simulation the system truly occupies every point in the space continuum, and relevant dynamic processes occur at every point in space and at every instant in time. The space variables must therefore be retained explicitly in formulating the mathematical model. Just as in the case of lumped parameter systems, conservation principles are invoked to permit the derivation of the governing equations. Because of the multiplicity of independent variables, these equations are partial differential equations of the general form

$$\frac{\partial}{\partial t}\left(a\frac{\partial u}{\partial t}\right) + b\frac{\partial u}{\partial t} = \sum_{k=1}^{m}\frac{\partial}{\partial x_k}\left[\sum_{i=1}^{m} c_{i_k}\frac{\partial u}{\partial x_i} + d_k u\right] + \sum_{k=1}^{m} e_k\frac{\partial u}{\partial x_k} + fu \tag{1.2}$$

where the parameters a, b, c, d, e, f may be functions of the space and time variables as well as of u.

Often a considerable number of the parameters of Eq. (1.2) may be taken to be zero, so that simplified forms of this equation result. The solution of Eq. (1.2) and most of its simplified forms arising in the analysis of systems existing in nature is never easy. Further simplifying assumptions are always required to permit computer implementation, and even then major obstacles arise.

1.5. THE EVOLUTION OF MODELING AND SIMULATION

A mathematical model of a system is only useful insofar as such a model provides solutions. Modern computers and computational methods are concerned with strategies to obtain these solutions in an efficient manner. The most elegant method of solving a partial differential equation, of course, is via a direct mathematical analysis that leads to an analytical solution. When the conditions of a problem make a direct analysis difficult or impossible, a solution may in general be found by experimental or computational methods.

Experimental methods enjoyed widespread popularity before the advent of modern digital computers. Most of these methods are based on the physical analogies. An analogy between two things is said to exist whenever there is a resemblance not necessarily of the two things themselves, but of two or more attributes, circumstances, or effects. For instance, concepts such as continuity and conservation are relevant in many fields of study and suggest the existence of certain similarities among them. Although perfect analogies do not exist, even crude analogies often provide useful insights. For example, transonic and supersonic flows can be studied by using suitably designed fluid-flow analogs. In fact, the Hele-Shaw model, quite popular until World War II, can be considered to be regarded as a *hydrodynamic* computer that exploited the analogy between perfectly inviscid, two-dimensional fluid flow and a variety of phenomena characterized by the Laplace equation. A variety of other physical (or direct) analogies led to such exotic "computational tools" as soap films and rubber membranes, however, the electroconductive analogs were the most widely used experimental analog devices in the precomputer era. An important reason for their popularity is the flexibility with which electrical quantities can be measured and controlled.

The electroconductive analogs can be subdivided into the continuous-space, continuous-time or CSCT variety and the discrete-space, continuous-time or DSCT variety. The conductive sheet (using the Teledeltos paper) and the conductive liquid (using the electrolytic tank) belong to the CSCT category. Resistance networks, resistance-capacitance networks, and other network analyzers belong to the DSCT variety. These electroconductive analogs were in widespread use at least until the early 1950s.

Analogies can also be exploited in an indirect manner by taking advantage of the similarity of the mathematical equations describing different phenomena. Once an appropriate mathematical model is formulated, the dynamic behavior of the physical system can be simulated on a computer by solving the relevant mathematical equation. The analog computer (more correctly, the differential analyzer) is one such device that is highly suitable for simulating systems characterized by ordinary differential equations. In an analog computer, electronic circuit modules are constructed each with the capability of performing certain basic mathematical operations, such as addition, negation, multiplication, division, temporal integration, and a few other operations, all in real time. These modules, for example, can be used to form the sum of a pair of voltages or produce a voltage that is proportional to the time integral of a single voltage. The real-time solution capabilities of analog computers permitted the engineer to use them as "live models" in an interactive design atmosphere. Also, it may be safely stated that the analog computer represents one of the earliest versions of the parallel computers. As the problem size increases, one simply has to use a bigger computer, and the solution always appears instantaneously.

Perhaps the most serious limitation of analog computers, aside from their limited accuracy, is their inability to solve equations for functions of more than one independent variable (that is, partial differential equations). An analog computer is basically an ideal tool to solve ordinary differential equations. To solve a partial differential equation (herein after referred to as PDE) on an analog computer, one has to transform the given equation into a system of ordinary differential equations (or ODEs). This transformation, often called the method of lines, entails the discretization of all but one of the independent variables. The nature of the resulting ODEs and the methods of solving these ODEs are not totally independent of the type of the original PDE (that is, elliptic, parabolic, or hyperbolic) and the type of variable that was left continuous (time or space). Although this approach resulted in a variety of analog and hybrid computer methods, the speed advantage of the analog approach soon eroded with the increasing availability of interactive digital computers and of special-purpose simulation languages. The present trend is unmistakably toward a digital computer orientation in modeling and simulation. Although references to supercomputers and networks of microcomputers will be made, the primary thrust of this book is on the use of the traditional Von Neumann type of digital computers.

What is a Von Neumann-type digital computer? It is a stored program computer in which programs and data are stored in the memory unit, and the programs are executed one instruction at a time in a sequential fashion. This type of computer consists of five basic functional subsystems: input unit, output unit, memory unit, control unit and the arithmetic-logic unit (ALU). It is customary to call the combination of the ALU and the control unit by the name central processing unit (CPU) or, simply, the processing unit. Communication with the outside world is through the input-output units in a sequential fashion. Many of the small computers of today are of this basic type.

As computer technology advanced on both the hardware and software fronts, demands for better performance (in terms of higher speeds, throughputs and resource utilization) continued to persist in an inexorable fashion. An oft-cited classic example is the computational demands of a weather forecasting model. In an article on modeling the weather (*Datamation*, May 1978), Francis J. Balint, of the National Oceanic and Atmospheric Administration, estimates that doubling the resolution level of a model would lead to a fourfold increase in main memory, eightfold increase in auxiliary storage, and a sixteen-fold increase in computing power. In spite of all the latest developments, Balint maintains that a continuing problem that modelers face is the difficulty to produce a useful forecast well in advance of the weather occurrence.

The unrelenting demand for better performance can be met by improving the Von Neumann computer from four different angles. The first approach is to increase the speed at which the basic components of a computer operate. However, physical and technological limitations are already being reached, and any improvements in this area are likely to be evolutionary only. A second means of improving performance is through advancements in computer architecture. A number of new ideas, such as the use of overlapping operations and input-output concurrency have begun to modify the basic Von Neumann architecture. Although a great deal of concurrency can be obtained through concepts such as buffering and channel independence, generally, programs have to wait for completion of some input-output action since the external devices are several orders of magnitude slower than the CPU. This led to the concept of ***multiprogramming***. In multiprogramming, one is concerned with time and resource sharing of a computer system by two or more programs resident simultaneously in the main memory unit.

The third approach is to use more than one computer to solve a large problem. Definitions in this emerging area have not yet been standardized. Array processing, systolic processing, synchronous multiple processing, asynchronous multiple processing, and distributed processing are some of the terms being used to describe various facets of this emerging area. For simplicity, the term multiple processing is used here as a catch-all term. In multiple processing, we are concerned with the simultaneous processing of two or more portions of the same program by two or more processing units. The next conceptual step is to combine multiprogramming and multiprocessing toward the very general idea of ***parallel processing***. Some of the more relevant ideas in this area are discussed briefly in the following section.

Finally, some schools of thought are proposing an entirely new type of machine architecture, the ***data-flow machine***. To understand the basic principle behind a data-flow computer, it is necessary to realize that the conventional (Von Neumann) computer is essentially a ***control-flow*** computer. That is, conventional computers, even those capable of concurrent activity, are implicitly synchronous and sequential and require any desired concurrency to be indicated explicitly. These can be contrasted with data-flow computers, which allow concurrent operations to be activated as soon as their input data are available. This topic is relatively new and any further discussion here is not justified.

1.6. SUPERCOMPUTERS AND MICROCOMPUTERS IN FIELD SIMULATION

The spectrum of new ideas in digital computer technology represents not only new advances in hardware and software design but also new innovations in arranging various building blocks. However, technological and design improvements alone are never sufficient; innovative architectural features are often necessary to improve performance. This usually takes the form of some kind of parallelism or overlap so that, despite the finite time taken for a certain operation, if several are performed in parallel using additional hardware, the apparent time taken for each operation is reduced. Several architectural approaches are possible to achieve this parallelism. In ***array processing***, one uses a collection of identical processors with each processor executing the same instruction sequence, but using different data. In ***vector processing***, the basic arithmetic operations are broken into a set of elementary steps. Each elementary step is then implemented in hardware, and the resulting arithmetic unit operates as a production line or ***pipeline***. Vector and array processing principles are widely used in the construction of supercomputers. In obtaining high rates of arithmetic operations, supercomputers sacrifice a certain amount of flexibility. Despite its apparent power, a given machine may not perform well on some problems because the idiosyncrasies of the problem prohibit the machine's maximum performance from being realized. Added to this, the high cost of these machines makes their availability limited to a few.

Another approach to realize high-performance parallel structures is to implement them using low-cost microprocessors in a network configuration with simple and regular communication paths. During the past several years there have been many proposals for such networks for the solution of partial differential equations. Network approaches appear particularly suitable for the treatment of field problems in which the space domain in one, two, or three dimensions is discretized with the aid of finite difference approximations. A separate microcomputer can then be dedicated to one or a group of adjacent grid points, and very substantial increases in speed over conventional digital methods can potentially be attained. Nonetheless, by 1980 none of the ambitious proposals formulated had been implemented or had progressed beyond the small prototype stage.

While there are some major technical problems that had yet to be fully solved (optimal ways of interconnecting the microprocessors, system reliability, and software), the fashioning of large network simulators appeared to be definitely within the state of the art. A major reason for the lack of progress in this area would appear to be related to the economics of computer systems and the fact that no computer manufacturer has a real stake in the development of such systems. In conventional computer systems, the manufacturer acquires components at a relatively small fraction of the system cost and then assembles these using architectures of his own design as well as a variety of engineering innovations. Viewed from the perspective of economics, the proposed microcomputer networks are not to be regarded as the successors or descendants of the supercomputers. CRAY-I, STAR-100, ASC, and even ILLIAC-IV were produced according to lines that have become conventional in the computer field. That is, a large amount of system engineering was superimposed upon the basic cost of the components. And it is for these reasons that major manufacturers took an interest in their development. The proposed microcomputer networks, on the other hand, can really be regarded as the descendants of the analog network analyzers that played a very important role in field simulations in the 1930s, 1940s, and 1950s. In the network analyzer approach, the field under study is first discretized using finite difference approximations, and each grid point is represented by a simple network of resistors, capacitors, and sometimes inductors. In time, these network analogs became the most powerful and widely used technique for solving elliptic, parabolic, and hyperbolic differential equations.

Although network analyzers were in very wide use during the post-World War II years, there emerged no manufacturers of such networks. With the exception of some units produced for highly specialized applications (for example aircraft vibration studies), most networks were designed and implemented by the user. In effect, each user or very narrow class of users designed networks most suitable for their special purposes and became expert in their use. This may well be the road that will be followed in applying microcomputers to field simulation. In other words, rather than implementing a single computer system to treat the many varieties of Navier-Stokes equations, a network might be fashioned to model a specific petroleum reservoir, or the atmosphere of a specific urban area, or a specific important aquifer. By this approach, the problem of providing for specified boundary and initial conditions, singularities, nonuniform field parameter, and so on, can be immeasurably reduced both from the hardware and software viewpoints.

SUGGESTIONS FOR FURTHER READING

This chapter touched upon a number of topics spanning various activities of modeling and simulation. For conciseness, only a representative sample of reference material is suggested here.

1. *For a discussion of the basic assumptions of paradigms of mathematical models, consult*
   T. S. KUHN, *The Structure of Scientific Revolutions.* Chicago: University of Chicago Press, 1962.

2. *A philosophical discussion on causality can be found in*
   M. BORN, *The Natural Philosophy of Cause and Chance.* London: Clarendon Press, 1969.

3. *The emphasis in this book is on modeling and simulation of field problems, that is, problems characterized by PDEs. For a discussion on modeling and simulation of problems characterized by ODEs, consult:*
   F. J. RICCI, *Analog/Logic Programming and Simulation.* Rochelle Park, N.J.: Hayden Book Co., Inc., 1972.

   G. A. KORN and J. V. WAIT, *Digital Continuous Systems Simulation.* Englewood Cliffs, N.J.: Prentice-Hall, Inc., 1978.

4. *An extensive discussion of analog methods to solve PDEs can be found in:*
   W. J. KARPLUS, *Analog Simulation: Solution of Field Problems.* New York: McGraw-Hill Book Co., 1958.

   D. VITKOVITCH (ed.), *Field Analysis: Experimental and Computational Methods.* New York: Van Nostrand Reinhold Co., 1966.

5. *A tutorial introduction to parallel computation can be found in the Special Issue on Parallel Processors and Processing*, ACM Computing Surveys, *Vol. 9, No. 1 March 1977.*

6. *The status of supercomputers as of 1979 can be found in the brief survey article by*
   J. C. KNIGHT, "The Current Status of Super Computers," *Intl. J. of Computers and Structures*, Vol. 10, pp. 401–409, 1979.

7. *A tutorial introduction and comparative description of a number of large-scale vector and array processors as well as multiprocessing systems can be found in the following two IBM Research Reports:*

G. PAUL, *Large Scale Vector/Array Processors*, IBM Research Report RC 7306 (#31432), dated September 1978, IBM Research Division, Yorktown Heights, N.Y.

M. SATYANARAYANAN, *A Survey of Multiprocessing Systems*, IBM Research Report RC 7346 (#31610), dated October 1978, IBM Research Division, Yorktown Heights, N.Y.

8. *A brief tutorial introduction to data-flow computers can be obtained from:*

J. B. DENNIS, D. P. MISUNAS, and P. S. THIAGARAJAN, *Data Flow Computer Architecture*, MIT Project MAC, Computation Structures Group, Memo 104, August 1974.

P. C. TRELEVAN, "Exploiting Program Concurrency in Computing Systems," *Computer*, January 1979, pp. 42–50.

2

The Physical Basis for Partial Differential Equations Characterizing Fields

2.1. INTRODUCTION

Although this text is devoted to the study of techniques for solving partial differential equations, using digital computers, the objective is not to treat any and every partial differential equation which may occur to mathematicians. Rather the scope is limited to those partial differential equations that are of interest to physical scientists and engineers, because the equations characterize physical phenomena. The solution of these equations therefore constitutes an essential element in scientific and engineering analysis. Since the mathematical models employed to describe physical processes in virtually every area of physics take the form of partial differential equations, the range of application of these equations is very wide indeed.

It is not necessary to have an intimate familiarity with a physical area in order to devise algorithms and computer programs for the solution of the governing partial differential equations. However, a general perspective over the link between mathematics and physics is virtually indispensable for a number of reasons. Some understanding of the physical basis for mathematical models is necessary in order to communicate effectively with the specialists of the application areas who bring a problem to the numerical analyst or computer scientist for solution. Such communication is vital, because the choice of the algorithms and of the general computational approach is often influenced by the objective of the simulation and the requirements of the ultimate user of the solution.

Additionally, a common feature of most digital computer solutions of complex problems, including partial differential equations, is that they are prone to a variety of sources of error. So numerous and pervasive are these error sources that only very rarely are correct solutions obtained the first time a problem is placed on the computer. Usually, the time expended in locating and correcting errors far exceeds the time spent in programming and in obtaining the final solution. Errors that affect the final quality of computer solutions fall into the following categories:

1. Errors in the formulation of the problem: that is, errors in the construction of the mathematical model characterizing the physical system under analysis.
2. Errors in numerical analysis: that is, errors in translating the partial differential equations into algorithms suitable for digital computer treatment.
3. Errors in program design: that is, errors in translating the numerical algorithms into a computer language.
4. Errors in coding: that is, errors in preparing the program and input data (for example, typing and keypunch errors, syntax errors).
5. Truncation errors: that is, errors resulting from the approximation of continuous independent variables by discretized variables.
6. Round-off errors: that is, errors due to the fact that variables are represented by a limited number of digits.
7. Errors due to malfunction of the computer system.

An important aid in the elimination of these errors, that is, to "debug" the program, is a general understanding of what the solution should look like. In fact, the more physical insight that can be brought to bear upon the preparation and the checkout of computer programs, the more likely is it that the final computer output constitutes a valid and credible solution of the problem.

It is the purpose of this chapter to provide a measure of such insight by demonstrating how partial differential equations are derived from physical principles. The discussion is facilitated by first introducing certain unifying concepts that link the various areas of physics. This is followed by derivations of the principal partial differential equations that characterize field problems.

2.2. UNIFYING CONCEPTS

The problems of engineering and applied physics can be classified into two broad categories: those that apply to lumped systems, and those that apply to distributed systems. Familiar examples of lumped systems appear particularly in mechanics and electric circuit theory. The differential equations characterizing lumped systems have only one independent variable, time; hence they are ordinary differential equations. By contrast, in distributed parameter systems or ***fields***, spatial dimensions comprise an integral part of the mathematical model. Mathematically, the independent problem variables may include three space variables as well as the time variable.

The concept of "field" has wide-ranging applications in physics and engineering, including such diverse areas as electrodynamics, fluid mechanics, solid mechanics, heat transfer, acoustics, optics, and many more. In each of these disciplines, the dependent variables (the physical quantities characterizing the excitation and response of the field) are defined as continuous functions in space and time. The intrinsic properties of the medium, which determine the parameters of the governing equations, are likewise defined continuously for every point in the time-space continuum. The partial differential equations characterizing the field can then be derived by applying basic physical laws and by following the procedure described in the rest of this chapter. The formulation of the partial differential equations characterizing distributed parameter systems is greatly simplified by recognizing certain conceptual similarities in the independent variables, the dependent variables, the parameters, and the basic laws governing the behavior of systems belonging to the many diverse areas of physics. This unified view serves as a convincing explanation for the otherwise surprising observation that three basic partial differential equations and their modified forms serve as the mathematical models for virtually all areas of physics and engineering.

The independent variables in the mathematical models of distributed parameter systems always include space variables and sometimes the time variable. The space variables identify location of any point within the field with respect to a reference point—the center of the coordinate system. Depending upon the dimensionality of the field, one, two, or three space variables may be involved. For convenience in dealing with symmetries within the field and with regularly shaped field boundaries, one of a variety of different coordinate systems may be employed. The Cartesian coordinate system with its orthogonal x, y, and z axes is by far the most widely used coordinate system. Where there is symmetry within the field about a straight line, the cylindrical coordinate system has important advantages; where there is symmetry about a point, the spherical coordinate system is the most appropriate. In all coordinate systems, the space variables are zero at the center of the coordinate system, and the distance and direction of any point in the field with respect to the center of the coordinate system is readily determined from the coordinates of that point. For simplicity, most of the discussion in this text is limited to space variables expressed in the Cartesian or rectangular coordinate system. Cylindrical and spherical coordinates are briefly discussed in Section 2.7. The time variable appears as an independent variable in mathematical models only for systems in the transient state. Many field problems are concerned only with the static or steady state, in which case the space variables are the only dependent variables. However, whenever the time variable appears in mathematical models, a specific instant of time is arbitrarily designated as $t = 0$, and a solution is sought only for times $0 \leq t$.

The dependent variables of the mathematical model are the physical quantities that are measured by instruments and that constitute the excitations and responses of the system. These are generally plotted as functions as space and/or time. A survey of the diverse areas of physics reveals that these dependent variables fall into two major categories: the potential functions (also referred to as across variables) and the flux functions (also referred to as through variables). An ***across variable*** relates the condition at one point within the field to that at some other field point or arbitrary reference point. The measuring instrument recording an across field variable must be applied simultaneously to two separated points, and the magnitude specified for this variable usually represents the difference in value between the two points. Thus, for example, temperature, which is always referred from absolute zero or the freezing point of water, is the across variable in heat transfer systems; pressure or velocity potential is the across variable in fluid mechanic systems, while electric voltage or potential is the across variable in electrostatic and electrodynamic systems. A ***through variable*** on the other hand requires only one point within the field for its specification and represents a measure of the flux traversing an elemental cross section of the field. Thus, heat flux in a thermal system, flow rate in a fluid mechanics system, and current density in an electrodynamic system are all through variables. Mathematically, the across variables are generally scalars, while the through variables are vectors.

The parameters of the mathematical model (the numerical coefficients in the characterizing equations) are derived from the intrinsic physical properties of the material constituting the field, and in the case of linear systems are independent of any specific existing excitations or responses. They may be measured by obtaining a sample of this material and by performing a laboratory experiment. In such an experiment an excitation is applied to the sample and the resulting response measured. The nature and magnitude of the field parameter are determined by the relationship between these measured excitations and responses. A survey of the many areas of physics in which partial differential equations arise reveals that the field of parameters invariably falls into one of only three classes.

1. Dissipators or dampers.
2. Reservoirs of potential.
3. Reservoirs of flux.

Where the material comprising a field is a dissipator, the excitation-response relationship between potential and the flux is one of proportionality. Consider a field described by a single space variable oriented in the x direction, such as, for example, a thin wire in an electrical system. A potential difference, Δp, is applied across the two ends of a small portion or sample of this system, and the resulting flux, f, is measured. For a dissipative material,

$$\Delta p = -D(\Delta x) f \tag{2.1}$$

where D is the dissipativity per unit length, and Δx is the distance between the two ends of the sample. The minus sign indicates that positive flux flows in the direction of decreasing potential, that is, from a high potential to a low potential. If Δx is made to approach zero, Eq. (2.1) becomes

$$\frac{dp}{dx} = -Df \tag{2.2}$$

where the term on the left is termed the potential gradient. The parameter D is a measure of the extent to which the material comprising the field dissipates energy by converting it into heat (or in the case of heat transfer systems by leading to an increase in entropy). This parameter is termed resistivity in electrical systems, viscosity in fluid flow systems, thermal resistivity in heat transfer systems, and appears in virtually all physical systems.

The material comprising a field acts as a reservoir if it is capable of storing matter or energy temporarily. In the case of reservoirs of potential, this storage occurs whenever a potential difference is placed across a sample of the material; the greater the potential difference, the more is stored. For example, an electrical capacitor stores electrical energy whenever a voltage difference is applied across its terminals. When a resistor or a short circuit is then placed across these terminals, the capacitor produces a current in a direction such as to oppose any change in potential. For a sample of a one-dimensional field with a reservoir of potential property, the relationship between the potential, p, and the flux, f, is

$$\Delta f = -E_p \, \Delta x \frac{dp}{dt} \tag{2.3}$$

where E_p is the potential reservoir property per unit length, and Δx is the length of the sample. Letting Δx approach zero leads to

$$\frac{\partial f}{\partial x} = -E_p \frac{\partial p}{\partial t} \tag{2.4}$$

Distributed capacitance or capacitivity is the reservoir of potential parameter in electrical fields, compressibility plays the same role in fluid flow systems, and thermal capacity or specific heat is E_p in heat transfer systems.

In the case of reservoir of flux, the material comprising the field acts to store energy whenever a flux is made to flow through it. For example, in electric circuits an inductor stores energy in its magnetic field whenever a current flows through it; if the terminals of the inductor are short-circuited, a voltage is produced with a polarity tending to oppose any change in current. For a field sample of length Δx, the relationship between flux and potential is

$$\Delta p = -E_f \, \Delta x \frac{df}{dt} \tag{2.5}$$

where E_f is a measure of the flux reservoir property per unit length. Again letting Δx approach zero,

$$\frac{\partial p}{\partial x} = -E_f \frac{\partial f}{\partial t} \tag{2.6}$$

Distributed inductance or inductivity is the reservoir of flux parameter in distributed electrical systems, while density or inertia play the same role in fluid flow systems. On the other hand, heat transfer systems never display a reservoir of flux property.

The pertinent across and through variables as they appear in the more important areas of physics are summarized in Table 2.1 together with the parameters or system characteristics necessary to specify completely the properties of the system. It is shown in this chapter that the presence or absence of one or more of the three basic parameter types has far-reaching importance in determining the behavior of the system and its

Table 2.1

Physical Area	*Across Variable* p *(potential)*	*Through Variable* f *(flux)*	*Parameters*		
			Dissipators D	*Reservoirs of Flux* E_f	*Reservoirs of Potential* E_p
Electrodynamics	Voltage	Current	Resistivity	Inductivity	Capacivity
Electrostatics	Electric potential	Flux	—	—	Dielectric permittivity
Magnetics	Potential, MMF	Flux	Reluctance	Permeability	—
Electromagnetics	Potential, EMF	Flux	Conductivity	Permeability	Dielectric permittivity
Statics (mechanical)	Displacement	Force	—	Spring constant	—
Dynamics (mechanical)	Displacement or velocity	Force	Viscous damping	Spring constant	Mass (inertia)
Elasticity	Strain	Stress	Viscous damping	Young's modulus	Inertia
Fluid mechanics	Velocity potential (pressure)	Flow rate	Viscosity	Inertia (density)	Compressibility
Particle diffusion	Concentration	Mass transfer rate	Diffusivity	Inertial forces	Compressibility
Heat transfer	Temperature	Heat flux	Thermal resistance	—	Thermal capacitance

mathematical description. Where only one parameter is present (in nonnegligible magnitude), the mathematical model takes the form of an elliptic type of equation, such as Laplace's or Poisson's equations; where dissipation and either reservoir of potential or reservoir of flux (but not both) are present, the governing equation is parabolic. Where both types of reservoirs are present, the field is characterized by a hyperbolic partial differential equation.

The basic physical principles that account for the analogous nature of field problems in the diverse areas of physics are known as the principles of conservation and continuity. These principles occur in various forms in all the branches of physics. The conservation principle applies to that quantity whose transport is measured by the through variable. According to this principle, the total amount of this quantity existing within the field at any time subsequent to an arbitrary initial instant, $t = 0$, must be equal to the algebraic sum of the net amount added (or subtracted) by external excitations plus the amount initially present within the field at time $t = 0$. For example, in electrical systems, electric charge is conserved; similarly, in dynamic mechanical systems momentum is conserved, and in fluid flow systems mass is conserved.

The principle of continuity also deals with through variables and specifies that the through variable is continuous and must emanate from a source (internal or external excitation) and return to the same or some other source. Generally, the conservation principle implies the continuity principle, and vice versa. Usually it is merely a matter of convention or terminology that has caused one principle to assume a more prominent role in a specific area.

2.3. PROCEDURE FOR DERIVING THE CHARACTERIZING EQUATIONS

The structural similarity of the mathematical models characterizing the many different disciplines within physics and engineering is largely due to the similarity in the methodology employed to derive the characterizing equations. As shown previously, regardless of the physical discipline, the characterizing equations involve space and time as the independent variables, potential or flux as the dependent variable, and parameters falling into three distinct classes. Moreover, the basic physical laws governing the diverse disciplines all take the form of conservation principles. The partial differential equations constituting the mathematical model can furthermore be derived using a procedure that is virtually independent of the physical application area. This procedure takes the following steps:

1. *Identify those field parameters that are present with nonnegligible magnitudes.* A field may contain all the three parameter types, dissipation, reservoir of potential, and reservoir of flux, or only one or two of these parameters may be present. Sometimes all three parameters are present, but one or two of these may be of negligible importance in determining the potential and flux distribution. For example, a wire may contain inductance and capacitance, but these may be ignored in determining the voltage distribution along the wire, so that only the resistivity is of significance.

2. *Decide whether the problem should be formulated in one, two, or three space dimensions and which coordinate system is appropriate.* This decision is generally made on the basis of the geometry of the system and the purposes of the modeling effort. Clearly, all physical systems exist in three space dimensions, and representation in one or two space dimensions therefore inevitably entails approximations. In general, a rectangular coordinate system is preferred, unless there is symmetry about a line or a point.

3. *Select a typical elemental region.* The geometry of this element depends upon the number of space dimensions as well as upon the coordinate system selected. In rectangular coordinates, if the system can be characterized adequately in one space dimension, the elemental field portion is a line segment Δx in length; if two space dimensions are required, the elemental portion will be a rectangle of dimensions Δx by Δy; if three dimensions are needed, the elemental portion will be a rectangular solid Δx by Δy by Δz. This is illustrated in Figure 2.1. A similar approach is employed in other coordinate systems. As examples, elemental field portions in cylindrical and spherical coordinates are shown in Figure 2.2. These elements are considered typical of the field as a whole. That is, a description of the potential distribution and flux in an elemental portion of the field is considered to be representative of all regions within the field boundaries.

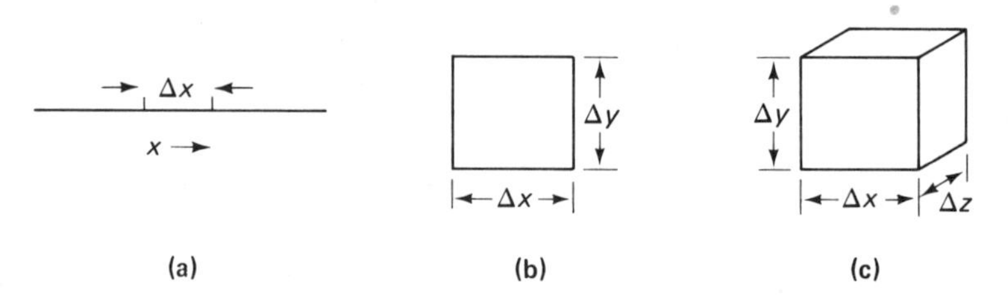

Figure 2-1 Elemental regions in Cartesian coordinates for one-, two- and three-dimensional systems

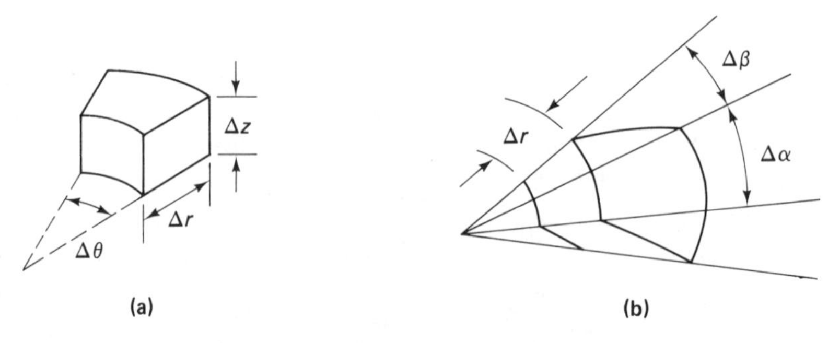

Figure 2-2 Elemental regions for three-dimensional systems: (a) cylindrical coordinates; (b) spherical coordinates

4. *Express the flux across each boundary of the typical element in terms of the potential, p, and its derivatives with respect to space and/or time.* For fields in one, two, and three space dimensions, this leads to two, four, and six equations, respectively. These equations are derived from Eqs. (2.2), (2.4), and (2.6) as determined by the types of field parameters existing in the field under study.

5. *Calculate the net flux into the element.* Where the field is purely dissipative, this involves only the summation of the flux across all the boundaries of the element; where reservoir parameters are present, the storage in the reservoir associated with the element must also be taken into account.

6. *Invoke the conservation principle applying to the physical discipline to which the system belongs.* This involves accounting completely for all flux into and out of the element.

7. *Let the dimensions of the element shrink to zero.* In rectangular coordinates, this means that Δx, Δy, and $\Delta z \longrightarrow 0$.

These steps result in a partial differential equation that characterizes the entire field. With the exception of the physical discipline of elasticity, the terms appearing in the characterizing differential equations are limited to first and second derivatives with respect to space and time.

A convenient and frequently used method for classifying the basic partial differential equations that characterize field problems follows from a consideration of the mathematical character of the solutions. This method of classification is only briefly outlined next to provide a link with more formal mathematical treatments. As demonstrated in the following sections of this chapter, the linear equations governing physical fields take the general form

$$A\frac{\partial^2 u}{\partial \alpha^2} + 2B\frac{\partial^2 u}{\partial \alpha\,\partial \beta} + C\frac{\partial^2 u}{\partial \beta^2} = D\frac{\partial u}{\partial \alpha} + E\frac{\partial u}{\partial \beta} + Fu + G \tag{2.7}$$

By letting the parameters A to G assume positive, negative, or zero magnitudes, all the equations derived in this chapter can be expressed as special cases of Eq. (2.7). The two independent variables α and β may both be space coordinates, or one may be a space coordinate and the other the time variable. The relative magnitudes of A, B, and C determine the nature of the equation.

If $AC > B^2$, as is the case if $B = 0$ and A and C are both positive, the equation is termed ***elliptic***. This definition applies even if another term $H(\partial^2 u/\partial \gamma^2)$, where H is a positive number, is added to the left side of Eq. (2.7), so that three-dimensional fields are included in this definition.

If $AC = B^2$, as is the case if B and either A or C are equal to zero, the equation is termed ***parabolic***. The addition of additional second-order derivatives such as $H(\partial^2 u/\partial \gamma^2)$ does not influence the basic parabolic character of the equation.

If $AC < B^2$, as is the case if B is positive or zero and either A or C is negative, the equation is termed ***hyperbolic***.

The classification of partial differential equations into these three categories is valuable because *the basic analytical and numerical methods for treating field problems are inherently different for the three types of equations.*

2.4. ELLIPTIC PARTIAL DIFFERENTIAL EQUATIONS—LAPLACE'S EQUATION

Elliptic partial differential equations characterize those distributed parameter systems in which no more than one of the three parameter types appears with nonnegligible magnitude. That is, all fields that are purely dissipative, purely reservoirs of potential, or purely reservoirs of flux are characterized by elliptic partial differential equations. In addition, fields containing dissipation as well as reservoir elements are characterized by elliptic equations when they have reached the quiescent or steady state. The most often encountered of the elliptic partial differential equations, and indeed of all partial differential equations of applied physics, is known as Laplace's equation. It is the basic equation of potential theory and plays a prominent role in virtually all major areas of physics and engineering. The derivation of this important equation will be illustrated for systems in one, two, and three space dimensions.

Consider first the system shown in Figure 2.3a, in which a thin conductor of flux, assumed to have constant dissipative properties D/unit length, is connected at

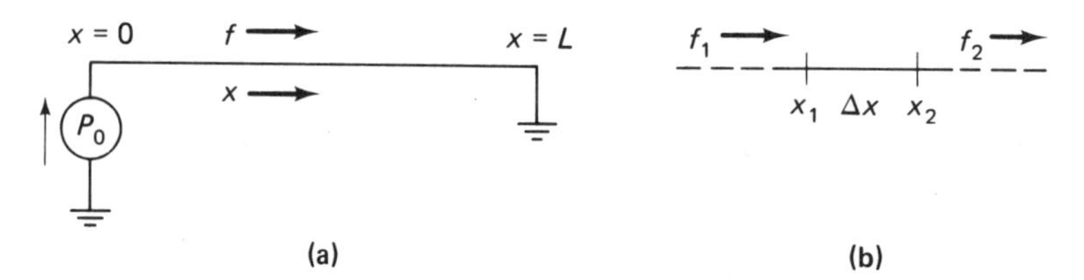

Figure 2-3 (a) One-dimensional field containing distributed dissipation but no reservoir properties; (b) elemental line segment

one end to a source of constant potential p_0, while its other extremity is connected to a source of zero potential (that is, ground). In accordance with the stepwise procedure outlined in the preceding section, we note that we are dealing with a system containing D but negligible E_p and E_f. We note further that this is a one-dimensional system oriented in the x direction. We now define an elemental line segment as shown in Figure 2.3b. The element is seen to have two boundaries located at x_1 and x_2. From Eq. (2.2), the potential-flux relationship at the two boundary points is

$$\left(\frac{dp}{dx}\right)_1 = -Df_1, \qquad \left(\frac{dp}{dx}\right)_2 = -Df_2 \tag{2.8}$$

Since the field is assumed to be purely dissipative, there can be no storage within the elemental line segment. Therefore, in accordance with the conservation principle

$$f_2 - f_1 = 0 \tag{2.9}$$

Solving Eqs. (2.8) for f_1 and f_2 and substituting in Eq. (2.9),

$$\frac{1}{D}\left(\frac{\partial p}{\partial x}\right)_2 - \frac{1}{D}\left(\frac{\partial p}{\partial x}\right)_1 = 0 \tag{2.10}$$

By multiplying both sides of Eq. (2.10) by D and by dividing both sides by Δx, this equation transforms to

$$\frac{\left(\frac{\partial p}{\partial x}\right)_2 - \left(\frac{\partial p}{\partial x}\right)_1}{\Delta x} = 0 \tag{2.11}$$

We now shrink the size of the elemental line segment of Figure 2.3b by letting $\Delta x \longrightarrow 0$, which yields

$$\frac{\partial^2 p}{\partial x^2} = 0 \tag{2.12}$$

This equation is known as Laplace's equation in one dimension. Its solution, either analytically or numerically, provides the function $p(x)$, the potential at all points within the field. Such a solution is only possible if the potential or flux at the boundaries of the field is specified. In the case of Figure 2.3a, these boundary conditions are

$$\text{at} \quad \begin{cases} x = 0, & p = p_0 \\ x = L, & p = p_L \end{cases} \tag{2.13}$$

As an example of a two-dimensional field, consider a rectangular sheet as shown in Figure 2.4a. This field may be a sheet made of an electrically conductive medium, a cross section of a fluid flow field, or a cross section of a thermal conductor. The differential element, a rectangle, is shown in Figure 2.4b.

The potential gradient across any side of the differential element is equal to the dissipative parameter, D, times the net flux, F, across the side (the net amount of

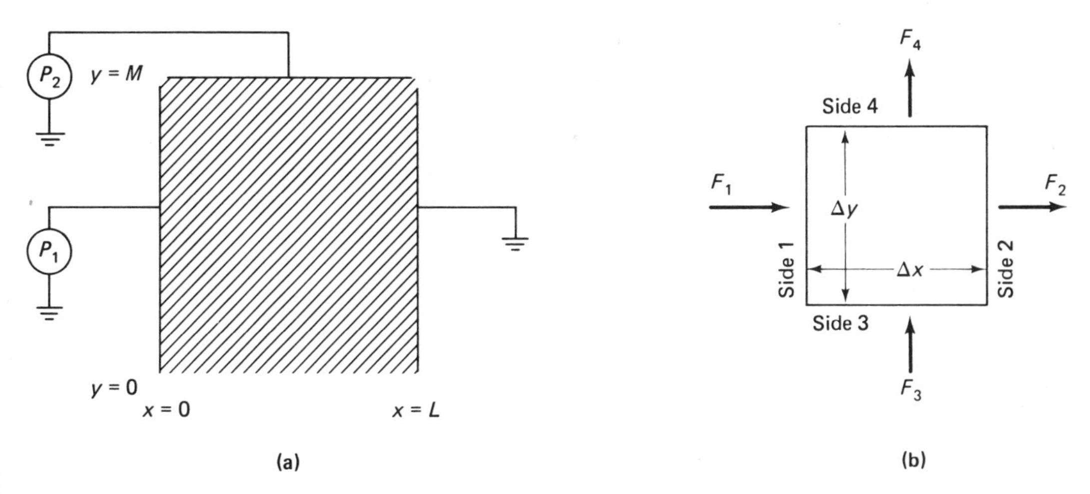

Figure 2-4 (a) Purely dissipative two-dimensional field; (b) elemental rectangle

material or energy passing over this line per second). In this case, F is the flux per unit length, f, times Δx or Δy. The potential gradients across the four sides of the element are then expressed as

$$\left(\frac{\partial p}{\partial x}\right)_1 = -\frac{F_1 D}{\Delta y}, \quad \left(\frac{\partial p}{\partial x}\right)_2 = -\frac{F_2 D}{\Delta y}, \quad \left(\frac{\partial p}{\partial y}\right)_3 = -\frac{F_3 D}{\Delta x}, \quad \left(\frac{\partial p}{\partial y}\right)_4 = -\frac{F_4 D}{\Delta x} \tag{2.14}$$

where the subscripts 1, 2, 3, and 4 refer to the left, right, bottom, and top sides of the differential element, respectively. Note that partial derivatives must be used in Eq. (2.14) since more than one independent variable is involved.

Solving for the flux components,

$$\begin{aligned} F_1 &= -\frac{\Delta y}{D}\left(\frac{\partial p}{\partial x}\right)_1, \quad & F_2 &= -\frac{\Delta y}{D}\left(\frac{\partial p}{\partial x}\right)_2 \\ F_3 &= -\frac{\Delta x}{D}\left(\frac{\partial p}{\partial y}\right)_3, \quad & F_4 &= -\frac{\Delta x}{D}\left(\frac{\partial p}{\partial y}\right)_4 \end{aligned} \tag{2.15}$$

According to the conservation principles, the net flux into the element, $(F_1 - F_2) + (F_3 - F_4)$ must equal zero. So

$$\frac{\Delta y}{D}\left[\left(\frac{\partial p}{\partial x}\right)_2 - \left(\frac{\partial p}{\partial x}\right)_1\right] + \frac{\Delta x}{D}\left[\left(\frac{\partial p}{\partial y}\right)_4 - \left(\frac{\partial p}{\partial y}\right)_3\right] = 0 \tag{2.16}$$

which can be rewritten as

$$\frac{1}{\Delta x}\left[\left(\frac{\partial p}{\partial x}\right)_2 - \left(\frac{\partial p}{\partial x}\right)_1\right] + \frac{1}{\Delta y}\left[\left(\frac{\partial p}{\partial y}\right)_4 - \left(\frac{\partial p}{\partial y}\right)_3\right] = 0 \tag{2.17}$$

Letting $\Delta x \longrightarrow 0$ and $\Delta y \longrightarrow 0$, Eq. (2.17) becomes

$$\frac{\partial^2 p}{\partial x^2} + \frac{\partial^2 p}{\partial y^2} = 0 \tag{2.18}$$

This equation is known as Laplace's equation in two dimensions and, together with the appropriate description of the excitations acting at the boundaries, characterizes completely the potential anywhere within the system. In the case of Figure 2.4a, these boundary conditions include constant potentials at three sides of the sheet; at the lower side, however, no excitation is applied so that there can be no flux across this boundary. In terms of the potential variable, this implies that the potential gradient in the direction normal to this side vanishes, since flux is directly proportional to the potential gradient. The appropriate boundary conditions are therefore

$$\begin{aligned} &\text{at}\begin{cases} x = 0 \\ 0 < y < M \end{cases} \quad p = p_1, \qquad &\text{at}\begin{cases} 0 < x < L \\ y = M \end{cases} \quad p = p_2 \\ &\text{at}\begin{cases} x = L \\ 0 < y < M \end{cases} \quad p = 0, \qquad &\text{at}\begin{cases} 0 < x < L \\ y = 0 \end{cases} \quad \frac{\partial p}{\partial y} = 0 \end{aligned} \tag{2.19}$$

A three-dimensional field is illustrated in Figure 2.5, which shows a rectangular solid composed of a dissipative material having uniform properties. Excitation is

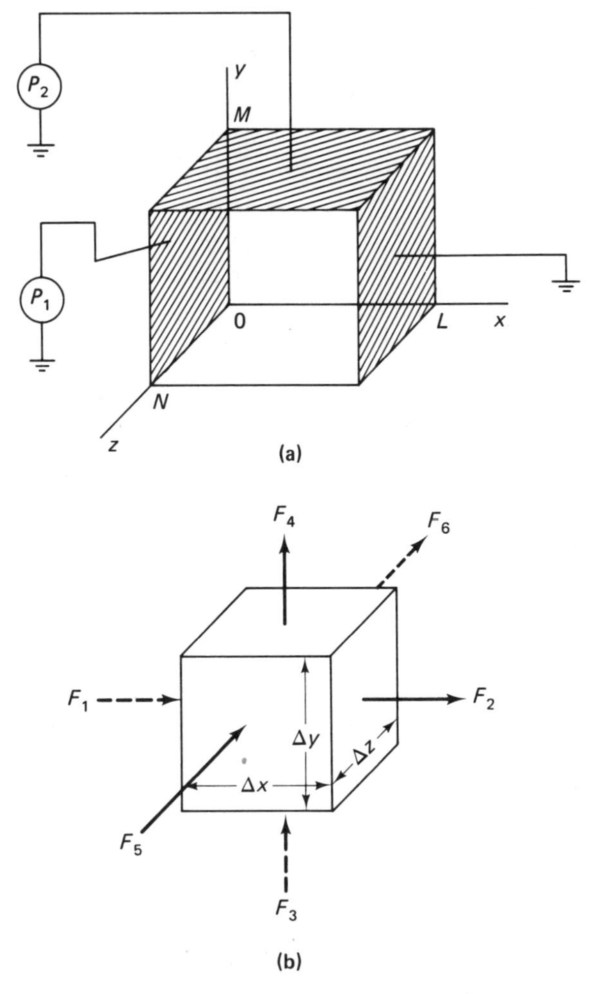

Figure 2-5 (a) Purely dissipative three-dimensional field; (b) elemental rectangular solid

again in the form of constant potentials applied to three entire sides: the left, right, and top surfaces of the solid. This system might represent an electrically conductive solid such as graphite, with highly conductive face plates covering three of its sides. The differential element, shown in Figure 2.5b, is a small block within this solid. Subscripts 1 through 6 refer to the left, right, bottom, top, front, and rear faces, respectively.

The analysis of this system follows lines similar to those used previously. The potential gradient across each face is first expressed in terms of the flux F across the face and the dissipative parameter D as

$$\begin{aligned}
\left(\frac{\partial p}{\partial x}\right)_1 &= -\frac{F_1 D}{\Delta z\,\Delta y} \\
\left(\frac{\partial p}{\partial x}\right)_2 &= -\frac{F_2 D}{\Delta z\,\Delta y} \\
\left(\frac{\partial p}{\partial y}\right)_3 &= -\frac{F_3 D}{\Delta x\,\Delta z} \\
\left(\frac{\partial p}{\partial y}\right)_4 &= -\frac{F_4 D}{\Delta x\,\Delta z} \\
\left(\frac{\partial p}{\partial z}\right)_5 &= -\frac{F_5 D}{\Delta x\,\Delta y} \\
\left(\frac{\partial p}{\partial z}\right)_6 &= -\frac{F_6 D}{\Delta x\,\Delta y}
\end{aligned} \tag{2.20}$$

In this case the flux, F, across each boundary surface of the element is equal to the local flux density (flux/unit area) times the area of the surface. Solving Eqs. (2.20) for $F_1, \ldots, F_6$ and invoking the conservation principle that

$$F_1 - F_2 + F_3 - F_4 + F_5 - F_6 = 0 \tag{2.21}$$

yields, upon rearrangement,

$$\frac{\Delta x\,\Delta y\,\Delta z}{D}\left\{\frac{\left(\frac{\partial p}{\partial x}\right)_2 - \left(\frac{\partial p}{\partial x}\right)_1}{\Delta x} + \frac{\left(\frac{\partial p}{\partial y}\right)_4 - \left(\frac{\partial p}{\partial y}\right)_3}{\Delta y} + \frac{\left(\frac{\partial p}{\partial z}\right)_6 - \left(\frac{\partial p}{\partial z}\right)_5}{\Delta z}\right\} = 0 \tag{2.22}$$

Since

$$\begin{aligned}
\frac{\partial^2 p}{\partial x^2} &\cong \frac{\left(\frac{\partial p}{\partial x}\right)_2 - \left(\frac{\partial p}{\partial x}\right)_1}{\Delta x} \\
\frac{\partial^2 p}{\partial y^2} &\cong \frac{\left(\frac{\partial p}{\partial y}\right)_4 - \left(\frac{\partial p}{\partial y}\right)_3}{\Delta y} \\
\frac{\partial^2 p}{\partial z^2} &\cong \frac{\left(\frac{\partial p}{\partial z}\right)_6 - \left(\frac{\partial p}{\partial z}\right)_5}{\Delta z}
\end{aligned} \tag{2.23}$$

as $\Delta x \longrightarrow 0$, $\Delta y \longrightarrow 0$, $\Delta z \longrightarrow 0$, Equation (2.22) becomes

$$\frac{\partial^2 p}{\partial x^2} + \frac{\partial^2 p}{\partial y^2} + \frac{\partial^2 p}{\partial z^2} = 0 \tag{2.24}$$

This equation is known as Laplace's equation in three dimensions and together with the boundary conditions applicable to this specific problem constitutes a complete characterization of the potential distribution within the field. In this case the boundary conditions at the six faces of the solid of Figure 2.5a are

$$\text{at}\begin{cases} x = 0 \\ 0 < y < M \\ 0 < z < N \end{cases} \quad p = p_1, \qquad \text{at}\begin{cases} 0 < x < L \\ y = M \\ 0 < z < N \end{cases} \quad p = p_2$$

$$\text{at}\begin{cases} x = L \\ 0 < y < M \\ 0 < z < N \end{cases} \quad p = 0, \qquad \text{at}\begin{cases} 0 < x < L \\ y = 0 \\ 0 < z < N \end{cases} \quad \frac{\partial p}{\partial y} = 0$$

$$\text{at}\begin{cases} 0 < x < L \\ 0 < y < M \\ z = 0 \end{cases} \quad \frac{\partial p}{\partial z} = 0 \tag{2.25}$$

$$\text{at}\begin{cases} 0 < x < L \\ 0 < y < M \\ z = N \end{cases} \quad \frac{\partial p}{\partial z} = 0$$

Laplace's equation, Eqs. (2.12), (2.18), and (2.24) can be expressed more compactly by introducing the Laplacian operator ∇^2, which merely signifies "the sum of the second derivatives with respect to all Cartesian space variables of interest." So regardless of the number of dimensions, Laplace's equation becomes

$$\nabla^2 p = 0 \tag{2.26}$$

which is also often written in vector notation as

$$\text{div}\,[\text{grad}\, p] = \nabla \cdot (\nabla p) = 0 \tag{2.27}$$

The boundary excitations in the preceding examples were specified to be sources of potential. Since sources of the flux variable also occur in engineering work, *there are actually two types of boundary conditions: a specified potential or a specified flux.* It has already been shown that, since no flux flows across an unexcited boundary, the potential gradient normal to this boundary is equal to zero. If a nonzero flux is specified at a boundary, the potential gradient normal to this boundary must be directly proportional to this variable. The boundary conditions may therefore be expressed in a more general form as

$$\begin{aligned} p &= k_1 \\ \frac{\partial p}{\partial n} &= k_2 \end{aligned} \tag{2.28}$$

where k_1 and k_2 are specified constants or functions of space variables and may be positive, negative, or zero, and where n is the direction normal to the boundary. This subject is considered in considerable detail in Chapter 3.

2.5. ELLIPTIC PARTIAL DIFFERENTIAL EQUATIONS: MODIFIED FORMS

The mathematical model of a distributed parameter system characterized by an elliptic partial differential equation assumes the simple and compact form of Eq. (2.26) only if a number of assumptions explicitly or implicitly made in its derivation are satisfied. In this section, three modified forms of Laplace's equation are considered. Each of these frequently arises in physics and engineering. They include modifications due to nonuniformities in the field parameter, due to moving coordinate systems, and due to distributed sources of flux.

An interesting feature of Eqs. (2.12), (2.18), and (2.24) is that the dissipative parameter D does not appear explicitly. It follows therefore that the potential distribution within the field is independent of the magnitude of this parameter. This is not unexpected. For example, in Figure 2.3a it is obvious that the potential at the midpoint of the conductor is exactly equal to one-half the potential difference between the two end points, and that this is true regardless of the magnitude of the dissipativity, provided of course that D is the same at all points within the system. Implicit in the derivation of Eq. (2.26) is the assumption that the characteristics of the field are uniform and isotropic. Where D is not independent of the space parameters, some modification in the previous derivations must be effected. If the field consists, for example, of two sections of different dissipative properties, these two sections can be treated as two separate fields. In that case the conservation principle demands that the flux across the junction of the two sections be the same on both sides of the junction and therefore specifies a boundary condition. This really then involves two separate field problems, each governed by Eq. (2.26).

More generally, the dissipative parameter D can be specified as a function of the space coordinates either by an analytic expression or numerically in graphical or tabular form. Referring to the derivation of Laplace's equation in one space dimension, and letting $D = D(x)$, the voltage gradients at the two ends of the line segment of Figure 2.3b are

$$f_1 = -\frac{1}{D_1}\left(\frac{dp}{dx}\right)_1, \qquad f_2 = -\frac{1}{D_2}\left(\frac{dp}{dx}\right)_2 \tag{2.29}$$

where D_1 and D_2 are the local dissipativities at x_1 and x_2, respectively. Equation (2.11) therefore becomes

$$\frac{1}{\Delta x}\left(\frac{1}{D_2}\left(\frac{dp}{dx}\right)_2 - \frac{1}{D_1}\left(\frac{dp}{dx}\right)_1\right) = 0 \tag{2.30}$$

Now when $\Delta x \longrightarrow 0$

$$\frac{\partial}{\partial x}\left(\frac{1}{D}\frac{\partial p}{\partial x}\right) = 0 \tag{2.31}$$

The functional relationship between D and x is therefore embedded within the second derivative. In a similar manner, the derivations for nonuniform fields in two and three space dimensions can readily be shown to lead to

$$\frac{\partial}{\partial x}\left(\frac{1}{D}\frac{\partial p}{\partial x}\right) + \frac{\partial}{\partial y}\left(\frac{1}{D}\frac{\partial p}{\partial y}\right) = 0$$

$$\frac{\partial}{\partial x}\left(\frac{1}{D}\frac{\partial p}{\partial x}\right) + \frac{\partial}{\partial y}\left(\frac{1}{D}\frac{\partial p}{\partial y}\right) + \frac{\partial}{\partial z}\left(\frac{1}{D}\frac{\partial p}{\partial z}\right) = 0 \tag{2.32}$$

Or in the more compact notation

$$\nabla\left(\frac{1}{D}\nabla p\right) = 0 \tag{2.33}$$

The above derivations and Eq. (2.33) apply also to fields in which D is a function of the field potential p. Under these conditions, the field is recognized to be nonlinear, and certain convenient analytical techniques (such as the superposition theorem) are inapplicable.

A second important modification of Laplace's equation occurs in the characterizations of fields in which the medium through which the flux flows is itself moving. This is equivalent to considering that the origin of the coordinate system, with respect to which all positions in space are measured, is moving at a specified velocity. Consider, for example, the study of heat transfer in the static or steady state, a phenomenon governed by Laplace's equation. Suppose now that the medium comprising the field is a liquid. While the fluid is at rest, the temperature distribution within this field is governed by the basic Laplace equation, Eq. (2.26). If now the fluid that comprises the field is made to flow, say, in the x direction, the transfer of heat is influenced not only by the temperature gradients, as in Eq. (2.26), but also by the velocity of the fluid particles themselves. Equation (2.26) must therefore be modified to take this phenomenon into account. For problems in one, two, and three space dimensions, respectively, it can be shown that this modification takes the form

$$\frac{d^2p}{dx^2} = v_x \frac{dp}{dx}$$

$$\frac{\partial^2 p}{\partial x^2} + \frac{\partial^2 p}{\partial y^2} = v_x \frac{\partial p}{\partial x} + v_y \frac{\partial p}{\partial y} \tag{2.34}$$

$$\frac{\partial^2 p}{\partial x^2} + \frac{\partial^2 p}{\partial y^2} + \frac{\partial^2 p}{\partial z^2} = v_x \frac{\partial p}{\partial x} + v_y \frac{\partial p}{\partial y} + v_z \frac{\partial p}{\partial z}$$

where v_x, v_y, and v_z are velocities of the coordinate systems in the x, y, and z directions, respectively. These terms may be constant, or they may themselves be functions of the space variables or of the potential p.

Another important modification of Laplace's equation is required if distributed sources or sinks of flux are active within the field. Consider, for example, a conductor made of metal. Under steady-state conditions the temperature distribution within this medium will be governed by Laplace's equation, Eq. (2.26). However, if an electrical current is made to flow through this material, heat will be generated at every point within the field (the I^2R loss) and will therefore affect the temperature distribution.

A similar phenomenon occurs if one considers the steady-state temperature distribution within the core of a nuclear reactor in which heat is generated by distributed radioactive particles. Under these conditions, the differential elements of Figures 2.3b, 2.4b, and 2.5b for one-, two-, and three-dimensional fields, respectively, must be modified by the inclusion of an additional flux input representing the distributed internal source. The application of the conservation principle then leads to

$$f_1 - f_2 = f_i$$
$$F_1 - F_2 + F_3 - F_4 = -f_i\,\Delta x\,\Delta y \tag{2.35}$$
$$F_1 - F_2 + F_3 - F_4 + F_5 - F_6 = -f_i\,\Delta x\,\Delta y\,\Delta z$$

where f_i is the additional flux per unit length, area, or volume, respectively, flowing into the element as a result of the distributed source. Laplace's equation then takes the general form

$$\nabla\left(\frac{1}{D}\nabla p\right) = -kf_i \tag{2.36}$$

for $D = D(x, y, z)$. Also, f_i can itself be a function of the space coordinates or of p. This equation is known as Poisson's equation.

The boundary conditions associated with the modified forms of Laplace's equation described in this section are identical to those for the basic form of Laplace's equation as expressed by Eq. (2.28).

2.6. ELLIPTIC PARTIAL DIFFERENTIAL EQUATIONS: GENERAL PROPERTIES AND APPLICATIONS

A number of general conclusions may be drawn from the foregoing analysis. First, it is apparent that Laplace's equation does not contain the independent time variable. This is a direct result of the fact that only one of the three types of parameters (dissipation) was assumed to be present within the system. The direct consequence of the absence of the other two types of parameters (reservoirs of potential and flux) is that, if the excitation is a function of time, the response everywhere within the system will have exactly the same transient characteristics, and that steady state is achieved instantaneously at all points in the field. For example, if the potential applied to one of the boundaries of the field shown in Figures 2.3a, 2.4a, and 2.5a is suddenly changed (that is, a step change), the potential, p, at all points within the field will change at the same instant to new values. This can also be shown to be true if the field contains only potential reservoirs or only flux reservoirs. The key point is that, if a field is composed entirely of one of the three basic element types, time is not an independent variable. The time variable also disappears, and Laplace's equation applies in fields that contain reservoirs as well as dissipation, provided that enough time has elapsed since a previous change in excitation so that steady-state conditions have been reached.

A second important feature of all the foregoing models, with the exception of those characterized by Eq. (2.36), is that the excitations are applied only at the boundaries or extremities of the fields. No potential or flux sources appear within the field. According to the conservation principle, therefore, all the energy or matter entering or leaving the field must be accounted for by external sources, and all lines of flux must terminate at boundaries. This implies directly that there can be no maxima or minima of potential within the field. Since the presence of such a potential maximum would imply that some point within the field has a higher potential than all neighboring points, and since flux always flows from a high potential to the low potential, such a maximum would mean that matter or energy emanates in all directions from this point. In vector analysis such a phenomenon is known as ***divergence***, and fields governed by Laplace's equation must have zero divergence. Likewise, a minimum of potential would imply a sink of matter or energy and again would violate the continuity and conservation principles. This means also that potential gradients cannot change polarity within the field.

The reason for the wide applicability of Laplace's and Poisson's equations is that systems containing only one of the three basic types of elements occur in a great many physical areas and that the conservation and continuity principles are applicable to these systems. A complete review of all the areas of application of these powerful equations would require the inclusion of lengthy and intricate analyses of many highly specialized and esoteric problems. Therefore, only the most often encountered and important applications will be summarized here.

Three basic body force fields are encountered in classical physics: the fields generated by masses, electric charges, and magnetic poles. All regions of free space that do not contain these field excitations are systems governed by Laplace's equation. Gravitational, electrostatic, and magnetic fields are therefore subject to the same type of analysis as that performed previously. The across variable in each of these three areas is termed a potential function and has a significance analogous to that of the potential, p, employed previously. Thus the symbol ϕ is frequently used to indicate a general potential function.

The concept of potential functions can also be extended to certain fluid flow systems where ϕ is termed the "velocity potential" and is defined by

$$\begin{aligned} \frac{\partial \phi}{\partial x} &= -v_x \\ \frac{\partial \phi}{\partial y} &= -v_y \\ \frac{\partial \phi}{\partial z} &= -v_z \end{aligned} \tag{2.37}$$

where v_x, v_y, and v_z represent the velocity of a fluid particle in the x, y, and z directions, respectively. Since Laplace's equation applies only to systems containing one type of element, a careful examination of the fluid system must precede the application of that equation. If the liquid particles are forced through a medium containing very small pore channels such as compacted sand, the forces upon the fluid due to viscous friction are much larger than the inertial forces; and if, furthermore, the fluid is assumed to be incompressible so there can be no storage of fluid particles, the system is purely dissipative, and Laplace's equation may be applied.

When an incompressible liquid flows in an open channel or a pipe, it is frequently permissible to assume that it is an "ideal" fluid having negligible viscosity. Under these conditions, the inertial (kinetic energy) forces are the only ones that have to be considered, and this system is again governed by Laplace's equation. The through variable in a fluid flow system is the translational velocity of the fluid particles. Since fluid particles may also rotate about their own axes, it is necessary that the flow be irrotational if Laplace's equation is to be applicable.

In mechanics, Laplace's equation is used to describe the static deflection of elastic membranes or sheets having negligible mass. These systems act as pure springs (reservoirs of potential). In problems of heat transfer, it is generally not practicable to describe a system entirely by means of one of the three parameter types. Heat flow systems in which the excitation varies with time are, therefore, not governed by Laplace's equation. If steady-state or static conditions have been reached, however, the reservoirs of potential have acquired all the heat that they can store and therefore exercise no further influence upon the heat conduction phenomenon. Under these conditions, the temperature within the system is a potential function subject to the same equation and analogous boundary conditions as those described for the preceding systems.

Poisson's equation arises in the above-mentioned disciplines wherever there are distributed sources. The gravitational field within a solid (such as the earth), the magnetic field within a magnet, and the electrostatic field within a vacuum containing a space charge are all examples of such a situation. The equation also arises in mechanics in the study of torsional stresses and strains, in heat transfer where distributed sources of heat are present, and in many other disciplines.

In the equations derived in this chapter, the potential, p, is the dependent variable. The solutions of these equations yield values of the potential as a function of x, y, and z. Digital computers usually generate solutions in tabular form, providing the potentials at discretely spaced points in the x, y, z domain. On occasion, the solution of problems in two space dimensions and for cross sections of a three-dimensional field are provided as plots of lines of equal potential, similar to topological contour maps. Sometimes a solution for the flux distribution is preferred to a solution for the field potentials. In accordance with Eq. (2.2), flux, f, is proportional to the potential gradient. The flux vector at any point in the field is therefore perpendicular to the line of constant potential, since the potential gradient is zero parallel to such a line. This implies that the lines of constant flux are orthogonal to the lines of equal potential at all points in a field. Once the values for the potential have been determined in a field problem, the solution for the flux at all points in the field can therefore be obtained very easily. Alternatively, flux can be employed as the dependent variable in the derivation of the field equations. This results in partial differential equations that are structurally similar to those already derived for potentials.

2.7. THE LAPLACIAN IN CURVILINEAR COORDINATE SYSTEMS

The purpose of any coordinate system is to specify uniquely the location of any point within a field with respect to some arbitrary reference point, termed the origin. The number of items of information that must be employed, that is, the number of independent space variables, is equal to the number of dimensions of the field under study. To this point the discussion has been limited to the Cartesian coordinate system in which the distance from three mutually perpendicular linear axes intersecting at (0, 0, 0) determines the coordinates of a point within the field, as shown in Figure 2.6a.

Another widely used coordinate system is the cylindrical coordinate system shown in Figure 2.6b. The z-coordinate is specified in the same manner as in the Cartesian coordinate system. The position in the xy-plane, however, is given in terms of the polar coordinates r and θ. If a straight line normal to the z-axis is drawn from the z-axis to the point in question, the length of that line is equal to r, and the angle

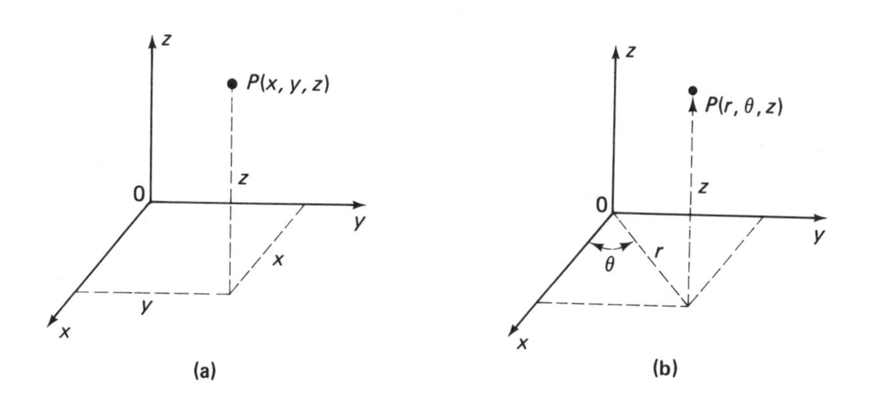

Figure 2-6 Most widely used spatial coordinate systems: (a) Cartesian coordinates; (b) cylindrical coordinates;

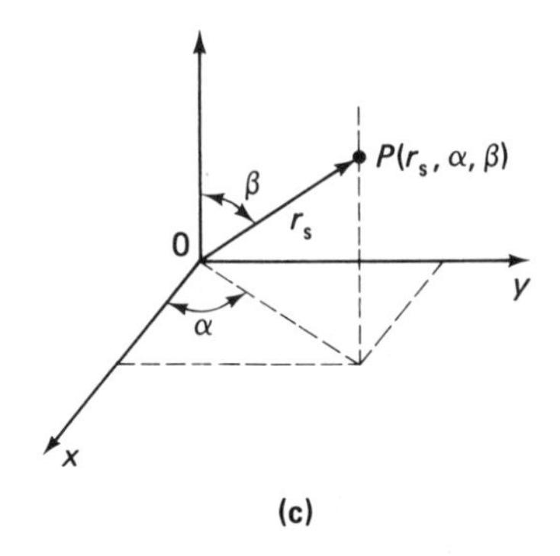

Figure 2-6 (*continued*) (c) spherical coordinates

that this line makes with the x-axis is equal to θ. Thus the three space coordinates in this system are z, r, and θ. This coordinate system is particularly useful if the potential distribution is symmetrical about a straight line, the z-axis. In that case, the angle θ need not be specified, since the potential is constant for any given r and z. The number of independent space variables can therefore be reduced from three to two. Similarly, if the potential in the cylindrical field is the same for every cross section normal to the axis of the cylinder, the z-coordinate need not be specified, and r and θ are the only independent space variables.

The third major coordinate system is the spherical coordinate system, illustrated in Figure 2.6c. Here a line is drawn from the origin to the point under consideration. The length of this line is equal to r_s, the angle that this line makes with the z-axis is identified as β, and the angle that the projection of this line on the xy-plane makes with the x-axis is identified as α. A linear dimension and two angles therefore comprise the specification of a point in spherical coordinates. Spherical coordinates are particularly appropriate where the potential function is symmetrical about a point in space, since α and β need not be specified under these circumstances, so r_s becomes the only independent variable.

To express Laplace's equation, Eqs. (2.26), in cylindrical and spherical coordinates, it is only necessary to make simple trigonometric transformations upon x, y, and z in Eqs. (2.12), (2.18), or (2.24). The appropriate transformation equation and the expressions for the Laplacian in three dimensions are presented in Table 2.2.

2.8. PARABOLIC PARTIAL DIFFERENTIAL EQUATIONS: BASIC AND MODIFIED FORMS

Ranking with Laplace's equation as one of the most important and fundamental equations of applied physics is the parabolic equation known variously as the diffusion or conduction equation. This equation is also first derived for a simple one-dimensional system such as that shown in Figure 2.7. The field in this case contains distributed

Table 2.2. Relation between principal coordinate systems

Coordinate System	To Convert to: Cartesian	Cylindrical	Spherical	$\nabla^2\phi$
Cartesian x, y, z	$x = x$ $y = y$ $z = z$	$r = \sqrt{x^2 + y^2}$ $\theta = \cos^{-1} \dfrac{x}{\sqrt{x^2 + y^2}}$ $z = z$	$r_s = \sqrt{x^2 + y^2 + z^2}$ $\alpha = \cos^{-1} \dfrac{x}{\sqrt{x^2 + y^2}}$ $\beta = \cos^{-1} \dfrac{z}{\sqrt{x^2 + y^2 + z^2}}$	$\dfrac{\partial^2\phi}{\partial x^2} + \dfrac{\partial^2\phi}{\partial y^2} + \dfrac{\partial^2\phi}{\partial z^2}$
Cylindrical r, θ, z	$x = r\cos\theta$ $y = r\sin\theta$ $z = z$	$r = r$ $\theta = \theta$ $z = z$	$r_s = \sqrt{r^2 + z^2}$ $\alpha = 0$ $\beta = \cos^{-1} \dfrac{z}{\sqrt{r^2 + z^2}}$	$\dfrac{1}{r}\dfrac{\partial}{\partial r}\left(r\dfrac{\partial\phi}{\partial r}\right) + \dfrac{1}{r^2}\dfrac{\partial^2\phi}{\partial\theta^2} + \dfrac{\partial^2\phi}{\partial z^2}$
Spherical r_s, α, β	$x = r_s \sin\beta\cos\alpha$ $y = r_s \sin\beta\sin\alpha$ $z = r_s\cos\beta$	$r = r_s\sin\beta$ $\theta = \alpha$ $z = r_s\cos\beta$	$r_s = r_s$ $\alpha = \alpha$ $\beta = \beta$	$\dfrac{1}{r^2}\dfrac{\partial}{\partial r}\left(r^2\dfrac{\partial\phi}{\partial r}\right) + \dfrac{1}{r^2\sin^2\beta}\dfrac{\partial^2\phi}{\partial\alpha^2} + \dfrac{1}{r^2\sin\beta}\dfrac{\partial}{\partial\beta}\left(\sin\beta\dfrac{\partial\phi}{\partial\beta}\right)$

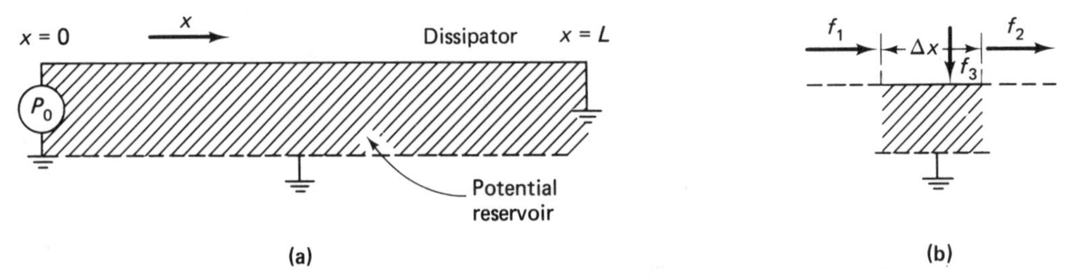

Figure 2-7 (a) One-dimensional field containing dissipation and potential reservoir properties; (b) elemental line segment

dissipation, just as in Figure 2.3, but also distributed reservoir properties (potential or flux). Figure 2.7a may describe an electrical resistor (dissipator) coupled to ground by means of a dielectric (potential reservoir), or it may represent a thermal conductor manifesting thermal resistance (dissipation) and heat capacity (potential reservoir). The differential element is shown in Figure 2.7b.

In addition to the flux f_1 and f_2 along the conductor there is now also flux f_3, which enters the reservoir. The conservation principle now specifies that

$$f_1 - f_2 - f_3 = 0 \tag{2.38}$$

If the storage properties per unit length are expressed by E_p, the flux serving to "fill" the reservoir comprised by the differential element can be expressed in accordance with Eq. (2.3) as

$$f_3 = f_1 - f_2 = E_p \, \Delta x \frac{\partial p}{\partial t} \tag{2.39}$$

where p refers to the average potential (with respect to ground) of the differential element, and $E_p \, \Delta x$ is the total reservoir capacity of this element. The difference between the flux f_1 and f_2 expressed in terms of the rate of change of flux per unit length is

$$f_1 - f_2 = -\frac{\partial f}{\partial x} \Delta x \tag{2.40}$$

if higher-order terms of a Taylor series expansion are neglected. But in accordance with Eq. (2.2), the flux f can be expressed in terms of the potential gradient as

$$f = -\frac{1}{D} \frac{\partial p}{\partial x} \tag{2.41}$$

where again D represents the dissipation per unit length.

Differentiating Eq. (2.41) with respect to x yields

$$-\frac{\partial f}{\partial x} = \frac{\partial}{\partial x}\left(\frac{1}{D} \frac{\partial p}{\partial x}\right) \tag{2.42}$$

Combining Eqs. (2.39), (2.40), and (2.42) yields

$$\frac{\partial}{\partial x}\left(\frac{1}{D} \frac{\partial p}{\partial x}\right) = E_p \frac{\partial p}{\partial t} \tag{2.43}$$

If the dissipation D is not a function of x, that is, if the conductor is uniform,

$$\frac{\partial^2 p}{\partial x^2} = DE_p \frac{\partial p}{\partial t} \tag{2.44}$$

which is known as the diffusion equation in one dimension. A complete specification of the problem includes the potentials $p(0, t)$ and $p(L, t)$ at the two ends of the field for all time, subsequent to an initial time $t = 0$, as well as the potential $p(x, 0)$ everywhere along the field at the initial instant. This type of problem therefore requires more information than the one-dimensional Laplace's equation. Given Eq. (2.43) or (2.44) along with the necessary boundary and initial conditions, the potential everywhere along the field for all time $t > 0$ can be determined.

The corresponding two-dimensional system is a sheet made of a dissipative material of dissipativity D having reservoir capabilities described by E_p per unit area, as shown in Figure 2.8. Such a system might be a sheet of electrically resistive material

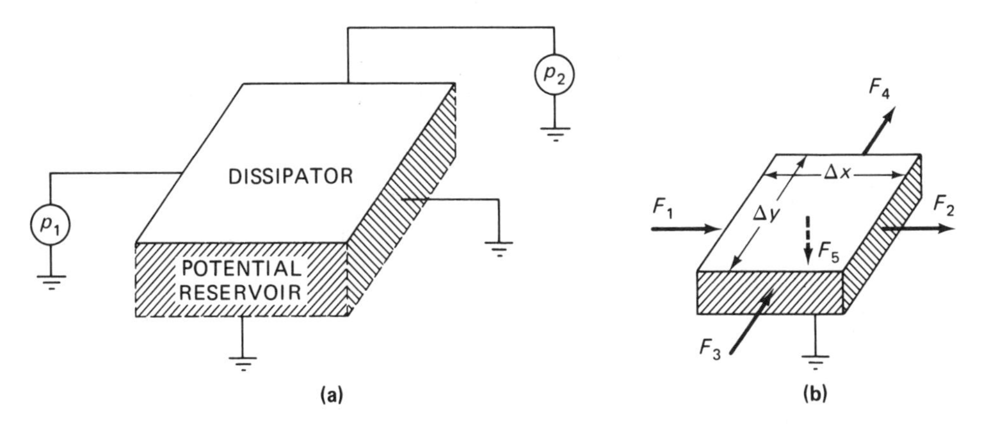

Figure 2-8 (a) Two-dimensional field containing dissipation and potential reservoir properties; (b) elemental rectangle

coupled to ground by a dielectric, or it may be a thin sheet of thermally conductive material. A comparison of the differential element shown in Figure 2.4b with that of Figure 2.8b suggests that the analysis of the latter system can be accomplished by modifying the equations pertaining to the Laplacian field to include the quantity stored in the reservoir. Equations (2.14) and (2.15) still apply. Now, however, according to the conservation principle

$$F_1 - F_2 + F_3 - F_4 = F_5 \tag{2.45}$$

But the flux F_5, serving to feed the reservoir, is related to the rate of change of potential and the total reservoir capacity of the element by

$$F_5 = E_p \, \Delta x \, \Delta y \frac{\partial p}{\partial t} \tag{2.46}$$

where p is again the average potential existing within the element. Inserting Eqs. (2.15) and (2.46) in Eq. (2.45) yields

$$\frac{\Delta x\,\Delta y}{D}\left(\frac{\left(\frac{\partial p}{\partial x}\right)_2-\left(\frac{\partial p}{\partial x}\right)_1}{\Delta x}+\frac{\left(\frac{\partial p}{\partial y}\right)_4-\left(\frac{\partial p}{\partial y}\right)_3}{\Delta y}\right)=E_p\,\Delta x\,\Delta y\,\frac{\partial p}{\partial t} \tag{2.47}$$

When $\Delta x \longrightarrow 0$ and $\Delta y \longrightarrow 0$, Eq. (2.45) becomes the diffusion equation in two dimensions,

$$\frac{\partial}{\partial x}\left(\frac{1}{D}\frac{\partial p}{\partial x}\right)+\frac{\partial}{\partial y}\left(\frac{1}{D}\frac{\partial p}{\partial y}\right)=E_p\frac{\partial p}{\partial t} \tag{2.48}$$

If D is not a function of x or y, Eq. (2.48) becomes

$$\frac{\partial^2 p}{\partial x^2}+\frac{\partial^2 p}{\partial y^2}=DE_p\frac{\partial p}{\partial t} \tag{2.49}$$

with necessary boundary conditions similar to those expressed by Eqs. (2.19). Now, however, an initial condition is required specifying $p(x, y, 0)$, the potential at every point within the field at $t = 0$.

The governing equation for a three-dimensional field containing dissipation and one type of reservoir can be derived in a similar fashion. In this case the differential element is a cube similar to that shown in Figure 2.5b, but containing, in addition, distributed reservoir properties E_p per unit cube. From the preceding development it may readily be inferred that in the case of a three-dimensional diffusion field

$$\frac{\partial}{\partial x}\left(\frac{1}{D}\frac{\partial p}{\partial x}\right)+\frac{\partial}{\partial y}\left(\frac{1}{D}\frac{\partial p}{\partial y}\right)+\frac{\partial}{\partial z}\left(\frac{1}{D}\frac{\partial p}{\partial z}\right)=E_p\frac{\partial p}{\partial t} \tag{2.50}$$

For uniform D,

$$\frac{\partial^2 p}{\partial x^2}+\frac{\partial^2 p}{\partial y^2}+\frac{\partial^2 p}{\partial z^2}=DE_p\frac{\partial p}{\partial t} \tag{2.51}$$

where in this case the boundary conditions are similar to those expressed in Eq. (2.19) plus the initial condition $p(x, y, z, 0)$, the potentials at all points within the field at the initial instant.

Introducing again the Laplacian operator, Eqs. (2.44), (2.49), and (2.51) may be generalized to read

$$\nabla^2 p = k\frac{\partial p}{\partial t} \tag{2.52}$$

where k is determined by the parameters of the system. The boundary conditions pertinent to this equation are the same as those for Laplace's equation, expressed by Eq. (2.28). In addition, an initial condition specifying the quantity stored in the field reservoir at the initial time must be furnished. If reservoirs of potential are present, the initial potential distribution $p(x, y, z, 0)$ must be given; if reservoirs of flux are present, the initial rate of change of potential $\partial p(x, y, z, 0)/\partial t$ may be furnished instead.

All the modifications discussed in Section 2.7 in connection with elliptic partial differential equations also apply to parabolic partial differential equations. Where the medium in which diffusion is taking place is itself in motion, that is, where the coordinate system is moving, the diffusion equation in one, two, and three dimensions becomes

$$\frac{\partial}{\partial x}\left(\frac{1}{D}\frac{\partial p}{\partial x}\right) = E_p\frac{\partial p}{\partial t} + v_x\frac{\partial p}{\partial x}$$

$$\frac{\partial}{\partial x}\left(\frac{1}{D}\frac{\partial p}{\partial x}\right) + \frac{\partial}{\partial y}\left(\frac{1}{D}\frac{\partial p}{\partial y}\right) = E_p\frac{\partial p}{\partial t} + v_x\frac{\partial p}{\partial x} + v_y\frac{\partial p}{\partial y} \tag{2.53}$$

$$\frac{\partial}{\partial x}\left(\frac{1}{D}\frac{\partial p}{\partial x}\right) + \frac{\partial}{\partial y}\left(\frac{1}{D}\frac{\partial p}{\partial y}\right) + \frac{\partial p}{\partial z}\left(\frac{1}{D}\frac{\partial p}{\partial z}\right) = E_p\frac{\partial p}{\partial t} + v_x\frac{\partial p}{\partial x} + v_y\frac{\partial p}{\partial y} + v_z\frac{\partial p}{\partial z}$$

where v_x, v_y, and v_z are the velocities of the medium in the x, y, and z directions, respectively. Such fields are said to involve combined mass transfer and diffusion.

The presence of distributed sources of flux in systems containing dissipation and one type of reservoir parameter entail the addition of a term proportional to the distributed source to the right-hand side of the diffusion equation. For nonuniform fields in one, two, and three space dimensions, the characterizing equations become

$$\frac{\partial}{\partial x}\left(\frac{1}{D}\frac{\partial p}{\partial x}\right) = E_p\frac{\partial p}{\partial t} - f_i$$

$$\frac{\partial}{\partial x}\left(\frac{1}{D}\frac{\partial p}{\partial x}\right) + \frac{\partial}{\partial y}\left(\frac{1}{D}\frac{\partial p}{\partial y}\right) = E_p\frac{\partial p}{\partial t} - f_i \tag{2.54}$$

$$\frac{\partial}{\partial x}\left(\frac{1}{D}\frac{\partial p}{\partial x}\right) + \frac{\partial}{\partial y}\left(\frac{1}{D}\frac{\partial p}{\partial y}\right) + \frac{\partial}{\partial z}\left(\frac{1}{D}\frac{\partial p}{\partial z}\right) = E_p\frac{\partial p}{\partial t} - f_i$$

where f_i has the units of flux/unit length, flux/unit area, and flux/unit volume, respectively.

2.9. PARABOLIC PARTIAL DIFFERENTIAL EQUATIONS: GENERAL PROPERTIES AND APPLICATIONS

In the foregoing derivation, the nature of the reservoir, whether it be a reservoir of a potential or of flux, was not specified. It was only assumed that one *or* the other type of parameter was present. In general terms, the preceding development leads to the conclusion that the diffusion equation, Eq. (2.52), is characteristic of systems that are dissipative and contain *one* type of reservoir parameter. This parameter may be of the potential or flux variety, but only one type may be present in the system. The presence of the combination of this parameter and the dissipative parameter implies that the response does not achieve its final value instantaneously as in the case of fields governed by Laplace's equation. The dissipative parameter is sometimes known as a ***damping term*** and acts in combination with the reservoir parameter to delay the attainment of final values. The constant k, determined by the system parameters, is a measure of this delay. Time is therefore very definitely an independent variable in such field problems.

The presence of only one rather than both types of storage elements implies that the final values of potential in a field, in response to a sudden change in excitations at the boundaries, are approached monotonically; that is, there is no change in polarity of potential gradients nor overshooting of the final value of the field potentials. For example, if in Figure 2.7a the initial condition specifies that the potential is everywhere equal to zero, and if the potential source p_0 takes the form of a step of magnitude Q at time $t = 0$, the final or steady-state potential everywhere along the system, $p(x, \infty) = Q(1 - x/L)$, is approached asymptotically at all points along the field. At no time is the potential anywhere in the field higher than the final value, and at no time and at no place is the rate of change of potential negative. For this reason the diffusion equation is sometimes known as the equalization equation. The rate at which this final value is approached is determined by the dissipative and storage parameters. While Laplace's equation for uniform fields is independent of the field characteristics, the same is therefore not true of the diffusion equation. The parameters of the field must be furnished to permit the prediction of the transient field potentials.

It is apparent that if steady-state conditions exist, that is, if none of the excitations is a function of time and if sufficient time has elapsed since any previous change in excitation, the $\partial p/\partial t$ term in the diffusion equation vanishes, and the Laplacian is equal to zero. Laplace's equation may therefore be considered a special or degenerate case of the diffusion equation. This implies, as pointed out in the preceding section, that field problems governed by the diffusion equation are described by Laplace's equation if static or steady-state conditions exist.

The diffusion equation finds application in all areas in which there arise problems involving two types of elements only one of which is a reservoir. In heat transfer problems, the system under study is generally a three-dimensional field containing thermal resistance and thermal capacitance, which act, respectively, as dissipative and reservoir of potential parameters. Thermal resistance is not an energy dissipative parameter in the same sense as an electrical resistor, in that no energy is actually converted. Nevertheless, from considerations of entropy it may be reasoned that the two resistances play analogous roles in their respective areas, in that both act to delay or dampen the achievement of final response values. Hence the diffusion equation permits the determination of the temperature everywhere within the field for all time after the "initial time," provided that the temperature or heat flux at the field boundaries, as well as the temperature distribution within the field, at the "initial time" is specified.

As the name implies, the diffusion equation also describes the diffusion of one type of fluid particles in a space occupied by a different fluid. For example, it describes the diffusion of carbon monoxide in motionless air or of ink in stagnant water. In such problems, attention is usually focused on the concentration of one of the two fluids, and this concentration or density, denoted by ρ, comprises the dependent variable of the system. The flux or flow of particles (through variable) is then related to the gradient of the density of the particles (across variable) by the diffusion constant (dissipative or damping term), which is related in turn to the freedom with which molecules are able to move within the field. At the same time there occurs a storage of fluid particles in every volume element of the field. This storage is proportional to the density of the particles. Hence every volume element of the field comprises a reservoir of potential. The diffusion equation under these conditions is generally expressed as

$$\nabla^2 \rho = \frac{1}{k}\frac{\partial \rho}{\partial t} \tag{2.55}$$

where ρ is the density of the particles and k is termed the diffusivity. These same considerations also apply to the absorption of fluid particles by solids and to the drying of porous solids saturated with liquids.

In Section 2.2 it was pointed out that, when a fluid flows through an open channel or porous medium, the inertia of the fluid particles corresponds to a reservoir of flux, the viscous forces to dissipation, and the compressibility effects to a potential reservoir. In problems of irrotational incompressible flow in which viscous forces and inertial forces arise, the diffusion equation can therefore be applied to predict the velocity potential or pressure at points within the flow stream. The same is true for problems involving the analysis of viscous flow of compressible fluids when inertial forces can be neglected, as is the case in gas flow in a porous medium.

In electromagnetics, Maxwell's equations reduce to the form of the diffusion equation in fields that have conductivity but in which either the permeability or dielectric constant may be neglected. Some caution is required in the application of these equations, however, since the potential function in this case is a vector having both magnitude and direction, while it is a scalar in the other field problems previously discussed.

2.10. HYPERBOLIC PARTIAL DIFFERENTIAL EQUATIONS: THE WAVE EQUATION

The third fundamental equation of physics is termed the wave equation and describes the familiar phenomenon of wave motion. In this case the one-dimensional system, such as that shown in Figure 2.9a, has distributed parameters that are reservoirs of

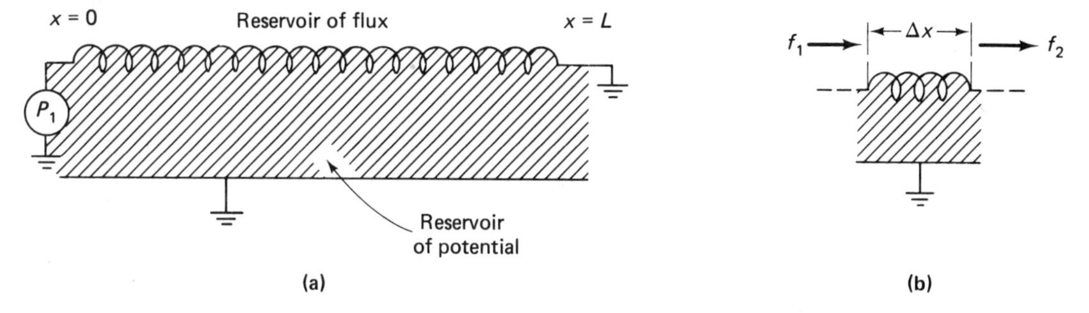

Figure 2-9 (a) One-dimensional field containing flux reservoir and potential reservoir properties; (b) elemental line segment

potential as well as of flux, but no dissipation is present. Such a system may represent an electric transmission line having negligible resistance but appreciable inductivity and dielectric coupling to ground. Referring to the differential element shown in Figure 2.9b, in accordance with Eq. (2.4), the rate of change of flux along the length of the system is expressed as

$$-\frac{\partial f}{\partial x} = E_p \frac{\partial p}{\partial t} \tag{2.56}$$

where E_p is descriptive of the potential reservoir capabilities per unit length. Similarly, the potential gradient is related to the rate of change of flux by parameter according to

$$-\frac{\partial p}{\partial x} = E_f \frac{\partial f}{\partial t} \tag{2.57}$$

where E_f represents the flux storage capability per unit length. Differentiating Eq. (2.56) with respect to time and Eq. (2.57) with respect to x yields

$$\begin{aligned} -\frac{\partial^2 f}{\partial x\,\partial t} &= E_p \frac{\partial^2 p}{\partial t^2} \\ -\frac{\partial^2 f}{\partial x\,\partial t} &= \frac{1}{E_f}\frac{\partial^2 p}{\partial x^2} \end{aligned} \tag{2.58}$$

Combining these two equations yields

$$\frac{\partial^2 p}{\partial x^2} = E_p E_f \frac{\partial^2 p}{\partial t^2} \tag{2.59}$$

which is known as the wave equation in one dimension. To predict the transient potential along the one-dimensional field, the potentials at $x = 0$ and $x = L$ must be known for all time, as in Laplace's equation and the diffusion equation, and in addition two initial conditions are required. These initial conditions correspond to the potential and the flux at every point in the system at the initial instant time, $t = 0$. This may correspond to a specification of the potential $p(x, 0)$ and the rate of change of potential $\partial p(x, 0)/\partial t$.

Two-dimensional and three-dimensional fields governed by the wave equation possess distributed potential and flux reservoirs in a plane and in a three-dimensional region, respectively. The preceding analysis is readily extended to such systems and leads to

$$\frac{\partial^2 p}{\partial x^2} + \frac{\partial^2 p}{\partial y^2} = E_p E_f \frac{\partial^2 p}{\partial t^2} \tag{2.60}$$

for a two-dimensional system, and

$$\frac{\partial^2 p}{\partial x^2} + \frac{\partial^2 p}{\partial y^2} + \frac{\partial^2 p}{\partial z^2} = E_p E_f \frac{\partial^2 p}{\partial t^2} \tag{2.61}$$

for the three-dimensional case. In more general notation, Eqs. (2.59), (2.60), and (2.61) become

$$\nabla^2 p = k \frac{\partial^2 p}{\partial t^2} \tag{2.62}$$

In general, the wave equation is applicable to all those systems that include reservoirs both of potential and of flux but do not have dissipative or damping parameters. The presence of these two types of reservoirs implies an interchange of matter or energy. None of the energy applied to the system or present in the system initially can be lost. It can only be changed back and forth from kinetic to potential energy. This is what gives rise to the familiar wave or vibration patterns. In the diffusion equation, the constant k is identified with the time constant, the rate at which steady-state conditions or equilibrium conditions are approached. The significance of the constant k in the wave equation is different, since evidently there can be no equilibrium in the same sense in such a system. In the wave equation, the constant k determines rather the velocity with which a disturbance is propagated. For example, in Figure 2.9a, the smaller the $E_p E_f$ product, the smaller the constant k, and the more rapidly felt is the full effect of a sudden change of the excitation $p(t)$ from one end of the system to the other end.

In electrical systems the wave equation arises where inductance and capacitance, but no resistance, are present. This is approximately the case in the study of "ideal" transmission lines. In dynamics, pure wave motion occurs if appreciable inertial mass forces and spring forces are present, and if viscous damping may be neglected. The vibration of a drum head with negligible damping is an example of such motion. Vibrating strings may likewise exhibit these properties. In fluid dynamics, flux storage is associated with the inertia of individual fluid particles, while the storage of potential in a fluid implies compression of the fluid. If the viscous forces within the fluid system are negligible compared to the inertial forces, and if the fluid is compressible, the wave equation can therefore be expected to apply. Wave motion in fluids is frequently observed as sound waves in air or in water. The wave equation is therefore of great importance in the field of acoustics. In electromagnetics the wave equation applies to those systems in which the conductivity is negligible but which contain appreciable permeability and dielectric properties. This is the case in a vacuum and in most non-metallic materials.

2.11. HYPERBOLIC PARTIAL DIFFERENTIAL EQUATIONS: THE DAMPED WAVE EQUATION

To this point, only systems containing two types of parameters have been considered. Many systems contain both types of energy reservoirs as well as dissipation of appreciable magnitude. To derive the governing equations, the system shown in Figure 2.9a must be modified to include dissipation. For full generality it is necessary to provide for two types of dissipation: series dissipation D_s, which effects a dissipation of energy by virtue of the flow of flux, and parallel dissipation D_p, which leads to a loss of system energy by "leakage." In electric transmission line theory, for example, such a situation arises in parallel transmission lines having appreciable series resistance, as well as inductance, and appreciable leakage conductance between the two lines as well as dielectric coupling. In this case, in applying the conservation principle to express the change of flux along the x-coordinate (for example, f_1 to f_2 in Figure 2.9b), it is necessary to include energy flow into the potential energy reservoir as well as through the parallel dissipative element. Accordingly,

$$-\frac{\partial f}{\partial x} = \frac{1}{D_p}p + E_p\frac{\partial p}{\partial t} \tag{2.63}$$

where E_p is again the potential reservoir capability per unit length and D_p is the parallel dissipation per unit length. Similarly, the potential gradient is expressed as

$$-\frac{\partial p}{\partial x} = D_s f + E_f\frac{\partial f}{\partial t} \tag{2.64}$$

where E_f is again the flux reservoir capability per unit length and D_s is the series dissipation per unit length.

Equations (2.63) and (2.64) are two equations in two unknowns: p and f. The simultaneous solution of these two equations for p yields

$$\frac{\partial^2 p}{\partial x^2} = E_p E_f\frac{\partial^2 p}{\partial t^2} + \left(\frac{E_f}{D_p} + D_f E_p\right)\frac{\partial p}{\partial t} + \frac{D_s}{D_p}p \tag{2.65}$$

Equation (2.65) is sometimes known as the telegraph equation since it is descriptive of the voltage distribution along telegraph transmission lines. It is apparent that if both $1/D_p$ and D_s in Eq. (2.65) are equal to zero, the equation reduces to the wave equation. If, on the other hand, E_f and $1/D_p$ are zero, the diffusion equation results. If any three of the four types of parameters E_f, E_p, $1/D_p$, or D_s are equal to zero, Eq. (2.65) becomes Laplace's equation. In more general terms, Eq. (2.65) can be written as

$$\nabla^2 p = k_1\frac{\partial^2 p}{\partial t^2} + k_2\frac{\partial p}{\partial t} + k_2 p \tag{2.66}$$

and requires the same boundary and initial conditions as the wave equation.

The characteristic response of systems governed by Eq. (2.66) to step function excitations at a boundary is one of damped oscillation. That is, an equilibrium condition is gradually approached. This approach is not necessarily monotonic, as in the case of the diffusion equation, but may involve overshooting and oscillation about the equilibrium state. In that respect this equation corresponds to the characterizing ordinary differential equation for a lumped system exhibiting underdamped, critically damped, or overdamped behavior.

In addition to electrical transmission lines, the damped wave equation finds application in all those physical systems in which all three types of parameters are present. In dynamic fields containing appreciable mass, spring forces, and viscous damping, this equation governs the motion of points within the system. A vibrating string or elastic sheet generally responds to a sudden blow or excitation with sinusoidal oscillations, which gradually die out or reduce to zero. In fluid dynamic systems where the fluid is compressible and both viscous forces and inertial forces are appreciable, the damped wave equation again applies. Likewise, electromagnetic field problems for systems containing appreciable permeability, dielectric properties, and conductivity are governed by Eq. (2.66).

2.12. BIHARMONIC EQUATIONS

A special class of equations arises in the theory of elasticity. These equations are generally similar to the elliptic, parabolic, and hyperbolic partial differential equations derived previously, but contain space derivatives of fourth order instead of second order. This complication arises from the fact that in stress analysis the through variable (the stress) is not a vector quantity as in the preceding examples but is rather described by a tensor. To specify completely a vector quantity, such as current or heat flux, it is necessary to know only its magnitude and direction. In the case of a tensor quantity, additional information must be supplied. For the stress tensor, six components must be known before the stress is completely defined. Three of these are the same as for vectors, the components in the x, y, and z directions. Three other components are necessary to define a plane to which the stress is referred. In stress analysis one is actually concerned with two types of stresses, the normal stress and the shear stress; accordingly, a specification of the total stress contains more information than the specification of current in electrodynamics or heat flux in heat transfer systems.

The basic laws of elasticity corresponding to the general conservation principles are the equations of equilibrium and compatibility. In the general application of these equations to relate the stress and the strain in an elastic body, it is convenient to define a stress function ϕ according to

$$\begin{aligned} \frac{\partial^2 \phi}{\partial x^2} &= \sigma_y \\ \frac{\partial^2 \phi}{\partial y^2} &= \sigma_x \\ \frac{\partial^2 \phi}{\partial x\,\partial y} &= \rho_{xy} \end{aligned} \tag{2.67}$$

where σ_x and σ_y are the normal stresses in the x and y directions, respectively, and ρ_{xy} is the corresponding shear stress. Under static conditions, equilibrium and compatibility then lead to the biharmonic equation, which takes the form

$$\frac{\partial^4 \phi}{\partial x^4} = 0 \tag{2.68}$$

in one dimension, and

$$\frac{\partial^4 \phi}{\partial x^4} + 2\frac{\partial^4 \phi}{\partial x^2\,\partial y^2} + \frac{\partial^4 \phi}{\partial y^4} = 0 \tag{2.69}$$

for two dimensions. These equations are elliptic equations approximately analogous to Laplace's equation in other systems. The left-hand side of Eqs. (2.68) and (2.69) can be abbreviated by the biharmonic operator $\nabla^4\phi$ so that these equations can be expressed compactly as

$$\nabla^4 \phi = 0 \tag{2.70}$$

In stress problems the weight of the beam or elastic plate being studied corresponds to the internal distributed sources as described in Section 2.5. Where the weight of the elastic member is appreciable, Eq. (2.70) is modified to read

$$\nabla^4 \phi = w \tag{2.71}$$

where w is the weight per unit length or area. This expression is approximately analogous to Poisson's equation.

Under transient conditions, spring forces, characterized by Young's modulus, come into play and vibrations described by

$$\nabla^4 \phi = k \frac{\partial^2 \phi}{\partial t^2} \tag{2.72}$$

arise. In yet more general formulations, Eq. (2.72) may be modified by the addition of terms proportional to $\partial \phi / \partial t$ and ϕ.

2.13. SUMMARY

The purpose of the preceding discussion is to demonstrate how the partial differential equations characterizing physical systems can be deduced directly from physical considerations. This development is limited to fields governed in some manner by the basic principles of conservation and continuity. The development of the appropriate partial differential equations and boundary conditions then proceeds as follows:

1. Identify the across and through variables appropriate to the system, recognizing that the across variable is generally the algebraic difference between two scalar quantities and that the through variable is generally a vector.

2. Examine the characteristics of the field to determine the types of parameters that are present and that cannot be neglected in the analysis. The possible types of parameters fall into three categories: reservoirs of potential, reservoirs of flux, and dissipators.

3. Write the pertinent partial differential equation by noting the combination of parameter types present. Here the symbol ∇^2 is useful in specifying compactly that the dependent variable is to be differentiated twice with respect to each pertinent Cartesian space variable and that the sum of the second derivatives is to be taken.

4. Modify the basic partial differential equation to take into account any internal distributed sources that may be present.

5. Specify the appropriate boundary and initial conditions. The boundary conditions must specify completely and uniquely the potential or potential gradient at every extremity or boundary of the field. This specification usually takes the form either of a constant potential or a constant potential gradient. The initial conditions must specify the energy stored by every energy reservoir element within the field. If in addition to the Laplacian term the partial differential equation includes a first derivative with respect to time, one initial condition is necessary at each point in the field. If the equation includes a second derivative with respect to time, two initial conditions are required.

Table 2.3 constitutes a summary of the concepts introduced in this chapter.

Table 2.3.

CLASS OF PDE	DESCRIPTIVE NAME	TYPICAL EXAMPLES	PARAMETERS CHARACTERIZING FIELD	NATURE OF TRANSIENT SOLUTION
Elliptic	Laplace's	$\nabla^2 p = 0$	D or E_p or E_f	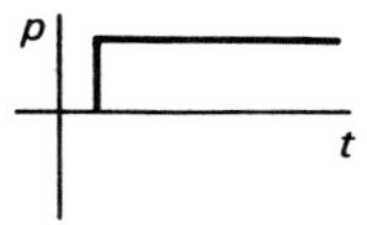
		$\nabla(\sigma \nabla p) = 0$		
	Poisson's	$\nabla^2 p = \text{k}$		
Parabolic	Diffusion	$\nabla^2 p = k \frac{\partial p}{\partial t}$	$D + (E_p$ or $E_f)$	
	Moving coordinate	$\nabla^2 p + k_1 \frac{\partial p}{\partial x} = k_2 \frac{\partial p}{\partial t}$		
Hyperbolic	Wave	$\nabla^2 p = k \frac{\partial^2 p}{\partial t^2}$	$E_p + E_f$	
	Damped wave	$\nabla^2 p = k_1 \frac{\partial^2 p}{\partial t^2} + k_2 \frac{\partial p}{\partial t} + k_3 p$	$D + E_p + E_f$	
Biharmonic	Static beam	$\nabla^4 p = 0$	E_p	
	Beam vibration	$\nabla^2 p = k \frac{\partial^2 p}{\partial t^2}$	$E_p + E_f$	
	Loaded beam	$\nabla^4 p = w$	E_p	

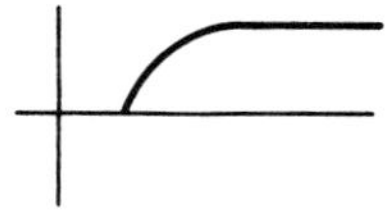

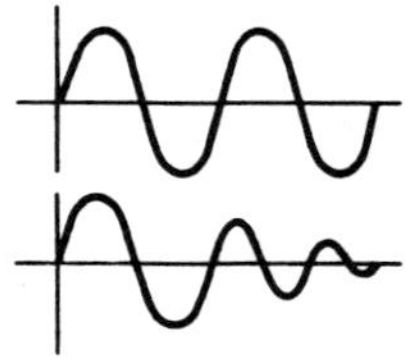

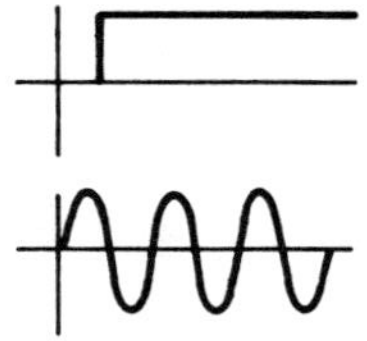

BC: $p = k; \ \frac{\partial p}{\partial n} = k$

IC: $(p)_1 = f(x, y, z, 0) \ \left(\frac{\partial p}{\partial t}\right) = f(x, y, z, 0)$

SUGGESTIONS FOR FURTHER READING

1. *The following are three famous books that present detailed analyses of the mathematical models that arise in all branches of classical physics.*

   P. N. MORSE and H. FESHBACH, *Methods of Theoretical Physics*. New York: McGraw-Hill Book Co., 1953 (2 volumes).

   A. N. TYCHONOV and A. A. SAMARSKI, *Partial Differential Equations of Mathematical Physics* (English translation). San Francisco: Holden-Day, Inc., 1964 (2 volumes).

   R. COURANT and D. HILBERT, *Methods of Mathematical Physics*. New York: John Wiley & Sons, Inc., (Interscience Division) 1962 (2 volumes).

2. *The following are briefer treatments of the solution of partial differential equations by analytic methods.*

   F. JOHN, *Partial Differential Equations*, 3rd ed. New York: Springer-Verlag New York, Inc., 1978.

   H. S. WEINBERGER, *A First Course in Partial Differential Equations*. New York: John Wiley & Sons, Inc., 1965.

   W. F. AMES, *Nonlinear Partial Differential Equations in Engineering*. New York: Academic Press, Inc., 1965.

   D. L. POWERS, *Boundary Value Problems*. New York: Academic Press, Inc., 1972.

3. *For a discussion of a variety of problems of contemporary interest, the following could serve as a useful starting point.*

   C. A. BREBBIA, (ed.), *Applied Numerical Modeling*, Proceedings of the First International Conference held at the University of Southampton, 11-15th July, 1977. New York: John Wiley & Sons, Inc., 1978.

EXERCISES

1. In Section 2.4 it is shown that in fields consisting of only one of the three basic parameter types, the potential distribution is characterized by Laplace's equation in one, two, and three space dimensions. Show that in such fields, the flux f is characterized by $\nabla^2 f = 0$.

2. By a derivation similar to that in Section 2.8, show that a one-dimensional field with distributed flux reservoir and dissipator properties is characterized by the diffusion equation

$$\frac{\partial^2 p}{\partial x^2} = k_f \frac{\partial p}{\partial t}$$

3. Derive the one-dimensional parabolic equation for systems manifesting both diffusion and dispersion:

$$\frac{\partial^2 p}{\partial x^2} = k_1 \frac{\partial p}{\partial t} + k_2 \frac{\partial p}{\partial x}$$

   from basic physical considerations. You may use either a fluid flow field or a heat transfer field for illustration.

4. Use the procedure described in Section 2.3 to derive the one-dimensional wave equation

$$\frac{\partial^2 p}{\partial x^2} = k \frac{\partial^2 p}{\partial t^2}$$

5. What conditions must exist in a *fluid flow* field in order that each of the following equations apply (where ϕ is velocity potential)? What is the physical significance of the coefficient k, and which of the three basic parameters are present in the field in each case?

(a) $\dfrac{\partial^2\phi}{\partial x^2} = k\dfrac{\partial^2\phi}{\partial t^2}$

(b) $\dfrac{\partial^2\phi}{\partial x^2} = k\dfrac{\partial\phi}{\partial x}$

(c) $\dfrac{\partial^2\phi}{\partial x^2} = k(1 + x)$

(d) $\dfrac{\partial^2\phi}{\partial x^2} = k\dfrac{\partial\phi}{\partial t}$

(e) $\dfrac{\partial}{\partial x}\left(kx\dfrac{\partial\phi}{\partial x}\right) = 0$

6. What conditions (parameters, excitations, and so on) must exist in a *heat transfer* field in order that each of the following equations be applicable. Be specific as to the significance of each concept in the heat transfer area.

(a) $\dfrac{\partial^2\phi}{\partial x^2} + \dfrac{\partial^2\phi}{\partial y^2} = 0$

(b) $\dfrac{\partial^2\phi}{\partial x^2} + \dfrac{\partial^2\phi}{\partial y^2} = k\dfrac{\partial^2\phi}{\partial t^2}$

(c) $\dfrac{\partial^2\phi}{\partial x^2} + \dfrac{\partial^2\phi}{\partial y^2} + \alpha\dfrac{\partial\phi}{\partial y} = k\dfrac{\partial\phi}{\partial t}$

(d) $\dfrac{\partial}{\partial x}\left(x^2\dfrac{\partial\phi}{\partial x}\right) = k\dfrac{\partial\phi}{\partial t}$

(e) $\dfrac{\partial^2\phi}{\partial x^2} + \dfrac{\partial^2\phi}{\partial y^2} = k\phi^\alpha$

7. Consider the following four partial differential equations:

(a) $\dfrac{\partial}{\partial x}\left(\dfrac{1}{x}\dfrac{\partial\phi}{\partial x}\right) + \dfrac{\partial}{\partial y}\left(y^2\dfrac{\partial\phi}{\partial y}\right) = 0$

(b) $\dfrac{\partial^2\phi}{\partial x^2} - \dfrac{t}{x}\dfrac{\partial\phi}{\partial t} = 0$

(c) $\dfrac{\partial^2\phi}{\partial x^2} + x^2\dfrac{\partial\phi}{\partial x} = 0$

(d) $\dfrac{\partial^2\phi}{\partial x^2} - \cos\phi + x = 0$

For each of these equations, describe a physical field problem that would be characterized by that equation. Indicate clearly what parameters are nonlinear, time varying, or nonuniform, what internal excitation exists, and any other special features. Your examples should be drawn from the following areas: heat transfer, fluid dynamics, or electrostatics.

8. For each of the following three physical situations:

- Write the governing partial differential equation (PDE).
- State whether the equation is elliptic, parabolic, hyperbolic, or biharmonic.

- Describe the boundary conditions and intial conditions that must be specified to permit the solution of the PDE.

(a) An *incompressible* fluid is flowing through a *porous* medium, which has the shape of a rectangular solid (three dimensions). The fluid conductivity in the x and y directions is everywhere equal to 5 (in appropriate units); the fluid conductivity in the z direction is given by $D = (10 + z + \frac{1}{2}z^2)$, where $z = 0$ is the bottom of the field. (In fluid flow in a porous medium the dissipative parameter is the viscosity and is inversely proportional to the fluid conductivity; compressibility is the reservoir of potential parameter; the inertia of the fluid is the reservoir of flux parameter and is negligible.)

(b) It is desired to study the transient temperature distribution in a square flat plate. The thermal conductivity is everywhere equal to 7 (in appropriate units); the heat capacity (specific heat) is determined by the relation $E_p = (1 + e^{T-T_0})$, where T is the temperature and $T_0 = 32°\text{F}$; the plate contains distributed radioactive matter that generates heat uniformly throughout the plate but that is decaying in time, so that the heat flux generated is given by $i = 10e^{-0.1t}$.

(c) It is desired to determine the steady-state temperature distribution in a long thin cylindrical rod of length 100, which is a core in a nuclear reactor. Distributed nonuniformly throughout the rod is radioactive matter, and heat flux is generated by the radioactive particles. The particles are densest at $x = 0$ and their density decreases linearly with x as x goes from 0 to 100. The thermal conductivity and heat capacity of the rod are uniform.

9. Consider a one-dimensional fluid-flow field characterized by each of the following equations. For each equation:

- State whether the equation is elliptic, parabolic, or hyperbolic.
- State whether the field is uniform.
- State whether there are internal distributed sources of fluid.
- State whether viscosity (dissipative parameter) is present in nonnegligible magnitude.
- Assume that the potential at $x = 10$ is maintained at zero and that a unit step excitation of potential is applied at $x = 0$ as shown. In as much detail as possible, sketch the transient potential $\phi(5, t)$ at $x = 5$. Assume zero initial conditions.

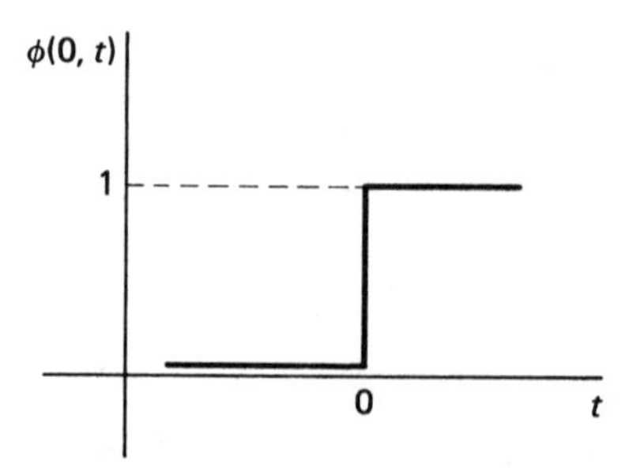

- With reference to the response function plotted above, indicate the significance of the parameter k. What is the effect upon this curve of doubling the magnitude of this parameter? Indicate by a sketch.

(a) $\dfrac{\partial}{\partial x}\left[(a + kx)\dfrac{\partial \phi}{\partial x}\right] = 0, \qquad 0 \leq x \leq 10$

(b) $\dfrac{\partial^2 \phi}{\partial x^2} = k\dfrac{\partial^2 \phi}{\partial t^2} + k\dfrac{\partial \phi}{\partial t}, \qquad 0 \leq x \leq 10$

(c) $\dfrac{\partial}{\partial x}\left(a\dfrac{\partial \phi}{\partial x}\right) = k\dfrac{\partial \phi}{\partial t} + k\phi^2$

10. Consider the following simultaneous partial differential equations:

$$\frac{\partial^2 \phi}{\partial x^2} + \frac{\partial^2 \phi}{\partial y^2} + \frac{\partial^2 \phi}{\partial z^2} = k\frac{\partial \phi}{\partial t} + g\psi^2$$

$$\frac{\partial^2 \psi}{\partial x^2} + \frac{\partial^2 \psi}{\partial y^2} + \frac{\partial^2 \psi}{\partial z^2} = ht^2 + n$$

where g, h, k, and n are constant and

$$0 \leq t \leq 1$$
$$0 \leq x \leq 1$$
$$0 \leq y \leq 1$$
$$0 \leq z \leq 1$$
$$\phi = \psi = 0 \text{ on all boundaries for } 0 \leq t \leq 1$$
$$\phi = \psi = 0 \text{ everywhere at } t = 0$$

These equations comprise the mathematical model for complex physical and chemical reactions arising in certain particle diffusion situations. Note that there are two dependent variables ϕ and ψ and that the two partial differential equations are coupled by the term $g\psi^2$. Using the concepts developed in this chapter:

(a) State whether the equations are elliptic, parabolic, or hyperbolic.
(b) State whether either of the equations is nonlinear, time varying, or nonuniform?
(c) Discuss the nature of the modifying factors on the right-hand side of the two equations.
(d) In as much detail as possible, discuss the solution dynamics; that is, predict the general nature (shape) of the solutions, $\phi\ (t, x, y)$ and $\psi\ (t, x, y)$.

11. The problem of lighting the landing deck of an aircraft carrier for night landings in a heavy fog is to be treated by simulation. In particular, it is desired to determine the optimum number, the optimum positions, and adequate intensities of an array of lights to be placed along the edges of the deck so as to best assist pilots coming in for a landing at various altitudes and approach angles. The simulation should permit a determination of what the pilot sees as he approaches the carrier under various fog conditions (assume that the fog layer density is uniform in the x and y directions but inversely proportional to z, and that it has different densities on different nights). By making many simulation runs for different approaches and for different light configurations, a preferred lighting arrangement is to be selected by trial and error. Construct the mathematical model using partial differential equations, including all pertinent equations, initial conditions, boundary conditions, parameters, and so on.

12. A silicon wafer the size of a penny may contain anywhere from 100 to 1000 junction devices. Each junction device, in the course of its operation, generates heat. This heat, if not dissipated quickly, can destroy the functioning of the devices. Mathematical models are required to study various strategies of heat dissipation. Keeping the physical nature of the problem in perspective, discuss how you would formulate this problem. What methods do you suggest to solve this mathematical formulation? Discuss.

13. You are asked to formulate a mathematical model for the following physical problem taken from the area of air pollution and relating to the diffusion of automobile exhaust through a one-dimensional transparent tunnel. An automobile engine is located at one end of the tunnel and emits nitrogen dioxide (NO_2) in a time-varying manner. A wind of velocity u is blowing through the tunnel at a constant rate. As a result, the exhaust fumes are carried through the tunnel in the x direction. While in the tunnel, light quanta $h\mu$ due to sunlight effect a chemical change resulting in the formation of ozone (O_3) and nitrous oxide (NO). At the same time, reverse reactions involving organic reactants (XO) act to change the NO back to NO_2. The basic chemical reactions in operations are

$$NO_2 + O_2 + h\mu \longrightarrow NO + O_3$$
$$NO + O_3 \longrightarrow NO_2 + O_2$$
$$NO + XO \longrightarrow NO_2 + X$$
$$O_3 + X \longrightarrow XO + O_2$$

The rate of formation or disappearance of the three major constituants is expressed approximately as

$$\frac{\partial(NO_2)}{\partial t} = -a(NO_2) + b(O_3)(NO) + c(NO)(XO)$$

$$\frac{\partial(NO)}{\partial t} = a(NO_2) - b(O_3)(NO) - d(NO)(XO)$$

$$\frac{\partial O_3}{\partial t} = a(NO_2) - b(O_3)(NO) - e(O_3)$$

where the parameters a, b, c, d, and e are rate constants.

To simulate a particle diffusion system of this type, it is necessary to recognize that the concentration of NO_2, for example, at every point in the system is determined not only by the mass flux in the principal direction (the x direction), but also by the chemical generation and disappearance of NO_2. The rate of change of the NO_2 concentration at a point depends, therefore, upon the concentration of NO_2, NO, and O_3 at the same point in the system. The complete simulation therefore demands that the particle diffusion of the three constituents NO_2, NO, and O_3 be represented by three simultaneous, coupled diffusion equations with mass transfer.

Write the three simultaneous partial differential equations applying to this system. You may assume that the source of NO_2 is located at $x = 0$, and that it emits pure NO_2 at a rate expressed as $f(t)$. Be sure to include the effects of the photochemical reactions in the mathematical model, and be sure to specify all necessary boundary and initial conditions. You may assume that the tunnel is 100 feet long and that it is open at the far end, so that the concentration of all three constituants drops to zero at that point.

14. Circuits are distinguished from fields by the fact that the flux is restricted to ideal channels or connecting paths. As a result, the time variable is the only independent variable in circuit problems, whereas field problems have space as well as time independent variables. The same types of parameter types arise in lumped and distributed parameter systems. In electric circuits, resistors, capacitors, and inductors comprise the dissipators, reservoirs of potential, and reservoirs of flux, respectively. As a result, the mathematical models characterizing circuits containing one, two, or three of the basic element types manifest a striking similarity to the partial differential equations characterizing fields with corresponding parameter types. Explore the similarity of the mathematical models for circuits and fields by preparing a table similar to Table 2.3 for circuits containing various combinations of parameter types.

3

Posing the Mathematical Problem

3.1. INTRODUCTION

So far the emphasis has been on deriving the partial differential equations from physical principles. The next step is to formulate a problem that is mathematically meaningful and correct. This chapter is concerned with the various aspects of stating a problem in a correct mathematical language.

A word about the transition in notation from Chapter 2 to Chapter 3. In Chapter 2 the letters and symbols used to represent the variables were chosen such that they serve as convenient mnemonics to the physical variables they represent. Thus t was used to represent time, p for potential, V for voltage, and so on. Because a mathematical model is an abstraction of a real system, it is useful to represent all equations with a uniform notation. Toward this end, we strived to use the letter u for the dependent variable in a partial differential equation. Similarly, the letter y was used in an ordinary differential equation. Minor variations from this system do occur, and their meaning should be clear from the context.

3.2. LINEAR AND NONLINEAR EQUATIONS

A majority of the equations or mathematical models in the preceding sections were derived under the implicit assumption of linearity. Phrases such as "assume the fluid is inviscid," "for constant thermal conductivity," "in a homogeneous, isotropic medium," "given a perfect insulator," often give the impression that everything is ideal, isotropic, frictionless, rigid, inviscid, incompressible, and so forth. Many problems of practical interest often fail to satisfy one or more of these assumptions, the most critical of them being the assumption about linearity.

The question of linearity is often discussed in the language of linear operators. An operator is a mathematical rule that when applied on a function produces another function. For example, the expression

$$\frac{\partial u}{\partial x} + \frac{\partial^2 u}{\partial x\, \partial y} + \frac{\partial^3 u}{\partial y^3}$$

can be compactly written as $L[u]$, where the differential operator L is defined by

$$L \triangleq \frac{\partial}{\partial x} + \frac{\partial^2}{\partial x\, \partial y} + \frac{\partial^3}{\partial y^3} \tag{3.1}$$

A ***linear operator*** L satisfies the following relations:

$$L[au] = aL[u] \tag{3.2}$$

$$L[u + v] = L[u] + L[v] \tag{3.3}$$

$$L[au + bv] = aL[u] + bL[v] \tag{3.4}$$

where a and b are constants. Differential operators such as $\partial/\partial t$, $\partial^2/\partial x^2$, and so on, can be readily verified to be linear.

To appreciate the importance of this linearity assumption while solving partial differential equations, consider a simple linear heat conduction equation

$$\frac{\partial^2 T}{\partial x^2} = \alpha \frac{\partial T}{\partial t} \tag{3.5}$$

where α is a constant. Let T_1 and T_2 be solutions of Eq. (3.5); that is,

$$\frac{\partial^2 T_1}{\partial x^2} = \alpha \frac{\partial T_1}{\partial t}, \qquad \frac{\partial^2 T_2}{\partial x^2} = \alpha \frac{\partial T_2}{\partial t} \tag{3.6}$$

By virtue of the linearity of the operators $\partial^2/\partial x^2$ and $\partial/\partial t$, it follows that $T_1 + T_2$ (in fact, $aT_1 + bT_2$, where a and b are arbitrary constants) is also a solution. This statement is the ***principle of superposition***, which was responsible for the great success of linearized models. According to this principle, elementary solutions can be combined to yield more flexible solutions, which can be made to satisfy auxiliary conditions (such as boundary conditions) that describe a particular physical phenomenon.

What will happen to the superposition principle if an equation is not linear? Consider the heat conduction equation

$$\frac{\partial}{\partial x}\left[k(T)\frac{\partial T}{\partial x}\right] = c\frac{\partial T}{\partial t} \tag{3.7}$$

where c is a constant and k depends upon temperature T. The situation now is very different from that of Eq. (3.5). For concreteness, let us assume that

$$k(T) = k_0 T \tag{3.8}$$

where k_0 is a constant. Now Eq. (3.7) becomes

$$T\frac{\partial^2 T}{\partial x^2} + \left(\frac{\partial T}{\partial x}\right)^2 = \frac{c}{k_0}\frac{\partial T}{\partial t} \tag{3.9}$$

Suppose T_1 and T_2 are solutions of Eq. (3.9). Can we say that $T_1 + T_2$ is also a solution? To see this, $T_1 + T_2$ is substituted in Eq. (3.9). After some algebra, one gets

$$\left[T_1 \frac{\partial^2 T_1}{\partial x^2} + \left(\frac{\partial T_1}{\partial x}\right)^2 - \frac{c}{k_0}\frac{\partial T_1}{\partial t}\right] + \left[T_2 \frac{\partial^2 T_2}{\partial x^2} + \left(\frac{\partial T_2}{\partial x}\right)^2 - \frac{c}{k_0}\frac{\partial T_2}{\partial t}\right]$$
$$+ T_1 \frac{\partial^2 T_2}{\partial x^2} + T_2 \frac{\partial^2 T_1}{\partial x^2} + 2\frac{\partial T_1}{\partial x}\frac{\partial T_2}{\partial x} \stackrel{?}{=} 0 \qquad (3.10)$$

The first two bracketed terms vanish by virtue of Eq. (3.9). The last three terms are not zero. Therefore, $T_1 + T_2$ is not a solution. The principle of superposition fails because Eq. (3.7) is not linear.

When a partial differential equation is not linear, it can be either ***quasilinear*** or ***nonlinear***. If the coefficients, such as k of Eq. (3.7), are functions only of the dependent variable and not of its derivatives, then the equation is quasilinear. Otherwise, the equations are nonlinear. This subtle distinction will not be observed unless a useful purpose is served in a specific context.

Representing nonlinear equations using the compact Laplacian operator ∇^2 is not straightforward because the symbol ∇^2 is essentially reserved by tradition to represent a linear operator. However, extension of the use of div and grad operators to quasilinear and nonlinear equations is straightforward. For instance, a three-dimensional analog to Eq. (3.7) can now be written as

$$\text{div}\,[k(T)\,\text{grad}\,T] = c\frac{\partial T}{\partial t} \qquad (3.11)$$

or

$$\nabla[k(T)\,\nabla T] = c\frac{\partial T}{\partial t} \qquad (3.12)$$

The left side of Eq. (3.12) is often called the ***pseudo-Laplacian***.

3.3. INITIAL AND BOUNDARY CONDITIONS

In Chapter 2 attention was focused on deriving partial differential equations from physical principles. The specification of a differential equation is complete only after some associated auxiliary conditions are also specified. These auxiliary conditions may appear either as initial conditions, boundary conditions, or both. The terms "boundary conditions" and "initial conditions" were used in the preceding sections with no formal explanation. To clarify their meaning, a short review on how one uses these terms in connection with ordinary differential equations is first presented.

Consider an nth-order ordinary differential equation

$$y^{(n)} = f(x, y, y^{(1)}, \ldots, y^{(n-1)}) \qquad (3.13)$$

where x represents a space coordinate. The solution of such an equation contains n arbitrary integration constants. To obtain a specific solution, one has to evaluate these arbitrary constants by imposing a number of auxiliary conditions on the problem. These conditions may concern the values of the solution y and its derivatives up to the $(n-1)$st order at a certain point, usually the beginning point of the interval within which the solution is desired; it may concern the values of the solution and its derivatives at the beginning point and at the end point of the considered interval.

Since x is considered here as a space coordinate, and the conditions imposed on the solution and its derivatives are in general conditions that have to be satisfied at the beginning and end point of the considered interval—the boundary of the interval—we call such conditions ***boundary conditions***. In the case where the independent variable x represents the time, we will call the conditions imposed on the solution and its derivatives at the beginning point of the considered time interval ***initial conditions*** and those imposed at the end point ***end conditions*** (the latter are very unusual). Boundaries do not necessarily have to be finite points; the end point of a considered time interval may very well be the point ∞. The same holds true for space intervals, as the reader might well imagine.

The reader should be cautioned that the nomenclature introduced here is by no means standard. For example, the term "boundary conditions" is frequently used to denote all auxiliary conditions. This use is certainly justified by the argument that the beginning and end points of a time interval also constitute a boundary of an interval, even though not in the strict geometrical sense.

The situation described changes rather radically when one considers general solutions of partial differential equations. For instance, it can readily be verified that

$$u = f(x) + g(y) \tag{3.14}$$

in a solution of

$$\frac{\partial^2 u}{\partial x\, \partial y} = 0 \tag{3.15}$$

where f and g are arbitrary differentiable functions. Similarly, as long as f and g are twice differentiable, one can easily verify that

$$u = f(x + y) + g(x - y) \tag{3.16}$$

is a solution of

$$\frac{\partial^2 u}{\partial x^2} - \frac{\partial^2 u}{\partial y^2} = 0 \tag{3.17}$$

Thus one is led to suspect that a general solution to a partial differential equation contains *arbitrary functions* in analogy to the appearance of *arbitrary constants* in the solution of ordinary differential equations.

The fact that the solutions of partial differential equations contain arbitrary functions indicates that one has to impose rather severe conditions on a solution that is supposed to describe a specific process completely. However, there is some flexibility in the interpretation of how and where these conditions are to be applied. First, the term "boundary" is not necessarily a boundary in the strict geometric sense. It could be a space-time boundary such as a rectangle in the xt plane. Such boundaries require a different treatment. For example, the boundary for the problem in Eq. (2.43) is an open polygon, $t = 0, x = 0, x = 1$ in the xt plane. The auxiliary conditions specified on the boundary defined by $t = 0$ are referred to as initial conditions. Second, one has to decide on the number of boundary conditions. Having in mind that the solution of an ordinary differential equation of the second order requires two boundary conditions, one suspects that the number of boundary conditions required to completely determine a solution depends upon the highest ordered derivative term for each dimension in the equation. When the boundaries are parallel to the coordinate axes, extension of the preceding rule to partial differential equations is easy. For instance, Laplace's equation on a square, such as Eq. (2.18), has second-order derivatives in x and y and, therefore, requires two sets of boundary conditions: one set specifies the conditions on the boundaries parallel to the x-axis and the other set specifies the conditions on the boundaries parallel to the y-axis. This statement requires modification when a field is bounded by a curve. Then specification of suitable conditions everywhere along the curve would be required.

Therefore, while solving partial differential equations, one has to pay careful attention to what is specified along the boundary of the region. For concreteness, consider $Lu = g$ as the equation to be solved over a region Ω whose boundary is denoted by Γ. If the problem is to determine u in Ω such that u satisfies the condition

$$u = f \qquad \text{on } \Gamma \tag{3.18}$$

where f is a prescribed continuous function on Γ, then such a problem is called the ***Dirichlet problem*** or the ***first boundary value problem***. Stated in words, a Dirichlet problem reads as follows: Find the solution of a partial differential equation inside a region Ω, given the value of the solution along the boundary Γ of the region Ω. The task of finding the temperature distribution in the interior of a plate when the temperature is prescribed at all points on the boundary of the plate is an example of a Dirichlet problem.

If the problem is to determine u in Ω such that the normal derivative $\partial u/\partial n$ satisfies the condition

$$\frac{\partial u}{\partial n} = f \qquad \text{on } \Gamma \tag{3.19}$$

where f is a prescribed function continuous on Γ, then such a problem is called the ***Neumann problem*** or the ***second boundary value problem***. The symbol $\partial u/\partial n$ denotes the directional derivative of u along the outward normal to Γ. As an aid in visualization, it is useful to observe that $\partial u/\partial n \equiv 0$ across an impermeable or insulating boundary.

Notice that the right sides of Eqs. (3.18) and (3.19) are nonzero. Such boundary conditions are said to be ***inhomogeneous***. An inhomogeneous boundary condition can be further classified as ***time varying*** and ***time invariant***. For example, boundary conditions such as

$$u(0, t) = 0, \qquad \alpha u_x(1, t) = 0$$

are homogeneous, while

$$u(0, t) = 3, \qquad \alpha u_x(0, t) = e^{-t}$$

are examples of inhomogeneous boundary conditions, the first being time invariant and the second time varying. The symbol u_x stands for $\partial u/\partial x$.

Problems involving a combination of Dirichlet and Neumann types of boundary conditions are referred to as the ***third boundary value problem***. A typical example of this type of boundary condition is

$$u(x_0, t) + \alpha u_x(x_0, t) = \gamma(t), \qquad \alpha \neq 0 \tag{3.20}$$

A simple physical interpretation to Eq. (3.20) can be given as follows. Suppose we have a uniform rod of conductivity κ extending from $x = 0$ to $x = L$, whose left end is in contact with a heat reservoir at temperature $\gamma(t)$, and let $u(x, t)$ be the temperature of the rod. If there is a thin film (such as oxide film, lubricating oil film) on the end of the rod, then the statement

$$u(0, t) = \gamma(t) \tag{3.21}$$

is not strictly correct. Rather the problem should be treated as one of conduction in a composite rod. Toward this end, let L_0 be the thickness of the film and let the temperature inside the film be $u_0(x, t)$. Assuming perfect thermal contact between the rod and the film, the ***jump conditions*** that must be satisfied at $x = 0$ are

$$u(0, t) = u_0(0, t) \tag{3.22}$$

$$\kappa \frac{\partial u}{\partial x}(0, \ ) = \kappa_0 \frac{\partial u_0}{\partial x}(0, t) \tag{3.23}$$

Since the temperature at the other end of the film is that of the heat reservoir, we have

$$u_0(-L_0, t) = \gamma(t)$$

Since the thickness L_0 of the film is very small, $\partial u_0/\partial x$ can be approximated by

$$\frac{\partial u_0}{\partial x}(0, t) \approx \frac{u_0(-L_0, t) - u_0(0, t)}{-L_0}$$

$$\approx \frac{\gamma(t) - u(0, t)}{-L_0}$$

Substituting this in Eq. (3.23), one gets

$$\kappa \frac{\partial u}{\partial x}(0, t) = -\frac{\kappa_0}{L_0}[\gamma(t) - u(0, t)]$$

or

$$u(0, t) - \frac{L_0 \kappa}{\kappa_0} \frac{\partial u}{\partial x}(0, t) = \gamma(t) \tag{3.24}$$

which is of the same form as the boundary condition of the third kind given in Eq. (3.20).

If the problem is to determine u in Ω such that a Dirichlet type of boundary condition is required to be satisfied on a portion Γ_1 of the boundary Γ, and a Neumann type of boundary condition on the rest Γ_2 of Γ, then the problem is called ***mixed*** or ***fourth boundary value problem***. That is, in this problem the boundary conditions take the form

$$\begin{aligned} u &= f_1 \qquad \text{on } \Gamma_1 \\ \frac{\partial u}{\partial n} &= f_2 \qquad \text{on } \Gamma_2 \end{aligned} \tag{3.25}$$

and

$$\Gamma = \Gamma_1 + \Gamma_2$$

The preceding four problems are collectively referred to as the ***interior boundary value problems***. These differ from ***exterior boundary value problems*** in two respects. First, for problems of the latter variety, part of the boundary is at infinity. Second, solutions of the exterior problems must satisfy an additional requirement, that the solution be bounded at infinity.

Specification of initial conditions is quite analogous to the case of ordinary differential equations. For concreteness, by an ***initial-value problem*** of a vibrating string we mean solution of a problem such as

$$\frac{\partial^2 u}{\partial t^2} = c^2 \frac{\partial^2 u}{\partial x^2} \tag{3.26}$$

satisfying the initial conditions

$$\begin{aligned} u(x, t_0) &= f(x) \\ \frac{\partial u}{\partial t}(x, t_0) &= g(x) \end{aligned} \tag{3.27}$$

where $f(x)$ is the initial displacement and $g(x)$ is the initial velocity along the string. Problems of this kind, in a slightly general context, are called ***Cauchy problems***. Then the data on the right side of Eq. (3.27) is called Cauchy data. Notice that nothing is mentioned about the length of the vibrating string and how the string is supported at both ends. Suppose the string is of finite length stretched over the interval [0, 1]; then in addition to the initial conditions in Eq. (3.27), one has to specify the boundary conditions, such as

$$\begin{aligned} u(0, t) &= p(t) \\ u(1, t) &= q(t) \end{aligned} \qquad t \geq 0 \tag{3.28}$$

Because both initial and boundary conditions are specified here, Eqs. (3.26) to (3.28) are also referred to as ***initial-boundary value problems***.

3.4. EIGENVALUE PROBLEMS

So far attention has been focused primarily on initial and boundary value problems. Quite frequently, interesting physical problems lead to eigenvalue problems. The concept of an eigenvalue is of great importance in both pure and applied mathematics. A physical system, such as a pendulum, a vibrating string, or a rotating shaft, has connected with it certain numbers characteristic of the system, that is, the period of the pendulum, the frequencies of the various overtones of the string, the critical angular velocities at which the shaft will buckle, and so forth. Other interesting eigenvalue problems arise in the study of vibration of structures, determination of the modal shape of an electromagnetic field in a waveguide, and in describing the behavior of shallow-water waves in lakes and harbors. The German word *eigen* means "characteristic," and the hybrid word eigenvalues stands for the characteristic numbers referred to.

To fix the ideas firmly, consider a homogeneous linear initial-boundary value problem

$$\frac{\partial u}{\partial t} = \frac{1}{\sigma(x)}\frac{\partial}{\partial x}\left[\kappa(x)\frac{\partial u}{\partial x}\right] + q(x)u, \qquad a < x < b,\ t > 0 \tag{3.29a}$$

$$\begin{aligned} u(a, t) &= 0 \\ u(b, t) &= 0 \end{aligned} \qquad t > 0 \tag{3.29b}$$

$$u(x, 0) = f(x), \qquad a < x < b$$

describing the cooling of a nonuniform rod with a source whose strength is proportional to the local temperature. If the solution $u(x, t)$ can be written as $u(x, t) = y(x) \cdot \psi(t)$, that is, if the solution is *separable*, then the above Eq. (3.29a) can be equivalently written as (see Exercise 2) two ordinary differential equations. The equation describing $y(x)$, for instance, looks like

$$\begin{aligned} &\frac{1}{\sigma(x)}\frac{d}{dx}\left[\kappa(x)\frac{dy}{dx}\right] + q(x)y + \lambda y = 0, \qquad a < x < b \\ &\qquad\qquad y(a) = 0, \qquad y(b) = 0 \end{aligned} \tag{3.30}$$

which is of the form

$$Ly = \lambda y \tag{3.31}$$

where the differential operator L is defined as

$$L = -\frac{1}{\sigma(x)}\frac{d}{dx}\left[\kappa(x)\frac{d}{dx}\right] - q(x) \tag{3.32}$$

Problems of the kind characterized by Eq. (3.31) are called ***eigenvalue problems***. The particular problem posed in Eq. (3.30) is called the ***Strum-Liouville problem***. If $\sigma(x) > 0$ and $\kappa(x) > 0$ everywhere in $[a, b]$, and if $\sigma(x)$, $\kappa(x)$, and $q(x)$ are real-valued continuous functions in $[a, b]$, and if $\kappa(x)$ has a continuous first derivative in $[a, b]$, then the Strum-Liouville problem is said to be ***regular***. If either the interval $[a, b]$ is infinite or if the conditions for regularity are not satisfied, then the problem is ***singular***. It is needless to stress that singular problems require specialized treatment.

The general nature of the eigenvalue problem can be demonstrated by considering an extremely simplified version of Eq. (3.30):

$$\frac{d^2y}{dx^2} + \lambda y = 0 \tag{3.33a}$$

$$y(0) = 0, \qquad y(b) = 0 \tag{3.33b}$$

This simple linear boundary value problem can be solved rather easily using the ***method of basic solutions***. Toward this end, Eq. (3.33) is first solved for the rather specialized set of initial conditions,

$$\begin{aligned} y_1(0) &= 1 \qquad & y_2(0) &= 0 \\ y_1'(0) &= 0 \qquad & y_2'(0) &= 1 \end{aligned} \tag{3.34}$$

The resulting set of solutions, called the ***basic set***, can be readily obtained to be

$$y_1(x) = \cos x\sqrt{\lambda}$$
$$y_2(x) = \frac{1}{\sqrt{\lambda}}\sin x\sqrt{\lambda} \tag{3.35}$$

Any linear combination of this set is a general solution of Eq. (3.33). Thus, a general solution of Eq. (3.33) is

$$y(x) = A\cos x\sqrt{\lambda} + \frac{B\sin x\sqrt{\lambda}}{\sqrt{\lambda}} \tag{3.36}$$

where the arbitrary constants A and B are determined by the specified boundary conditions. For instance, the boundary condition $y(0) = 0$ requires that $A = 0$. Thus the solution of the eigenvalue problem in Eq. (3.33) is $y(x) = B(\sin x\sqrt{\lambda})/\sqrt{\lambda}$. The second boundary condition, $y(b) = 0$, could be satisfied by setting $B = 0$. But this results in the trivial solution $y(x) = 0$. For a nontrivial solution, λ must be chosen such that

$$\frac{\sin b\sqrt{\lambda}}{\sqrt{\lambda}} = 0$$

That is, a nontrivial solution to Eq. (3.33) can be obtained by choosing λ as

$$\lambda_n = \left(\frac{n\pi}{b}\right)^2, \qquad n = 1, 2, \ldots, \tag{3.37}$$

The ***eigenfunctions*** belonging to these eigenvalues are

$$y_n(x) = \sin x\sqrt{\lambda_n} = \sin\frac{n\pi x}{b}, \qquad n = 1, 2, \ldots \tag{3.38}$$

On a number of occasions, the operator L satisfies the ***self-adjoint*** property. In such cases all the eigenvalues are real. Indeed, the eigenvalues of self-adjoint Strum-Liouville problems can be arranged such that

$$\lambda_0 < \lambda_1 < \lambda_2 < \cdots < \lambda_n < \cdots$$

and $\lim \lambda_n \longrightarrow \infty$ as $n \longrightarrow \infty$. However, in practical problems, one ordinarily seeks to determine the first five or ten eigenvalues and the corresponding eigenfunctions.

The nature of the eigenvalue problem that arises in lumped and distributed parameter systems is generally the same with some minor differences. Consider, for example, a lumped parameter system such as a system of n coupled particles in one dimension. The state of this system can be completely described by the n coordinates of a position vector $\mathbf{x}(t)$. However, if we are interested in a distributed parameter system, such as the vibration of a string of length L, then the state of the string at any time t can no longer be determined by a finite set of n numbers, but only by a function $u(x, t)$ defined over the interval $0 \leq x \leq L$. For this reason the problem of n coupled particles is said to be ***finite dimensional***, and the string problem is said to be ***infinite dimensional***. In mechanics, the former problem is often referred to as one having n ***degrees of freedom*** and the latter as having ***infinite degrees of freedom***.

The interrelationship between infinite dimensional and finite dimensional problems is of importance in a computational context. For example, the ***Rayleigh-Ritz method*** (also known as the ***Ritz method***) consists of reducing an infinite dimensional problem to a sequence of increasing finite dimensional problems. A slight generalization of the Ritz method leads to the ***Galerkin method*** (also known as the ***Babunov-Galerkin method***). Further discussion on the Ritz and Galerkin methods can be found in subsequent chapters. In the case of n degrees of freedom, every possible motion of the system is a superposition of n spatial motions, the ***eigenvibrations***, whose frequencies, which are equal to the square roots of the *eigenvalues*, are completely determined by the physical characteristics of the system. In the case of distributed parameter systems, where there are infinitely many degrees of freedom, it can be shown that an eigenvibration, that is, a vibration of form $u(x, t) = y(x) \cdot \psi(t)$, is possible only for the functions $y(x)$, determined up to an arbitrary constant factor, of a certain infinite sequence $y_1(x), y_2(x), \ldots, y_n(x), \ldots$, called the ***eigenfunctions*** of the system. The frequencies of these eigenvibrations are called ***eigenfrequencies*** and are completely determined by the system.

The eigenvalue problem, as stated here, can be associated with a variety of differential operators. Eigenvalue problems associated with Laplace's operator, that is, $L = (\partial^2/\partial x^2) + (\partial^2/\partial y^2)$, and so on, for instance, occur in the design of nuclear reactors, whether stable (for power production) or unstable (atom bombs). In this case, one is generally interested in the smallest eigenvalue and the associated eigenfunction. If L is a biharmonic operator, that is, $L = (\partial^4/\partial x^4) + (\partial^4/\partial y^4)$, and so on, then λ is proportional to a frequency of vibration of an ideal elastic plate. In certain simple cases, one can get upper and lower bounds for the eigenvalues using various mathematical and physical principles. Therefore, in the context of computation it is natural to inquire whether the methods under investigation can furnish such upper and lower bounds, and whether such bounds can be made arbitrarily close without excessive labor.

3.5. VARIATIONAL FORMULATIONS

Field problems arising in engineering and applied physics can be formulated in two different ways. In the first approach, the problem is formulated in terms of differential equations and some auxiliary conditions. A differential equation only describes the local behavior of a variable, that is, the behavior of a typical infinitesimal region. Starting from a given point, a differential equation permits the construction of many possible solutions in a stepwise fashion. The auxiliary conditions are invoked to choose a particular solution. In the second approach, a variational, extremum principle valid over the whole region is postulated, and the correct solution is the one minimizing some quantity J that is defined by suitable integration of the unknown quantities over the whole domain. A quantity such as J that is a function of unknown functions is known as a ***functional***.

Both these procedures are mathematically equivalent; one is based on differentiation and the other on integration. Conceptually, integration is a deductive process, a process easier to visualize than differentiation, which is essentially an inductive generalization. Computationally, integration is a smoothing operation, whereas differentiation is a "noisy" operation. Finally, in the integral (variational) formulation all the necessary information is absorbed into one equation; there is no need for the specification of separate auxiliary conditions. Therefore, it is extremely useful to study partial differential equations using variational formulation. A more extensive treatment of this philosophy appears in Chapters 8 and 9. The present concern is only to study some of the fundamental aspects of the variational approach by studying first a problem in one dimension. Toward this goal consider

$$J = \int_{x_1}^{x_2} F(x, u, u', u'')\, dx \tag{3.39}$$

where the variable u and its first two derivatives with respect to x, that is, u' and u'', are functions of x. The quantities F and J are called ***functionals*** because they are functions of other functions. In many engineering problems some well-defined physical meaning can be attached to J, for example, the potential energy of a deformable solid. Notice that the integral in Eq. (3.39) is defined over $[x_1, x_2]$. Typically, the values of u or u' are specified at the boundaries. For concreteness, let

$$\begin{aligned} u(x_1) &= u_1 \\ u(x_2) &= u_2 \end{aligned} \tag{3.40}$$

Now let us focus on the task of minimizing J. For instance, one can try a trial solution $\bar{u}(x)$ that satisfies the given boundary conditions in Eq. (3.40). Then one can proceed to correct the value of $\bar{u}(x)$ until J achieves a minimum value. This is the gist of the ***calculus of variations***. To develop a strategy for the modification of $\bar{u}(x)$, let $\bar{u}(x)$ be a solution in the neighborhood of the exact solution $u(x)$ so that one is allowed to write

$$\bar{u}(x) = u(x) + \delta u \tag{3.41}$$

where δu, the difference between the exact and tentative solutions, is called the variation in u. The δ in δu can be treated as an operator, similar to the differential operator, d. Fortunately, this operator commutes with differential and integral operators, permitting one to write

$$\delta \int F\, dx = \int (\delta F)\, dx \tag{3.42a}$$

$$\frac{\delta(du)}{dx} = \frac{d(\delta u)}{dx} \tag{3.42b}$$

Also, the total differential δF can be written down, in a way analogous to the total derivative df, as

$$\delta F = \frac{\partial F}{\partial u}\delta u + \frac{\partial F}{\partial u'}\delta u' + \frac{\partial F}{\partial u''}\delta u'' \tag{3.42c}$$

Now, taking variations on both sides of Eq. (3.39), one gets

$$\delta J = \int_{x_1}^{x_2} \delta F\, dx = \int_{x_1}^{x_2} \left(\frac{\partial F}{\partial u}\, \delta u + \frac{\partial F}{\partial u'}\, \delta u' + \frac{\partial F}{\partial u''}\, \delta u'' \right) dx \tag{3.43a}$$

In ordinary calculus a function $f(x)$ is maximized or minimized by setting $df = 0$. Similarly, a necessary condition for making the functional J stationary is

$$\delta J = 0 \tag{3.43b}$$

Integrating the last two terms in Eq. (3.43a) by parts, one gets

$$\int_{x_1}^{x_2} \frac{\partial F}{\partial u'}\, \delta u'\, dx = \left[\frac{\partial F}{\partial u'}\, \delta u \right]\Bigg|_{x_1}^{x_2} - \int_{x_1}^{x_2} \frac{d}{dx}\left(\frac{\partial F}{\partial u'} \right) \delta u\, dx \tag{3.44a}$$

$$\int_{x_1}^{x_2} \frac{\partial F}{\partial u''}\, \delta u''\, dx = \left[\frac{\partial F}{\partial u''}\, \delta u' \right]\Bigg|_{x_1}^{x_2} - \left[\frac{d}{dx}\left(\frac{\partial F}{\partial u''} \right) \delta u \right]\Bigg|_{x_1}^{x_2}$$

$$+ \int_{x_1}^{x_2} \frac{d^2}{dx^2}\left(\frac{\partial F}{\partial u''} \right) \delta u\, dx \tag{3.44b}$$

Combining Eqs. (3.43) and (3.44), we get

$$\delta J = \int_{x_1}^{x_2} \left[\frac{\partial F}{\partial u} - \frac{d}{dx}\left(\frac{\partial F}{\partial u'} \right) + \frac{d^2}{dx^2}\left(\frac{\partial F}{\partial u''} \right) \right] \delta u\, dx$$

$$+ \left[\frac{\partial F}{\partial u'} - \frac{d}{dx}\left(\frac{\partial F}{\partial u''} \right) \right] \delta u \Bigg|_{x_1}^{x_2}$$

$$+ \left[\left(\frac{\partial F}{\partial u''} \right) \delta u'' \right]\Bigg|_{x_1}^{x_2} \equiv 0 \tag{3.45}$$

Since δu is an arbitrary variation, vanishing of δJ therefore requires that the individual terms vanish. That is,

$$\frac{\partial F}{\partial u} - \frac{d}{dx}\left(\frac{\partial F}{\partial u'} \right) + \frac{d^2}{dx^2}\left(\frac{\partial F}{\partial u''} \right) = 0 \tag{3.46a}$$

$$\left[\frac{\partial F}{\partial u'} - \frac{d}{dx}\left(\frac{\partial F}{\partial u''} \right) \right] \delta u \Bigg|_{x_1}^{x_2} = 0 \tag{3.46b}$$

$$\left[\left(\frac{\partial F}{\partial u''} \right) \delta u' \right]\Bigg|_{x_1}^{x_2} = 0 \tag{3.46c}$$

Equation (3.46a) is the governing differential equation for the problem and is called the ***Euler equation*** or the ***Euler-Lagrange equation***. The conditions specified in Eq. (3.46b) and (3.46c) give the boundary conditions. For instance, if the conditions

$$\frac{\partial F}{\partial u'} - \frac{d}{dx}\left(\frac{\partial F}{\partial u''} \right) \Bigg|_{x_1}^{x_2} = 0 \quad \text{and} \quad \frac{\partial F}{\partial u''} \Bigg|_{x_1}^{x_2} = 0 \tag{3.47}$$

called the ***natural boundary conditions***, are satisfied, then Eqs. (3.46b) and (3.46c) are satisfied. Then Eq. (3.47) gives the ***free boundary conditions***. If the natural boundary conditions are not satisfied, then in order to satisfy Eqs. (3.46b) and (3.46c) one must force

$$\delta u(x_1) = 0, \qquad \delta u(x_2) = 0 \tag{3.48}$$
$$\delta u'(x_1) = 0, \qquad \delta u(x_2) = 0$$

For this reason, Eq. (3.48) is called ***forced boundary conditions*** or ***geometric boundary conditions***.

Example. Consider a beam on an elastic foundation with one end rigidly supported and the other end propped on an elastic spring support. From beam theory it is a well-known fact that the total potential energy of the beam is

$$J = \int_0^l \frac{1}{2} EI(u'')^2 \, dx + \int_0^l \frac{k_f u^2}{2} \, dx + \frac{k_s}{2} u^2(l) - \int_0^l pu \, dx$$

where k_f is the foundation reaction spring constant per unit length, k_s is the spring constant for the flexible end support, E and I are beam parameters, and u is the deflection. Using the procedure discussed previously,

$$\delta J = \int_0^l (EIu'') \, \delta u'' \, dx + \int_0^l k_f u \, \delta u \, dx$$
$$+ k_s u(l) \, \delta u(l) - \int_0^l p \, \delta u \, dx = 0$$

Using the geometric boundary conditions given (for a rigid support), that is,

$$u(x = 0) = u'(x = 0) = 0$$

and integrating by parts, one gets

$$\int_0^l [(EIu'')'' - p + k_f u] \, \delta u \, dx + (EIu'' \, \delta u')\Big|_{x=l} - [(EIu'')' - k_s u] \, \delta u\Big|_{x=l} = 0$$

Setting each term individually to zero, the governing differential equation is

$$\frac{\partial^2}{\partial x^2}\left(EI \frac{\partial^2 u}{\partial x^2}\right) = P - k_f u$$

The remaining terms give the natural boundary conditions:

$$EI \frac{\partial^2 u}{\partial x^2}\Big|_{x=l} = 0$$

and

$$\frac{\partial}{\partial x}\left(EI \frac{\partial^2 u}{\partial x^2}\right)\Big|_{x=l} + k_s u(l) = 0$$

Thus, given a variational statement, the transition to an equivalent governing differential equation is a simple matter, as demonstrated here. The reverse process, however, is more involved. Given a differential equation, one may or may not be able to find an equivalent variational formulation. Under very specialized conditions involving linear, positive definite, self-adjoint operators, one can proceed from a differential equation to a variational formulation. This is an important point to remember while formulating finite element procedures.

EXTENSION TO HIGHER DIMENSIONS

The idea presented so far can be generalized to higher dimensions. Before going into a derivation of the general case, it is instructive to state the final result. For example, the well-known Euler theorem states that if the integral

$$J_1(u) = \iiint_\Omega F\left(x, y, z, u, \frac{\partial u}{\partial x}, \frac{\partial u}{\partial y}, \frac{\partial u}{\partial z}\right) dx\, dy\, dz \tag{3.49}$$

is to be minimized over a bounded region Ω, then the necessary and sufficient conditions for this minimum to be reached is that the unknown function $u(x, y, z)$ should satisfy the differential equation

$$\frac{\partial}{\partial x}\left(\frac{\partial F}{\partial(\partial u/\partial x)}\right) + \frac{\partial}{\partial y}\left(\frac{\partial F}{\partial(\partial u/\partial y)}\right) + \frac{\partial}{\partial z}\left(\frac{\partial F}{\partial(\partial u/\partial z)}\right) = \frac{\partial F}{\partial u} \tag{3.50}$$

within the same region, provided u satisfies the same boundary conditions in both cases. Taking this general result for granted, one can quickly verify, for instance, that the differential equation

$$\frac{\partial}{\partial x}\left(k_x \frac{\partial u}{\partial x}\right) + \frac{\partial}{\partial y}\left(k_y \frac{\partial u}{\partial y}\right) + \frac{\partial}{\partial z}\left(k_z \frac{\partial u}{\partial z}\right) + \alpha = 0 \tag{3.51}$$

subject either to the boundary conditions

$$u = u_b \tag{3.52}$$

or the conditions

$$k_x \frac{\partial u}{\partial x} l_x + k_y \frac{\partial u}{\partial y} l_y + k_z \frac{\partial u}{\partial z} l_z + q + \alpha u = 0 \tag{3.53}$$

is equivalent to the minimization of

$$J(u) = \iiint_\Omega \left[\frac{1}{2}\left\{k_x\left(\frac{\partial u}{\partial x}\right)^2 + k_y\left(\frac{\partial u}{\partial y}\right)^2 + k_y\left(\frac{\partial u}{\partial z}\right)^2\right\} - Qu\right] dx\, dy\, dz \tag{3.54}$$

subject to u obeying the same boundary conditions. In Eq. (3.53), l_x, l_y, and l_z are the direction cosines of the outward normal to the boundary surface.

In general, it is easy to apply the boundary condition in Eq. (3.52), whereas Eq. (3.53) is difficult to implement. To overcome this difficulty, it is generally preferable not to impose Eq. (3.53), but to add to the functional $J_1(u)$ in Eq. (3.49) another integral pertaining to the boundary, which, upon minimization, automatically yields the boundary condition. In this problem this integral is simply

$$\int_{\Gamma_1} \left(qu + \frac{1}{2} au^2\right) d\Gamma_1 \tag{3.55}$$

where Γ_1 is that portion of the boundary on which boundary conditions such as Eq. (3.53) are imposed. Thus, in the general case, it is customary to rewrite Eq. (3.49) as

$$J(u) = \int_\Omega F(x, y, x, u, u_x, u_y, u_z)\, d\Omega + \int_{\Gamma_1} \left(qu + \frac{1}{2}\alpha u^2\right) d\Gamma_1 \tag{3.56}$$

where $u_x = \partial u/\partial x$, and so on.

As before, one can proceed to get

$$\delta J = \int_\Omega \left(\frac{\partial F}{\partial u}\,\delta u + \frac{\partial F}{\partial u_x}\,\delta u_x + \frac{\partial F}{\partial u_y}\,\partial u_y + \frac{\partial F}{\partial u_z}\,\delta u_z\right) d\Omega + \int_{\Gamma_1} (q\,\delta u + \alpha u\,\delta u)\,d\Gamma_1 = 0 \tag{3.57}$$

$$= \int_\Omega \left(\frac{\partial F}{\partial u}\,\delta u + \frac{\partial F}{\partial u_x}\frac{\partial}{\partial x}(\delta u) + \cdots\right) d\Omega + \int_{\Gamma_1} (q\,\delta u + \alpha u\,\delta u)\,d\Gamma_1 = 0 \tag{3.58}$$

Integrating the second term under the first integral by parts with respect to x,

$$\int \frac{\partial F}{\partial u_x}\frac{\partial}{\partial x}(\delta u)\,d\Omega = \int_{\Gamma_1} \frac{\partial F}{\partial u_x}\,\delta u l_x\,d\Gamma - \int_\Omega \frac{\partial}{\partial x}\left(\frac{\partial F}{\partial u_x}\right)\delta u\,d\Omega$$

where l_x is the direction cosine of the outward drawn normal to the surface with the x-axis. Performing similar operations on other terms of Eq. (3.58), one gets

$$\delta J = \int_\Omega \delta u \left\{\frac{\partial F}{\partial u} - \frac{\partial}{\partial x}\left(\frac{\partial F}{\partial u_x}\right) - \frac{\partial}{\partial y}\left(\frac{\partial F}{\partial u_y}\right) - \frac{\partial}{\partial z}\left(\frac{\partial F}{\partial u_z}\right)\right\} d\Omega + \int_{\Gamma_1} \delta u \left\{q + \alpha u + l_x \frac{\partial F}{\partial u_x} + l_y \frac{\partial F}{d u_y} + l_z \frac{\partial F}{\partial u_z}\right\} d\Gamma_1 = 0 \tag{3.59}$$

Therefore, the Euler equation within Ω is

$$\frac{\partial F}{\partial u} - \frac{\partial}{\partial x}\left(\frac{\partial F}{\partial u_x}\right) - \frac{\partial}{\partial y}\left(\frac{\partial F}{\partial u_y}\right) - \frac{\partial}{\partial z}\left(\frac{\partial F}{\partial u_z}\right) = 0 \tag{3.60}$$

and on the boundary Γ_1, the condition is

$$l_x \frac{\partial F}{\partial u_x} + l_y \frac{\partial F}{\partial u_y} + l_z \frac{\partial F}{\partial u_z} + q + \alpha u = 0 \tag{3.61}$$

Note that Eqs. (3.50) and (3.60) are identical and Eqs. (3.53) and (3.61) are equivalent. Notice also that Eq. (3.53), for $k_x = k_y = k_z = k_1$ and $q = \alpha = 0$, can be written as

$$\left(k_1 \frac{\partial}{\partial x}\mathbf{i} + k_1 \frac{\partial}{\partial y}\mathbf{j} + k_1 \frac{\partial}{\partial z}\mathbf{k}\right) \cdot (l_x\mathbf{i} + l_y\mathbf{j} + l_z\mathbf{k})u = 0$$

or

$$\text{grad } u \cdot \mathbf{n} = 0 \tag{3.62}$$

where $\mathbf{n}$ is the unit normal to the boundary Γ_1. This is what is normally referred to as the normal derivative boundary condition.

VARIATIONAL FORMULATION OF CERTAIN EIGENVALUE PROBLEMS

The variational approach, discussed so far, can be profitably used in characterizing eigenvalues also in variational terms, that is, as certain maxima or minima. Consider, for example, the equation which follows:

$$I_1(u) = \iint_\Omega \left(\frac{1}{2}a\left(\frac{\partial u}{\partial x}\right)^2 + b\left(\frac{\partial u}{\partial x}\right)\left(\frac{\partial u}{\partial y}\right) + \frac{1}{2}c\left(\frac{\partial u}{\partial y}\right)^2 - \frac{1}{2}fu^2\right) dx\, dy \qquad (3.63)$$

As before, we are interested in rendering $I_1(u)$ stationary with respect to all sufficiently smooth functions u in a certain class K. Suppose the functions u of K are required to satisfy Dirichlet boundary conditions and are further subjected to the constraint

$$I_2(u) = \frac{1}{2}\iint_\Omega u^2\, dx\, dy = \text{constant} \qquad (3.64)$$

Then the eigenvalue problem can be stated as one of finding the nonzero functions u in K for which there exists a constant λ such that $I_1(u) - \lambda I_2(u)$ is stationary. Writing

$$J(u) = I_1(u) - \lambda I_2(u)$$
$$= \iint_\Omega \left[\frac{1}{2}a\left(\frac{\partial u}{\partial x}\right)^2 + b\left(\frac{\partial u}{\partial x}\right)\left(\frac{\partial u}{\partial y}\right) + \frac{1}{2}c\left(\frac{\partial u}{\partial y}\right)^2 - \frac{1}{2}fu^2 - \frac{1}{2}\lambda u^2\right] dx\, dy \qquad (3.65)$$

and comparing this with Eq. (3.55), it is evident that a function u in K that renders Eq. (3.63) stationary must satisfy the condition

$$-L(u) = \lambda u \qquad \text{in } \Omega \qquad (3.66)$$

and the natural boundary condition

$$\beta\frac{\partial u}{\partial n} + \delta\frac{\partial u}{\partial s} = 0 \qquad \text{on } \Gamma - \Gamma_1 \qquad (3.67)$$

where Γ_1 is a part of Γ, where u is required to satisfy a Dirichlet type of boundary condition, and $L(u)$ is given by

$$L(u) = \frac{\partial}{\partial x}\left(a\frac{\partial u}{\partial x} + b\frac{\partial u}{\partial y}\right) + \frac{\partial}{\partial y}\left(b\frac{\partial u}{\partial x} + c\frac{\partial u}{\partial y}\right) + fu \qquad (3.68)$$

The problem in Eq. (3.66) is the eigenvalue problem for the operator $-L$.

3.6. WELL-POSED AND ILL-POSED PROBLEMS

The partial differential equations encountered in practical problems are often very complicated, and the practitioner is often unskilled in handling difficult mathematical questions of existence, uniqueness, convergence, stability, and the like. The time pressures on the engineering profession usually induce the engineer to assume the existence and uniqueness of the solution. A considerable degree of intuition and empiricism, therefore, enters the practical or engineering approach to get working solutions to PDEs. This zeal for a practical solution can be justified, at least in part, if the problem is correctly formulated and properly posed.

Roughly, a problem is said to be ***well posed*** or ***properly posed*** (in the Hadamard sense) if a unique solution exists and depends continuously on the data. A problem that is not well posed is said to be ***ill posed***, ***nonwell posed***, or ***improperly posed***. The criteria given for well-posedness is widely accepted because it is physically reasonable. The existence and uniqueness criteria are an affirmation of determinism. That is, experiments could be repeated with an expectation that consistent data lead to consistent solutions. Continuous dependence on data is an expression of stability; that is, small changes in the auxiliary data should produce a correspondingly small change in the solutions.

Until recently, the prevailing attitude toward ill-posed problems is that there is no serious need to consider such problems since they are of no physical interest. For instance, it was widely believed for many years that problems which exhibit discontinuous dependence on the data do not correspond to any *real formulations*; that is, they do not arise in the study of natural phenomena. In other words, there is something wrong with the mathematical model and not with the physical problem it portrays. This attitude supports the belief that ill-posed problems are essentially ill-conceived problems. Now it is widely recognized that this attitude toward improperly posed problems is erroneous.

Typical ill-posed problems in the study of partial differential equations are the initial boundary value problem for the backward heat equation, the Cauchy problem for the Laplace's equation, and the Dirichlet problem for the wave equation. A simple example of the backward heat equation consists in determining the initial temperature distribution $u(x, y, 0) = \xi$ on a plate whose edges are maintained at a specified temperature g, in order that at a given time $t = T$ the temperature distribution $u(x, y, T)$ attains a specified temperature χ. That is, it is required to solve

$$\nabla^2 u = \frac{\partial u}{\partial t} \qquad \text{in } \Omega \times (0, T) \tag{3.69a}$$

with

$$\begin{aligned} u &= g^* \qquad \text{on } \Gamma \times (0, T) \\ u(x, y, T) &= \chi^* \end{aligned} \tag{3.69b}$$

where g^* and χ^* are, respectively, the measured values of g and χ. This problem can be quickly recognized as a type of ***control problem*** where the control appears in an initial condition. Another type of control problem of immense practical interest is one in which the control appears in the boundary conditions. This is a control problem in the true sense because in problems of this kind one can deliberately exert control in the course of evolution.

A second type of ill-posed problem is a boundary value problem with over-abundant data on one part of the boundary and insufficient data on another part of the boundary. Problems of this kind occur if a portion of the boundary where data should be taken becomes inaccessible. An example of this situation can be found in geological problems involving oil reservoirs or groundwater aquifers. Due to physical and topographic peculiarities, a portion of the boundary becomes inaccessible for the measurement of boundary conditions. Worst yet is the case where the analyst has no idea about where exactly the boundary of an underground aquifer is. A more formal example belonging to this category is a Laplace's equation with Cauchy data on a portion Γ_1 of the boundary and no data on the rest of the boundary Γ_2. That is, find u with

$$\nabla^2 u = 0 \qquad \text{in } \Omega \tag{3.70a}$$

$$u = g^* \qquad \text{on } \Gamma_1$$

$$\frac{\partial u}{\partial n} = h^* \qquad \text{on } \Gamma_1 \tag{3.70b}$$

and $\Gamma = \Gamma_1 + \Gamma_2$.

A third type of improperly posed problem is a class of inverse problems that arise when certain coefficients in the equation or in the boundary conditions are unknown. This is the system identification or parameter estimation problem. In addition, a number of inverse eigenvalue problems have appeared in the literature. The more important ones of this category are the inverse Strum-Liouville problems and the problem posed by the now famous question, "Can you hear the shape of a drum?"

SUGGESTIONS FOR FURTHER READING

1. *For an extensive discussion of the eigenvalue problem, consult:*

   G. POLYA and G. SZEGO, *Isoparametric Inequalities in Mathematical Physics.* Princeton, N.J.: Princeton University Press, 1951.

   G. POLYA, "Estimates for eigenvalues," in *Studies in Mathematics and Mechanics Presented to Richard von Mises.* New York: Academic Press, Inc., 1954.

   S. H. GOULD, *Variational Methods for Eigenvalue Problems.* Toronto: University of Toronto Press, 1957.

2. *For an excellent treatment of the variational formulations, see:*

   R. COURANT and D. HILBERT, *Methods of Mathematical Physics*, Vol. 1. New York: John Wiley & Sons, Inc. (Interscience Division), 1953.

   L. V. KANTOROVIC and V. I. KRYLOV, *Approximate Methods of Higher Analysis.* The Netherlands: P. Noordhoff Ltd., 1958.

3. *For a discussion of a number of inverse problems arising in mathematical physics, consult:*

   M. M. LAVRENTIEV, V. G. ROMANOV, and V. G. VASILIEV, *Multidimensional Inverse Problems for Differential Equations*, Lecture Notes in Mathematics, Vol. 167, Berlin: Springer-Verlag, 1970.

4. *For an excellent discussion on improperly posed problems, see:*

   L. E. PAYNE, *Improperly Posed Problems in Partial Differential Equations*, CBMS Regional Conference Series, Vol. 22, Philadelphia: Society of Industrial and Applied Mathematics, 1975.

EXERCISES

1. For each of the following equations, state whether the PDE is linear, quasilinear, or nonlinear. If it is linear, state whether it is homogeneous or nonhomogeneous, and give its order. Note the notation used: $u_x = \partial u/\partial x$, $u_{xy} = \partial^2 u/\partial x \partial y$, and so on.

(a) $u_{xx} + xu_y = y$
(b) $uu_x - 2xyu_y = 0$
(c) $u_x^2 + uu_y = 1$
(d) $u_{xx}^2 + u_x^2 + \sin u = e^y$

2. Consider the PDE

$$a(x, y)u_{xx} + c(x, y)u_{yy} + d(x, y)u_x + e(x, y)u_y + f(x, y)u = 0$$

Let the solution be of the form

$$u(x, y) = X(x)Y(y)$$

where X and Y are, respectively, functions of x and y alone and are twice differentiable. Under the given separable conditions, show that the given PDE can be written as two ODEs of the form

$$\alpha_1(x)\frac{d^2u}{dx^2} + \alpha_2(x)\frac{du}{dx} + [\alpha_3(x) + \lambda]u = 0$$

Now introduce

$$p(x) = \exp\left[\int \frac{\alpha_2}{\alpha_1}\,dx\right], \qquad q(x) = \frac{\alpha_3}{\alpha_1}p, \qquad s(x) = \frac{1}{\alpha_1}p$$

and show that the given ODE can be written as the Strum-Liouville equation

$$Lu + \lambda s(x)u = 0$$

where L is the operator $L = \dfrac{d}{dx}\left(p\dfrac{d}{dx}\right) + q$.

3. Show that the eigenvalues of the Strum-Liouville system

$$u'' + \lambda u = 0, \qquad 0 \leq x \leq \pi$$
$$u(0) = 0, \qquad u'(\pi) = 0$$

are

$$\lambda_n = \frac{(2n-1)^2}{2^2}, \qquad n = 1, 2, 3, \ldots$$

and the corresponding eigenfunctions are $\sin\dfrac{(2n-1)}{2}x$, $n = 1, 2, 3, \ldots$.

4. The set of all eigenvalues of a Strum-Liouville system is called the ***spectrum*** of the system. Show that the spectrum of the Strum-Liouville system

$$u'' + \lambda u = 0, \qquad -\pi \leq x \leq \pi$$

with ***perodic boundary conditions***

$$u(\pi) = u(-\pi) \qquad \text{and} \qquad u'(\pi) = u'(-\pi)$$

is given by $\lambda_n = n^2$, $n = 1, 2, 3, \ldots$.

5. Consider the eigenvalue problem

$$Lu \triangleq -\left(\frac{\partial^2 u}{\partial x^2} + \frac{\partial^2 u}{\partial y^2}\right) = \lambda u \qquad \text{in } \Omega$$
$$u = 0 \qquad \text{on } \Gamma$$

where $\Omega = \{(x, y), 0 < x < 1, 0 < y < 1\}$ and Γ is the boundary of Ω. Show that the eigenvalues of the operator L are of the form

$$\lambda_{m,p} = (m^2 + p^2)\pi^2, \qquad m = 1, 2, \dots, p = 1, 2, \dots$$

and the eigenvectors are of the form

$$u_{m,p} = 2 \sin m\pi x \sin p\pi x$$

6. Show that

$$\frac{\partial}{\partial x}\left(k_x \frac{\partial u}{\partial x}\right) + \frac{\partial}{\partial y}\left(k_y \frac{\partial u}{\partial y}\right) + Q = 0$$

is the Euler equation of the functional

$$J(u) = \iint_\Omega \left\{\frac{1}{2}\left[k_x\left(\frac{\partial u}{\partial x}\right)^2 + k_y\left(\frac{\partial u}{\partial y}\right)^2\right] - Qu\right\} d\Omega$$

7. Show that

$$L(u) = (au_x + bu_y)_x + (bu_x + cu_y)_y + fu$$

is the Euler equation of the functional

$$J(u) = \iint_\Omega \left(\frac{1}{2}au_x^2 + bu_x u_y + \frac{1}{2}cu_y^2 - \frac{1}{2}fu^2 + gu\right) dx\, dy + \int_\Gamma \left(\frac{1}{2}\alpha u^2 + \gamma u\right) ds$$

8. Are the following problems well posed or not? Given your reasons.

(a) Determine the unknown functions $u = u(x, t)$ and $f = f(x)$ that satisfy

$$\begin{aligned} u_{tt} - u_{xx} &= f(x), && 0 < x < 1, \quad t < 0 \\ u(x, 0) &= \phi(x), && 0 \leq x \leq 1 \\ u_t(x, 0) &= \psi(x), && 0 \leq x \leq 1 \\ u(0, t) &= \mu_1(t), && t \geq 0 \\ u(1, t) &= \mu_2(t), && t \geq 0 \end{aligned}$$

(b)

$$\begin{aligned} \nabla^2 u + u_t &= 0 && \text{in } \Omega \times (0, T) \\ u &= 0 && \text{on } \Gamma \times (0, T) \\ u(x, 0) &= f(x) \end{aligned}$$

(c)

$$\begin{aligned} &\nabla^2 w + w_t = \epsilon^2 \nabla^4 w = 0 && \text{in } \Omega \times (0, T) \\ &w = 0, \quad \nabla^2 w = 0 && \text{on } \Gamma \times (0, T) \\ &w(x, 0) = f(x) \end{aligned}$$

(d)

$$\begin{aligned} &u_{xx} + u_{yy} = 0 \\ &u(x, 0) = 0, \qquad u_y(x, 0) = n^{-1} \sin nx \end{aligned}$$

TRANSFORMATIONS

4

Finite Difference Approximations

4.1. INTRODUCTION

The mathematical models characterizing distributed parameter systems take the form of partial differential equations, including elliptic, parabolic, hyperbolic, and biharmonic partial differential equations. As such they contain terms such as

$$\frac{\partial^4 u}{\partial x^4}, \quad \frac{\partial^2 u}{\partial x^2}, \quad \frac{\partial^2 u}{\partial x\,\partial y}, \quad \frac{\partial u}{\partial x}, \quad \frac{\partial^2 u}{\partial t^2}, \quad \text{and} \quad \frac{\partial u}{\partial t}$$

If a digital computer is to be employed to implement the mathematical model, that is, to solve the partial differential equations, the mathematical model must first be transformed so as to require only arithmetic operations of addition, subtraction, multiplication, and division in its solution. A number of approaches to the deriving of suitable digital computer algorithms for the treatment of partial differential equations have been developed over the years. The most widely used of these are the finite difference method and the finite element method.

The aim of both these methods is to reduce a given continuum problem into a discrete mathematical model suitable for solution on a digital computer. Toward this end various discretization schemes have been proposed both by engineers and mathematicians. Among the various discretization strategies, the finite difference and the finite element methods are perhaps the most general and therefore the most popular. The finite difference method, developed primarily by mathematicians, is a very general procedure and can be used directly if the governing differential equations of a physical system are available. The rudiments of the method are easy to learn. Therefore, the finite difference method enjoyed widespread and unchallenged popularity for a long time both in the context of analog simulation and digital simulation.

However, in the mid-1950s, the finite element method began to appear in the engineering literature. Although the roots of the finite element method have been traced to much older mathematical literature, much of the credit for the early development of the method as a practical tool rightfully belongs to the engineers. The engineering community addressed the question of discretization at an intuitive level rather than at a formal mathematical level. As many physical systems of engineering interest can be viewed as interconnected components, the engineers attempted to create analogies between finite portions of a continuum with discrete components or "finite elements" of the physical system. Thus the starting point of the traditional finite element method was the physical system itself, and the discretization process was intuitive and based on physical arguments. Nevertheless, the finite element method, as practiced by the engineering community, almost always led to acceptable numerical solutions.

Both the finite difference and the finite element methods are considered to be total discretization schemes because derivatives no longer appear in the discretized model. In addition to these total discretization schemes, some semidiscretization methods also play an important role in solving field problems. Historically, people interested in analog computer simulation were the first to be attracted to these semidiscretization methods. Now, after a period of neglect, these schemes are gaining ground among people interested in developing general-purpose software packages (see Chapter 11) for solving partial differential equations. These semidiscretization schemes, called the *method of lines*, discretize all but one independent variable. This procedure converts a given partial differential equation into a system of ordinary differential equations. These equations take the form of initial or boundary value problems depending upon the type of the original partial differential equation and the manner in which time and space variables are treated. These ordinary differential equations are then solved using one of the standard methods of solving them.

Although all these three discretization methods will be discussed in this book, the purpose of this chapter is to focus on the finite difference method.

4.2. DISCRETE REPRESENTATION OF VARIABLES, FUNCTIONS, AND DERIVATIVES

The basic objective of the finite difference method is to represent the time-space continuum by a set of discretely spaced points. A separate algebraic equation approximating the partial differential equation is derived for every one of these points. The solution of the partial differential equation is found by solving these algebraic equations. Therefore, the first step in obtaining these algebraic equations is to define a set of discretely spaced points. This set of points, variously referred to as *net*, *mesh*, *grid*, *lattice*, or *node* points, is superimposed upon the field, and attention is limited to these points. Approximations for the variables and derivatives are then obtained in terms of the function values at the node points.

Consider first the discrete representation of the independent continuous variable x, which lies between 0 and 1 shown in Figure 4.1. This interval is first replaced by a set of equally spaced points, and the variable x is defined only at these grid points.

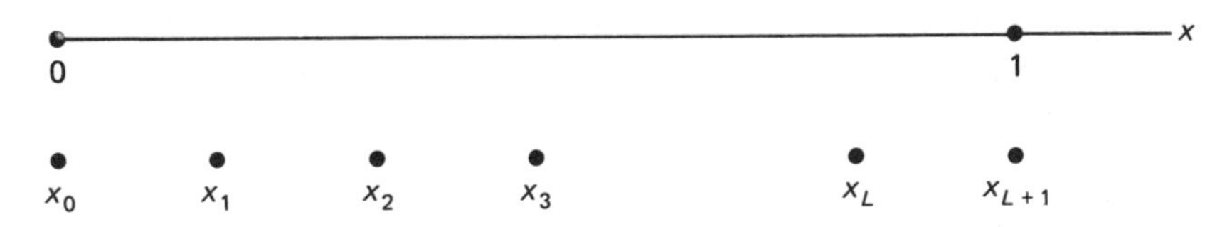

Figure 4.1 Discrete representation of a variable x

The value of the variable at each of these grid points is denoted by a subscript l and is defined as

$$x_l = l(\Delta x) \tag{4.1}$$

where Δx, called the finite difference, is the distance between successive grid points. The subscript l takes integral values from 0 to $L + 1$, where L is the total number of internal grid points in [0, 1]. Some useful relations between adjacent points are

$$\begin{aligned} x_{l+1} &= x_l + \Delta x \\ x_{l-1} &= x_l - \Delta x \end{aligned} \tag{4.2}$$

Now any function $u(x)$, defined on [0, 1], can be evaluated at $x = x_l$, $l = 0, 1, \ldots, L + 1$ and denoted by

$$u(x_l) = u_l \tag{4.3}$$

To determine the discrete representation of derivatives, it is useful to review what a derivative means. Derivatives represent rates of change. A derivative contains information about the local variation of a function. Therefore, a difference approximation to a derivative should attempt to couple information at the neighboring points of a grid. This coupling can be achieved by using the Taylor series expansion of the function about a neighboring point. Thus, expanding $u(x + \Delta x)$ about the point x,

$$u(x + \Delta x) = u(x) + \frac{du}{dx}\Delta x + \frac{d^2u}{dx^2}\frac{(\Delta x)^2}{2!} + \frac{d^3u}{dx^3}\frac{(\Delta x)^3}{3!} + \cdots \tag{4.4}$$

Using the nomenclature for the discrete version, this Taylor series can be rewritten as

$$u_{l+1} = u_l + \left(\frac{du}{dx}\right)_l (\Delta x) + \left(\frac{d^2u}{dx^2}\right)_l \frac{(\Delta x)^2}{2!} + \left(\frac{d^3u}{dx^3}\right)_l \frac{(\Delta x)^3}{3!} + \cdots \tag{4.5}$$

In a similar manner, the value of $u(x - \Delta x)$ can be written as u_{l-1} and expanded thus:

$$u_{l-1} = u_l - \left(\frac{du}{dx}\right)_l (\Delta x) + \left(\frac{d^2u}{dx^2}\right)_l \frac{(\Delta x)^2}{2!} - \left(\frac{d^3u}{dx^3}\right)_l \frac{(\Delta x)^3}{3!} + \cdots \tag{4.6}$$

When Eq. (4.5) is solved for the first derivative, one gets

$$\left(\frac{du}{dx}\right)_l = \frac{u_{l+1} - u_l}{(\Delta x)} - \left(\frac{d^2u}{dx^2}\right)_l \frac{(\Delta x)}{2!} - \left(\frac{d^3u}{dx^3}\right)_l \frac{(\Delta x)^2}{3!} + \cdots \tag{4.7a}$$

$$= \frac{u_{l+1} - u_l}{(\Delta x)} + \mathrm{O}(\Delta x) \tag{4.7b}$$

where $\mathrm{O}(\Delta x)$ denotes terms containing first and higher powers of (Δx). If this series is truncated after the first term, what is left is called a finite difference approximation (or analog) to the first derivative at the grid point l. That is,

$$\left(\frac{du}{dx}\right)_l \simeq \frac{u_{l+1} - u_l}{(\Delta x)} \tag{4.8}$$

The error involved in this approximation is said to be of the order (Δx), because (Δx) occurs in the first term of the truncated series. Therefore, the truncation error in Eq. (4.8) is said to be of first order, and Eq. (4.8) itself is said to be first order correct. As the preceding approximation involves the use of function values at the point in question, that is, l, and at the next point, $l + 1$, it is referred to as a ***forward difference*** approximation.

Similarly, one can derive a ***backward difference*** approximation from Eq. (4.6) as

$$\left(\frac{du}{dx}\right)_l = \frac{u_l - u_{l-1}}{(\Delta x)} + \mathrm{O}(\Delta x) \tag{4.9}$$

$$\simeq \frac{u_l - u_{l-1}}{(\Delta x)} \tag{4.10}$$

which is also first order correct. Even though both forward and backward difference analogs were derived for the grid point l, they are not centered at the grid point l. The forward difference analog is said to be centered at $(l + \frac{1}{2})$ and the backward analog at $(l - \frac{1}{2})$.

A more accurate analog to the first derivative, which is centered at the grid point l, is obtained by subtracting Eq. (4.6) from Eq. (4.5) and evaluating (du/dx). Then

$$\left(\frac{du}{dx}\right)_l = \frac{u_{l+1} - u_{l-1}}{2(\Delta x)} + \mathrm{O}(\Delta x)^2 \tag{4.11}$$

$$\simeq \frac{u_{l+1} - u_{l-1}}{2(\Delta x)} \tag{4.12}$$

This is a ***central difference*** analog and is second order correct.

To obtain a finite difference approximation to the second derivative, Eqs. (4.5) and (4.6) are added, and the resulting equation is solved for $(d^2u/dx^2)_l$ to yield

$$\left(\frac{d^2u}{dx^2}\right)_l = \frac{u_{l+1} - 2u_l + u_{l-1}}{(\Delta x)^2} + \mathrm{O}(\Delta x)^2 \tag{4.13}$$

The first term on the right side is a finite difference approximation to the second derivative. That is,

$$\left(\frac{d^2u}{dx^2}\right)_l \simeq \frac{1}{(\Delta x)^2}(u_{l-1} - 2u_l + u_{l+1}) \tag{4.14}$$

This approximation to the second derivative is called the ***second central difference***. The error involved in using this analog is of the order of $(\Delta x)^2$.

This procedure can be readily extended to more than one space dimension. For example, the Laplacian in two space dimensions can be approximated by

$$\left(\frac{\partial^2 u}{\partial x^2} + \frac{\partial^2 u}{\partial y^2}\right)_{l,m} \simeq \frac{1}{(\Delta x)^2}(u_{l-1,m} - 2u_{l,m} + u_{l+1,m}) + \frac{1}{(\Delta y)^2}(u_{l,m-1} - 2u_{l,m} + u_{l,m+1}) \tag{4.15}$$

where

$$u_{l,m} = u(l\,\Delta x, m\,\Delta y) \tag{4.16}$$

Treatment of time derivative terms is no different from the preceding. Mathematically, the symbol t in partial differential equations is just another variable and may therefore be subjected to the same finite difference operations as the space derivatives. The continuous time variable is divided into discrete intervals, and attention is focused upon the function values existing at instants of time spaced Δt units apart from each other. For example, for $du(t)/dt$, the forward, backward, and central difference approximations at time $j(\Delta t)$ are, respectively,

$$\left(\frac{du}{dt}\right)_j \simeq \frac{u_{j+1} - u_j}{(\Delta t)}, \qquad \text{forward or single-step forward} \tag{4.17}$$

$$\left(\frac{du}{dt}\right)_j \simeq \frac{u_j - u_{j-1}}{(\Delta t)}, \qquad \text{backward or single-step backward} \tag{4.18}$$

$$\left(\frac{du}{dt}\right)_j \simeq \frac{u_{j+1} - u_{j-1}}{2(\Delta t)}, \qquad \text{central or multistep forward} \tag{4.19}$$

where u_j is defined as

$$u_j = u(t_j) \qquad \text{and} \qquad t_j = j(\Delta t) \tag{4.20}$$

Similarly, the second time derivative can be approximated by

$$\left(\frac{d^2 u}{dt^2}\right)_j \simeq \frac{u_{j-1} - 2u_j + u_{j+1}}{(\Delta t)^2} \tag{4.21}$$

These approximations, when appropriately combined with the approximations for the space derivatives, provide discretized analogs to many partial differential equations. However, improper discretization of the time variable could lead to the ***computational instability***, which will be discussed in Chapter 6.

A COMPACT NOTATION FOR DIFFERENCE OPERATORS

Writing the various difference expressions again and again can become a tedious affair. Several shorthand notations are available for this purpose. For example, the forward, backward and central difference operators, D_{fx}, D_{bx}, and D_{cx}, can be defined as

$$D_{fx}u_{l,j} = \frac{1}{(\Delta x)}(u_{l+1,j} - u_{l,j}), \qquad \text{forward difference} \tag{4.22}$$

$$D_{bx}u_{l,j} = \frac{1}{(\Delta x)}(u_{l,j} - u_{l-1,j}), \qquad \text{backward difference} \tag{4.23}$$

$$D_{cx}u_{l,j} = \frac{1}{2}(D_{fx} + D_{bx})u_{l,j}$$

$$= \frac{1}{2(\Delta x)}(u_{l+1,j} - u_{l-1,j}), \qquad \text{central difference} \tag{4.24}$$

Using these definitions as building blocks, higher-order operators, such as δ_x^2 and δ_x^4 can be defined. For example,

$$\delta_x^2 u_{l,j} = D_{bx}D_{fx}u_{l,j} = \frac{1}{(\Delta x)^2}(u_{l+1,j} - 2u_{l,j} + u_{l-1,j}) \tag{4.25}$$

Similarly,

$$\delta_y^2 u_{l,j} = D_{by}D_{fy}u_{l,j} = \frac{1}{(\Delta y)^2}(u_{l,m+1} - 2u_{l,m} + u_{l,m-1}) \tag{4.26}$$

Notice also that

$$D_{fx}D_{bx}u_{l,j} = D_{bx}D_{fx}u_{l,j} \tag{4.27}$$

More on the use of these operators can be found, for example, in Mitchell's book.[1]

4.3. A MODEL PROBLEM

To illustrate the ideas presented so far, consider Laplace's equation

$$\frac{\partial^2 u}{\partial x^2} + \frac{\partial^2 u}{\partial y^2} = 0 \tag{4.28}$$

on the unit square $0 \leq x \leq 1, 0 \leq y \leq 1$, with the boundary condition

$$u(x, y) = g(x, y) \tag{4.29}$$

That is, Eqs. (4.28) and (4.29) constitute a Dirichlet problem. To apply the finite-difference method, a uniform square mesh of size $\Delta x = \Delta y = \frac{1}{3}$ is superimposed on this unit square. The intersections of this grid (the grid points) with the interior and boundary of the unit square are numbered in some convenient fashion, as shown in Figure 4.2. Instead of attempting to find the solution $u(x, y)$ satisfying Eq. (4.28), for *all* $0 < x < y < 1$ and the boundary conditions in Eq. (4.29), attention is now focused on the interior grid points, and an approximate solution is sought at these grid points.

By virtue of Eq. (4.15), a finite-difference approximation to Eq. (4.28) can be written, at any interior grid point (l, m), as

$$\left(\frac{\partial^2 u}{\partial x^2} + \frac{\partial^2 u}{\partial y^2}\right)_{l,m} \simeq (u_{l-1,m} + u_{l+1,m} - 4u_{l,m} + u_{l,m-1} + u_{l,m+1}) = 0 \tag{4.30}$$

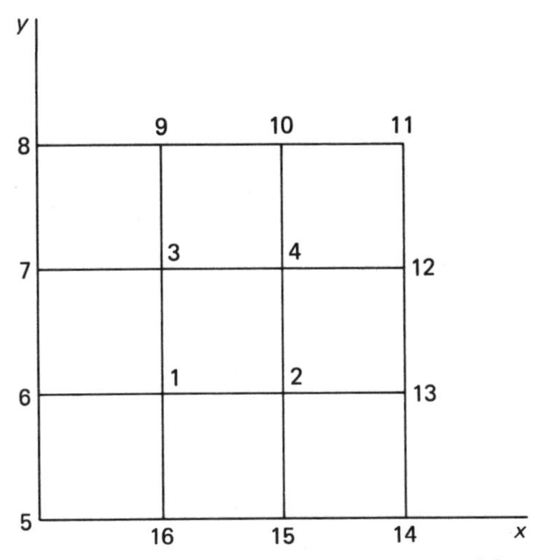

Figure 4.2 Finite difference grid for the model problem

At this point, it is convenient to relabel the subscripts so that the notation in Eq. (4.30) conforms with the notation in Figure 4.2. That is,

$$u_1 = u(\tfrac{1}{3}, \tfrac{1}{3}), \qquad u_2 = u(\tfrac{2}{3}, \tfrac{1}{3}), \qquad u_3 = u(\tfrac{1}{3}, \tfrac{2}{3}), \qquad u_4 = u(\tfrac{2}{3}, \tfrac{2}{3})$$

Similarly, g_5 stands for the value of the specified boundary condition at the origin, g_{11} at the point (1, 1), and so forth. With this notation Eq. (4.30), when applied to the four interior points, yields

$$\begin{aligned} 4u_1 - u_2 - u_3 - 0 &= g_6 + g_{16} \\ -u_1 + 4u_2 - 0 - u_4 &= g_{13} + g_{15} \\ -u_1 - 0 + 4u_3 - u_4 &= g_7 + g_9 \\ -0 - u_2 - u_3 + 4u_4 &= g_{10} + g_{12} \end{aligned} \tag{4.31}$$

which can be written in matrix notation as

$$\begin{bmatrix} 4 & -1 & -1 & 0 \\ -1 & 4 & 0 & -1 \\ -1 & 0 & 4 & -1 \\ 0 & -1 & -1 & 4 \end{bmatrix} \begin{bmatrix} u_1 \\ u_2 \\ u_3 \\ u_4 \end{bmatrix} = \begin{bmatrix} g_6 + g_{16} \\ g_{13} + g_{15} \\ g_7 + g_9 \\ g_{10} + g_{12} \end{bmatrix} \tag{4.32}$$

Now suppose $g(x, y)$ is specified as zero on all sides of the square except the side defined by $y = 1$, where $g(x, y) = 4500x(1 - x)$. Then the right side of Eq. (4.32) can be evaluated to yield

$$\begin{bmatrix} 4 & -1 & -1 & 0 \\ -1 & 4 & 0 & -1 \\ -1 & 0 & 4 & -1 \\ 0 & -1 & -1 & 4 \end{bmatrix} \begin{bmatrix} u_1 \\ u_2 \\ u_3 \\ u_4 \end{bmatrix} = \begin{bmatrix} 0 \\ 0 \\ 1000 \\ 1000 \end{bmatrix} \tag{4.33}$$

which can be solved directly to yield

$$u_1 = 125, \qquad u_2 = 125, \qquad u_3 = 375, \qquad u_4 = 375$$

In the preceding discussion, a square mesh of size $\Delta x = \Delta y = \frac{1}{3}$ has been used, and the matrix obtained was of the order $(3 - 1)^2 = 4$. If one uses, say, $\Delta x = \Delta y = \frac{1}{20}$, then one would obtain a matrix of the order $(20 - 1)^2 = 361$. However, out of $361^2 = 130{,}321$ elements in the matrix, fewer than 1800 are nonzero. Indeed, regardless of the grid size chosen, there will be no more than five nonzero elements in any row or column of the resulting matrix. Such matrixes with relatively small numbers of nonzero entries are called ***sparse***.

Most of the matrixes encountered while solving PDEs not only exhibit sparsity but also exhibit a regular pattern of occurrence of the nonzero entries. Although not obvious at the outset, closer inspection reveals that the matrix in Eq. (4.33) has the ***block tridiagonal*** structure with two 2 by 2 submatrixes on the main diagonal and a $-\mathbf{I}$ on the two off diagonals. While solving elliptic and parabolic equations in one space dimension, and on several other occasions, we will also encounter the simple tridiagonal structure. Numerical solution of sparse systems exhibiting these specialized structural patterns is relatively easy because of the availability of specialized algorithms for these systems. [2] This subject will be treated in more detail in Chapter 10.

4.4. ACCURACY OF APPROXIMATIONS

John von Neumann and H. H. Goldstein have identified four major sources of errors that are nearly unavoidable when describing physical systems.

1. Mathematical models of nature are generally derived with several assumptions. A system that is nonlinear and time varying is perhaps represented by a linear and constant coefficient differential equation. A distributed parameter system can often be approximated by a lumped parameter model. Errors of this nature are called ***modeling errors***.
2. Most descriptions of natural phenomena require measured data. Errors introduced at this stage are called ***measurement errors***.
3. Most descriptions of field problems involve an infinite continuum. In the context of numerical analysis, one can deal only with a finite number of terms from limiting processes which are normally described by infinite series. Errors introduced in this context are called ***truncation errors***.
4. Finally, when computations are performed, they can be done only with a finite precision on a computer. This results from a limited size of registers in the arithmetic unit of the computer. Errors introduced by dropping off of digits are called ***roundoff errors***.

Of the four types of errors, the treatment of truncation and roundoff errors falls in the realm of the numerical analyst.

ROUNDOFF ERRORS

In computer calculations roundoff errors reflect the fact that only a finite number of bits are available to calculate the differences used to approximate derivatives. The expectation is that numbers will round down as often as up, with no accumulating error. However, it is possible that all numbers would be rounded up, although the probability of such an occurrence is small. The expected error is proportional to the standard deviation expected for such random events. But we know

$$\sigma \propto \sqrt{N}$$

where σ is the standard deviation and N is the number of such calculational roundoffs.

As calculations use a finer mesh, the truncation error decreases but the roundoff error increases owing to the increased number of computations performed. The economics of computers, however, indicate that the truncation error is the limiting factor for most problems. That is, most calculations in large electronic computers have considerably larger truncation errors than roundoff errors. This can be understood by the following example. Let the truncation error be proportional to grid size and time step for a time-dependent problem with one spatial dimension. Reduce the truncation error a factor of 2 by halving Δx and Δt. The problem then requires twice as much memory and four times as many calculations. Four times as many calculations means an increase in roundoff error by a factor of 2. Reducing the truncation error by a factor of 2 relative to its original value requires an overall improvement in roundoff by a factor of 4. Thus, if we want to maintain the same relative value of roundoff error and truncation error, only two more binary bits per number are required. For a factor of 2 improvement in truncation error and roundoff error, the increase in memory for improving truncation error is much larger than that required for improving roundoff error.

When roundoff error is large enough to seriously affect the solution, it can usually be detected readily. If insignificant changes in the physical system result in large changes in the solution, roundoff error is the likely culprit.

TRUNCATION ERRORS

The truncation error refers to the error introduced by the truncation of the series expansion of a function. In most computer calculations, the truncation error is the dominant error. Intuitively, one would assume that the accuracy of any finite difference solution could be decreased by increasing the grid size, since, evidently, if the grid spacing is reduced to zero, the discretized equivalent becomes identical to the continuous field. This intuition, though correct, does not give a total picture. Consider, for instance, a wire of uniform resistivity, 100 feet long, having a 100-volt battery connected to one end while the other end is grounded. This system is governed by Laplace's equation, so that Eq. (4.14) is applicable. If a net spacing of $\Delta x = 50$ feet is employed, u_1 refers to the potential at the midpoint of the wire while u_0 and u_2

represent the potentials at the end points. Then

$$\frac{d^2u}{dx^2} = \frac{1}{(\Delta x)^2}(u_0 + u_2 - 2u_1) = 0 \tag{4.34}$$

or

$$u_1 = \tfrac{1}{2}(u_0 + u_2) = \tfrac{1}{2}(100 + 0) = 50 \text{ volts} \tag{4.35}$$

As is readily verified, this calculation has provided precisely the correct value for the potential at the midpoint of the wire. Despite the fact that an extraordinarily large net spacing has been used, there is no error in the solution. To gain better insight into this problem, two methods, one based on Taylor series and the other based on Fourier series, are discussed next.

Taylor Series Method. While deriving finite difference analogs, such as the second central difference used in the preceding example, some of the higher-order terms in a Taylor series expansion were neglected. For instance, the Taylor series expansions shown in Eq. (4.5), and rearranged in Eq. (4.7a), can be rewritten as

$$\left(\frac{du}{dx}\right)_l = \frac{u_{l+1} - u_l}{(\Delta x)} + \alpha(\Delta x) + \beta(\Delta x)^2 + \cdots$$

where $\alpha, \beta, \ldots$ are functions of x and derivatives (second and higher order) of u evaluated at the grid point l. In other words, the error in the forward difference approximation

$$\left(\frac{du}{dx}\right)_l \simeq \frac{u_{l+1} - u_l}{(\Delta x)}$$

is proportional to (Δx) and higher powers of (Δx). If (Δx) is small, the error will be dominated by the lowest power of (Δx) in the error term, and for this case it is the first power. The preceding approximation is therefore first order accurate. To reduce the truncation error at a given grid point, we must therefore use a finer mesh, that is, a smaller value for (Δx). Alternatively, the truncation error can be reduced by using a higher-order approximation that uses information from many more neighboring grid points. Either of these choices can be very expensive. It is significant to note that truncation error does not give the *size* of the error. Rather, it indicates how rapidly the error vanishes as the grid is refined.

Fourier Series Method. Consider a continuous function $u(x)$ defined everywhere on the interval [0, 1]. A discrete representation of this function as

$$u_l = u(x_l), \qquad l = 0, 2, \ldots, L - 1 \tag{4.36}$$

is evidently an incomplete description of the function $u(x)$. Information about the function in the interval (x_l, x_{l+1}), for all l, is lost. Nevertheless, for $x_l < \tilde{x} < x_{l+1}$, the value of $u(\tilde{x})$ can be obtained by using an interpolation formula such as

$$\tilde{u} = \epsilon u_{l+1} + (1 - \epsilon)u \tag{4.37}$$

where

$$\epsilon = \frac{\tilde{x} - x_l}{x_{l+1} - x_l} \tag{4.38}$$

and $\tilde{u}$ is an approximation to $u(\tilde{x})$. It is evident that if u changes very rapidly over the interval (x_l, x_{l+1}), then $\tilde{u}$ is not likely to be a good approximation to $u(\tilde{x})$. For a slowly varying function, $\tilde{u}$ would be a sufficiently good approximation to $u(\tilde{x})$. In physics and engineering, it is customary to visualize a rapidly varying function as consisting of high-frequency or short-wave components, and vice versa for a slowly varying function. Thus, qualitatively, one can see that finite difference methods are generally good for "long-wavelength" phenomena and poor for "short-wavelength" phenomena, and the approximation gets better as more grid points are included in the domain of interest. These ideas will now be quantified using techniques of Fourier analysis.

Making the usual assumptions of Fourier analysis, that is, the function u is periodic, say, over $X = [0, 1]$ and satisfies the Dirichlet conditions, the function u can be represented by an infinite Fourier series

$$u(x) = \sum_{k=-\infty}^{\infty} g_k e^{i2\pi kx/X} \tag{4.39}$$

where

$$g_k = \frac{1}{X}\int_0^1 u(x)e^{-i2\pi kx/X}\,dx \tag{4.40}$$

Equation (4.39) says that any reasonably behaved function $u(x)$ can be analyzed into an infinite set of Fourier components or modes, where g_k is the amplitude of the kth harmonic or mode with a wavelength X/k.

The same type of analysis can be applied to the discrete approximation of $u(x)$. Since u_l are finite, the set $\{u_l\}$ can be represented as a vector of dimension L, and therefore $\{u_l\}$ may only be written as the sum of a finite set of L Fourier modes. Under the standard assumptions of Fourier analysis, one can write

$$u_l = \sum_{k=0}^{L-1} \hat{g}_k e^{i2\pi kl/L} \tag{4.41}$$

where the amplitude of each mode is specified by

$$\hat{g}_k = \frac{1}{L}\sum_{l=0}^{L-1} u_l e^{-i2\pi kl/L} \tag{4.42}$$

Note that in going from Eq. (4.39) to Eq. (4.41), X is replaced by $L\,\Delta X$ and x is replaced by $l\,\Delta x$. As no wavelength smaller than Δx can be defined on the mesh, the infinite series becomes a finite series. Thus, in going from $u(x)$ to u_l only a finite set of Fourier modes (or wavelengths) is chosen, and no phenomena below the cutoff wavelength Δx are described. Thus, $\{u_l\}$ is a long-wavelength approximation to $u(x)$.

This digression on Fourier analysis is sufficient to enable one to estimate the accuracy of a difference approximation to a derivative. Consider, for instance, a central difference approximation to a first derivative,

$$\left(\frac{du}{dx}\right) \simeq \frac{u_{l+1} - u_{l-1}}{2(\Delta x)} \tag{4.43}$$

The function $u(x)$ on the left side can be expressed as a Fourier series, and for a Fourier mode

$$u = ge^{ikx} \tag{4.44}$$

the operation of the left side of Eq. (4.43) yields

$$\frac{du}{dx} = ikge^{ikx} = iku \tag{4.45}$$

For the same Fourier mode, the operation on the right side yields

$$\frac{u_{l+1} - u_{l-1}}{2(\Delta x)} = \frac{ge^{ikx_{l+1}} - ge^{ikx_{l-1}}}{2(\Delta x)} \tag{4.46}$$

Recalling that $x_{l+1} = x_l + \Delta x$ and $x_{l-1} = x_l - \Delta x$, the right side of Eq. (4.46) becomes

$$\begin{aligned}\frac{u_{l+1} - u_{l-1}}{2(\Delta x)} &= \frac{g}{2(\Delta x)}[e^{ik(x_l+\Delta x)} - e^{ik(x_l-\Delta x)}] \\ &= \frac{g}{(\Delta x)} e^{ikx_l} \frac{1}{2}(e^{ik\Delta x} - e^{ik\,\Delta x}) \\ &= \frac{iu}{(\Delta x)} \sin k(\Delta x)\end{aligned} \tag{4.47}$$

Since, for small $k(\Delta x)$, $\sin k(\Delta x)$ is approximately $k(\Delta x)$, the right side of Eq. (4.47) approaches the right side of Eq. (4.45). Hence, the formula in Eq. (4.43) is a good approximation as long as the wave number k is small (that is, the wavelength $2\pi/k$ is large). The longer the wavelength (that is, the smoother the function $u(x)$), the better is the approximation in Eq. (4.43). An estimate of the error in the approximation can be obtained by expanding the right side of Eq. (4.47). For example, for small $k(\Delta x)$, we have

$$\begin{aligned}\frac{du}{dx} \simeq \frac{u_{l+1} - u_{l-1}}{2(\Delta x)} &= \frac{iu}{(\Delta x)}\left[k(\Delta x) - \frac{k(\Delta x)^3}{3!} + O(k^5(\Delta x)^5)\right] \\ &= iuk\left[1 - \frac{k^2(\Delta x)^2}{6} + O(k^4(\Delta x)^4)\right] \\ &= \left[1 - \frac{k^2(\Delta x)^2}{6} + O(k^4(\Delta x)^4)\right]\frac{du}{dx}\end{aligned} \tag{4.48}$$

Therefore, the approximation in Eq. (4.43) is second-order correct. A similar analysis leads to the conclusion that a central difference approximation to a second derivative is also second-order correct. This is left to the student as an exercise.

4.5. HIGHER-ORDER DIFFERENCE APPROXIMATIONS

So far discussion has been confined to the derivation of three-point formulas for the first derivative and five-point formulas for the second derivative. As the problems become more complex, these standard second-order-accurate methods are not suitable to achieve the required accuracy in the solution. One method of increasing the accuracy of the difference schemes is to increase the number of grid points participating in approximating the derivative. This strategy obviously would require more storage, but with modern computers this is generally not a constraint.

For illustrative purposes, consider the derivation of a four-point approximation to the first derivative du/dx centered at the grid point l. As before, we proceed by writing Taylor series expansions for u_{l-2}, u_{l-1}, u_{l+1}, and u_{l+2} about the grid point l.

$$u_{l-2} = u_l - 2hu_l' + \frac{4}{2!}h^2u_l'' - \frac{8}{3!}h^3u_l''' + \frac{16}{4!}h^4u_l^{(\text{iv})} + \cdots$$

$$u_{l-1} = u_l - hu_l' + \frac{1}{2!}h^2u_l'' - \frac{1}{3!}h^3u_l''' + \frac{1}{4!}h^4u_l^{(\text{iv})} + \cdots$$

$$u_{l+1} = u_l + hu_l' + \frac{1}{2!}h^2u_l'' + \frac{1}{3!}h^3u_l''' + \frac{1}{4!}h^4u^{(\text{iv})} + \cdots$$

$$u_{l+2} = u_l + 2hu_l' + \frac{4}{2!}h^2u_l'' + \frac{8}{3!}h^3u_l''' + \frac{16}{4!}h^4u_l^{(\text{iv})} + \cdots$$

Multiplying these four equations respectively by $+1$, -8, $+8$, and -1 and adding, one gets

$$u_l' = \frac{1}{12h}(u_{l-2} - 8u_{l-1} + 8u_{l+1} - u_{l+2}) + O(h^4) \tag{4.49}$$

which is fourth-order accurate. Note that the two-point scheme shown in Eq. (4.12) is only second-order accurate. Extension of this idea to other derivatives is straightforward.

The usual objection to fourth-order and higher-order schemes comes from the additional grid points (besides the standard set) necessary to achieve the higher order accuracy. This introduces some difficulties of having to consider additional fictitious grid points while treating points on or near the boundary. One way to avoid the use of fictitious grid points is to introduce the concept of *one-sided approximations* (of the same order of accuracy as the rest). Such one-sided approximations are used in Chapter 13 to solve a Navier-Stokes equation. The additional nodes almost always destroy the simple tridiagonal form produced by second-order schemes.

A somewhat circuitous method, [3] attributed to Kreiss, attains fourth-order accuracy while maintaining compactness (that is, it retains the tridiagonal form for the coefficient matrix). This method is briefly presented here primarily to illustrate the variety of avenues one can explore in this area. Kreiss's idea is to first approximate the first and second derivatives as

$$u_l' = \left(\frac{D_{cx}}{1 + \frac{1}{6}h^2 D_{fx} D_{bx}}\right) u_l \tag{4.50}$$

$$u_l'' = \left(\frac{D_{fx} D_{bx}}{1 + \frac{1}{12}h^2 D_{fx} D_{bx}}\right) u_l \tag{4.51}$$

where D_{fx}, D_{bx}, and D_{cx} are defined in Eqs. (4.22) to (4.24). In practice, Eqs. (4.50) and (4.51) are used as follows. Set the derivatives to some other function, say,

$$u_l' = F_l, \qquad u_l'' = S_l. \tag{4.52}$$

Then solve for these two new functions F and S from relations obtained after multiplying by the denominator in either Eq. (4.50) or (4.51). These relations are

$$\frac{1}{6}F_{l+1} + \frac{2}{3}F_l + \frac{1}{6}F_{l-1} = \left(\frac{1}{2h}\right)(u_{l+1} - u_{l-1}) \tag{4.53}$$

$$\frac{1}{12}S_{l+1} + \left(\frac{5}{6}\right)S_l + \frac{1}{12}S_{l-1} = \left(\frac{1}{h^2}\right)(u_{l+1} - 2u_l + u_{l-1}) \tag{4.54}$$

These two equations can be solved for F and S by inverting a tridiagonal matrix using Thomas's algorithm described in Chapter 10. With the addition of the differential equation defining u, we have a set of three equations in three unknowns, y, F, and S.

The accuracy of the schemes shown in Eqs. (4.53) and (4.54) can be obtained by Taylor expansions. For example, it can be shown (show!) that

$$F_l = u_l' - (\tfrac{1}{180})h^4u^{(v)} \tag{4.55}$$

$$S_l = u_l'' - (\tfrac{1}{240})h^4u^{(vi)} \tag{4.56}$$

4.6. TREATMENT OF NORMAL DERIVATIVE BOUNDARY CONDITIONS

In the model problem of Section 4.3 a Dirichlet type of boundary condition was discussed. In many problems, the value of the normal derivative of the solution, $\partial u/\partial n$, rather than the solution u itself, is usually specified. For example, while modeling an aquifer with an impermeable rock boundary, the condition that no water flows across the boundary is stated by $\partial u/\partial n = 0$, where u is the hydraulic potential. In electrical generators, the current flowing in the electrical windings generates heat that is radiated into the environment. This process is usually represented by an equation such as $\partial u/\partial n = g(x, y, u)$.

For concreteness, consider Laplace's equation in the model problem once again:

$$\frac{\partial^2 u}{\partial x^2} + \frac{\partial^2 u}{\partial y^2} = 0, \qquad 0 \le x \le 1, \qquad 0 \le y \le 1 \tag{4.57}$$

with

$$\frac{\partial u}{\partial y} = 10 \qquad \text{on the side } y = 1 \tag{4.58}$$

$$u = 0 \qquad \text{on all other sides} \tag{4.59}$$

That is, a Dirichlet type of boundary condition is applied on three sides of a square and a Neumann type on one side. If a square mesh of size $\Delta x = \Delta y = \frac{1}{3}$ is superimposed as before, the equation analogous to Eq. (4.32) would be

$$\begin{bmatrix} 4 & -1 & -1 & 0 \\ -1 & 4 & 0 & -1 \\ -1 & 0 & 4 & -1 \\ 0 & -1 & -1 & 4 \end{bmatrix} \begin{bmatrix} u_1 \\ u_2 \\ u_3 \\ u_4 \end{bmatrix} = \begin{bmatrix} g_6 + g_{16} \\ g_{13} + g_{15} \\ g_7 + g_? \\ g_? + g_{12} \end{bmatrix} \tag{4.60}$$

In Eq. (4.60) the locations meant for the entries g_9 and g_{10} are filled by $g_?$ because they represent the values of u at the grid points 9 and 10, which are not specified; instead, the values of the first derivative, $\partial u/\partial y$, at these points are specified.

One method of handling this problem is to introduce fictitious grid points, x_{-9} and x_{-10}, outside the region as shown in Figure 4.3. Now $\partial u/\partial y$ at the point 9 can be approximated using any of the analogs presented earlier. For instance, a second-order-correct analog at the point 9 can be written as

$$\left(\frac{\partial u}{\partial y}\right)_9 \simeq \frac{u_{-9} - u_3}{(\Delta x)^2} \tag{4.61}$$

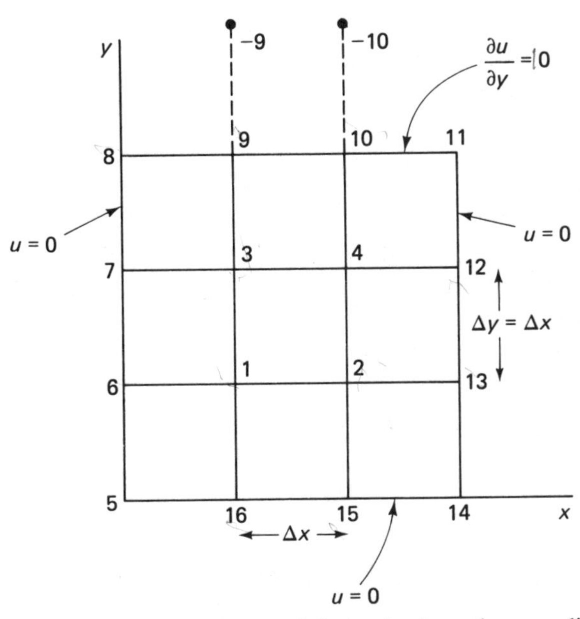

Figure 4.3 Treatment of a normal derivative boundary condition via the use of fictitious grid points

solving for the fictitious value

$$u_{-9} \simeq u_3 + (\Delta x)^2 \left(\frac{\partial u}{\partial y}\right)_9 \tag{4.62}$$

substituting this in the third equation of Eq. (4.60)

$$-u_1 + 4u_3 - u_4 = g_7 + u_3 + (\Delta x)^2 \left(\frac{\partial u}{\partial y}\right)_9 \tag{4.63}$$

which upon rearranging becomes

$$\begin{aligned} -u_1 + 3u_3 - u_4 &= g_7 + (\Delta x)^2 \left(\frac{\partial u}{\partial y}\right)_9 \\ &= 0 + (\tfrac{1}{3})^2(10) \end{aligned} \tag{4.64}$$

by virtue of the boundary conditions specified in Eq. (4.59). A similar treatment at node 10 yields

$$-u_2 - u_3 + 3u_4 = \left(\tfrac{10}{9}\right) \tag{4.65}$$

Now, Eq. (4.60) can be rewritten as

$$\begin{bmatrix} 4 & -1 & -1 & 0 \\ -1 & 4 & 0 & -1 \\ -1 & 0 & 3 & -1 \\ 0 & -1 & -1 & 3 \end{bmatrix} \begin{bmatrix} u_1 \\ u_2 \\ u_3 \\ u_4 \end{bmatrix} = \begin{bmatrix} 0 \\ 0 \\ \frac{10}{9} \\ \frac{10}{9} \end{bmatrix} \tag{4.66}$$

APPROXIMATIONS WITHOUT USING FICTITIOUS GRID POINTS

It should be observed at this point that use of fictitious grid points is merely a convenient way to obtain a difference approximation at the boundary. The use of fictitious points in no way implies that the differential equation is being applied outside the region of its applicability. Indeed, difference approximations for points on the boundary can be obtained, without recourse to fictitious points, by using Taylor series. This is demonstrated next.

Expressing the value of u at the grid points 3 and 1 in terms of the value at the point 9 (see Figure 4.3),

$$u_3 = u_9 - (\Delta y)\left(\frac{\partial u}{\partial y}\right)_9 + \frac{(\Delta y)^2}{2!}\left(\frac{\partial^2 u}{\partial y^2}\right)_9 - \cdots \tag{4.67}$$

$$u_1 = u_9 - (2\Delta y)\left(\frac{\partial u}{\partial y}\right)_9 + \frac{(2\Delta y)^2}{2!}\left(\frac{\partial^2 u}{\partial y^2}\right)_9 - \cdots \tag{4.68}$$

Multiplying Eq. (4.67) by 4 and subtracting Eq. (4.68),

$$4u_3 - u_1 = 4u_9 - u_9 - [4(\Delta y) - 2(\Delta y)]\left(\frac{\partial u}{\partial y}\right)_9 + O(\Delta y^3) \tag{4.69}$$

solving for $\left(\frac{\partial u}{\partial y}\right)_9$

$$\left(\frac{\partial u}{\partial y}\right)_9 = \frac{u_1 - 4u_3 + 3u_9}{2(\Delta y)} - O(\Delta y^2) \tag{4.70}$$

Thus

$$\left(\frac{\partial u}{\partial y}\right)_9 \simeq \frac{u_1 - 4u_3 + 3u_9}{2(\Delta y)} \tag{4.71}$$

is a second-order-correct analog to the derivative on the boundary. Solving for u_9, one gets

$$u_9 = \frac{1}{3}\left[4u_3 - u_1 + 2(\Delta y)\left(\frac{\partial u}{\partial y}\right)_9\right] \tag{4.72}$$

which can be readily identified as the missing value of g_9 in Eq. (4.60). Invoking the boundary conditions in Eq. (4.59), the third equation of Eq. (4.60) becomes

$$-u_1 + 4u_3 - u_4 = 0 + \tfrac{1}{3}[4u_3 - u_1 + 2(\tfrac{1}{3})(10)] \tag{4.73}$$

which, upon rearranging, is

$$-\tfrac{2}{3}u_1 + \tfrac{8}{3}u_3 - u_4 = \tfrac{20}{9}$$

or

$$-2u_1 + 8u_3 - 3u_4 = \tfrac{20}{3} \tag{4.74}$$

A similar analysis at node 10 yields

$$-2u_2 - 3u_3 + 8u_4 = \tfrac{20}{3} \tag{4.75}$$

Now Eq. (4.60) can be written as

$$\begin{bmatrix} 4 & -1 & -1 & 0 \\ -1 & 4 & 0 & -1 \\ -2 & 0 & 8 & -3 \\ 0 & -2 & -3 & 8 \end{bmatrix} \begin{bmatrix} u_1 \\ u_2 \\ u_3 \\ u_4 \end{bmatrix} = \begin{bmatrix} 0 \\ 0 \\ \frac{20}{3} \\ \frac{20}{3} \end{bmatrix} \tag{4.76}$$

This, of course, is not the same as Eq. (4.66).

4.7. TREATMENT OF CURVED AND IRREGULAR BOUNDARIES

If the boundary of the field does not coincide with the grid lines of the net, the finite difference equations corresponding to nodes in the immediate vicinity of the boundary must be modified slightly. Consider the model problem with a slightly modified boundary, as shown in Figure 4.4. The difference approximations to the first and second derivatives at a grid point such as point 4 should now be modified.

Using Taylor's series

$$u_B = u_4 + (a\,\Delta x)\left(\frac{\partial u}{\partial x}\right)_4 + \frac{(a\,\Delta x)^2}{2!}\left(\frac{\partial^2 u}{\partial x^2}\right)_4 + O(\Delta x)^3 \tag{4.77}$$

$$u_3 = u_4 - (\Delta x)\left(\frac{\partial u}{\partial x}\right)_4 + \frac{(\Delta x)^2}{2!}\left(\frac{\partial^2 u}{\partial x^2}\right) + O(\Delta x)^3 \tag{4.78}$$

Eliminating $(\partial^2 u/\partial x^2)_4$ from Eqs. (4.77) and (4.78) gives

$$\left(\frac{\partial u}{\partial x}\right)_4 = \frac{1}{(\Delta x)}\left\{\frac{1}{a(1+a)}u_B - \frac{1-a}{a}u_4 - \frac{a}{(1+a)}u_3\right\} \tag{4.79}$$

which is accurate to the order of $(\Delta x)^2$. Similarly, the elimination of $(\partial u/\partial x)_4$ leads to

$$\left(\frac{\partial^2 u}{\partial x^2}\right)_4 = \frac{1}{(\Delta x)^2}\left\{\frac{2}{a(1+a)}u_B + \frac{2}{(1+a)}u_3 - \frac{2}{a}u_4\right\} \tag{4.80}$$

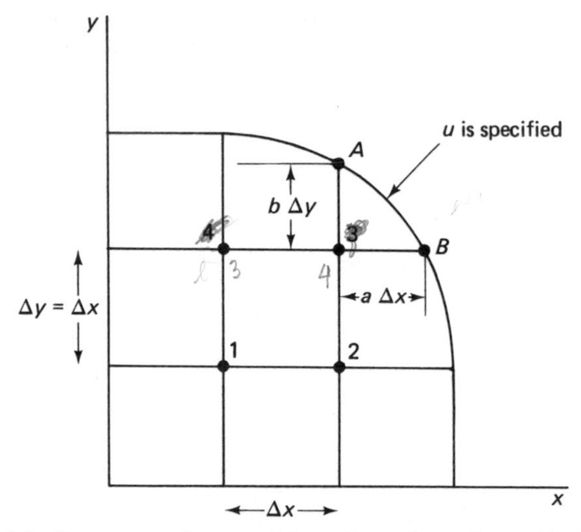

Figure 4.4 Treatment of a curved boundary when u is specified on the boundary

which is accurate to the order of only (Δx). Similarly,

$$\left(\frac{\partial^2 u}{\partial y^2}\right)_4 = \frac{1}{(\Delta y)^2}\left\{\frac{2}{b(1+b)}u_A + \frac{2}{(1+b)}u_2 - \frac{2}{b}u_4\right\} \tag{4.81}$$

which is accurate only to the order of $(\Delta y) = (\Delta x)$. Therefore, even though $(\partial^2 u/\partial x^2)$ can be approximated at the grid points 1, 2, and 3 with an accuracy of the order of $(\Delta x)^2$, the overall accuracy of the approximation is only of the order of (Δx) due to the irregularity of the boundary near the grid point 4. With Eqs. (4.80) and (4.81) properly substituted, Eq. (4.32) becomes, for $\Delta x = \Delta y$,

$$\begin{bmatrix} 4 & -1 & -1 & 0 \\ -1 & 4 & 0 & -1 \\ -1 & 0 & 4 & -1 \\ 0 & \dfrac{-2}{b(1+b)} & \dfrac{-2}{a(1+a)} & \dfrac{2}{a}+\dfrac{2}{b} \end{bmatrix} \begin{bmatrix} u_1 \\ u_2 \\ u_3 \\ u_4 \end{bmatrix} = \begin{bmatrix} g_6 + g_{16} \\ g_{13} + g_{15} \\ g_7 + g_9 \\ \dfrac{2}{b(1+b)}g_{10} + \dfrac{2}{a(1+a)}g_{12} \end{bmatrix} \tag{4.82}$$

NORMAL DERIVATIVES SPECIFIED ON CURVED BOUNDARIES

The preceding treatment requires a little more refinement if the normal derivative $\partial u/\partial n$ is specified on an irregular boundary. Once again, there are several possible ways to handle this situation. In all these procedures it is generally assumed that the boundary has a continuously turning tangent; that is, corners are excluded.

One procedure for treating this case starts with an expression for the directional derivative

$$\left.\frac{\partial u}{\partial n}\right|_\alpha = \frac{\partial u}{\partial x}\cos\alpha + \frac{\partial u}{\partial y}\sin\alpha \tag{4.83}$$

where the angle α with the x-axis specifies the direction. Now consider a boundary as shown in Figure 4.5, where $\partial u/\partial n$ is known on the boundary. That is, $\partial u/\partial n$ is known at points A, B, and C in the figure. Expanding $u(x, y)$ about the point 4, which is assumed to be the origin for the time being, in Taylor series,

$$u(x, y) = u_4 + x\left(\frac{\partial u}{\partial x}\right)_4 + y\left(\frac{\partial u}{\partial y}\right)_4 + \frac{x^2}{2}\left(\frac{\partial^2 u}{\partial x^2}\right)_4 + \left(\frac{\partial^2 u}{\partial x\,\partial y}\right)_4 xy + \left(\frac{\partial^2 u}{\partial y^2}\right)_4 \frac{y^2}{2} + \cdots \tag{4.84}$$

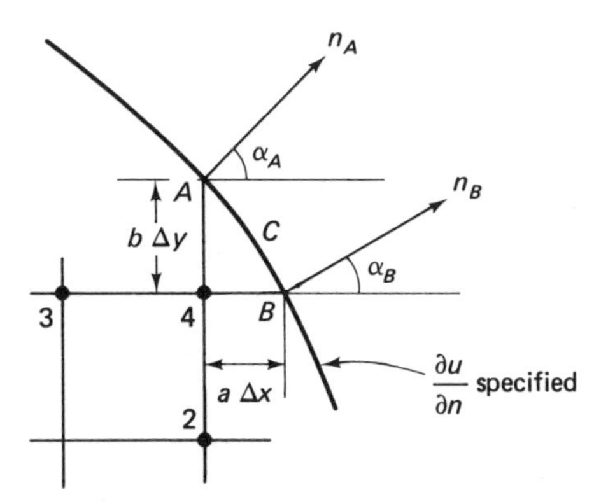

Figure 4.5 Treatment of a curved boundary when the normal derivative of u is specified on the boundary

Differentiating with respect to x and y,

$$\frac{\partial u}{\partial x} = \left(\frac{\partial u}{\partial x}\right)_4 + x\left(\frac{\partial^2 u}{\partial x^2}\right)_4 + y\left(\frac{\partial^2 u}{\partial x\,\partial y}\right)_4 + \cdots \tag{4.85a}$$

$$\frac{\partial u}{\partial y} = \left(\frac{\partial u}{\partial y}\right)_4 + x\left(\frac{\partial^2 u}{\partial x\,\partial y}\right)_4 + y\left(\frac{\partial^2 u}{\partial y^2}\right)_4 + \cdots \tag{4.85b}$$

Using Eq. (4.85) in conjunction with Eq. (4.83) and specializing for the point, say, A,

$$\left(\frac{\partial u}{\partial n}\right)_A = \left(\frac{\partial u}{\partial x}\right)_4 + (b\,\Delta y)\left(\frac{\partial^2 u}{\partial x\,\partial y}\right)_4 \cos\alpha_A + \left(\frac{\partial u}{\partial y}\right)_4 + (a\,\Delta x)\left(\frac{\partial^2 u}{\partial x\,\partial y}\right)_4 \sin\alpha_A \tag{4.86}$$

Analogous expressions can be obtained by specializing at the points B and C. Adding to these three equations the two Taylor series for u_2 and u_3 gives five equations from which expressions for the derivatives $(\partial u/\partial x)$, $(\partial u/\partial y)$, $(\partial^2 u/\partial x^2)$, $(\partial^2 u/\partial y^2)$, and $(\partial u/\partial x\,\partial y)$ at the point 4 can be obtained. These computations are left as an exercise.

4.8. USE OF GRADED GRIDS

In many problems, the field is not uniform in its character. There are usually parts of the field where the solution u varies slowly, and in parts it varies rapidly. The entire interest in the solution may be in the behavior of the solution u near one corner. To reveal the solution in sufficient detail in the region of interest, it is often necessary to use a net with a very small value of Δx and Δy. If this small value of mesh size were maintained throughout the field, it would usually result in far more equations than a computer can handle. These considerations have led to the use of ***graded nets***.

The major problem with graded nets is in obtaining the equations at the interfaces between different nets. To keep adjoining nets reasonably compatible, it is customary to double or halve the mesh size at each interface as shown in Figure 4.6.

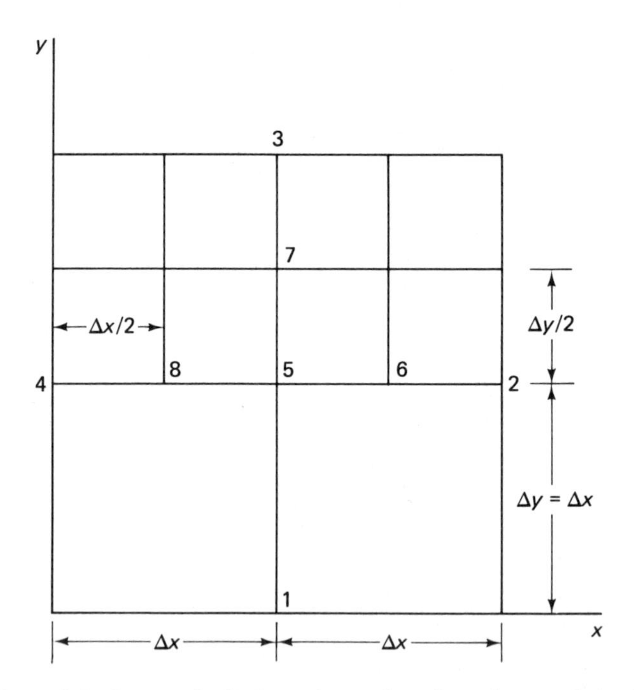

Figure 4.6 One method of treating an interface via a graded grid

The procedure for setting up difference approximations at interior points such as 1 or 7 is straightforward. But the points 5 and 6 are representative of the two types of points on the interface, and they are troublesome to handle without loss of accuracy. Many methods are available to handle such points.

For instance, for the model problem, the value u_5 can be written first as the average of u_1, u_2, u_3, and u_4, which are obtained by covering the entire region with a coarse grid. Then the value at u_6 is obtained as the average of u_5 and u_2. That is,

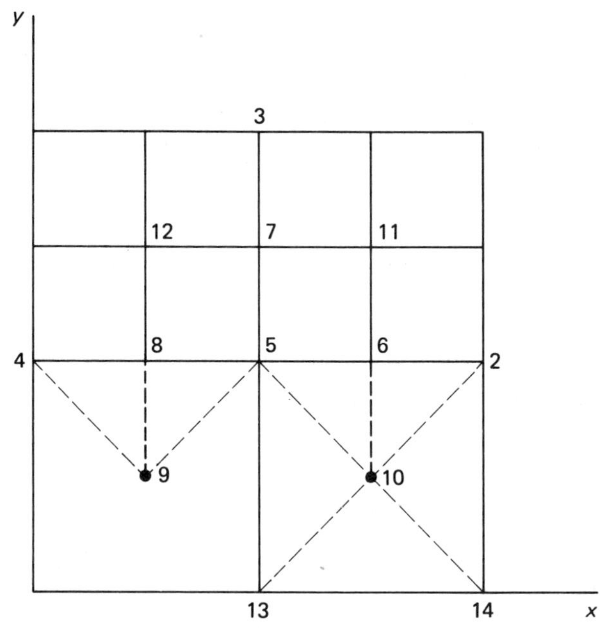

Figure 4.7 Another method of treating an interface with a graded grid

$$u_5 \simeq \tfrac{1}{4}(u_1 + u_2 + u_3 + u_4) \tag{4.87}$$

$$u_6 \simeq \tfrac{1}{2}(u_5 + u_2) \tag{4.88}$$

Alternatively, one can introduce new grid points at the midpoints of the large squares just south of the interface as shown in Figure 4.7. Now the point 5 is treated as an ordinary point of the coarse grid with neighbors 1, 2, 3, and 4 as before. The point 6 is treated as an ordinary point of the fine grid with neighbors 2, 11, 5, and 10. This requires the value of u at the grid point 10, which is obtained by using the formula

$$u_{10} \simeq \frac{1}{\sqrt{2}(\Delta x)} [u_2 + u_5 + u_{13} + u_{14}] \tag{4.89}$$

It can be readily verified that either of the preceding methods, when applied to the model problem, will lead to coefficient matrixes that are not symmetric.

4.9. DIFFERENCE APPROXIMATIONS IN POLAR COORDINATES

Partial differential equations defined on circular boundaries can usually be solved more conveniently in polar coordinates than Cartesian coordinates, because they avoid the use of awkward difference analogs near the curved boundary.

As an example of this, consider Laplace's equation in polar coordinates,

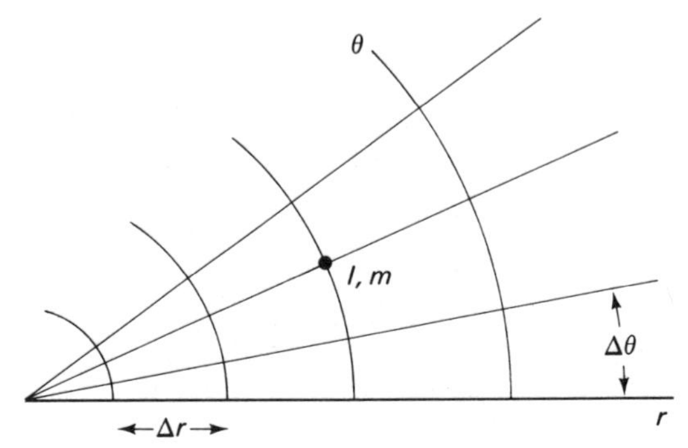

Figure 4.8 Finite difference grid in polar coordinates

$$\frac{\partial^2 u}{\partial r^2} + \frac{1}{r}\frac{\partial u}{\partial r} + \frac{1}{r^2}\frac{\partial^2 u}{\partial \theta^2} = 0 \tag{4.90}$$

Now a grid in the $r\theta$ plane is defined by concentric circles $r = l(\Delta r)$ and the radial lines $\theta = m(\Delta\theta)$, as shown in Figure 4.8. Now, a difference analog at the grid point (l, m) can be written as

$$\frac{u_{l+1,m} - 2u_{l,m} + u_{l-1,m}}{(\Delta r)^2} + \frac{1}{l(\Delta r)}\frac{u_{l+1,m} - u_{l-1,m}}{2(\Delta r)} + \frac{1}{(l\,\Delta r)^2}\frac{u_{l,m+1} - 2u_{l,m} + u_{l,m-1}}{(\Delta\theta)^2} = 0 \tag{4.91}$$

If these equations are written in detail for $l = 1, 2, \ldots, L$ and $m = 1, 2, \ldots, M$, and if the values of u at the boundaries are assumed to be known, then Eq. (4.91) in vector matrix notation is

$$\mathbf{Au} = \mathbf{b} \tag{4.92}$$

where

$$\mathbf{u} = (u_{1,1}, u_{1,2}, \ldots, u_{1,M}, u_{2,1}, u_{2,2}, \ldots, u_{2,M}, \ldots, u_{L,M})^T \tag{4.93}$$

and $\mathbf{A}$ is a matrix of coefficients. Notice that the matrix is not symmetric.

The procedure described causes no difficulty as long as $r \neq 0$. For $r = 0$, the second and third terms of Eq. (4.90) appear to become infinitely large. This complication does not arise in the Cartesian coordinate system. Therefore, if a solution to the problem at $r = 0$ is desired, either the problem should be treated in Cartesian coordinate system by using the Cartesian equivalent of Eq. (4.90), that is,

$$\frac{\partial^2 u}{\partial x^2} + \frac{\partial^2 u}{\partial y^2} = 0 \tag{4.94}$$

or the region around the point $r = 0$ should be given special treatment. A straightforward method of implementing the second option is to construct a circle of radius Δr centered at $r = 0$ and treat this circular region separately. For example, if one designates the points of intersection of the circle with the axes by 1, 2, 3, 4, and the origin by 0, then

$$\nabla^2 u = \frac{(u_1 + u_2 + u_3 + u_4 - 4u_0)}{(\Delta r)^2} + O\{(\Delta r)^2\} \tag{4.95}$$

Rotation of the axes through a small angle clearly leads to a similar equation. Repetition of this type of rotation and addition of all such equations then gives

$$\nabla^2 u = \frac{4(u_M - u_0)}{(\Delta r)^2} + O\{(\Delta r)^2\} \tag{4.96}$$

where u_M is the mean value of u around the circle

When a two-dimensional problem in cylindrical coordinates possesses circular symmetry, then $\partial^2 u/\partial\theta^2 = 0$ and Eq. (4.90) simplifies to

$$\frac{\partial^2 u}{\partial r^2} + \frac{1}{r}\frac{\partial u}{\partial r} = 0 \tag{4.97}$$

If the problem is also symmetrical with respect to the origin, then $\partial u/\partial r = 0$ at $r = 0$ and $(1/r)\,(\partial u/\partial r)$ assumes the indeterminate form of 0/0 at $r = 0$. Using Maclaurin series $u'(r)$ can be expanded as

$$\begin{aligned}\frac{\partial u}{\partial r} &= u'(r) = u'(0) + ru''(0) + \frac{1}{2}r^2u'''(0) + \cdots \\ &= ru''(0) + \tfrac{1}{2}r^2u'''(0) + \cdots\end{aligned} \tag{4.98}$$

because $u'(r) = 0$ at $r = 0$:

$$\frac{1}{r}\frac{\partial u}{\partial r} = u''(0) + \frac{1}{2}ru'''(0) + \cdots \tag{4.99}$$

$$\lim_{r\to 0}\frac{1}{r}\frac{\partial u}{\partial r} = u''(0) \tag{4.100}$$

Therefore, a $r = 0$, Eq. (4.97) can be replaced by

$$\frac{\partial^2 u}{\partial r^2} + \frac{\partial^2 u}{\partial r^2} = 0 \tag{4.101}$$

This result can also be deduced from Eq. (4.94) because $\partial^2 u/\partial x^2 = \partial^2 u/\partial y^2$ owing to circular symmetry.

4.10. DIFFERENCE APPROXIMATIONS USING DIFFERENTIATION MATRIXES

Attention so far was confined to replacing derivatives by finite difference expressions. It is also possible to replace derivatives by rectangular matrixes with interesting consequences. For instance, consider a function $u(x)$. Up until now discretization of $u(x)$ was accomplished by superimposing a grid along the x-axis and confining attention to the grid points x_l, $l = 0, 1, 2, \ldots, L + 1$. The derivatives at the grid points were approximated in terms of the values at the neighboring grid points. Thus, to approximate higher and higher order derivatives, one has to include grid points farther and farther from the immediate grid points. Intuitively, it appears that a better accuracy can be obtained if one confines attention to function values in the immediate neighborhood of the grid point in question.

For concreteness, consider the problem of approximating du/dx by confining attention to function values *within* the interval $[x_l, x_{l+1}]$. Evidently, one can write

$$u_{l+1} = u_l + (x_{l+1} - x_l)u_l' + \frac{(x_{l+1} - x_l)^2}{2!} u_l'' + \cdots$$

or

$$u_l' = \frac{1}{h}(u_{l+1} - u_l) + O(h), \qquad h = x_{l+1} - x_l$$

$$\simeq \frac{1}{h}[-1, +1]\begin{bmatrix} u_l \\ u_{l+1} \end{bmatrix} = \mathbf{D}^{(1)}\mathbf{u} \tag{4.102}$$

That is, the derivative at the point l can be approximated by using the matrix $\boldsymbol{D}^{(1)}$ as an operator on the vector $\mathbf{u}$ whose components are the values of the function $u(x)$ evaluated at $x = x_l$ and $x = x_{l+1}$.

A better approximation to the first derivative can be obtained by using the values of u at three points in $[x_l, x_{l+1}]$. Suppose the two end points and the midpoint $x_{l+1/2}$ are used for this purpose. Now the following Taylor series can be written

$$u_{l+1/2} = u_l + \frac{h}{2} u_l' + \frac{(h/2)^2}{2!} u_l'' + \cdots$$

$$u_{l+1} = u_l + hu_l' + \frac{h^2}{2!} u_l'' + \cdots$$

Multiplying the first by 4 and subtracting the second from the first,

$$4u_{l+1/2} - u_{l+1} = 4u_l - u_l + (2h - h)u_l' + O(h^3)$$

or

$$u_l' = \frac{1}{h}(-3u_l + 4u_{l+1/2} - u_{l+1}) = O(h^2)$$

$$\simeq \frac{1}{h}[-3 \quad +4 \quad -1]\begin{bmatrix} u_l \\ u_{l+1/2} \\ u_{l+1} \end{bmatrix} \tag{4.103}$$

Similarly, one can write

$$u_{l+1}' \simeq \frac{1}{h}[1 \quad -4 \quad 3]\begin{bmatrix} u_l \\ u_{l+1/2} \\ u_{l+1} \end{bmatrix} \tag{4.104}$$

These results can be combined into

$$\mathbf{u}' = \begin{bmatrix} u_l' \\ u_{l+1}' \end{bmatrix} = \frac{1}{h}\begin{bmatrix} -3 & +4 & -1 \\ 1 & -4 & +3 \end{bmatrix}\begin{bmatrix} u_l \\ u_{l+1/2} \\ u_{l+1} \end{bmatrix} = \mathbf{D}^{(2)}\mathbf{u} \tag{4.105}$$

Thus, given a function value at three points, called the ***knots***, the two slopes associated with these points can be obtained by using the matrix $\mathbf{D}^{(2)}$ as an operator on the vector $\mathbf{u}$. In general, given a function value at $(n + 1)$ points, corresponding to an nth-order polynomial, an approximation to its first derivative at n points can be obtained by using a suitably defined matrix $\mathbf{D}^{(n)}$ as an operator on the vector $\mathbf{u} = (u_1, u_2, \ldots, u_{n+1})^T$. The matrixes $\mathbf{D}$, thus defined, are called ***differentiation matrixes***. Further details of this method can be found in the literature. [4]

4.11. REVIEW OF SOME BASIC IDEAS FROM FUNCTIONAL ANALYSIS

In ordinary analysis we work with the real or complex number system. Functional analysis is based on the use of linear spaces, which are generalizations of these number systems. From a conceptual standpoint, a linear space can be visualized as a collection of, for instance, vectors. A simple example of a linear space is the set or collection of two-component vectors $\mathbf{u}$, denoted by

$$R^2 = \{\mathbf{u} : \mathbf{u} = (u_1, u_2)\}$$

where u_1, u_2 are real numbers. Similarly, a linear space consisting of a collection of n-component vectors can be denoted by R^n. Another example of an important linear space is furnished by the set $C[0, 1]$ of real functions $u = u(t)$, which are continuous on the closed interval $0 \leq t \leq 1$. From a conceptual standpoint, the vectors $\mathbf{u} = (u_1, u_2)$, which form R^2, and the functions $u = u(t)$ belonging to $C[0, 1]$ are not as different as they may seem at first. The vector may be regarded as being a table of values for the function $u = u(t)$, which is defined only on the set of coordinates $i = 1$, 2 with

$$u(t_i) = u_i, \qquad i = 1, 2$$

being the components of the vector $\mathbf{u}$. On the other hand, the function $u = u(t)$ could be regarded as being a vector for which each number t in the interval $0 \leq t \leq 1$ is a coordinate. Thus $u = u(t)$ could be considered as a vector with infinitely many components. However, during the computation of solutions, one is limited to the use of a finite set of numbers, and this makes it impossible, in general, to give exact representations of elements of infinite dimensional spaces. Therefore, a central problem area of computational mathematics is concerned with finding suitable finite representations of elements of infinite-dimensional spaces and the interpretation of the results of finite computations in such spaces.

One method of obtaining an element of a finite-dimensional space that corresponds to a given element of an infinite-dimensional space is called ***projection***. To see the nature of this projection operation, consider a power series expansion of a function $f(x)$,

$$f(x) = \alpha_0 + \alpha_1 x + \alpha_2 x^2 + \cdots$$

where $\alpha_0, \alpha_1, \ldots$ are scalars. Evidently, $f(x)$ is a linear combination of the set of functions $\{x^n\}$, $n = 0, 1, 2, \ldots$. In a computational process, $f(x)$ is usually approximated by a polynomial such as

$$f(x) \longrightarrow p_n(x) = \alpha_0 + \alpha_1 x + \cdots + \alpha_n x^n$$

In such a situation, one says that $p_n(x)$ is said to be the projection of $f(x)$ onto an n-dimensional subspace. Evidently, one can get many different projections of $f(x)$ by choosing different elements from the set $\{x\}^n$ in the construction of $p_n(x)$. That is, one can replace $f(x)$ by using one of many suitable projections. Indeed, the finite difference approximations presented in the preceding sections can be viewed as projections from the space Ω to the ***grid space*** Ω_n.

This general viewpoint leads to the very constructive conclusion that one can construct approximating difference schemes using several techniques. As was mentioned earlier, the most satisfactory schemes are generally obtained for equations with sufficiently smooth coefficients and smooth solutions. These particular schemes have been the subject of ever-increasing interest, because the speed with which new complicated problems are emerging has had a definite bearing on the evolutionary pace of the computational means. It seems reasonable therefore to seek approximate solutions of specified accuracy not at the expense of a formal increase in the dimensionality of the subspaces involved (that is, not by decreasing the grid size), but rather by constructing more accurate approximations of the original problem using a priori information about the smoothness of the solution. This point of view turned out to be quite useful. The Ritz and Galerkin variational methods derived from this perspective have proved themselves to be very powerful indeed.

REFERENCES

1. A. R. MITCHELL, *Computational Methods for Partial Differential Equations*. New York: John Wiley & Sons, Inc., 1969.

2. D. M. YOUNG, *Iterative Solution of Large Linear Systems*. New York: Academic Press, Inc., 1971.

3. R. S. HIRSH, "Higher-order Accurate Difference Solutions of Fluid Mechanics Problems by a Compact Differencing Technique", *J. Computational Phys.*, Vol. 19, pp. 90–109, 1975.

4. A. RAEFSKY and V. VEMURI, "A Numerical Method for Boundary Value Problems", *International Journal of Computers and Electrical Engineering.*, Vol. 5, pp. 85–104, 1978.

EXERCISES

1. Derive a finite difference approximation to $\partial^3 u/\partial x^3$. Where is the resulting expression centered? What is the order of accuracy of the approximation?

2. For the model problem discussed in Section 4.3, choose $\Delta x = \Delta y = \frac{1}{40}$. What is the ratio of the nonvanishing entries of **A** to the total number?

3. Consider the regular hexagonal and triangular grids shown in the accompanying figures.

Using Taylor series method, derive, (a) four-point difference approximation on the hexagonal grid, and (b) seven-point approximation on the triangular grid to solve.

$$\frac{\partial^2 u}{\partial x^2} + \frac{\partial^2 u}{\partial y^2} = 0$$

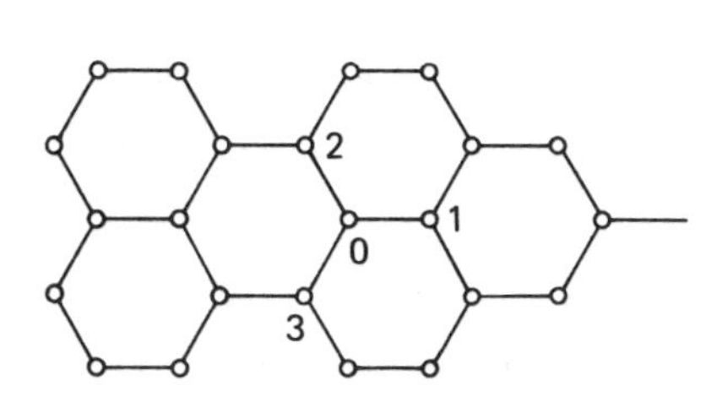

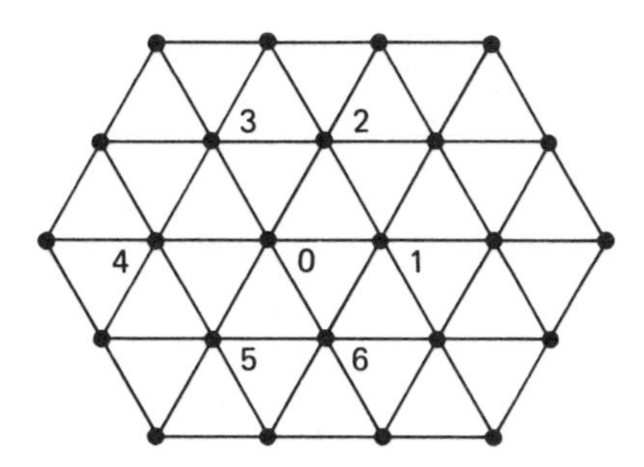

4. Consider the problem of finding five-point difference approximations to Laplace's equation in cylindrical coordinates,

$$\frac{\partial^2 u}{\partial r^2} + \frac{1}{r}\frac{\partial u}{\partial r} + \frac{\partial^2 u}{\partial z^2} = 0$$

over a uniform mesh $h = \Delta r = \Delta z$. Using Taylor series and central differences, show that the resulting coefficient matrix is nonsymmetric.

5. Consider

$$(x+1)\frac{\partial^2 u}{\partial x^2} + (y+1)\frac{\partial^2 u}{\partial y^2} - u = 1$$

in $0 \le x \le 1$, $0 \le y \le 1$ with the boundary conditions $u(0, y) = y$, $u(1, y) = y^2$, $u(x, 0) = 0$, $u(x, 1) = 1$. With $\Delta x = \Delta y = \frac{1}{3}$, write down the finite difference approximation in matrix form.

6. Derive a finite difference approximation to the operator

$$\frac{\partial}{\partial x}\left(A(x, y)\left(\frac{\partial u}{\partial x}(x, y)\right)\right)$$

at a point (lh, mh) where $h = \Delta x = \Delta y$. Repeat the problem for the operator

$$\frac{\partial}{\partial y}\left(C(x, y)\frac{\partial u}{\partial y}(x, y)\right)$$

7. Consider a simple ordinary differential equation such as

$$\frac{-d^2 y(x)}{dx^2} + \sigma(x)y(x) = f(x), \qquad 0 < x < 1$$

The interval $0 \le x \le 1$ is divided into ten equal parts and a discrete approximation to the preceding is written as $\mathbf{Ax} = \mathbf{b}$. How many equations does this represent for the following sets of boundary conditions?

(a) $y(0) = \alpha$, $y(1) = \beta$ (b) $y(0) = \alpha$, $y'(1) = \beta$ (c) $y'(0) = \alpha$, $y'(1) = \beta$

8. Using the Fourier analysis method described in Section 4.4, show that the approximation

$$\frac{d^2u}{dx^2} = \frac{u_{l+1} - 2u_l + u_{l-1}}{(\Delta x)^2}$$

is (a) only second order correct, and (b) only good for small $k(\Delta x)$ where the wavelength of interest is much larger than (Δx).

9. Most of the discussion in this chapter has been confined to the deriving of three-point difference approximations to the first derivative and five-point approximations to the second derivative.

(a) Discuss how you would proceed to approximate $\partial u/\partial x$ using five-point and seven-point approximations.

(b) Show that $\partial u/\partial x$ can be approximated by the seven-point formula

$$\left(\frac{\partial u}{\partial x}\right)_{x=x_l} = \frac{-u_{l-3} + 9u_{l-2} - 45u_{l-1} + 45u_{l+1} - 9u_{l+2} + u_{l+3}}{60h}$$

(c) Show that $\partial u/\partial x$ can be approximated at $x = x_l$ by the third-order unsymmetric approximation

$$\left(\frac{\partial u}{\partial x}\right)_{x=x_l} = \frac{1}{6h}(-2u_{l-1} - 3u_l + 6u_{l+1} - u_{l+2})$$

Results of this problem are useful in the development of the method of lines discussed in Chapter 6.

10. Consider the problem of approximating $\partial^2 u/\partial x^2$ at $x = x_l$ using a five-point formula. This problem becomes difficult when a grid point falls on or adjacent to a boundary. Resolve this difficulty by extending the solution to two fictitious grid points beyond the boundary by cubic extrapolation. Also show that the formulas for cubic extrapolation at the boundaries are

$$u_{-1} = 4u_0 - 6u_1 + 4u_2 - u_3$$

$$u_{-2} = 4u_{-1} - 6u_0 + 4u_1 - u_2$$

Analogous formulas hold good for u_{N+1} and u_{N+2}.

11. Discuss the advantages and disadvantages of higher-order formulas.

12. Consider a function $u(x)$ defined on $0 \le x \le 1$. Discretize $0 \le x \le 1$ such that $0 = x_0 < x_1 \ldots x_l < x_{l+1} \ldots x_n = 1$. Now consider a typical interval of size $h = x_{l+1} - x_l$. Using a procedure similar to that used in Section 4.10, derive an expression for the differentiation matrix for du/dx such that $\mathbf{u}'$ is related to the values u_l, $u_{l+1/3}$, $u_{l+2/3}$, and u_{l+1} via $\mathbf{D}^{(3)}$.

13. Repeat the procedure described in Section 4.9 and derive a family of matrixes to approximate d^2u/dx^2. [*Hint:* Note that $d^2u/dx^2 = d/dx\,(du/dx)$.]

14. Consider the problem of solving

$$\frac{du}{dx} + u = f(x), \quad x \in [x_l, x_{l+1}]$$

Suppose du/dx is approximated by using the differentiation matrix $\mathbf{D}^{(2)}$ defined in Eq. (4.105). Suppose $u(x)$ and $f(x)$ are replaced by

$$\mathbf{u} = (u_l, u_{l+1/2}, u_{l+1})^T \quad \text{and} \quad \mathbf{f} = (f_l, f_{l+1/2}, f_{l+1})^T$$

Write down the matrix analog to the differential equation. Can you solve the matrix equation? If so, how? If not, why not? Discuss your answer.

5

Boundary Value Problems

5.1. A GENERAL BOUNDARY VALUE PROBLEM

The basic aim of this chapter is to describe a general boundary value problem associated with elliptic partial differential equations and to show how their solution by finite difference methods leads to systems of algebraic equations with some specialized properties.

Let Ω be a bounded plane region with boundary Γ, and let $g(x, y)$ be a given function defined and continuous on Γ. The problem is to find a function $u(x, y)$ defined in $\Omega + \Gamma$ and continuous in $\Omega + \Gamma$ such that $u(x, y)$ is twice differentiable in Ω and satisfies the linear second-order partial differential equation

$$Lu \triangleq A\frac{\partial^2 u}{\partial x^2} + C\frac{\partial^2 u}{\partial y^2} + D\frac{\partial u}{\partial x} + E\frac{\partial u}{\partial y} + Fu = G \qquad (5.1)$$

in Ω. Here A, C, D, E, F, and G are analytic functions of x and y in Ω and satisfy $A \geq 0$, $C \geq 0$, and $F \leq 0$. The function u is required to satisfy the condition

$$\alpha(x, y)u(x, y) + \beta(x, y)\frac{\partial u}{\partial n}(x, y) = g(x, y) \qquad \text{on } \Gamma \qquad (5.2)$$

This problem is often referred to as the ***general boundary value problem***. If $\beta = 0$ in Equation (5.2), the problem is a ***generalized Dirichlet problem***. If, in addition, L in Eq. (5.1) is Laplacian operator, the problem is a ***Dirichlet problem***.

Consider a generalized Dirichlet problem. A first step in arriving at a finite difference analog to this problem is to superimpose, say, a square grid with $\Delta x = \Delta y = h$. Attention is now confined to the set of points Ω_h defined by the grid in Ω. These are ***interior points***. Without much ado, let us assume that some of the grid points fall exactly on Γ, and this set is denoted by Γ_h. These are ***boundary points***. If u and its derivatives are now approximated at each interior point (l, m) by, say, central difference approximations, then one gets

$$\alpha_1 u_{l-1,m} + \alpha_2 u_{l+1,m} + \alpha_3 u_{l,m-1} + \alpha_4 u_{l,m+1} - \alpha_0 u_{l,m} = t_{l,m} \tag{5.3}$$

where

$$\begin{aligned} \alpha_1 &= A_{l,m} - \tfrac{1}{2}hD_{l,m}, \qquad & \alpha_3 &= C_{l,m} - \tfrac{1}{2}hE_{l,m} \\ \alpha_2 &= A_{l,m} + \tfrac{1}{2}hD_{l,m}, \qquad & \alpha_4 &= C_{l,m} + \tfrac{1}{2}hE_{l,m} \\ \alpha_0 &= 2(A_{l,m} + C_{l,m} - \tfrac{1}{2}h^2F_{l,m}), \qquad & t_{l,m} &= h^2G_{l,m} \end{aligned} \tag{5.4}$$

In Eqs. (5.3) and (5.4), subscript (l, m) means that the subscripted quantities are evaluated at the grid point whose coordinates are (lh, mh).

Equation (5.3) written for each interior grid point represents a system of as many equations as there are interior grid points. This set of equations can be compactly written as

$$\mathbf{Au} = \mathbf{b} \tag{5.5}$$

The entries on the main diagonal of $\mathbf{A}$, that is, the a_{ii}'s, are the α_0's of Eq. (5.4), and the off-diagonal entries, the a_{ij}'s, are the negatives of the α_i's for $i = 1$ to 4. From Eq. (5.4) it is evident that, since $F \leq 0$,

$$a_{i,i} \geq 0, \tag{5.6}$$

$$a_{i,j} \leq 0, \qquad i \neq j \tag{5.7}$$

for all interior points, provided that h is sufficiently small, that is, provided

$$0 < h \leq \min \left\{ \frac{2A_{l,m}}{|D_{l,m}|}, \frac{2C_{l,m}}{|E_{l,m}|} \right\} \tag{5.8}$$

where the minimum is taken over all interior and boundary grid points. Inspection of Eqs. (5.3) and (5.4) also reveals that, since $F \leq 0$,

$$\alpha_0 \geq \sum_{i=1}^{4} \alpha_i \tag{5.9}$$

or, equivalently,

$$a_{i,i} \geq \sum_{\substack{\text{all } j \\ i \neq j}} |a_{i,j}| \tag{5.10}$$

In Eq. (5.10), the strict inequality should hold good for some i. These properties, observed by mere inspection, have profound computational significance while solving Eq. (5.5) using systematic iterative methods, described at length in Chapter 10.

Many useful concepts that play a crucial role in solving $\mathbf{Au} = \mathbf{b}$, such as ***diagonal dominance***, ***irreducibility***, and ***property* A**, can be related to the elementary observations made here. A matrix $\mathbf{A}$ is said to be *diagonally dominant* if Eq. (5.10) holds for all i.

The matrix is strictly diagonally dominant if the strict inequality holds for all i. The concept of irreducibility can be informally explained as follows. A matrix **A** is reducible if it can be partitioned such that a subset of equations can be solved separately; otherwise, **A** is irreducible. Property A can also be informally stated as follows. With reference to a two-space dimensional problem, let the grid points be labeled alternately as black (B) and white (W) in a chessboard fashion as follows:

$$
\begin{array}{ccccc}
B & W & B & W & \cdots \\
W & B & W & B & \cdots \\
B & W & B & W & \cdots \\
W & B & W & B & \cdots \\
\cdot & \cdot & \cdot & \cdot & \cdots \\
\cdot & \cdot & \cdot & \cdot & \cdots
\end{array}
$$

If the difference equation approximating the differential equation at point B can be solved for the function value at B exclusively in terms of the function values at those points labeled W, and vice versa, then matrix **A** has property A.

5.2. THE SELF-ADJOINT PROPERTY

If Eq. (5.3) is written out in full for a few of the grid points, it becomes evident that **A** is not in general a symmetric matrix. Symmetric matrixes have practical advantages from a computational viewpoint because the inverse of a symmetric matrix is symmetric, thereby reducing computational effort and storage requirements. If the differential operator L in Eq. (5.1) is *self-adjoint*, one can rightfully expect a symmetric **A**, because **A** is a discrete analog of L, and symmetry of a matrix is analogous to self-adjointness of a differential operator.

A differential operator L is said to be self-adjoint if, for any two sufficiently smooth functions ϕ and ψ,

$$I = \iint_{\Omega} [\phi L(\psi) - \psi L(\phi)]\, d\Omega \tag{5.11}$$

is a function of the values of ϕ, ψ and their derivatives *on* Γ *alone*. Somewhat more important is the concept of the self-adjointness of a *boundary value problem*. A boundary value problem, such as the one described by Eqs. (5.1) and (5.2), is said to be self-adjoint if the value of I in Eq. (5.11) vanishes for any two sufficiently smooth functions ϕ and ψ that satisfy the corresponding homogeneous boundary conditions, that is,

$$
\begin{aligned}
\alpha\phi + \beta\frac{\partial\phi}{\partial n} &= 0 \\
\alpha\psi + \beta\frac{\partial\psi}{\partial n} &= 0
\end{aligned} \tag{5.12}
$$

For instance, Laplace's equation with Dirichlet boundary conditions is self-adjoint. To demonstrate this, one has simply to prove for a two-dimensional case that

$$\iint \phi\left(\frac{\partial^2}{\partial x^2} + \frac{\partial^2}{\partial y^2}\right)\psi \, dx \, dy = \iint \psi\left(\frac{\partial^2}{\partial x^2} + \frac{\partial^2}{\partial y^2}\right)\phi \, dx \, dy \tag{5.13}$$

Toward this end, the left side operation is performed using integration by parts:

$$\int u \, dv = uv - \int v \, du \tag{5.14}$$

For simplicity, only the first term on the left side of Eq. (5.13) is considered.

$$\iint \phi \frac{\partial^2}{\partial x^2}\psi \, dx \, dy = \int \left[\int \phi \frac{\partial^2 \psi}{\partial x^2} dx\right] dy \tag{5.15}$$

The inner integral on the right side of Eq. (5.15) is

$$\int \phi \frac{\partial^2 \psi}{\partial x^2} dx = \int \phi \frac{\partial}{\partial x}\left(\frac{\partial \psi}{\partial x}\right) dx = \int \phi d\left(\frac{\partial \psi}{\partial x}\right)$$

which can be integrated by parts by setting $u = \phi$ and $v = (\partial\psi/\partial x)$ to yield

$$\int \phi \frac{\partial^2 \psi}{\partial x^2} dx = \phi \frac{\partial \psi}{\partial x} - \int \frac{\partial \psi}{\partial x}\frac{\partial \phi}{\partial x} dx \tag{5.16}$$

The first term on the right when evaluated on the boundary vanishes and can be dropped from further consideration. The second term on the right side of Eq. (5.16) can be further integrated by parts by setting $u = (\partial\phi/\partial x)$ and $v = \psi$ to yield

$$-\int \frac{\partial \psi}{\partial x}\frac{\partial \phi}{\partial x} dx = -\psi \frac{\partial \phi}{\partial x} + \int \psi \frac{\partial}{\partial x}\left(\frac{\partial \phi}{\partial x}\right) dx \tag{5.17a}$$

$$= \int \psi \frac{\partial^2 \phi}{\partial x^2} dx \tag{5.17b}$$

Equation (5.17b) follows from Eq. (5.17a) because the first term on the right side of Eq. (5.17a) vanishes on the boundary. Combining Eqs. (5.15) to (5.17a), one has

$$\iint \phi \frac{\partial^2 \psi}{\partial x^2} dx \, dy = \iint \psi \frac{\partial^2 \phi}{\partial x^2} dx \, dy \tag{5.18}$$

which proves that $\partial^2/\partial x^2$ is self-adjoint. Similarly, $\partial^2/\partial y^2$ is self-adjoint. Therefore, Laplace's equation with Dirichlet boundary conditions is self-adjoint. Indeed, it can be proved that any boundary value problem $\nabla^2 u = g$ with boundary conditions such as those in Eq. (5.12) is a self-adjoint problem.

The reader may well wonder at this point what types of differential operators are not self-adjoint and what types of boundary value problems associated with a self-adjoint operator L are not self-adjoint. Crudely, operators containing first-order derivatives are not likely to be self-adjoint because, obviously,

$$\int \phi \frac{d\psi}{dx} dx \neq \int \psi \frac{d\phi}{dx} dx$$

As an example of the latter case, the problem

$$Lu = \frac{d^2u}{dx^2} = 0, \qquad a \leq x \leq b \tag{5.19a}$$

with

$$u(a) = u_a \qquad \text{and} \qquad u'(a) = u'_a \tag{5.19b}$$

is not a self-adjoint boundary value problem. Proof of this assertion is left as an exercise. Similarly, it can be proved (show the proof) that an elliptic equation of the form of Eq. (5.1) is self-adjoint if

$$\frac{\partial A}{\partial x} = D \qquad \text{and} \qquad \frac{\partial C}{\partial y} = E \tag{5.20}$$

Even if these conditions are not satisfied, one can sometimes obtain the self-adjoint form

$$\frac{\partial}{\partial x}\left(A\frac{\partial u}{\partial x}\right) + \frac{\partial}{\partial y}\left(C\frac{\partial u}{\partial y}\right) + Fu = G \tag{5.21}$$

by multiplying both sides of Eq. (5.1) by an "integrating factor" $\mu(x, y)$ so that

$$\frac{\partial}{\partial x}(\mu A) = D, \qquad \frac{\partial}{\partial y}(\mu C) = E \tag{5.22}$$

are satisfied. Such a function $\mu(x, y)$ exists if and only if

$$\frac{\partial}{\partial y}\left(\frac{D - \frac{\partial A}{\partial x}}{A}\right) = \frac{\partial}{\partial x}\left(\frac{E - \frac{\partial C}{\partial y}}{C}\right) \tag{5.23}$$

If Eq. (5.23) is satisfied, Eq. (5.1) is said to be ***essentially self-adjoint***.

5.3. DISCRETIZING A SELF-ADJOINT PROBLEM

Treatment of this section parallels that of Section 5.1 except for the fact that the differential operator L here is assumed to be self-adjoint. Consider the self-adjoint differential equation

$$Lu = \frac{\partial}{\partial x}\left(A\frac{\partial u}{\partial x}\right) + \frac{\partial}{\partial y}\left(C\frac{\partial u}{\partial y}\right) + Fu = G \tag{5.24a}$$

in Ω. Here, as before, A, C, F, and G are analytic functions of x and y in Ω and satisfy $A \geq 0$, $C \geq 0$, and $F \leq 0$. The function u, as before, is required to satisfy the boundary condition

$$u(x, y) = g(x, y) \qquad \text{on } \Gamma \tag{5.24b}$$

That is, we have a generalized Dirichlet problem. By following steps similar to those in Section 5.2, a discrete analog to Eq. (5.24) can be obtained. For instance, the differential operators $(\partial/\partial x)(A\, \partial u/\partial x)$ and $(\partial/\partial y)(C\, \partial u/\partial y)$ can be replaced by

$$\frac{\partial}{\partial x}\left(A\frac{\partial u}{\partial x}\right) \simeq h^{-2}\{A_{l+1/2,m}(u_{l+1,m} - u_{l,m}) - A_{l-1/2,m}(u_{l,m} - u_{l-1,m})\}$$
$$\frac{\partial}{\partial y}\left(C\frac{\partial u}{\partial y}\right) \simeq h^{-2}\{C_{l,m+1/2}(u_{l,m+1} - u_{l,m}) - C_{l,m-1/2}(u_{l,m} - u_{l,m-1})\} \tag{5.25}$$

Substituting Eq. (5.25) in (5.24), one gets

$$\hat{\alpha}_1 u_{l-1,m} + \hat{\alpha}_2 u_{l+1,m} + \hat{\alpha}_3 u_{l,m-1} + \hat{\alpha}_4 u_{l,m+1} - \hat{\alpha}_0 u_{l,m} = t_{l,m} \tag{5.26}$$

where

$$\begin{aligned} \hat{\alpha}_1 &= A_{l-1/2,m}, \qquad \hat{\alpha}_3 = C_{l,m-1/2} \\ \hat{\alpha}_2 &= A_{l+1/2,m}, \qquad \hat{\alpha}_4 = C_{l,m+1/2} \\ \hat{\alpha}_0 &= (\hat{\alpha}_1 + \hat{\alpha}_2 + \hat{\alpha}_3 + \hat{\alpha}_4 + h^2F_{l,m}), \qquad t_{l,m} = h^2G_{l,m} \end{aligned} \tag{5.27}$$

Equation (5.26) written for each interior grid point represents a system of equations, which is written as

$$\hat{\mathbf{A}}\mathbf{u} = \mathbf{b} \tag{5.28}$$

Unlike $\mathbf{A}$ of Eq. (5.5), the matrix $\hat{\mathbf{A}}$ is necessarily symmetric. For, by Eqs. (5.26) and (5.27), the coefficient of $u_{l+1,m}$ in the equation written at grid point (l, m), that is, $A_{l+1/2,m} = A(x + h/2, y)$, is the same as the coefficient of $u_{l'-1,m}$ in the equation written at grid point (l', m) where $l' = l + 1$. This latter coefficient is

$$A_{l'-1/2,m} = A_{l+1-1/2,m} = A_{l+1/2,m} \tag{5.29}$$

In all other respects, statements made in Section 5.1 are also valid here.

This brief discussion is sufficient to motivate one to look for alternate discretization schemes that would always lead to a symmetric matrix whenever L is self-adjoint. One such procedure, the integration technique, is discussed in the following two sections. Another procedure, based on the variational principle, will be discussed in Section 5.6.

5.4. DIFFERENCE APPROXIMATIONS USING INTEGRATION METHOD

To illustrate the rudiments of this method, R. S. Varga [1] considers the simple self-adjoint ordinary differential equation

$$-\frac{d^2y(x)}{dx^2} + \sigma(x)y(x) = g(x), \qquad a < x < b \tag{5.30}$$

subject to the boundary conditions

$$y(a) = \alpha, \qquad y(b) = \beta \tag{5.31}$$

Here $\sigma(x)$ and $g(x)$ are given real continuous functions on $a \leq x \leq b$ with $\sigma(x) \geq 0$. As in the finite differencing procedure, a uniform grid of size $h = (b - a)/(L + 1)$ is superimposed on the interval $a \leq x \leq b$, and the grid points are denoted by

$$x_l = a + l\frac{(b-a)}{(L+1)} = a + lh, \qquad 0 \leq l \leq L+1 \tag{5.32}$$

as shown in Figure 5.1.

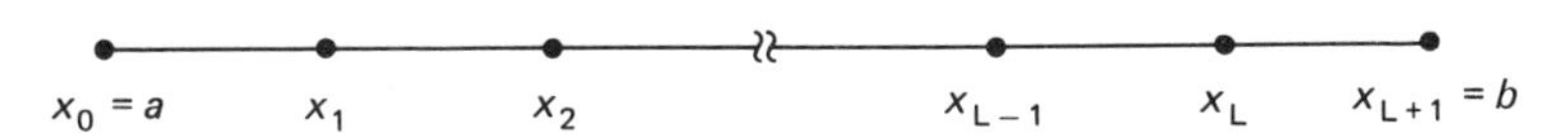

Figure 5.1 One-dimensional grid used in the integration method

Suppose Eq. (5.30) is integrated with respect to x from η_1 to η_2 with $a \leq \eta_1 \leq \eta_2 \leq b$; one gets

$$-\frac{dy(\eta_2)}{dx} + \frac{dy(\eta_1)}{dx} + \int_{\eta_1}^{\eta_2} \sigma(x)y(x)\,dx = \int_{\eta_1}^{\eta_2} g(x)\,dx \tag{5.33}$$

Equation (5.33) can be evaluated at any of the grid points. Toward this end, η_1 and η_2 are conveniently chosen as

$$\eta_1 = x_{l-1/2} = x_l - \frac{h}{2}, \qquad \eta_2 = x_{l+1/2} = x_l + \frac{h}{2} \tag{5.34}$$

In the same vein, the derivatives and integrals are approximated by

$$\frac{dy}{dx}(x_{l+1/2}) \simeq \frac{y_{l+1} - y_l}{h} \tag{5.35}$$

and

$$\int_{x_{l-1/2}}^{x_{l+1/2}} f(x)\,dx \simeq f_l(x_{l+1/2} - x_{l-1/2}) = f_l h \tag{5.36}$$

It should be noted that Eqs. (5.35) and (5.36) represent one of many possible approximations. With these approximations, Eq. (5.33) becomes

$$(-y_{l-1} + 2y_l - y_{l+1}) + h^2\sigma_l y_l = h^2 g_l \tag{5.37}$$

Or, in matrix notation

$$\mathbf{Ay} = \mathbf{b} \tag{5.38}$$

The elementary idea presented here can be refined and generalized to higher dimensions, nonuniform grids, and interface conditions. Toward this end, consider

$$-\frac{d}{dx}\left\{p(x)\frac{dy}{dx}\right\} + q(x)\frac{dy(x)}{dx} + \sigma(x)y(x) = g(x) \tag{5.39}$$

for $a < x < b$, subject to the boundary conditions

$$\alpha_1 y(a) - \beta_1 y'(a) = \gamma_1, \qquad \alpha_2 y(b) + \beta_2 y'(b) = \gamma_2 \tag{5.40}$$

where

$$p(x) > 0, \qquad q(x) > 0, \qquad \alpha_i, \beta_i \geq 0, \qquad \alpha_i + \beta_i > 0 \tag{5.41}$$

In many applications $p(x)$, $q(x)$, $\sigma(x)$, and $f(x)$ are only *piecewise continuous* and may even possess a finite number of discontinuities in $a \leq x \leq b$. The points of discontinuities often represent ***interfaces***. To accommodate this situation, p, q, σ, and f are assumed to be only piecewise continuous. However, it is necessary to require continuity of $y(x)$ and $\{p(x)\, dy/dx\}$ as the former has to be twice differentiable and the latter once. Differentiability of $\{p(x)\, dy/dx\}$ implies

$$p(\xi+)\frac{dy(\xi+)}{dx} = p(\xi-)\frac{dy(\xi-)}{dx}, \qquad a < \xi < b \tag{5.42}$$

where the notations $f(\xi+)$ and $f(\xi-)$ denote, respectively, the right and left hand limits of $f(x)$ at $x = \xi$.

As a first step of discretizing this problem, a grid, not necessarily uniform, defined by

$$a = x_0 < x_1 < x_2 < \cdots x_{L+1} = b \tag{5.43}$$

is superimposed on $a \leq x \leq b$. The only requirement in the selection of this grid is that the points of discontinuities, if any, of p, q, σ, and g be a subset of the preceding set of grid points. As in the previous case, Eq. (5.39) is integrated, but now within the limits $(x_l, x_{l+1/2})$, to give

$$\begin{aligned} -p_{l+1/2}\frac{dy_{l+1/2}}{dx} + p(x_l^+)\frac{dy(x_l^+)}{dx} + \int_{x_l}^{x_{l+1/2}} \frac{dy}{dx}\,dx \\ + \int_{x_l}^{x_{l+1/2}} \sigma(x)y(x)\,dx = \int_{x_l}^{x_{l+1/2}} g(x)\,dx \end{aligned} \tag{5.44}$$

If the grid point l is not a boundary point, Eq. (5.39) can also be integrated within the limits $x_{l-1/2} \leq x \leq x_l$ to yield

$$\begin{aligned} -p(x_l^-)\frac{dy(x_l^-)}{dx} + p_{l-1/2}\frac{dy_{l-1/2}}{dx} + \int_{x_{l-1/2}}^{x_l} q(x)\frac{dy}{dx}\,dx \\ + \int_{x_{l-1/2}}^{x_l} \sigma(x)y(x)\,dx = \int_{x_{l-1/2}}^{x_l} g(x)\,dx \end{aligned} \tag{5.45}$$

Adding Eqs. (5.44) and (5.45) and invoking the continuity condition in Eq. (5.42),

$$\begin{aligned} -p_{l+1/2}\frac{dy_{l+1/2}}{dx} + p_{l-1/2}\frac{dy_{l-1/2}}{dx} + \int_{x_{l-1/2}}^{x_{l+1/2}} q(x)\frac{dy}{dx}\,dx \\ + \int_{x_{l-1/2}}^{x_{l+1/2}} \sigma(x)y(x)\,dx = \int_{x_{l-1/2}}^{x_{l+1/2}} g(x)\,dx \end{aligned} \tag{5.46}$$

The derivatives can be approximated using expressions such as those in Eq. (5.35). As there is a possibility for the existence of an interface at the grid point l, care should be exercised in evaluating the integrals whose range of integration includes the grid point l. Therefore, the integrals are approximated using

$$\begin{aligned} \int_{x_{l-1/2}}^{x_{l+1/2}} f(x)\,dx &= \int_{x_{l-1/2}}^{x_l} f(x)\,dx + \int_{x_l}^{x_{l+1/2}} f(x)\,dx \\ &= f_l^-\left(\frac{h_{l-1}}{2}\right) + f_l^+\left(\frac{h_l}{2}\right) \end{aligned} \tag{5.47}$$

Now Eq. (5.46) can be discretized to yield

$$-p_{l-1/2}\left(\frac{y_{l+1}-y_l}{h_l}\right)+p_{l+1/2}\left(\frac{y_l-y_{l-1}}{h_{l-1}}\right)+q_l^+\left(\frac{y_{l+1}-y_l}{2}\right)$$
$$+q_l^-\left(\frac{y_l-y_{l-1}}{2}\right)+y_l\left(\frac{\sigma_l^-h_{l-1}+\sigma_l^+h_l}{2}\right)=\left(\frac{g_l^-h_{l-1}+g_l^+h_l}{2}\right) \tag{5.48}$$

Equation (5.48) is an approximation to Eq. (5.39) at the grid point l. Analogous equations hold good at all the interior grid points.

Now it is necessary to approximate the boundary conditions. For Dirichlet conditions, $\beta_1 = 0$ and $y_0 = \gamma_1/\alpha_1$ is a specified value. The general case, that is, when $\beta_1 \neq 0$, is a little more involved. At the left boundary, that is, at $l = 0$, one can write Eq. (5.44) as

$$-p_{1/2}\frac{dy_{1/2}}{dx}+p_0^+\frac{dy_0^+}{dx}+\int_{x_0}^{x_{1/2}} q(x)\frac{dy}{dx}\,dx$$
$$+\int_{x_0}^{x_{1/2}}\sigma(x)y(x)\,dx=\int_{x_0}^{x_{1/2}} g(x)\,dx \tag{5.49}$$

which can now be discretized to yield

$$-p_{1/2}\left(\frac{y_1-y_0}{h_0}\right)+p_0^+\left(\frac{\alpha_1 y_0-\gamma_1}{\beta_1}\right)+q_0^+\left(\frac{y_1-y_0}{2}\right)+\frac{\sigma_0^+y_0h_0}{2}=\frac{g_0^+h_0}{2} \tag{5.50}$$

The quantity in the parentheses of the second term is obtained by invoking the left boundary condition. Similarly, at the right boundary, one can approximate Eq. (5.45) by

$$-p_{L+1}\left(\frac{y_{L+1}-y_L}{h_L}\right)+p_{L+1/2}\left(\frac{\gamma_2-\alpha_2 y_{L+1}}{\beta_2}\right)+q_{L+1}^-\left(\frac{y_{L+1}-y_L}{2}\right)$$
$$+\frac{\sigma_{L+1}^-y_{L+1}h_{L+1}}{2}=\frac{g_{L+1}^-h_L}{2} \tag{5.51}$$

The various expressions can now be collected and written, for any interior point l, as

$$\alpha_1 y_{l-1}+\alpha_0 y_l+\alpha_2 y_{l+1}=t_l \tag{5.52}$$

where

$$\alpha_1=\frac{p_{l-1/2}}{h_{l-1}}+\frac{q_l^-}{2},\qquad \alpha_2=\frac{p_{l+1/2}}{h_l}-\frac{q_l^+}{2}$$
$$\alpha_0=\alpha_1+\alpha_2+\frac{\sigma_l^-h_{l-1}+\sigma_l^+h_l}{2},\qquad t_l=\frac{g_l^-h_{l-1}+g_l^+h_l}{2} \tag{5.53}$$

Equation (5.52) written for each interior grid point represents a system of equations

$$\mathbf{A}\mathbf{y}=\mathbf{b} \tag{5.54}$$

Equation (5.54) represents a discrete approximation to Eq. (5.39). The diagonal entries of **A** are positive and off-diagonal entries are nonpositive provided that h_l are all sufficiently small. Matrix **A** is also diagonally dominant and possesses property *A*. Another important fact to be noticed in the preceding derivation is the manner in which boundary conditions are handled *in conjunction* with the differential equation. Finally, inspection reveals that **A** is not symmetric. This is not surprising, because Eq. (5.39) is not self-adjoint. However, if $q(x) = 0$ and $p'(x)$ and $\sigma(x)$ are continuous, then Eq. (5.39) becomes self-adjoint, and matrix **A** becomes symmetric *even for a nonuniform grid.* The Taylor series method of finite differencing, discussed in Chapter 4, does not in general produce a symmetric matrix for arbitrary grid spacing even if the equation is self-adjoint.

5.5. INTEGRATION METHOD IN HIGHER DIMENSIONS

The integration method will now be extended to derive finite-difference approximations to second-order self-adjoint elliptic partial differential equations in two or more space variables. Here we once again follow R. S. Varga's work. [1]

As a concrete example, consider Ω to be the quarter-disk $0 < x < 1, 0 < y < 1, 0 < x^2 + y^2 < 1$ (see Figure 5.2), and consider

$$-\frac{\partial}{\partial x}\left(P\frac{\partial u}{\partial x}\right) - \frac{\partial}{\partial y}\left(P\frac{\partial u}{\partial y}\right) + \sigma u = g(x, y), \qquad x, y \in \Omega \tag{5.55}$$

with boundary conditions

$$\alpha u + \beta\frac{\partial u}{\partial n} = \gamma, \qquad x, y \in \Gamma \tag{5.56}$$

Here P, σ, α, β, and γ are all functions of x and y, and $\partial u/\partial n$ refers to outward pointing normal on Γ.

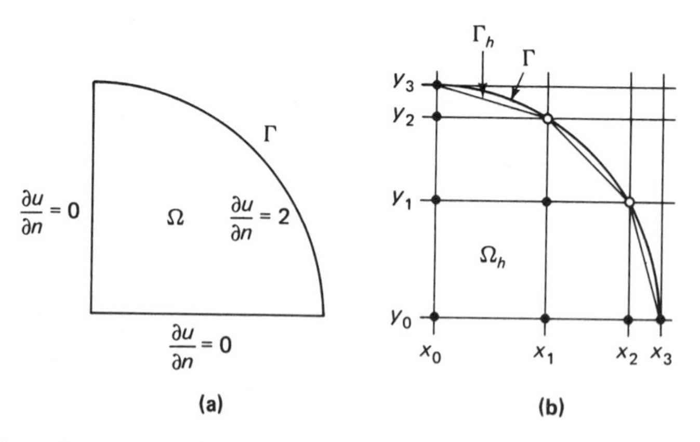

Figure 5.2 A two-dimensional region and a difference grid used in the integration method (Reproduced with permission from *Matrix Interative Analysis*, by R.S. Varga, © 1962, Prentice-Hall, Inc., Englewood Cliffs, N.J.)

As before, a first step for discretizing this problem is the definition of a grid. This is accomplished by drawing a system of straight lines parallel to the coordinate axes as shown in Figure 5.2b. The lines $x = x_i$ and $y = y_i$ are arranged such that some of the grid points fall precisely on the boundary. Therefore, the spacings $\Delta x_l = x_{l+1} - x_l$ and $\Delta y_l = y_{l+1} - y_l$ need not be uniform. The boundary Γ of Ω is now approximated by straight line segments Γ_h, which join the grid points lying on the boundary. The boundary conditions specified on Γ are now applied to Γ_h.

The next step is to associate with each grid point (l, m) an influence region r_{lm} defined as lying within the region defined by Ω_h and bounded by lines drawn parallel to the coordinate axes and passing through the midpoints of the grid lines passing through (l, m). Influence regions for two typical grid points are shown in Figure 5.3.

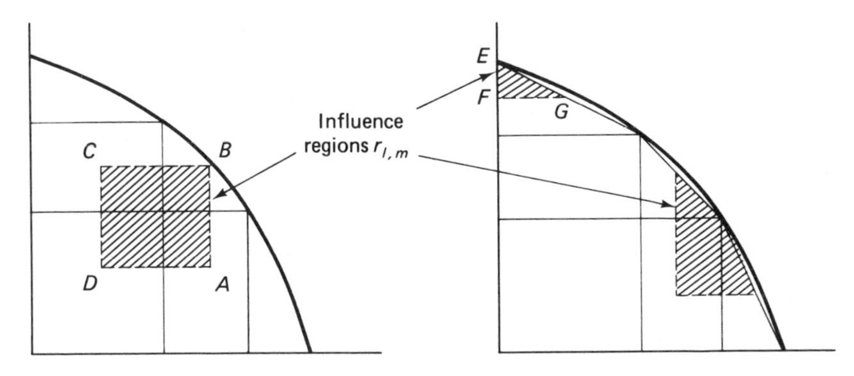

Figure 5.3 Definition of the influence regions

In the present case, the influence regions are either rectangular or polygonal (or triangular), according as the grid point is an internal point or a boundary point. For each grid point (l, m) for which $u_{l,m}$ is unknown, Eq. (5.55) is integrated over the corresponding influence region r_{lm}:

$$-\int_{r_{l,m}}\int \left\{\frac{\partial}{\partial x}\left[P\frac{\partial u}{\partial x}\right] + \frac{\partial}{\partial y}\left[P\frac{\partial u}{\partial y}\right]\right\} dx\, dy + \int_{r_{l,m}}\int \sigma\, dx\, dy = \int_{r_{l,m}}\int g\, dx\, dy \tag{5.57}$$

This integration can be performed by invoking Green's theorem, which states that for any two differentiable functions $T(x, y)$ and $S(x, y)$ defined in $r_{l,m}$

$$\int_{r_{l,m}}\int \left(\frac{\partial S}{\partial x} - \frac{\partial T}{\partial y}\right) dx\, dy = \int_{c_{l,m}} (T\, dx + S\, dy) \tag{5.58}$$

where $c_{l,m}$ represents the boundary of $r_{l,m}$, and the line integral on the right side is taken in the positive sense. If S and T are identified with $S = (P\, \partial u/\partial x)$ and $T = (P\, \partial u/\partial y)$, then Eq. (5.57) becomes

$$-\int_{c_{l,m}} \left\{\left(P\frac{\partial u}{\partial x}\right) dy - \left(P\frac{\partial u}{\partial y}\right) dx\right\} + \int_{r_{l,m}}\int \sigma u \, dx \, dy = \int_{r_{l,m}}\int g \, dx \, dy \quad (5.59)$$

Evaluation of the line integral, represented by the first term of Eq. (5.59), is as follows.

First, the influence region $r_{l,m}$ is considered to be rectangular in shape, as it would if (l, m) is an interior grid point (see Figure 5.4a). Then one can decompose the line integral around the perimeter of the rectangle as

$$\int_{c_{l,m}} = \int_{AB} + \int_{BC} + \int_{CD} + \int_{DA} \quad (5.60)$$

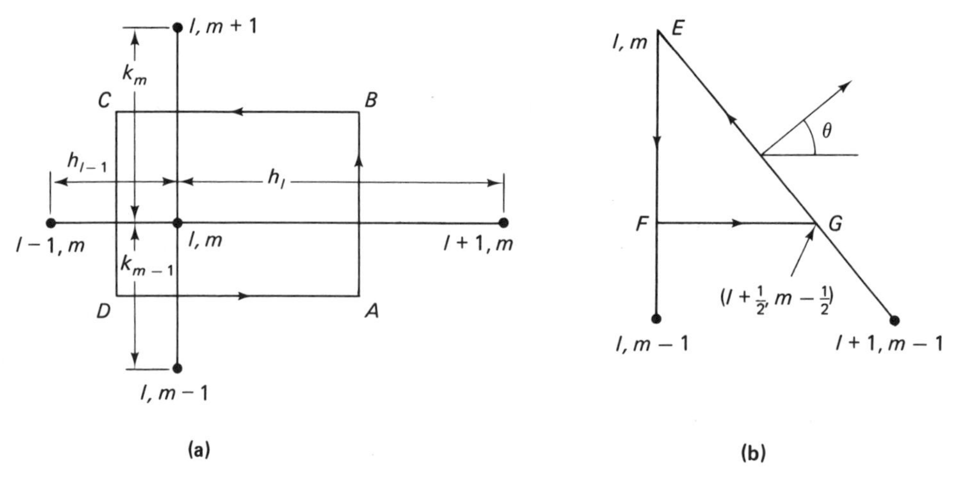

Figure 5.4 Contour for the line integrals surrounding the influence regions. (Reproduced with permission from *Matrix Interative Analysis*, by R. S. Varga, © 1962, Prentice-Hall, Inc., Englewood Cliffs, N.J.)

Now $(P\,\partial u/\partial x)$ and $(P\,\partial u/\partial y)$ are approximated at the midpoints of the sides of the rectangle $ABCD$ as

$$\begin{aligned}
P\frac{\partial u}{\partial x}\bigg|_{AB} &\simeq P_{l+1/2,m}\left(\frac{u_{l+1,m} - u_{l,m}}{h_l}\right) \\
P\frac{\partial u}{\partial y}\bigg|_{BC} &\simeq P_{l,m+1/2}\left(\frac{u_{l,m+1} - u_{l,m}}{k_m}\right) \\
P\frac{\partial u}{\partial x}\bigg|_{CD} &\simeq P_{l-1/2,m}\left(\frac{u_{l,m} - u_{l-1,m}}{h_{l-1}}\right) \\
P\frac{\partial u}{\partial y}\bigg|_{DA} &\simeq P_{l,m-1/2}\left(\frac{u_{l,m} - u_{l,m-1}}{k_{m-1}}\right)
\end{aligned} \quad (5.61)$$

Therefore, one can write

$$-\int_{AB}\left(P\frac{\partial u}{\partial x}\right) dy \simeq \left(\frac{k_{m-1} + k_m}{2}\right)\left\{P_{l+1/2,m}\left(\frac{u_{l,m} - u_{l+1,m}}{h_l}\right)\right\} \quad (5.62)$$

and similar expressions for the other parts of the line integral. Summing up,

$$-\int_{c_{l,m}} \left\{\left(P\frac{\partial u}{\partial x}\right) dy - \left(P\frac{\partial u}{\partial y}\right) dx\right\}$$

$$\simeq \left(\frac{k_{m-1}+k_m}{2}\right)\left\{P_{l+1/2,m}\left(\frac{u_{l,m}-u_{l+1,m}}{h_l}\right) + P_{l-1/2,m}\left(\frac{u_{l,m}-u_{l-1,m}}{h_{l-1}}\right)\right\}$$

$$+ \left(\frac{h_{l-1}+h_l}{2}\right)\left\{P_{l,m+1/2}\left(\frac{u_{l,m}-u_{l,m+1}}{k_l}\right) + P_{l,m-1/2}\left(\frac{u_{l,m}-u_{l,m-1}}{k_{l-1}}\right)\right\} \tag{5.63}$$

The double integrals in Eq. (5.59) are simply approximated by a formula such as

$$\int_{r_{l,m}}\int f_{l,m}\, dx\, dy = f_{l,m} \text{ (area of } r_{l,m}) \tag{5.64}$$

Whatever the shape of the influence region, the area of $r_{l,m}$ is generally easy to calculate. Therefore, for each grid point (l, m), Eqs. (5.63) and (5.64) can be substituted in Eq. (5.59) to yield a finite difference equation of the form

$$-\alpha_1 u_{l-1,m} - \alpha_2 u_{l+1,m} - \alpha_3 u_{l,m-1} - \alpha_4 u_{l,m+1} - \alpha_0 u_{l,m} = t_{l,m} \tag{5.65}$$

where

$$\alpha_1 = P_{l-1/2,m}\left(\frac{k_{m-1}+k_m}{2h_{l-1}}\right), \qquad \alpha_3 = P_{l,m-1/2}\left(\frac{h_{l-1}+h_l}{2k_{m-1}}\right)$$

$$\alpha_2 = P_{l+1/2,m}\left(\frac{k_{m-1}+k_m}{2h_l}\right), \qquad \alpha_4 = P_{l,m+1/2}\left(\frac{h_{l-1}+h_l}{2k_m}\right) \tag{5.66}$$

$$\alpha_0 = \alpha_1 + \alpha_2 + \alpha_3 + \alpha_4 + \sigma_{l,m}\left(\frac{h_{l-1}+h_l}{2}\right)\left(\frac{k_{m-1}+k_m}{2}\right)$$

$$t_{l,m} = g_{l,m}\left(\frac{h_{l-1}+h_l}{2}\right)\left(\frac{k_{m-1}+k_m}{2}\right)$$

at any internal grid point.

If a grid point lies on the boundary, then $r_{l,m}$ is more likely to be polygonal or triangular in shape. For concreteness, consider the triangular influence region EFG as shown in Figure 5.4b. The line integral in Eq. (5.59) is now integrated along the sides of the triangle EFG. That is,

$$\int_{c_{l,m}} = \int_{EF} + \int_{FG} + \int_{GE} \tag{5.67}$$

Evaluation of the integrals along EF and FG proceeds in a fashion similar to that described earlier. The more involved is the integration along GE. To accomplish this, the line GE is parametrically represented by the equations

$$x = x_{l+1/2} - \xi \sin\theta, \qquad y = y_{m-1/2} + \xi \cos\theta \tag{5.68}$$

where θ is the angle made by the outward normal to Γ_h with the positive x-axis. Then

$$\frac{\partial u}{\partial n} = \left(\frac{\partial u}{\partial x}\right)\cos\theta + \left(\frac{\partial u}{\partial y}\right)\sin\theta \tag{5.69}$$

Therefore, the line integral along GE is

$$-\int_{GE}\left\{\left(P\frac{\partial u}{\partial x}\right)dy-\left(P\frac{\partial u}{\partial y}\right)dx\right\}=-\int_{GE}\left\{\left(P\frac{\partial u}{\partial x}\right)\cos\theta+\left(P\frac{\partial u}{\partial y}\right)\sin\theta\right\}d\xi$$
$$=-\int_{GE}P\frac{\partial u}{\partial n}\,d\xi \tag{5.70}$$

The first step on the right side is obtained by using Eq. (5.68) and the second step by using Eq. (5.69). Now $(\partial u/\partial n)$ is obtained from the boundary condition specified in Eq. (5.56). Therefore,

$$-\int_{GE}\left\{\left(P\frac{\partial u}{\partial x}\right)dy-\left(P\frac{\partial u}{\partial y}\right)dx\right\}=-\int_{GE}P\left(\frac{\gamma(\xi)-\alpha(\xi)u(\xi)}{\beta(\xi)}\right)d\xi$$
$$\simeq -P_{l,m}\left(\frac{\gamma_{l,m}-\alpha_{l,m}u_{l,m}}{\beta_{l,m}}\right)\left(\frac{1}{2}\sqrt{h_l^2+k_{m-1}^2}\right) \tag{5.71}$$

The system of equations represented by Eqs. (5.65) and (5.71), when written out and rearranged, results in

$$\mathbf{Au}=\mathbf{b} \tag{5.72}$$

It can be readily verified that **A** is real with positive diagonal entries and non-positive off-diagonal entries and is strictly diagonally dominant. One can also verify that **A** is irreducible and consistently ordered.

5.6. DIFFERENCE APPROXIMATIONS USING VARIATIONAL METHOD

Consider an ordinary differential equation

$$-\frac{d^2y(x)}{dx^2}+\sigma(x)y(x)=g(x), \qquad a<x<b \tag{5.73}$$

subject to the boundary conditions

$$y(a)=\alpha, \qquad y(b)=\beta \tag{5.74}$$

These equations, incidentally, are the same as Eqs. (5.30) and (5.31). Using an argument analogous to the one used in Section 3.5, it can be easily shown that mathematically the preceding problem is equivalent to finding a function $w(x)$ (whose slope is at least piecewise continuous in $a<x<b$) that satisfies the boundary conditions and that minimizes

$$J(w)=\frac{1}{2}\int_a^b\left\{\left[\frac{dw(x)}{dx}\right]^2+\sigma(x)[w(x)]^2-2g(x)w(x)\right\}dx \tag{5.75}$$

To derive the difference equations, Eq. (5.75) is discretized over the interval $[a, b]$ as

$$J(w)=\frac{1}{2}\sum_{l=0}^{L}\int_{x_l}^{x_{l+1}}\left\{\left[\frac{dw(x)}{dx}\right]^2+\sigma(x)[w(x)]^2-2w(x)g(x)\right\}dx \tag{5.76}$$

and the integrals are approximated, as before, by

$$\int_{x_l}^{x_{l+1}} \left(\frac{dw(x)}{dx}\right)^2 dx \simeq \left(\frac{w_{l+1} - w_l}{(\Delta x)}\right)^2 (x_{l+1} - x_l) = \left(\frac{w_{l+1} - w_l}{(\Delta x)}\right)^2 \tag{5.77}$$

and

$$\int_{x_l}^{x_{l+1}} f(x)\, dx = \left(\frac{f_{l+1} + f_l}{2}\right)(x_{l+1} - x_l) = \left(\frac{f_{l+1} + f_l}{2}\right)(\Delta x) \tag{5.78}$$

Now Eq. (5.76) can be written as

$$J(w_1, w_2, \ldots, w_L) = \frac{1}{2}\sum_{l=0}^{L}\left\{\frac{(w_{l+1} - w_l)^2}{(\Delta x)} + \left(\frac{\sigma_{l+1}w_{l+1}^2 + \sigma_l w_l^2}{2}\right)(\Delta x) - (w_{l+1}g_{l+1} + w_l g_l)(\Delta x)\right\} \tag{5.79}$$

Introducing the following vector-matrix notation,

$$\mathbf{w} = (w_1, w_2, \ldots, w_L)^T$$

$$\mathbf{b} = (g_1 + \alpha/(\Delta x)^2, g_2, \ldots, g_{L-1}, g_L + \beta/(\Delta x)^2)^T$$

and

$$\mathbf{A} = \frac{1}{(\Delta x)^2}\begin{bmatrix} 2 + \sigma_1(\Delta x)^2 & -1 & & \\ -1 & 2 + \sigma_2(\Delta x)^2 & & \\ & & & -1 \\ & & -1 & 2 + \sigma_L(\Delta x)^2 \end{bmatrix}$$

one can express Eq. (5.79) as

$$\frac{2J(\mathbf{w})}{(\Delta x)} = \mathbf{w}^T\mathbf{A}\mathbf{w} - 2\mathbf{w}^T\mathbf{b} + \frac{\beta^2}{(\Delta x)^2} + \frac{\alpha^2}{(\Delta x)^2} + \frac{1}{2}\{\sigma(1)\beta^2 + \sigma(0)\alpha^2\} - (\beta g(1) + \alpha g(0)) \tag{5.80}$$

which is a quadratic form. Therefore, $J(\mathbf{w})$ can be minimized as a function of w_1, $w_2, \ldots, w_L$ by solving

$$\frac{\partial J(\mathbf{w})}{\partial w_i} = 0, \qquad 1 \leq i \leq L \tag{5.81}$$

which leads to the solution of a matrix equation of the type

$$\mathbf{A}\mathbf{y} = \mathbf{b} \tag{5.82}$$

where $\mathbf{y}$ is the minimizing value of $\mathbf{w}$. Note this equation is the same as Eq. (5.54).

The variational method can be readily extended to higher dimensions. The Taylor series method, the integration method, and the variational method are equivalent, and it is often difficult to decide on which method to use. The variational method depends upon the self-adjoint property of the differential equation and is not applicable in the general case. The integration method gives particularly simple discretization schemes to problems with nonuniform grids and internal interfaces. Both the integration method and variational method require more in the way of initial calculations. The Taylor series method is simple, easy to apply, and indispensable in deducing the order of accuracy of the approximations.

As an illustration of a very special case of the variational method, consider a function $w(x)$ as shown in Figure 5.5, where $w_1, w_2, \ldots, w_L$ are arbitrary real numbers and $w(x = a) = \alpha$ and $w(x = b) = \beta$. Then the value of $w(x)$ for any $x \in [x_l, x_{l+1}]$ can be written as

$$w(x) = w_l + \frac{w_{l+1} - w_l}{(\Delta x)}(x - x_l), \qquad 0 \leq l \leq L \tag{5.83}$$

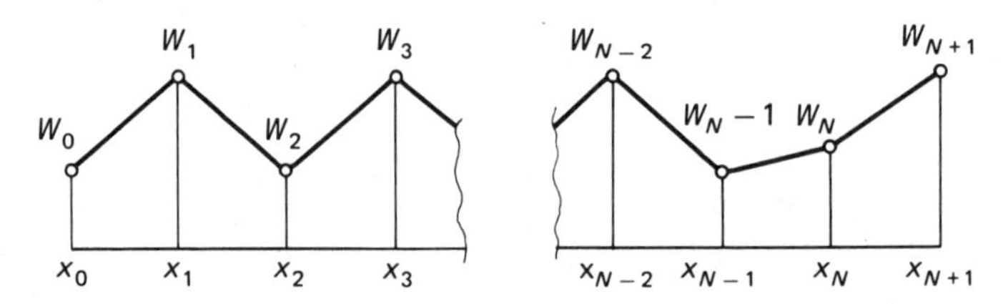

Figure 5.5 One-dimensional grid used with the variational method

Substituting this in Eq. (5.76), the first term of $\partial J/\partial w_l = 0$ becomes

$$\frac{\partial}{\partial w_l}\left[\frac{1}{2}\sum_{l=0}^{L}\int_{x_l}^{x_{l+1}}\left\{\frac{w_{l+1} - w_l}{(\Delta x)}\right\}^2 dx\right]$$

As the differentiation is with respect to w_l, and as w_l appears only in two terms under the summation sign, the preceding expression simplifies to

$$\left(\frac{w_l - w_{l+1}}{(\Delta x)}\right) + \left(\frac{w_l - w_{l-1}}{(\Delta x)}\right)$$

where the integration was performed with the help of Eq. (5.78). The second and third terms can likewise be simplified, and the system of equations represented by Eq. (5.81) can be written as

$$\left(\frac{w_l - w_{l+1}}{(\Delta x)}\right) + \left(\frac{w_l - w_{l-1}}{(\Delta x)}\right) + \int_{x_{l-1}}^{x_{l+1}} \sigma(x)w(x)t_l(x)\,dx$$

$$= \int_{x_{l-1}}^{x_{l+1}} g(x)t_l(x)\,dx, \qquad 1 \leq l \leq L \tag{5.84}$$

where $t_i(x)$ is the piecewise continuous function defined by

$$t_i(x_j) = \begin{cases} 1, & j = i \\ 0, & j \neq i \end{cases}, \qquad 0 \leq j \leq L + 1 \tag{5.85}$$

The system of equations represented by Eq. (5.84) can be written as

$$\hat{\mathbf{A}}\mathbf{w} = \hat{\mathbf{b}} \tag{5.86}$$

where **w** is same as the **y** of Eq. (5.82). However, $\hat{\mathbf{A}}$ and $\hat{\mathbf{b}}$ are not identical to **A** and **b** of Eq. (5.82). However, $\hat{\mathbf{A}}$ and **A** are real, symmetric, and tridiagonal.

5.7. SOLVING THE SYSTEM $Au = b$

It is clear, by now, that the solution of elliptic equations eventually becomes a problem of solving systems of algebraic equations. So far, linear equations only were considered. So the resulting system of equations is also linear, which is denoted by $\mathbf{Au} = \mathbf{b}$. In general, one may have to solve a nonlinear system of equations denoted by $\mathbf{f(u)} = \mathbf{b}$. Thus, the solution of elliptic systems by numerical methods, whether of the finite difference, Rayleigh-Ritz, or finite element type, ultimately depends on an ability to solve systems of simultaneous algebraic equations, either linear or nonlinear. The solution of nonlinear algebraic equations usually involves an iterative method, such as the Newton-Raphson method, that is based on a sequence of approximations to the system which hopefully converges to the desired solution. A simplified system of equations, usually linear, is often used at each step in the iteration process.

Solving the system of linear equations can itself be by a direct (noniterative) or iterative method. Sucesss of direct methods depends upon a variety of factors, some of which are (1) ***condition*** of the system, (2) need for ***pivoting*** during elimination, (3) the fact that pivoting is not necessary while solving positive definite systems, (4) matrix factorization techniques, and (5) ***sparsity*** and ***bandedness*** of matrixes. The Gaussian elimination method and the method based on the tridiagonal algorithm are typical examples of direct methods.

The use of iterative methods, often involving relaxation techniques, has found widespread application in the solution of both linear and nonlinear algebraic equations. Various relaxation techniques such as the Gauss-Seidel method and the successive overrelaxation (SOR) method, have been successfully employed in the modes of point, line, and block relaxations. The alternating direction iterative (ADI) methods, which are a special class of block relaxation methods have found widespread use in solving elliptic systems.

Some of the more important ideas associated with these methods are presented in Chapter 10. At this stage, students primarily interested in the methodology involved in getting solutions on a computer can go to Chapter 10. In this context it is extremely instructive to consider a simple problem and derive finite difference approximations to it using all the three techniques described earlier, using in one case a uniform grid and in another case a graded grid. Then it would be useful to study the resulting system $\mathbf{Au} = \mathbf{b}$ for symmetry, diagonal dominance, property A, and so on, in each case and compare the solutions in terms of computational time, storage requirements, convergence, and accuracy.

5.8. OTHER METHODS OF SOLVING ELLIPTIC EQUATIONS

The set of equations $\mathbf{Au} = \mathbf{b}$ obtained as a result of finite difference approximations is usually solved using classical iterative methods such as, say, the successive overrelaxation (SOR) method or the alternating direction iterative method (ADIM), or variants of these two popular methods. In addition, there are many other digital computer and analog computer oriented methods of solving elliptic equations. Some of the more important digital computer methods that fall outside the main stream of classical iterative methods are briefly described in this section.

THE METHOD OF LINES AND INVARIANT IMBEDDING

A variant of the finite difference method that has been found useful is the ***method of lines***. In this method all independent variables but one are discretized. This procedure leads to a system of ordinary differential equations in the remaining independent variable. As a specific example, a two-dimensional Laplace's equation on the unit square with Dirichlet boundary conditions is considered.

$$\begin{aligned} \frac{\partial^2 u}{\partial x^2} + \frac{\partial^2 u}{\partial y^2} = 0, &\qquad 0 \le x, y \le 1 \\ u(x, y) = g, &\qquad \text{on the boundary} \end{aligned} \tag{5.87}$$

If the y-variable is left continuous and the problem is discretized in the x-direction (see Figure 5.6), Eq. (5.87) can be written, for $l = 1, 2, \ldots, L - 1$, as

$$\frac{u_{l-1}(y) - 2u_l(y) + u_{l+1}(y)}{(\Delta x)^2} + \frac{d^2 u_l(y)}{dy^2} = 0 \tag{5.88a}$$

$$\begin{aligned} u_l(0) &= g_{1,l} \\ u_l(1) &= g_{2,l} \end{aligned} \tag{5.88b}$$

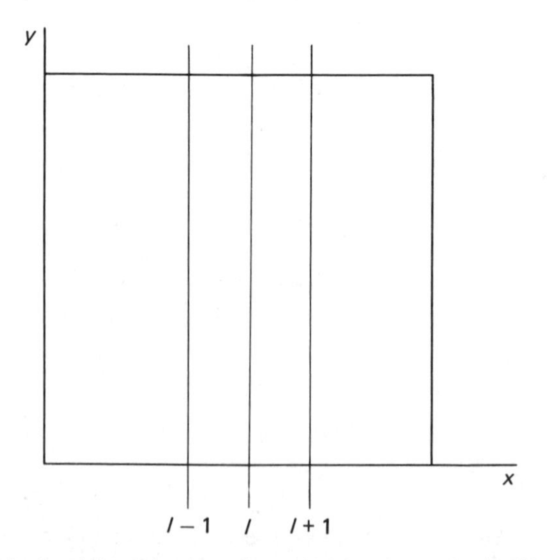

Figure 5.6 Semidiscretization of a region for the method of lines. Only three typical grid lines are shown

where g_1 and g_2 are determined from the boundary conditions. Equation (5.88) can be written in vector-matrix notation as

$$\ddot{\mathbf{u}} + \mathbf{Q}\mathbf{u} = \mathbf{r}(y) \tag{5.89a}$$

with

$$\mathbf{u}(0) = \mathbf{g}_1 \qquad \mathbf{u}(1) = \mathbf{g}_2 \tag{5.89b}$$

Here the matrix $\mathbf{Q}$ is a tridiagonal matrix and the vector $\mathbf{r}(y)$ is determined from boundary values. The values of $\mathbf{g}_1$ and $\mathbf{g}_2$ are also determined from the boundary values.

The system in Eq. (5.89) represents a set of ordinary differential equations with one condition specified at the boundary $y = 0$ and another condition specified at the boundary $y = 1$. Such problems are termed ***two-point boundary value problems*** (or TPBVPs). Solution of a general TPBVP associated with linear equations is a little bit involved. Several methods, such as the ***adjoint equation method***, ***sensitivity equation method***, or the ***complementary function method***, are available to solve a linear TPBVP. These methods can be extended to solve nonlinear TPBVPs as well. Equation (5.89), however, is difficult to solve owing to the inherent instability of the equation.

The invariant imbedding approach circumvents the preceding difficulties by reformulating Eq. (5.89) as a new and stable problem with specified initial conditions. The central idea of the invariant imbedding approach lies in treating $\mathbf{u}$ as a function of not only x, but also of the length of the interval. That is, $\mathbf{u}(x)$ is replaced by $\mathbf{u}(x, \xi)$, and when $\xi = 1$, one gets $\mathbf{u}(x, 1)$, the desired solution. In this procedure, it turns out that the solution of Eq. (5.89) can be written as

$$\mathbf{u}(x) = \mathbf{R}(x)\mathbf{v}(x) + \mathbf{b}(x)$$

where $\mathbf{R}(x)$ is the solution of a matrix Riccati equation, $\mathbf{b}(x)$ is the solution of an associated vector-matrix equation, and $\mathbf{v} = \dot{\mathbf{u}}$.

TENSOR PRODUCT METHOD WITH FAST FOURIER TRANSFORMS

In contrast to the iterative methods, the direct methods also play a useful role in solving elliptic boundary value problems. The Gaussian elimination method and the tridiagonal algorithm are examples of direct methods. The tensor product method is another. Central to the tensor product method is the assumption that the problem is separable. Then the solution can be expressed in terms of the tensor products (or direct products or Kronecker products) of solutions of lower-dimensional problems. This approach is conservative of storage and fast, particularly when it is used in conjunction with the fast Fourier transform (FFT) method. Major applications of this method so far have been to elliptic and parabolic equations defined over regular domains.

To illustrate the tensor product-FFT method, Poisson's equation

$$\frac{\partial^2 u}{\partial x^2} + \frac{\partial^2 u}{\partial y^2} = -g(x, y) \tag{5.90a}$$

$$u = 0 \qquad \text{on the boundary} \tag{5.90b}$$

in a rectangular region is considered. A finite difference grid of size $(\Delta x, \Delta y)$ is superimposed on the region. Using the usual five-point difference approximation, Eq. (5.90), at any grid point (l, m), becomes

$$\frac{u_{l-1,m} - 2u_{l,m} + u_{l+1,m}}{(\Delta x)^2} + \frac{u_{l,m-1} - 2u_{l,m} + u_{l,m+1}}{(\Delta y)^2} = -g_{l,m} \tag{5.91}$$

which when written in matrix notation becomes

$$[(\Delta x)^2]^{-1}\mathbf{A}\mathbf{u} + [(\Delta y)^2]^{-1}\mathbf{u}\mathbf{B} = \mathbf{g} \tag{5.92}$$

where $\mathbf{A}$ is an $L \times L$ symmetric tridiagonal matrix, and $\mathbf{B}$ is an $M \times M$ symmetric tridiagonal matrix. That is,

$$\mathbf{A} = \begin{bmatrix} -2 & 1 & & & \\ 1 & -2 & 1 & & \\ & & & & \\ & & & & 1 \\ & & & 1 & -2 \end{bmatrix}, \quad \mathbf{B} = \begin{bmatrix} -2 & 1 & & & \\ 1 & -2 & 1 & & \\ & & & & \\ & & & & 1 \\ & & & 1 & -2 \end{bmatrix}$$

The vector $\mathbf{u}$, of course, is of dimension (LM). The eigenvalues and eigenvectors of the preceding matrixes are well known. For instance, the eigenvalues and normalized eigenvectors of the $L \times L$ matrix $\mathbf{A}$ are, respectively,

$$\lambda_l = 2\left(1 - \cos\frac{l\pi}{L+1}\right), \qquad l = 1, 2, \ldots, L \tag{5.93}$$

and

$$q_{kl} = \left(\frac{2}{L+1}\right)^{1/2} \sin\left(\frac{kl\pi}{L+1}\right), \qquad k = 1, 2, \ldots, L \tag{5.94}$$

Thus, one may write

$$\mathbf{A} = \mathbf{Q}\boldsymbol{\Lambda}\mathbf{Q}^T \tag{5.95}$$

where $\boldsymbol{\Lambda}$ is a diagonal matrix with λ_l on its diagonal and $\mathbf{Q}$ is the matrix made up of the normalized eigenvectors. Therefore, it is also true that

$$\mathbf{Q}\mathbf{Q}^T = \mathbf{Q}^T\mathbf{Q} = \mathbf{I} \tag{5.96}$$

Substituting Eq. (5.95) in Eq. (5.92) and premultiplying through out by $\mathbf{Q}^T$ yields

$$[(\Delta x)^2]^{-1}\boldsymbol{\Lambda}\boldsymbol{\psi} + [(\Delta y)^2]^{-1}\boldsymbol{\psi}\mathbf{B} = f \tag{5.97}$$

where

$$\boldsymbol{\psi} = \mathbf{Q}^T\mathbf{u} \tag{5.98}$$

and

$$\mathbf{f} = \mathbf{Q}^T\mathbf{g} \tag{5.99}$$

If Eqs. (5.98) and (5.99) are written out in full, with the information contained in Eq. (5.94), it becomes obvious that $\boldsymbol{\psi}$ and $\mathbf{f}$ are Fourier transforms of $\mathbf{u}$ and $\mathbf{g}$, respectively. Thus, the sequence of steps in this procedure would be (1) calculate $\mathbf{f}$

using Eq. (5.99) and the known information about $\mathbf{g}$. This step is the Fourier analysis part. (2) Solve the tridiagonal system represented by Eq. (5.92) and obtain $\boldsymbol{\psi}$. (3) Solve for $\mathbf{u}$ using Eq. (5.98). This step is the Fourier synthesis part.

The speed advantage of the fast Fourier transforms would be lost if the eigenvalues and eigenvectors of $\mathbf{A}$ and $\mathbf{B}$ are not known analytically. Because $\mathbf{A}$ and $\mathbf{B}$ are tridiagonal, their eigenvalues and eigenvectors can be calculated on the computer at a price. One advantage of the tensor product method is that the required procedures, such as matrix multiplication, tridiagonal system solver, and FFT routines, can usually be found in user program libraries.

LINEAR PROGRAMMING METHOD

The system of equations

$$\mathbf{Au} = \mathbf{b} \tag{5.100}$$

obtained from a finite-difference approximation can also be solved using linear programming (or LP) method. In applying the LP method, one starts from an initial guess $\mathbf{u}^{(0)}$, which in general does not satisfy Eq. (5.100). For concreteness, let

$$\mathbf{Au}^{(0)} \leq \mathbf{b} \tag{5.101}$$

Defining a residual $\mathbf{r}$, Eq. (5.101) can be rewritten as

$$\mathbf{Au}^{(0)} + \mathbf{r} = \mathbf{b} \tag{5.102}$$

Now the problem of solving Eq. (5.100) can be stated as

$$\begin{aligned} &\text{minimize} \quad \mathbf{r} \\ &\text{subject to} \quad \mathbf{Au}^{(0)} + \mathbf{r} = \mathbf{b} \end{aligned} \tag{5.103}$$

which can be readily recognized as an LP problem in the standard format. One possible advantage of this method is that one can take advantage of the sparsity of $\mathbf{A}$ in conjunction with one of several decomposition algorithms available to solve large-scale LP problems.

DYNAMIC PROGRAMMING METHOD

The dynamic programming method is based on Bellman's ***principle of optimality***, which in turn has its roots in variational calculus. It is well known that many—perhaps most—elliptic partial differential equations are related to the minimization or maximization of an integral, which often represents the energy of a physical system (see Section 5.6). For instance, solving Laplace's equation

$$\frac{\partial^2 u}{\partial x^2} + \frac{\partial^2 u}{\partial y^2} = 0 \tag{5.104}$$

over the rectangle $0 \leq x \leq a$, $0 \leq y \leq b$ subject to the boundary conditions

$$u(x, y) = g(x, y) \tag{5.105}$$

is associated with the minimization of

$$J(u) = \iint \left[\left(\frac{\partial u}{\partial x} \right)^2 + \left(\frac{\partial u}{\partial y} \right)^2 \right] dx\, dy \tag{5.106}$$

subject to the boundary condition specified in Eq. (5.105).

The first step of the dynamic programming procedure, following Angel and Bellman,[2] is to superimpose a finite difference grid on the region of interest. This is achieved, for example, by letting

$$a = Lh \qquad \text{and} \qquad b = Mh$$

where L and M are integers and defining

$$u_{lm} = u(lh, mh), \qquad l = 0, 1, \ldots, L, \quad m = 0, 1, \ldots, M$$

Now, a discretized version of Eq. (5.106) can be written as

$$J(u) = \sum_{l=1}^{L} \sum_{m=1}^{M} [(u_{l,m} - u_{l,m-1})^2 + (u_{l,m} - u_{l-1,m})^2] \tag{5.107}$$

subject to the discretized boundary conditions specified by $u_{0,m}$, $u_{l,0}$, $u_{l,L}$, and $u_{l,M}$, which are calculated from Eq. (5.105).

The problem now is to minimize $J(u)$ over the set of all $u_{l,m}$ defined at all interior grid points. Since the primary interest is in finding the values $u_{l,m}$ that minimize $J(u)$, but not the minimum value of J itself, the constant terms in Eq. (5.107) can be removed without influencing the answer. Thus, removing terms of the form $(u_{l,M} - u_{l-1,M})^2$, which are fixed by the boundary conditions, Eq. (5.107) becomes

$$J(u) = \sum_{l=1}^{L} \left[\sum_{m=1}^{M-1} (u_{l,m} - u_{l-1,m})^2 + \sum_{m=1}^{M} (u_{l,m} - u_{l,m-1})^2 \right] \tag{5.108}$$

At this stage the discretization is complete. For notational convenience, the problem will now be put in vector-matrix format. Using a procedure similar to that used in Section 5.6, $J(u)$ is now expressed in a quadratic form. This procedure will facilitate the process of expressing the problem of minimizing J as a multistage decision process, which in turn can be handled efficiently by dynamic programming.

The following definitions are introduced. Let

$$\mathbf{u}_R = (u_{R1}, u_{R2}, \ldots, u_{RL})^T, \qquad R = 1, 2, \ldots, L-1 \tag{5.109}$$

Note $\mathbf{u}_0$ and $\mathbf{u}_L$ are known from the specified boundary conditions.

Let $\mathbf{Q}$ be a tridiagonal matrix of order $M-1$ with 2's on the main diagonal, -1's on the two off diagonals, and zeros everywhere else.

Let $\mathbf{r}_R$ be a set of $L-1$ vectors each of dimension $M-1$ that depend on the boundary conditions. That is,

$$\mathbf{r}_R = [r_{Rj}], \qquad \text{where } r_{Rj} = \begin{cases} u_{R,0}, & \text{for } j = 1 \\ u_{R,M}, & \text{for } j = M-1 \\ 0, & \text{otherwise} \end{cases} \tag{5.110}$$

$$s_R = u_{R0}^2 + u_{RM}^2$$

With this notation, Eq. (5.108) can be expressed in required quadratic form, using inner product notation, as

$$J(u) = \sum_{R=1}^{L} [\langle \mathbf{Q}\mathbf{u}_R, \mathbf{u}_R \rangle - 2\langle \mathbf{r}_R, \mathbf{u}_R \rangle + s_R + \langle \mathbf{u}_R - \mathbf{u}_{R-1}, \mathbf{u}_R, \mathbf{u}_{R-1} \rangle] \tag{5.111}$$

where the inner product is defined as

$$\langle \mathbf{x}, \mathbf{y} \rangle = \mathbf{x}^T\mathbf{y} = \sum_{l=1}^{L} x_l y_l$$

The minimization is now carried out over the vectors $\mathbf{u}_l$, $l = 1, 2, \ldots, L-1$. The direct approach for this minimization would be to differentiate Eq. (5.111) with respect to all u_{Rl}, $l = 1, 2, \ldots, L-1$. This, however, would result in the typical finite difference equations and consequently a linear system of order $(L-1) \times (M-1)$.

The dynamic programming method reformulates Eq. (5.111) as the sequence of variational problems defined by

$$f_R(\mathbf{v}) = \min_{\mathbf{u}_R, \mathbf{u}_{R+1}, \ldots, \mathbf{u}_{L-1}} \sum_{l=R}^{l=L} [\langle \mathbf{Q}\mathbf{u}_l, \mathbf{u}_l \rangle - 2\langle \mathbf{r}_l, \mathbf{u}_l \rangle + s_l + \langle (\mathbf{u}_l - \mathbf{u}_{l-1}), (\mathbf{u}_l - \mathbf{u}_{l-1}) \rangle] \tag{5.112}$$

for $R = 1, 2, \ldots, L-1$, and $\mathbf{v}$ is defined by

$$\mathbf{v} = \mathbf{u}_{R-1} \tag{5.113}$$

Comparing Eq. (5.112) with Eq. (5.111), it is clear that for $R = 1$ the following condition is satisfied:

$$f_1(\mathbf{u}_0) = \min_{\{\mathbf{u}_R\}} J(\mathbf{u}) \tag{5.114}$$

Each of the problems defined in Eq. (5.112) can be regarded as a solution to the discrete variational formulation of the given Laplace's equation over a smaller rectangle (the shaded region in Figure 5.7) with the boundary condition, $\mathbf{v}$, left unspecified.

Now the dynamic programming idea is brought into the problem by seeking a relation between $f_R(\mathbf{v})$ and $f_{R+1}(\mathbf{v})$. This is done by noting that the first four terms of

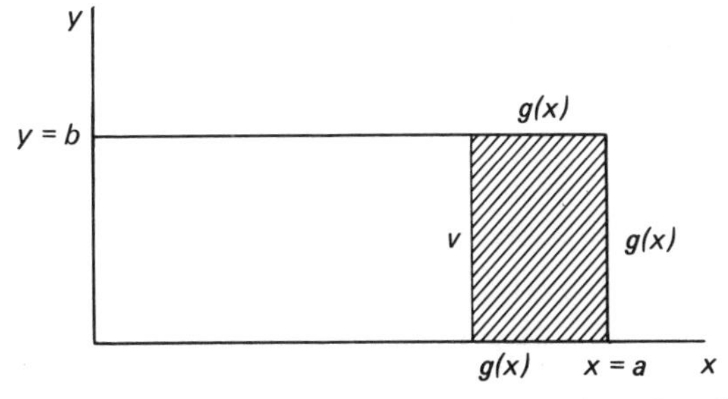

Figure 5.7 Truncated rectangle of interest (the shaded region) in the dynamic programming method

Eq. (5.112) depend on $\mathbf{u}_R$. Therefore, Eq. (5.112) can be rewritten as

$$f_R(\mathbf{v}) = \min_{\mathbf{u}_R}\Big[\langle \mathbf{Q}\mathbf{u}_R, \mathbf{u}_R\rangle - 2\langle \mathbf{r}_R, \mathbf{u}_R\rangle + s_R + \langle(\mathbf{u}_R - \mathbf{v}), (\mathbf{u}_R - \mathbf{v})\rangle$$
$$+ \min_{[\mathbf{u}_{R+1},\mathbf{u}_{R+2},\ldots,\mathbf{u}_{-L1}]} \sum_{l=R+1}^{L} \{\langle \mathbf{Q}\mathbf{u}_l, \mathbf{u}_l\rangle - 2\langle \mathbf{r}_l, \mathbf{u}_l\rangle + s_l \tag{5.115}$$
$$+ \langle(u_l - u_{l-1}), (u_l - u_{l-1})\rangle\}\Big]$$

Recognizing the similarity of the expression under the summation sign with Eq. (5.112), one can readily rewrite Eq. (5.115) as

$$f_R(\mathbf{v}) = \min_{\mathbf{u}_R} [\langle \mathbf{Q}\mathbf{u}_R, \mathbf{u}_R\rangle - 2\langle \mathbf{r}_R, \mathbf{u}_R\rangle + s_R + \langle(\mathbf{u}_R - \mathbf{v}), (\mathbf{u}_R - \mathbf{v})\rangle + f_{R+1}(\mathbf{u}_R)] \tag{5.116}$$

For $R = L$, Eq. (5.116) becomes

$$f_L(\mathbf{v}) = \langle \mathbf{Q}\mathbf{u}_L, \mathbf{u}_L\rangle - 2\langle \mathbf{r}_L, \mathbf{u}_L\rangle + s_L + \langle \mathbf{u}_L - \mathbf{v}, \mathbf{u}_L - \mathbf{v}\rangle \tag{5.117}$$

and is a known quantity because $\mathbf{u}_L$ is fixed by the specified boundary conditions. Starting from Eq. (5.117), one can proceed to solve Eq. (5.116) backward until $f_1(\mathbf{u}_0)$ is determined. However, Eqs. (5.116) and (5.117) are not quite convenient in a computational context. A computationally convenient set of equations can be obtained by taking advantage of the quadratic nature of $f_R(\mathbf{v})$. Because $f_R(\mathbf{v})$ is quadratic, it can be written as

$$f_R(\mathbf{v}) = \langle \mathbf{A}_R\mathbf{v}, \mathbf{v}\rangle - 2\langle \mathbf{b}_R, \mathbf{v}\rangle + c_R \tag{5.118}$$

where $\mathbf{A}_R$ is a symmetric matrix and the quantities $\mathbf{A}_R$, $\mathbf{b}_R$, and c_R are independent of $\mathbf{v}$. Substituting this quadratic form for $f_{R+1}(\mathbf{u}_R)$ in Eq. (5.115), we get

$$f_R(\mathbf{v}) = \min_{\mathbf{u}_R} [\langle \mathbf{Q}\mathbf{u}_R, \mathbf{u}_R\rangle - 2\langle \mathbf{r}_R, \mathbf{u}_R\rangle s_R$$
$$+ \langle(\mathbf{u}_R - \mathbf{v}), (\mathbf{u}_R - \mathbf{v})\rangle + \langle(\mathbf{A}_{R+1}\mathbf{u}_R), \mathbf{u}_R\rangle - 2\langle \mathbf{b}_{R+1}, \mathbf{u}_R\rangle + c_R] \tag{5.119}$$

Differentiating this with respect to $\mathbf{u}_R$ and equating $\partial f_R(x)/\partial \mathbf{u}_R$ to zero, the minimizing value of $\mathbf{u}_R$ can be found to be

$$\mathbf{u}_R = [\mathbf{Q} + \mathbf{A}_{R+1} + \mathbf{I}]^{-1}(\mathbf{v} + \mathbf{b}_{R+1} + \mathbf{r}_R) \tag{5.120}$$

Substituting this in Eq. (5.119), we get

$$f_R(\mathbf{v}) = \langle[\mathbf{I} - [\mathbf{I} + \mathbf{Q} + \mathbf{A}_{R+1}]^{-1}]\mathbf{v}, \mathbf{v}\rangle$$
$$- \langle 2[\mathbf{I} + \mathbf{Q} + \mathbf{A}_{R+1}]^{-1}(\mathbf{b}_{R+1} + \mathbf{r}_R), \mathbf{v}\rangle + c_{R+1} + s_{R+1} \tag{5.121}$$
$$- \langle[\mathbf{I} + \mathbf{Q} + \mathbf{A}_{R+1}]^{-1}(\mathbf{b}_{R+1} + \mathbf{r}_R), (\mathbf{b}_{R+1} + \mathbf{r}_R)\rangle$$

Comparing with Eq. (5.118), we get

$$\mathbf{A}_R = \mathbf{I} - [\mathbf{I} + \mathbf{Q} + \mathbf{A}_{R+1}]^{-1}$$
$$\mathbf{b}_R = [\mathbf{I} + \mathbf{Q} + \mathbf{A}_{R+1}]^{-1}(\mathbf{b}_{R+1} + \mathbf{r}_R)$$
$$= [\mathbf{I} - \mathbf{A}_R](\mathbf{b}_{R+1} + \mathbf{r}_R) \tag{5.122}$$
$$c_R = c_{R+1} + s_R - \langle[\mathbf{I} + \mathbf{Q} + \mathbf{A}_{R+1}]^{-1}(\mathbf{b}_{R+1} + \mathbf{r}_R), (\mathbf{b}_{R+1} + \mathbf{r}_R)\rangle$$
$$= c_{R+1} + s_R - \langle \mathbf{b}_R, (\mathbf{b}_{R+1} + \mathbf{r}_R)\rangle$$

From Eq. (5.117), we get the starting conditions

$$\mathbf{A}_L = \mathbf{I}, \qquad \mathbf{b}_L = \mathbf{u}_L \tag{5.123}$$

$$c_L = \langle[\mathbf{I} + \mathbf{Q}]\mathbf{u}_L, \mathbf{u}_L\rangle - 2\langle\mathbf{r}_L, \mathbf{u}_L\rangle + s_L$$

Since $\mathbf{v}$, by definition, is $\mathbf{u}_R$, Eq. (5.120) becomes

$$\mathbf{u}_R = [\mathbf{I} + \mathbf{Q} + \mathbf{A}_{R+1}]^{-1}(\mathbf{u}_{R-1} + \mathbf{b}_{R+1} + \mathbf{r}_R) \tag{5.124}$$

$$= [\mathbf{I} - \mathbf{A}_R]\mathbf{u}_{R-1} + \mathbf{b}_R \tag{5.125}$$

More thorough discussion of this method can be found in the literature.

MONTE CARLO METHODS

Another major approach to the solution of elliptic partial differential equations using digital computers is based upon the theory of probability. These methods serve not primarily to provide the potential distribution everywhere within a one-, two-, or three-dimensional field, but are rather used to determine the potential at one specific point. In principle the Monte Carlo method involves the selection of some point of interest within the field and the commencing of a series of ***random walks*** from this point. Each such walk involves a sequence of small steps, such that each step begins at the end of the preceding step; but the direction of each step is random. Eventually, each random walk will reach some point on the field boundary. The potential at this boundary point is recorded and a new walk is commenced. Provided enough random walks are taken, and provided each step in the walk is sufficiently small, it can be shown that the weighted average of the boundary intersections and the field points traversed will converge to the potential existing at the point of interest. So far the Monte Carlo method has been most useful for the solution of two-dimensional problems governed by Laplace's and Poisson's equations with Dirichlet boundary conditions (potential specified everywhere along the boundary), but the method is equally applicable to other types of field problems.

Consider first the application of the Monte Carlo method to a uniform two-dimensional field governed by Laplace's equation

$$\frac{\partial^2 u}{\partial x^2} + \frac{\partial^2 u}{\partial y^2} = 0 \tag{5.126}$$

As in the other methods described in this chapter, a finite difference grid is now specified. If a rectangular grid with $\Delta x = \Delta y = h$ is selected, the finite difference approximation for the potential at a typical node is given by

$$u_{l,m} = \tfrac{1}{4}(u_{l-1,m} + u_{l+1,m} + u_{l,m-1} + u_{l,m+1}) \tag{5.127}$$

For nonuniform fields or unequal net spacings, each of the terms in the right side of Eq. (5.127) has a different coefficient. The coefficients of the difference equation, which are all $\frac{1}{4}$ in this case, are interpreted as transition probabilities from one point to the neighboring points. Thus, Eq. (5.127) specifies that if the random walk has progressed to a point (l, m) there is an equal probability that any of the four neighboring points will be reached in the next step of the walk.

A point of interest is now selected within the field and may be assigned for convenience the coordinates (0, 0). All random walks are commenced at this point, and sequences of random numbers are employed to determine the direction of each step. Consider for example the use of two sequences A and B of random numbers, each containing the integers 0, 1, 2, 3, 4, 5, 6, 7, 8, and 9, and that one integer from each sequence is provided for each step of the calculation. The decision as to the direction of the corresponding step of the random walk can then be made in accordance with Table 5.1. Depending upon the combination of the two random numbers, therefore, the random walk will proceed for one step in the positive or negative x or y direction.

Table 5.1. A table illustrating the solution of random walk directions

Random Number from Sequence A	*Random Number from Sequence B*	*New x-coordinate*	*New y-coordinate*
0, 1, 2, 3, 4	0, 1, 2, 3, 4	l	$m + 1$
5, 6, 7, 8, 9	5, 6, 7, 8, 9	l	$m - 1$
0, 1, 2, 3, 4	5, 6, 7, 8, 9	$l + 1$	m
5, 6, 7, 8, 9	0, 1, 2, 3, 4	$l - 1$	m

This stepwise progress is repeated until a grid point lying on the field boundary is reached. The walk is then terminated, and the potential at that point, a specified boundary value, is recorded and placed in an accumulating register. A new random walk is then commenced from the same original point, (0, 0). It can be demonstrated, using certain theorems from the theory of probability, that as the number of such random walks becomes large the average of the boundary potentials at which each walk terminates will approach the true potential at the point (0, 0) of interest. The number of random walks required to produce such convergence to the solution depends upon the geometry of the field under study, the uniformity of the boundary conditions, the uniformity of the field parameters, and to a lesser extent upon the manner in which the finite difference grid is constructed. The average length of each random walk depends primarily upon the distance of the point under consideration from the field boundaries, and upon the net interval h.

A number of modifications and refinements are desirable and may be necessary in this procedure. Usually, it is impossible to specify in advance precisely how many random walks M have to be performed to achieve a sufficient convergence to the correct solution. For this reason, in place of specifying a fixed number of random walks, the desired statistical variant is specified. Additional computational steps must then be included to check whether this statistical specification of convergence has been met.

If the coefficients of the finite difference equation are not all identical, the transition probabilities in the four principal directions are different from each other. Under these conditions Table 5.1 must be replaced by a more complex table taking the specific transition probabilities into account. If the problem under consideration is governed by Poisson's equation instead of Laplace's equation, the term on the right side of this equation must be taken into account at each step in the random walk. For uniform fields and equal spacings in the x and y directions, a term $\frac{1}{4}h^2f(l, m)$ is added to the accumulator register at each step in the random walk. The Monte Carlo method is particularly suitable for the relatively quick determination of the potential at one or few points of interest. This method becomes inefficient if a solution is required everywhere in a region.

REFERENCES

1. R. S. VARGA, *Matrix Iterative Analysis.* Englewood Cliffs, N.J.: Prentice-Hall, Inc., 1963.
2. E. ANGEL and R. E. BELLMAN, *Dynamic Programming and Partial Differential Equations*, New York: Academic Press, Inc., 1972.

SUGGESTIONS FOR FURTHER READING

1. *There are many fine books in which elliptic boundary value problems are extensively treated. For a nice elementary introduction to the mechanics of discretization with elementary exercises and worked out solutions, see*

   G. D. SMITH, *Numerical Solution of Partial Differential Equations*, Oxford University Press, London, 1965.

2. *The self-adjoint property and other theoretical matters pertaining to truncation and round-off errors are treated in*

   G. E. FORSYTHE and W. R. WASOW, *Finite Difference Methods for Partial Differential Equations*, John Wiley & Sons, Inc., New York, 1960.

3. *For a good book dealing with discretization procedures, truncation error estimates, and the like; for elliptic parabolic and hyperbolic equations, consult*

   A. R. MITCHELL, *Computational Methods in Partial Differential Equations*, John Wiley & Sons, Inc., New York, 1969.

4. *For an excellent unified theoretical treatment of both finite difference and variational methods at a level suitable for advanced engineering and science students, consult*

   G. I. MARCHUK, *Methods of Numerical Mathematics*, Springer-Verlag New York, Inc., New York, 1975.

5. *The sections on integration and variational methods of deriving finite difference equations are based heavily on the material in*

   R. S. VARGA, *Matrix Iterative Analysis*, Prentice-Hall, Inc., Englewood Cliffs, N.J., 1963.

6. *Many of the important results concerning the basic problems of solving linear systems of algebraic equations can be found in*

   G. E. FORSYTHE and C. B. MOLER, *Computer Solution of Linear Algebraic Systems*, Prentice-Hall, Inc., Englewood Cliffs, N.J., 1967, and in

   L. FOX, *An Introduction to Numerical Linear Algebra*, Oxford University Press, New York, 1964.

7. *A thorough discussion on the methods of solving TPBVP's for ordinary differential equations can be found in*

S. M. ROBERTS and J. S. SHIPMAN, *Two-Point Boundary Value Problems: Shooting Methods*, American Elsevier, New York, 1972.

8. *For a discussion on the method of lines and invariant imbedding, consult*

G. M. WING, *Introduction to Transport Theory*, John Wiley & Sons, Inc., New York, 1962.

9. *For a thorough bibliography on tensor product-FFT method, refer to the review article by*

F. W. DORR, "The Direct Solution of the Discrete Poisson Equation on a Rectangle," *SIAM Review*, Vol. 12, No. 2, pp. 248–263, 1970.

10. *A discussion on linear programming method to solve elliptic equations can be found in*

J. R. CANNON, "The Numerical Solution of Dirichlet Problem for Laplace's Equation by Linear Programming," *J. Soc. Ind. Appl. Math.*, Vol. 12, pp. 233–237, 1964.

11. *The dynamic programming method is extensively treated in*

E. ANGEL and R. E. BELLMAN, *Dynamic Programming and Partial Differential Equations*, Academic Press, Inc., New York, 1972.

12. *For theoretical and practical aspects of the Monte Carlo method, see*

C. N. KLAHR, "A Monte Carlo Method for the Solution of Elliptic Partial Differential Equations," in *Mathematical Methods for Digital Computers*, A. Ralston and H. S. Wilf, eds., John Wiley & Sons, Inc., New York, 1960.

13. *An exhaustive discussion of solving field problems using analog techniques (i.e., nonnumerical methods) can be found in*

W. J. KARPLUS, *Analog Simulation: Solution of Field Problems*, McGraw-Hill Book Co., New York, 1958.

14. *Boundary value problems associated with the biharmonic operator are not treated in this chapter. For a tutorial introduction to this topic with extensive references to the literature, consult*

V. VEMURI, "Biharmonic Equations, Solution Methods," in *Encyclopedia of Computer Science and Technology*, Vol. 3, J. Belzer, A. G. Holzman and A. Kent, eds., Marcel Dekker, Inc., New York, 1976.

EXERCISES

1. Show that the problem in Eq. (5.19) is not self-adjoint.

2. Show that

$$\frac{\partial^2 u}{\partial x^2} + \frac{\partial^2 u}{\partial y^2} + x^2 \frac{\partial u}{\partial x} + y^2 \frac{\partial u}{\partial y} + u = 0$$

can be made self-adjoint by multiplying it throughout by $\mu(x, y) = \exp(x^3 + y^3)/3$. Show that the self-adjoint form is

$$\frac{\partial}{\partial x}\left[\exp((x^3 + y^3)/3)\frac{\partial u}{\partial x}\right] + \frac{\partial}{\partial y}\left[\exp((x^3 + y^3)/3)\frac{\partial u}{\partial y}\right] + \exp[(x^3 + y^3)/3]u = 0$$

3. Test whether the following are self-adjoint or essentially self-adjoint. Where possible, transform the given equation into the form of Eq. (5.21).

   (a) $\dfrac{\partial^2 u}{\partial x^2} + (k/y)\dfrac{\partial u}{\partial y} + \dfrac{\partial^2 u}{\partial y^2} = 0, \qquad k$ an integer

   (b) $\dfrac{\partial^2 u}{\partial x^2} + x\dfrac{\partial u}{\partial x} + y\dfrac{\partial^2 u}{\partial y^2} + xy\dfrac{\partial u}{\partial y} = 0.$

4. Repeat Problem 3 of Chapter 4 using the integration method.

5. Repeat Problem 4 of Chapter 4 using the integration method instead of the Taylor series method. Show that the resulting coefficient matrix is symmetric.

6. Consider a region Ω in the n-dimensional Euclidean space E_n. Now define the Hilbert space $L_2(\Omega)$ of all real, measurable, square integrable functions:

$$\int_\Omega f^2(x)\,dx < \infty$$

with the *inner product* defined by

$$\langle f, g\rangle \triangleq \int_\Omega fg\,d\Omega$$

Now define a subspace U of $L_2(\Omega)$ by imposing the following additional conditions which every $u \in U$ must satisfy: (i) $u = 0$ on the boundary Γ of Ω. (ii) u and its first and second derivatives are continuous on $\Omega + \Gamma$. Now consider an operator A that maps U into $L_2(\Omega)$. The operator A is said to be *positive* if

$$\langle Au, u\rangle > 0, \qquad u \neq 0, \qquad u \in U$$

   (a) Show that the operator $-\nabla^2$ is positive.

   The operator A is said to be *self-adjoint* if $A = A^*$, where the adjoint operator A^* is defined by the Lagrange identity

$$\langle Au, v\rangle = \langle u, A^*v\rangle$$

   (b) Show that $-\nabla^2$ is self-adjoint. (*Hint:* Use Green's first and second formulas.)

7. Consider the differential equation $Lu = g$, where L is a two-dimensional Laplacian operator defined on Ω. A square net of size h is superimposed on this region. Let L^h be a difference operator defined as follows

$$L^h = L_1 + L_2$$
$$L_1 = -D_{fx}D_{bx}$$
$$L_2 = -D_{fy}D_{by}$$

Show that L^h is a self-adjoint operator by showing

$$\langle L^h u, v\rangle = \langle u, L^h v\rangle$$

Hint: The discrete analogs of Green's first and second formulas are

$$-\sum_{l=1}^{n-1}(D_{fx}D_{bx}u)_{l,m}v_{l,m} = \sum_{l=1}^{n}(D_{bx}u)_{l,m}(D_{bx}v)_{l,m}$$

$$-\sum_{l=1}^{n-1}(D_{fx}D_{bx}u)_{l,m}v_{l,m} = \sum_{l=1}^{n-1}(D_{fx}D_{bx}v)_{l,m}u_{l,m}$$

8. Optimization of numerical processes and various theoretical estimates of algorithms often require knowledge of the norm of the operator L. The following relation is good for arbitrary operator L.

$$\|L\|^2 = \sup_{u \in U} \frac{\langle Lu, Lu \rangle}{\langle u, u \rangle}$$

If L is a positive semidefinite matrix on the Euclidean space, that is, if $L = \mathbf{A} \geq 0$ and if $\sigma \geq 0$, then show that

(a) $\|(\mathbf{I} + \sigma\mathbf{A})^{-1}\| \leq 1$

(b) $\|(\mathbf{I} - \sigma\mathbf{A})(\mathbf{I} + \sigma\mathbf{A})^{-1}\| \leq 1$

9. Given a $N \times N$ matrix $\mathbf{A}$, a directed graph of $\mathbf{A}$ can be constructed as follows. Choose any set of N distinct points in the plane and label them as $1, 2, \ldots, N$. For each $a_{lm} \neq 0$ of $\mathbf{A}$, draw a directed line from point l to point m. The matrix $\mathbf{A}$ is *irreducible* if for any two distinct points labeled there is an arrow from i to j or there is a path, consisting of a sequence of arrows, from i to j. Test the following matrices for irreducibility:

(a) $\begin{bmatrix} 0 & 1 & 0 & 1 \\ 1 & 0 & 1 & 0 \\ 0 & 1 & 0 & 1 \\ 1 & 0 & 0 & 1 \end{bmatrix}$ (b) $\begin{bmatrix} 1 & 1 & 1 \\ 0 & 0 & 1 \\ 0 & 0 & 1 \end{bmatrix}$ (c) $\begin{bmatrix} 2 & -1 & 0 \\ -1 & 2 & -1 \\ 0 & -1 & 2 \end{bmatrix}$

6

Initial Value Problems

6.1. GENERAL FORMULATION OF THE INITIAL VALUE PROBLEM

Problems to be considered in this chapter are initial value problems that arise in various branches of physics and engineering, such as heat flow, diffusion, fluid dynamics, magnetofluid dynamics, electromagnetic theory, wave mechanics, radiative transfer, and neutron transport. A steady state is not assumed to exist. Therefore, in the equations describing these phenomena, one of the independent variables, usually called t, has a timelike behavior. "Timelike" is used here to describe a class of problems that has boundary conditions specified at $t = t_0$ and a solution exists for $t \geq t_0$. Generally, this is a broader class of problems than the class that arises from specifying a solution in real time. Nevertheless, problems where t stands for real time are important, and for these problems the condition specified at $t = t_0$ is called the initial condition; therefore, such problems are called initial value problems. Well-posed initial value problems occur in both parabolic and hyperbolic types of equations and these two types will be treated together in this chapter.

6.2. AN EXPLICIT METHOD FOR DIFFUSION EQUATIONS

In this section, the main ideas are introduced by considering the familiar example of linear heat flow, or diffusion in one space dimension. If x denotes the coordinate along the length of a thin insulated rod in which heat can flow, and if $u = u(x, t)$ denotes the temperature at position x at time t, then the temperature satisfies the differential equation

$$K\frac{\partial^2 u}{\partial x^2} = a\frac{\partial u}{\partial t} \tag{6.1}$$

where K is the thermal conductivity of the material and a is the heat capacity per unit volume. For simplicity, both K and a are assumed to be constant. In the general case, they may be functions of x, or of x and t, or of x, t, u, and higher derivatives of u. In higher-dimensional anisotropic media, K and a depend upon the direction of flow and should be represented by tensors. However, discussion in this section is limited to linear, homogeneous, isotropic media so that K and a are constants and $K/a = \sigma > 0$.

Equation (6.1) is to be supplemented by boundary conditions at, say, $x = 0$ and $x = L$, which denote the ends of the rod. If the end $x = L$ is kept in melting ice, then at $x = L$, $u = 0$. If the end $x = 0$ is insulated, then at $x = 0$, $\partial u/\partial x = 0$. In addition, the initial temperature distribution along the length of the rod is required to be specified as an initial condition.

The first step in obtaining finite difference approximations to Eq. (6.1) involves, as usual, the definition of a finite difference grid in the xt plane as shown in Figure 6.1. Using a notation analogous to that used in Chapter 4, the approximation to $u(l\,\Delta x, j\,\Delta k)$ is denoted by $u_{l,j}$, and a typical point in the grid is denoted by (l, j). Figure 6.1 shows a typical grid point and its neighbors.

One of the simplest difference approximations to Eq. (6.1) is

$$\sigma\left[\frac{u_{l-1,j} - 2u_{l,j} + u_{l+1,j}}{(\Delta x)^2}\right] = \frac{u_{l,j+1} - u_{l,j}}{(\Delta t)} \tag{6.2}$$

where the space derivative at (l, j) has been replaced by a second central difference and the time derivative at (l, j) by a forward difference. [See Eqs. (4.14) and (4.17).] It is customary to show the grid points involved by a mnemonic scheme as shown in Figure 6.2. Such a figure is sometimes called a ***stencil.*** The points shown in this stencil are those points in the xt plane used in one application of the formula. The three points connected by the horizontal line are used in forming the second space difference. The two points connected by the vertical line are used in forming the time difference. Labels such as l and j are shown here only for clarity and generally will be omitted.

Equation (6.2) can be rearranged as

$$u_{l,j+1} = u_{l,j} + r(u_{l-1,j} - 2u_{l,j} + u_{l+1,j}) \tag{6.3}$$

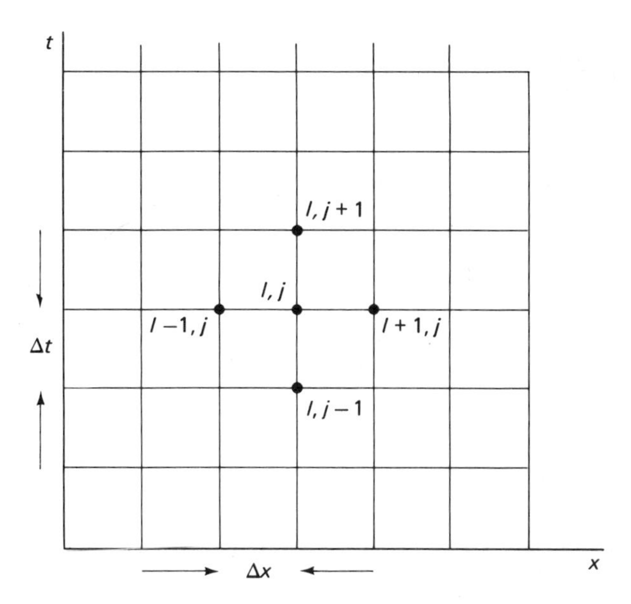

Figure 6.1 Finite difference grid in the xt domain

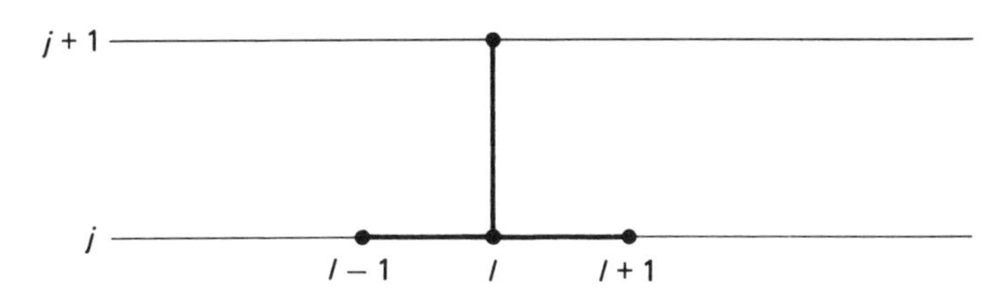

Figure 6.2 Computational stencil for a forward difference scheme

where

$$r = \frac{\sigma(\Delta t)}{(\Delta x)^2} \tag{6.4}$$

Thus, Eq. (6.3) gives a formula for the temperature $u_{l,j+1}$ in terms of the temperatures at the jth time level. At $j = 0$, the initial time level, the temperature along the rod is specified as an initial condition and is therefore known. Using this information in Eq. (6.3), one can calculate the temperature distribution along the rod at time level $j = 1$, then at time level $j = 2$ in terms of values at time level $j = 1$, and so forth. A formula such as Eq. (6.3), which expresses the unknown value at *one* grid point in terms of known values at other grid points, is called an ***explicit*** formula, and methods based on such formulas are ***explicit*** methods.

As a numerical illustration, consider the problem of solving the nondimensional equation

$$\frac{\partial^2 u}{\partial x^2} = \frac{\partial u}{\partial t} \tag{6.5}$$

with an initial temperature distribution, in nondimensional form, given by

$$u = \begin{cases} 2x, & 0 \le x \le \frac{1}{2} \\ 2(1-x), & \frac{1}{2} \le x \le 1 \end{cases} \tag{6.6}$$

and the boundary conditions given by

$$u = 0 \quad \text{at } x = 0 \text{ and } x = 1 \quad \text{for all time} \tag{6.7}$$

To solve this problem, let $\Delta x = \frac{1}{10}$ and $\Delta t = 1/1000$. So $r = \frac{1}{10}$ and Eq. (6.3) becomes

$$u_{l,j+1} = \tfrac{1}{10}(u_{l-1,j} - 8u_{l,j} + u_{l+1,j}) \tag{6.8}$$

Now Eq. (6.8) is solved for $j = 0$ and $l = 1, 2, \ldots, 9$, and the results appear in the row corresponding to $j = 0$ in Table 6.1. The table shows the solution only for grid points $l = 1, 2, \ldots, 5$, as the problem is symmetric with respect to $x = 1/2$ or $l = 5$. Now, the value of j is incremented by unity and the problem repeated (see Table 6.1).

Table 6.1. Solution of Equations (6.5) and (6.6) with $\Delta x = \frac{1}{10}$ and $\Delta t = 1/1000$

	$l=0$ $x=0$	$l=1$ 0.1	$l=2$ 0.2	$l=3$ 0.3	$l=4$ 0.4	$l=5$ 0.5
$(j=0)$ $t = 0.000$	0	0.2000	0.4000	0.6000	0.8000	1.0000
$(j=1)$ 0.001	0	0.2000	0.4000	0.6000	0.8000	0.9600
$(j=2)$ 0.002	0	0.2000	0.4000	0.6000	0.7960	0.9280
$(j=3)$ 0.003	0	0.2000	0.4000	0.5996	0.7896	0.9016
$(j=4)$ 0.004	0	0.2000	0.4000	0.5986	0.7818	0.8792
$(j=5)$ 0.005	0	0.2000	0.3999	0.5971	0.7732	0.8597
⋮						
$(j=10)$ 0.01	0	0.1996	0.3968	0.5822	0.7281	0.7867
⋮						
$(j=20)$ 0.02	0	0.1938	0.3781	0.5373	0.6486	0.6891

From *Numerical Solutions of Partial Differential Equations* by G. D. Smith, published by Oxford University Press, New York. © Oxford University Press, 1965, 2nd edition.

The exact (analytical) solution of Eq. (6.5) with the auxiliary conditions in Eqs. (6.6) and (6.7) is

$$u = \frac{-8}{\pi^2} \sum_{n=1}^{\infty} \frac{1}{n^2} \left(\sin \frac{1}{2} n\pi\right)(\sin n\pi x) \exp(-n^2\pi^2 t) \tag{6.9}$$

This expression can be evaluated at various grid points at various time levels. For instance, at $l = 3$, Eq. (6.9) is evaluated at various time levels, and the computed

Table 6.2. Comparison of the calculated solution from Table 6.1 (at $l = 3$) with the exact solution given in Equation (6.9)

	Finite Difference Solution ($l = 3$)	*Analytical Solution* ($l = 3$)	*Difference*	*Percentage Error*
$t = 0.005$	0.5971	0.5966	0.0005	0.08
$t = 0.01$	0.5822	0.5799	0.0023	0.4
$t = 0.02$	0.5373	0.5334	0.0039	0.7
$t = 0.10$	0.2472	0.2444	0.0028	1.1

solution from Table 6.1 is compared with the exact solution at $l = 3$. The results are summarized in Table 6.2.

The errors shown in Table 6.2 are considered acceptable. However, in going from $t = 0$ to $t = 0.1$, Eq. (6.8) was solved 100 times. A considerable saving in computer time can possibly be achieved by increasing Δt and consequently reducing the number of times Eq. (6.8) is solved. To test this hypothesis, Eq. (6.5) is solved once with $\Delta x = 1/10$, $\Delta t = 5/1000$, and once again with $\Delta x = 1/10$, $\Delta t = 1/100$; the results are displayed in Tables 6.3 and 6.4 and Figure 6.3.

Figure 6.3 shows a comparison of the computed solutions (dashed lines), such as those shown in Tables 6.3 and 6.4, with the exact solution (the continuous line). Inspection of Figure 6.3 reveals that the finite difference solution with $r = 0.48$ is not

Table 6.3. Solution of Equations (6.5) and (6.6) with $\Delta x = \frac{1}{10}$ and $\Delta t = 5/1000$

	$l = 0$ $x = 0$	1 0.1	2 0.2	3 0.3	4 0.4	5 0.5
$t = 0.000$	0	0.2000	0.4000	0.6000	0.8000	1.0000
0.005	0	0.2000	0.4000	0.6000	0.8000	0.8000
0.010	0	0.2000	0.4000	0.6000	0.7000	0.8000
0.015	0	0.2000	0.4000	0.5500	0.7000	0.7000
0.020	0	0.2000	0.3750	0.5500	0.6250	0.7000
⋮						
0.100	0	0.0949	0.1717	0.2484	0.2778	0.3071

Table 6.4. Solution of Equations (6.5) and (6.6) with $\Delta x = \frac{1}{10}$ and $\Delta t = 1/100$

	$l=0$ $x=0$	1 0.1	2 0.2	3 0.3	4 0.4	5 0.5
$t = 0.00$	0	0.2	0.4	0.6	0.8	1.0
0.01	0	0.2	0.4	0.6	0.8	0.6
0.02	0	0.2	0.4	0.6	0.4	1.0
0.03	0	0.2	0.4	0.2	1.2	−0.2
0.04	0	0.2	0.0	1.4	−1.2	2.6

From *Numerical Solutions of Partial Differential Equations* by G. D. Smith, published by Oxford University Press, New York.

as good as the earlier solutions; nevertheless, it is adequate. However, the solution in Figure 6.3b is completely meaningless. The phenomenon appearing in Figure 6.3b is not merely due to rounding errors, but is a property of the difference equation (6.8), or equivalently Eq. (6.3), and is called ***instability.*** In a situation like this, one would be tempted to reduce Δx in an attempt to reduce the oscillations in the solution. Indeed the errors would only be worse by reducing Δx, unless Δt is also suitably reduced (verify this!).

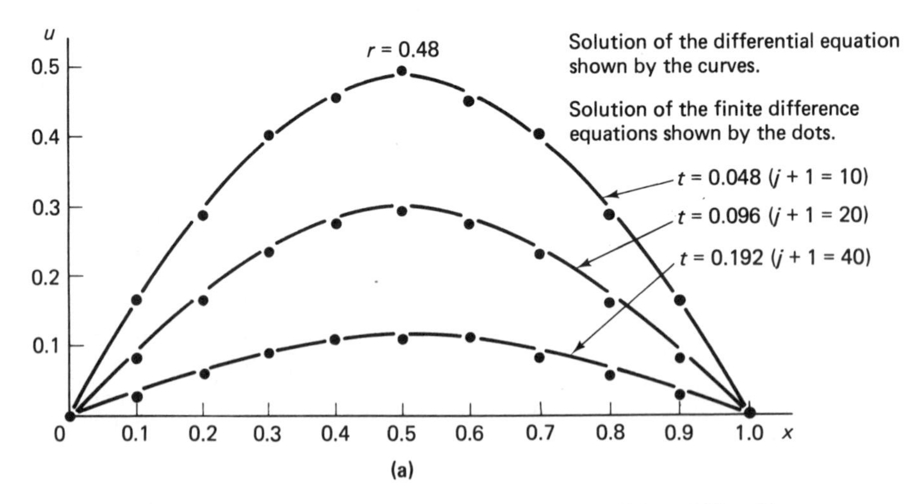

Figure 6.3 Demonstration of the phenomenon of instability (From *Numerical Solutions of Partial Differential Equations*, by G. D. Smith, published by Oxford University Press, Inc., New York, © Oxford University Press, 1965, 2nd edition)

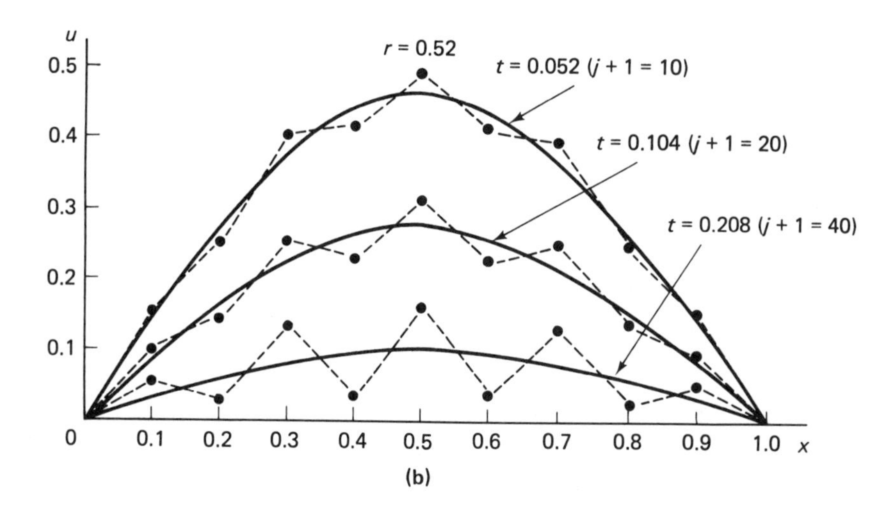

Figure 6.3 *(continued)*

6.3. CONSISTENCY, ACCURACY, EFFICIENCY, AND STABILITY

When a partial differential equation is approximated by a finite difference analog, one naturally expects that the difference scheme indeed represents the differential system in some sense. That is, a difference system is said to be ***consistent*** with a differential system when the former becomes identical with the later in the limit as $x \longrightarrow 0$ and $t \longrightarrow 0$. Clearly, consistency is a fundamental requirement.

Accuracy of a numerical solution depends on two major classes of errors: roundoff and truncation errors. ***Roundoff errors*** characterize the differences between the solution furnished by the computer and the exact solution of the difference equations. Roundoff errors arise as a result of finite precision arithmetic that can be carried out in a computer. In modern digital computers a sufficient number of significant digits are usually carried throughout all arithmetic operations to make any single rounding error negligible in magnitude. The real danger as far as roundoff errors are concerned results from the fact that solutions obtained in one cycle of calculations are used as "initial conditions" for obtaining the values of u for the subsequent time increments. There therefore exists the danger of error propagation and error growth as the solution proceeds over a large number of steps. This phenomenon is closely related to computational stability.

Truncation errors are caused by the approximations involved in representing differential equations. Truncation errors depend upon the spatial grid size (Δx) and step size (Δt). Magnitude of such errors can be estimated using Taylor series method, as was seen in Section 4.4.

Efficiency refers to the amount of computational work done by the computer in solving a problem over unit time-length. Usually there is a certain amount of tradeoff involved between efficiency and accuracy.

The final concept to be studied is stability. If $u(x, t)$ is the exact solution of the initial value problem, and $u_{l,j}$ is the solution of the finite difference equations, the error of the approximation at (l, j) is $(u_{l,j} - u(l\,\Delta x, j\,\Delta t))$, and one may ask two questions:

1. What is the behavior of $|u_{l,j} - u(l\,\Delta x, j\,\Delta t)|$ as $j \longrightarrow \infty$ for fixed Δx, Δt? That is, is the solution bounded as the time index $j \longrightarrow \infty$? This is the stability problem.
2. What is the behavior of $|u_{l,j} - u(l\,\Delta x, j\,\Delta t)|$ as Δt, $\Delta x \longrightarrow 0$ for a fixed value of $j\,\Delta t$? That is, is the difference scheme convergent?

For both questions, the number of cycles of calculation becomes infinite in the limit, and there is a possibility of unlimited amplification of errors. In a majority of pure initial value problems, the two questions lead to quite similar conclusions, but for problems in a bounded domain the results can be quite different.

From the discussion of the preceding section, it is clear that as the number of cycles of calculations become large there is a possibility for unlimited amplification of errors. If errors are amplified from time step to time step, the total accumulated error will quickly swamp the solution, making it worthless. The property of stability may therefore be stated as: a numerical method is stable if a small error at any stage produces a smaller cumulative error.

Various techniques are available for a quantitative treatment of stability of finite difference schemes. The best known among them is perhaps the linearized Fourier analysis method. An extension of this method, referred to as the ***spectral theory of difference operators***, can be used to study the effects of boundary conditions and the stability of certain equations with variable coefficients. The maximum principle and the energy method are other methods that are occasionally used to study the stability of finite difference schemes.

The concept of stability will now be studied quantitatively by first considering a single ordinary differential equation. For this simple case the dependent variable is a scalar, and it is easy to define an error, say ϵ_j, which occurs at time level j. In stability analysis, one is interested in the amplification of the error at time level $j + 1$. This can be done by writing

$$\epsilon_{j+1} = g\epsilon_j \tag{6.10}$$

where g is an amplification factor that is related to the integration scheme employed and therefore to the truncation errors of the integration scheme. For stability, then,

$$|\epsilon_{j+1}| \leq |\epsilon_j| \tag{6.11}$$

By virtue of Eq. (6.11)

$$|g\epsilon_j| \leq |\epsilon_j| \tag{6.12}$$

or

$$|g| \leq 1 \tag{6.13}$$

This idea can be generalized. For a system of N first-order ordinary differential equations, Eq. (6.12) can be rewritten as

$$\boldsymbol{\epsilon}_{j+1} = \mathbf{G}\boldsymbol{\epsilon}_j \tag{6.14}$$

where $\boldsymbol{\epsilon}_j$ is an N-vector and $\mathbf{G}$ is an $N \times N$ amplification matrix. If Eq. (6.14) is diagonalized,

$$\epsilon_{j+1,\mu} = g_\mu \epsilon_{j,\mu} \tag{6.15}$$

where g_μ is an eigenvalue of $\mathbf{G}$, and $\boldsymbol{\epsilon}_\mu$ is the associated eigenvector. The condition for stability must now be applied separately to the amplitude of each error eigenvector,

$$|\epsilon_{j+1,\mu}| \leq |\epsilon_{j,\mu}|, \qquad \text{for all } \mu$$

or

$$|g_\mu| \leq 1, \qquad \text{for all } \mu \tag{6.16}$$

The condition for stability, therefore, reduces to the requirement that the magnitude of each and every eigenvalue of the amplification matrix $\mathbf{G}$ must be smaller than or equal to unity. In general, the eigenvalues of the amplification matrix may be complex, in which case

$$|g_\mu| = \sqrt{g_\mu g_\mu^*} \tag{6.17}$$

where g_μ^* is the complex conjugate of g_μ.

The preceding background is sufficient to briefly consider a few techniques of analyzing the stability of some finite difference schemes.

FOURIER SERIES METHOD

This method, introduced by J. von Neumann, will be illustrated by analyzing Eq. (6.3) for stability. The idea is to derive an expression for the amplification factor for a Fourier mode in space. Following the procedure already used in Section 4.4, the function $u(x, t)$ can be expressed as a Fourier series, and for a Fourier mode in space, one can write, analogous to Eq. (4.44),

$$u(x, t) = \hat{u}(t)e^{ikx} \tag{6.18a}$$

or

$$u(l\,\Delta x, j\,\Delta t) = \hat{u}_j e^{ikx_l} \tag{6.18b}$$

Substituting in Eq. (6.3),

$$\hat{u}_{j+1}e^{ikx_l} = \hat{u}_j e^{ikx_l} + r\hat{u}_j(e^{ikx_{l+1}} - 2e^{ikx_l} + e^{ikx_{l-1}})$$

or

$$\begin{aligned}\hat{u}_{j+1} &= \hat{u}_j\{1 + 2r(e^{ik(\Delta x)} + e^{-ik(\Delta x)} - 1)\}\\ &= \hat{u}_j\{1 + 2r(\cos k(\Delta x) - 1)\}\end{aligned} \tag{6.19}$$

Thus, Eq. (6.19) relates the amplitude of a Fourier mode over consecutive time levels. Since there is only one dependent variable and only one partial differential equation, there is only one amplification factor g, not a matrix $\mathbf{G}$. By virtue of Eq. (6.12), and a little algebra, the amplification factor g is quickly recognized to be

$$\begin{aligned}g &= \{1 + 2r(\cos k(\Delta x) - 1)\}\\ &= 1 - 4r\sin^2\left(\frac{k(\Delta x)}{2}\right)\end{aligned} \tag{6.20}$$

Therefore, to satisfy the stability condition $|g| \leq 1$, one sets

$$\left| 1 - 4r \sin^2 \left(\frac{k(\Delta x)}{2}\right) \right| \leq 1 \tag{6.21}$$

This condition must hold good for every wave number k, or equivalently for every Fourier mode. So taking the maximum value of the sine function, the requirement for stability of the difference scheme in Eq. (6.3) is

$$-4r \geq -2 \qquad \text{or} \qquad r \leq \tfrac{1}{2}$$

which, in view of Eq. (6.4), translates to

$$\Delta t \leq \frac{1}{2} \frac{(\Delta x)^2}{\sigma} \tag{6.22}$$

Hence, to obtain stable numerical solutions, t should be chosen such that the preceding inequality is satisfied.

A useful physical interpretation can be given to this inequality. In a diffusion process, the time scale of interest is the diffusion time, defined as

$$\tau = \frac{(\Delta x)^2}{\sigma}$$

Therefore, the maximum permissible time step Δt_m is just one half the diffusion time associated with the mesh size (Δx). The quantity Δt_m can be visualized as the time for information to travel over a distance (Δx).

POSITIVE COEFFICIENT METHOD

The restriction shown in Eq. (6.22) leads to a simple stability test, which comes in handy. Rearranging Equation (6.3) as

$$u_{l,j+1} = Au_{l-1,j} + Bu_{l,j} + Cu_{l+1,j} \tag{6.23}$$

where $A = C = r$ and $B = (1 - 2r)$, one can readily see that the restriction in Eq. (6.22) implies that A, B, and C are positive and $A + B + C = 1$. However, it must be pointed out that the condition of a positive coefficient on $u_{l,j}$ cannot be used as a criterion for stability. It is only a reasonable consequence of the restriction established earlier. Nevertheless, the fact that $A, B, C > 0$ and $A + B + C = 1$ can be used to establish the boundedness of the solution. Using the maximum operation on both sides, Eq. (6.23) becomes

$$\begin{aligned}\operatorname*{Max}_l |u_{l,j+1}| &\leq (A + B + C) \operatorname*{Max}_l |u_{l,j}| = \operatorname*{Max}_l |u_{l,j}| \\ &\leq \operatorname*{Max}_l |u_{l,j-1}| \leq \cdots \leq \operatorname*{Max}_l |u_{l,0}| \end{aligned} \tag{6.24}$$

The solution is therefore bounded.

This procedure, in a slightly modified form, can be applied even if the coefficients of the differential equation are variable. However, the Fourier series method is more general and provides a practical insight.

ENERGY METHOD

The "energy method" is a name given to a group of techniques based on the use of certain energy-like quantities. In some cases, the energylike quantities do, in fact, correspond to physical energy in the system. In mathematical terminology, these quantities are called ***norms***. The key property that norms have in common with energy is that both are ***positive definite***. The energy method and the spectral theory of operators, referred to earlier, are complementary in the sense that the energy method can be used to obtain sufficient conditions for stability, whereas the spectral theory can be used to obtain necessary conditions.

The use of energy method depends mainly on one's ability to find a norm for a given difference scheme. As an illustration of how one would proceed, Eq. (6.3) is considered once again. For convenience, this equation along with its boundary conditions is reproduced here:

$$u_{l,j+1} = u_{l,j} + r(u_{l-1,j} - 2u_{l,j} + u_{l+1,j}) \tag{6.25a}$$

$$u_{0,j} = 0, \qquad u_{L,j} = 0, \qquad \text{for all } j \tag{6.25b}$$

Multiplying both sides by $(u_{l,j+1} + u_{l,j})$ and summing over all l, one gets after a little algebra

$$\sum_{l=1}^{L-1} (u_{l,j+1})^2 - \sum_{l=1}^{L-1} (u_{l,j})^2 = r\sum_{l=1}^{L-1} (u_{l,j+1} + u_{l,j})(u_{l-1,j} - 2u_{l,j} + u_{l,j}) \tag{6.26}$$

Defining

$$||u_j||^2 = (\Delta x) \sum_{l=1}^{L-1} (u_{l,j})^2 \tag{6.27}$$

and using the "summation by parts" formula,

$$\sum_{l=1}^{L-1} u_l(v_{l-1} - 2v_l + v_{l+1}) = u_L(v_L - v_{L-1}) - u_1(v_1 - v_0) - \sum_{l=1}^{L-1} (u_{l+1} - u_l)(v_{l+1} - v_L) \tag{6.28}$$

Eq. (6.26) reduces, after using the initial conditions in Eq. (6.25b), to

$$\frac{1}{(\Delta x)}\left[||u_{j+1}||^2 - ||u_j||^2\right] = -r\left[(u_{1,j} + u_{1,j+1})u_{1,j} + (\Delta x)||D_{fx}u_{l,j}||^2 + (\Delta x)^2 \sum_{l=1}^{L-1} (D_{fx}u_{l,j})(D_{fx}u_{l,j+1})\right] \tag{6.29}$$

where D_{fx} is the forward difference operator, defined earlier as $D_{fx}u_{l,j} = (u_{l+1,j} - u_{l,j})/(\Delta x)$ [see Eq. (4.22)].

Now, if S_j is defined as

$$S_j = ||u_j||^2 - \tfrac{1}{2}r[(u_{l,j})^2(\Delta x) + (\Delta x)^2||D_{fx}u_{l,j}||^2] \tag{6.30}$$

then, by virtue of Eq. (6.29), we can write

$$S_{j+1} - S_j = -\frac{r}{2}\left[(u_{1,j} + u_{1,j+1})^2 + ||D_{fx}u_j||^2(\Delta x) + ||D_{fx}u_{j+1}||^2(\Delta x) + 2(\Delta x)^3 \sum_{l=1}^{L-1} (D_{fx}u_{l,j})(D_{fx}u_{l,j+1})\right] \tag{6.31}$$

Except for the last term, every term inside the parentheses on the right side of Eq. (6.31) is positive. However, by Schwartz's inequality

$$\left|(\Delta x)^2 \sum_{l=1}^{L-1} (D_{fx}u_{l,j})(D_{fx}u_{l,j+1})\right| \leq ||D_{fx}u_j|| \cdot ||D_{fx}u_{j+1}||$$

$$\leq \tfrac{1}{2}(||D_{fx}u_j||^2 + ||D_{fx}u_{j+1}||^2)$$

Therefore

$$S_{j+1} - S_j \leq 0 \tag{6.32}$$

which proves that S_j is nonincreasing. From Eq. (6.30), it is obvious that $S_0 \leq ||u_0||^2$. Therefore, from Eq. (6.32), it follows that $S_j \leq ||u_0||^2$.

The final step is to show that $||u_j||^2 \leq (\text{constant}) \cdot S_j$. Starting from Eq. (6.30),

$$S_j = (\Delta x) \sum_{l=1}^{L-1} (u_{l,j})^2 - \frac{r}{2}(\Delta x)\left[(u_{1,j})^2 + \sum_{l=1}^{L-1} (u_{l+1,j} - u_{l,j})^2\right] \leq ||u_0||^2 \tag{6.33}$$

But, because $u_{L,j} = 0$, one can say that

$$\tfrac{1}{2}\left[(u_{1,j})^2 + \sum_{l=1}^{L-1} (u_{l+1,j} - u_{l,j})^2\right] \leq \sum_{l=1}^{L-1} (u_{l,j})^2 \tag{6.34}$$

Therefore, $S_j \leq ||u_0||^2$ is true if

$$r \leq 1 - \epsilon, \qquad \epsilon > 0 \tag{6.35}$$

Then Eq. (6.33) yields

$$S_j \geq (\Delta x) \sum_{l=1}^{L-1} (u_{l,j})^2 - (1 - \epsilon)(\Delta x) \sum_{l=1}^{L-1} (u_{l,j})^2 \tag{6.36}$$

$$\geq ||u_j||^2 - (1 - \epsilon)||u_j||^2$$

or

$$S_j \geq \epsilon ||u_j||^2 \tag{6.37}$$

That is, $||u_j||^2 \leq (\text{constant})S_j$, which means that the solution u_j is bounded as $j \longrightarrow \infty$, which in turn is the requirement for stability. Thus, the condition in Eq. (6.35), that is, $r \leq (1 - \epsilon)$, $\epsilon > 0$, or equivalently,

$$\frac{\sigma(\Delta t)}{(\Delta x)^2} \leq 1 \tag{6.38}$$

is the condition for stability.

6.4. IMPLICIT METHODS FOR THE DIFFUSION EQUATION

The stability condition, Eq. (6.22), places a serious limitation upon the magnitude of Δt. If a relatively small Δx has been chosen to minimize truncation errors in the approximation of the space derivatives, stability considerations may dictate the choice of a much smaller Δt than would be required, merely to keep the truncation errors involved in the approximation of the time derivative within acceptable bounds. Under these conditions, the admissible value of Δt may turn out to be so very small that an enormous number of cycles of calculations is required to complete a problem.

BACKWARD DIFFERENCE METHOD

The difficulty described can be avoided if Eq. (6.1) is approximated by

$$\sigma\frac{u_{l-1,j+1} - 2u_{l,j+1} + u_{l+1,j+1}}{(\Delta x)^2} = \frac{u_{l,j+1} - u_{l,j}}{(\Delta t)} \tag{6.39}$$

This equation is the same as Eq. (6.2) except for the second subscript on u on the left side. Equation (6.39) is said to have utilized a backward difference because, relative to the time level $j + 1$ at which space differences are expressed, the time derivative is approximated by a backward difference approximation. The mnemonic for this scheme is shown in Figure 6.4.

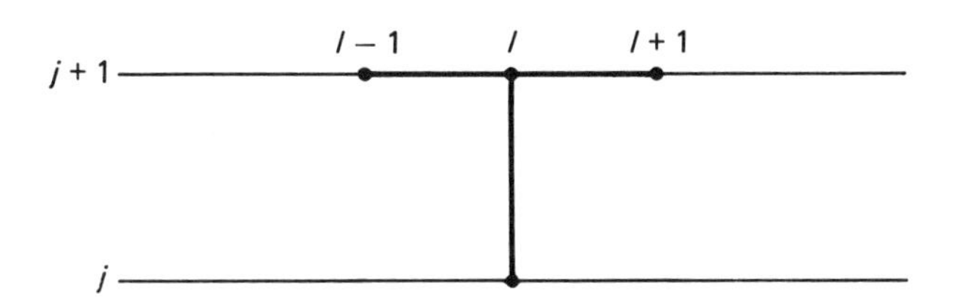

Figure 6.4 Computational stencil for a backward difference scheme

Equation (6.39) can be rearranged, as before, as

$$u_{l-1,j+1} - \left(2 + \frac{1}{r}\right)u_{l,j+1} + u_{l+1,j+1} = -\frac{1}{r}u_{l,j} \tag{6.40}$$

where $r = \sigma(\Delta t)/(\Delta x)^2$. Now it is no longer possible to solve for the values of u at time level $j + 1$ explicitly in terms of u values at time level j. To evaluate the unknowns, one must solve the set of simultaneous equations represented by Eq. (6.40), for $l = 1, 2, \ldots, L - 1$, using a method such as the tridiagonal algorithm (see Chapter 10).

CRANK-NICOLSON METHOD

It can be readily shown that Eq. (6.40) is stable under all circumstances. That is, the size of (Δt) can be chosen independently of the size of (Δx). The preceding implicit approximation, though efficient and simple to use, is only first-order correct in time, even though it is second-order correct in space. There are two ways to approach this problem. Since the space derivative in Eq. (6.39) is centered at the grid point $(l, j + 1)$, one can attempt to center the approximation for $\partial u/\partial t$ at the same point by writing

$$\left(\frac{\partial u}{\partial t}\right)_{l,j+1} = \frac{u_{l,j+2} - u_{l,j}}{2(\Delta t)} + O(\Delta t)^2 \tag{6.41}$$

which is second-order correct. However, it can be shown that use of the preceding approximation in the right side of Eq. (6.40) leads to an unconditionally unstable difference scheme. Alternatively, since the time derivative in Eq. (6.39) is centered at $(l, j + \frac{1}{2})$, one can attempt to center the approximation for $\partial^2 u/\partial^2 x$ at $(l, j + \frac{1}{2})$ by writing the approximation as the average of the second central differences at $(l, j + 1)$ and (l, j). Such an approximation, called the ***Crank-Nicolson approximation***, for Eq. (6.1) looks like

$$\frac{\sigma}{2}\left[\frac{u_{l-1,j+1} - 2u_{l,j+1} + u_{l+1,j+1}}{(\Delta x)^2} + \frac{u_{l-1,j} - 2u_{l,j} + u_{l+1,j}}{(\Delta x)^2}\right] = \frac{u_{l,j+1} - u_{l,j}}{(\Delta t)} \tag{6.42}$$

or, in the more compact and descriptive notation,

$$\frac{\sigma}{2}(\delta_x^2 u_{l,j+1} + \delta_x^2 u_{l,j}) = \frac{u_{l,j+1} - u_{l,j}}{(\Delta t)} \tag{6.43}$$

where δ_x^2 is the second central difference operator defined in Eq. (4.25). A mnemonic for this method is shown in Figure 6.5. Rearranging the terms, Eq. (6.42) can be written as

$$u_{l-1,j+1} - \left(2 + \frac{2}{r}\right)u_{l,j+1} + u_{l+1,j+1} = -u_{l-1,j} + \left(2 - \frac{2}{r}\right)u_{l,j} - u_{l+1,j} \tag{6.44}$$

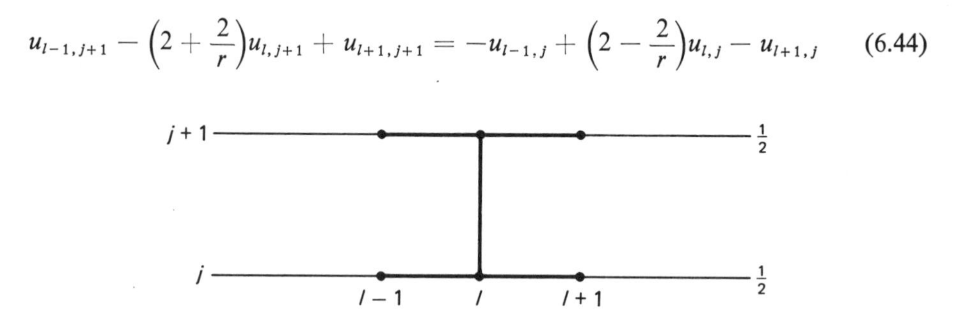

Figure 6.5 Computational stencil for the Crank-Nicolson scheme

One has to solve Eq. (6.44) at each time level j, for $l = 1, 2, \ldots, L - 1$. For $j = 0$, for example, the right side can be computed using initial condition data, and the left side contains three unknowns. Thus, to advance one time step, Eq. (6.44), representing a system of $L - 1$ equations, is solved using an algorithm such as the tridiagonal algorithm.

The Crank-Nicolson method requires more computation per time step than the backward difference method of Eq. (6.40), because evaluation of the right side of Eq. (6.44) requires more computation than evaluating the right side of Eq. (6.40). However, a larger value of (Δt) can be used in conjunction with the Crank-Nicolson method, as it is second-order correct. Therefore, a given elapsed time can be traversed in a fewer number of time steps by the Crank-Nicolson method.

Both the backward difference method and the Crank-Nicolson methods belong to a general family of difference schemes obtained by approximating $\partial^2 u/\partial x^2$ as a weighted average of second central differences at time levels j and $j + 1$. That is, Eq.

(6.1) is approximated as

$$\sigma\{\theta\delta_x^2 u_{l,j+1} + (1-\theta)\delta_x^2 u_{l,j}\} = \frac{u_{l,j+1} - u_{l,j}}{(\Delta t)} \tag{6.45}$$

where θ is a real constant in $0 \leq \theta \leq 1$.

When $\theta = 0$, this scheme becomes explicit, and Eq. (6.45) reduces to Eq. (6.2). When $\theta \neq 0$, the scheme becomes implicit, and for $\theta = 1$, Eq. (6.45) reduces to Eq. (6.39). For $\theta = \frac{1}{2}$, the equation reduces to the Crank-Nicolson method.

Equation (6.45) can be analyzed for stability by using a Fourier mode. Set $u(x, t) = \hat{u}(t) \exp(ikx)$ as before; Eq. (6.45), after some rearranging, becomes

$$\hat{u}_{j+1} = \hat{u}_j - r\theta(1 - \cos k(\Delta x))\hat{u}_{j+1} - r(1-\theta)(1 - \cos k(\Delta x))\hat{u}_j \tag{6.46}$$

Therefore, the amplification factor satisfies the equation

$$g = \frac{1 - 2r(1-\theta)(1 - \cos k(\Delta x))}{1 + 2r\theta(1 - \cos k(\Delta x))} \tag{6.47}$$

Inspection of Eq. (6.47) reveals that g is always real for all real k, and its value never exceeds $+1$. A plot of g as a function of $\xi = 2r(1 - \cos k(\Delta x))$ is shown in Figure 6.6 for various values of θ. If $1/2 \leq \theta \leq 1$, the value of g is always greater than -1; hence $|g| \leq 1$ and the difference scheme is unconditionally stable. If $0 \leq \theta \leq 1/2$, for stability, the vaue of ξ must be restricted by the value at which the curve intersects the line $g = -1$. Consequently, Eq. (6.45) is stable if

$$r = \frac{\sigma\,\Delta t}{(\Delta x)^2} \leq \frac{1}{2(1-2\theta)} \qquad \text{for } 0 \leq \theta < 1/2 \tag{6.48}$$

$$\text{no restriction} \qquad \text{for } 1/2 \leq \theta \leq 1$$

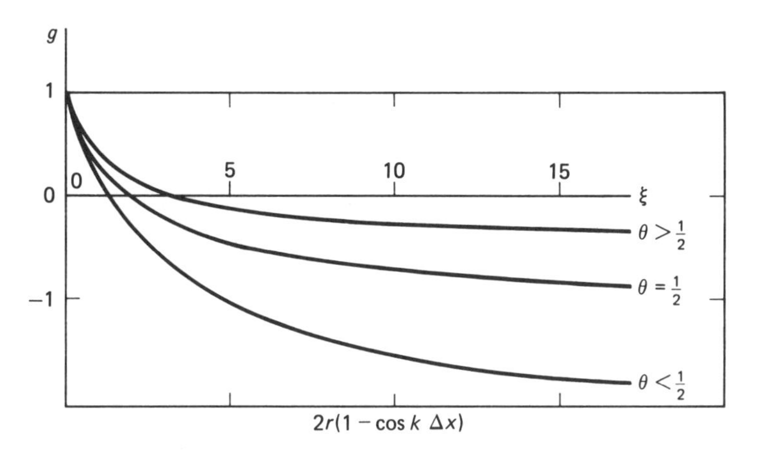

Figure 6.6 Growth factor g for the implicit difference equation (6.45) (Reproduced with permission from *Difference Methods for Initial Value Problems*, 2nd ed., by R. D. Richtmyer and K. W. Morton, Interscience Publishers, New York, © John Wiley & Sons)

The sufficiency of this condition can also be established by the energy method. This is left as an exercise.

DUFORT-FRANKEL METHOD

One of the most difficult decisions to make when solving initial value problems is whether to use an explicit or an implicit method. Explicit methods are easy to implement but are generally conditionally stable. Implicit methods, on the other hand, are generally always stable but require more involved computations. The Dufort-Frankel method is exceptional for being both explicit and unconditionally stable. These two desirable properties are achieved, however, at the expense of consistency, as will be seen presently.

The Dufort-Frankel scheme is different in the sense it employs three time levels. Equation (6.1) under this approximation becomes

$$\frac{\sigma(\Delta t)}{(\Delta x)^2}\{u_{l+1,j} - (u_{l,j+1} + u_{l,j-1}) + u_{l-1,j}\} = \frac{u_{l,j+1} - u_{l,j-1}}{2(\Delta t)} \tag{6.49}$$

This is readily obtained by replacing the $u_{l,j}$ term of the second central difference by the time average, $(u_{l,j+1} + u_{l,j-1})/2$, and the grid point (l, j) is not considered at all. Solving for $u_{l,j+1}$,

$$u_{l,j+1} = \left(\frac{1-2r}{1+2r}\right)u_{l,j-1} + \frac{2r}{1+2r}(u_{l+1,j} + u_{l-1,j}) \tag{6.50}$$

For this three-level formula, the amplification factor is obtained by solving the quadratic

$$g^2 - \frac{4r}{1+2r}\cos k(\Delta x)g - \frac{1-2r}{1+2r} = 0$$

or

$$g = \frac{1}{1+2r}\{2r\cos k(\Delta x) \pm \sqrt{(1 - 4r^2\sin^2 k(\Delta x))}\} \tag{6.51}$$

Case 1. If $4r^2 \sin^2 k(\Delta x) \leq 1$, which corresponds to small values of (Δt), the amplification factor is real and smaller than unity. Therefore, the Dufort-Frankel method is stable.

Case 2. If $4r^2 \sin^2 k(\Delta x) > 1$, which corresponds to large time steps, the amplification factor becomes complex and its magnitude is given by

$$|g| = \frac{1-2r}{1+2r} \tag{6.52}$$

which is also less than unity.

Therefore, for any value of (Δt) and for all k, the magnitude of g is less than unity. The complex nature of g tends to produce slight oscillations in the solution when the Dufort-Frankel method is used with large time steps. Though stable, this method is not very accurate when large time steps are used.

To examine the consistency of this scheme, each term in Eq. (6.49) is replaced by a Taylor series expansion, such as

$$u_{l,j+1} = u_{l,j} + (\Delta t)\left(\frac{\partial u}{\partial t}\right)_{l,j} + \cdots$$

Collecting terms and comparing with the original differential equation, an expression for the truncation error is obtained:

$$\frac{u_{l,j+1} - u_{l,j-1}}{2(\Delta t)} - \sigma \frac{u_{l+1,j} - u_{l,j+1} - u_{l,j-1} + u_{l-1,j}}{(\Delta x)^2} - \left(\frac{\partial u}{\partial t} - \sigma \frac{\partial^2 u}{\partial x^2}\right)_{l,j}$$
$$= \sigma\left(\frac{\Delta t}{\Delta x}\right)^2 \left(\frac{\partial^2 u}{\partial t^2}\right)_{l,j} + \mathrm{O}(\Delta t)^2 + \mathrm{O}(\Delta x)^2 + \mathrm{O}\frac{(\Delta t)^4}{(\Delta x)^2} \quad (6.53)$$

Consistency requires that $\Delta t/\Delta x \longrightarrow 0$ as $t \longrightarrow 0$. That is, Eq. (6.49) is consistent with Eq. (6.1) if and only if $\Delta t \longrightarrow 0$ faster than $\Delta x \longrightarrow 0$. If $\Delta t/\Delta x$ is kept fixed, say equal to M, then the first term of the right side of Eq. (6.53) would tend to $\sigma M \partial^2 u/\partial t^2$, rather than to zero. Then Eq. (6.49) would be consistent with the hyperbolic equation

$$\frac{\partial u}{\partial t} - \sigma \frac{\partial^2 u}{\partial x^2} + \sigma M^2 \frac{\partial^2 u}{\partial t^2}$$

rather than with the given parabolic equation in Eq. (6.1).

The term ***flexibility*** is sometimes used in the place of consistency.

OTHER METHODS

A number of other methods available for the treatment of parabolic equations are not treated here because of space limitations. However, some of the highlights of the more important methods are summarized in Table 6.5.

Table 6.5. Finite difference approximations to $\dfrac{\partial u}{\partial t} = \sigma \dfrac{\partial^2 u}{\partial x^2}$, σ = constant > 0

1. An explicit method 	1. $\dfrac{u_{l,j+1} - u_{l,j}}{\Delta t} = \sigma \dfrac{(\delta^2 u)_{l,j}}{\Delta x^2}$ $e = \mathrm{O}(\Delta t) + \mathrm{O}[(\Delta x)^2]$. Stable is $\sigma\, \Delta t/(\Delta x)^2$ = const. $\leq \frac{1}{2}$ as Δt, $\Delta x \longrightarrow 0$.
2. Crank-Nicolson method 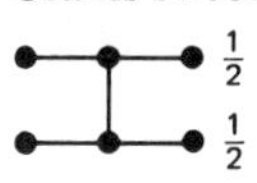	2. $\dfrac{u_{l,j+1} - u_{l,j}}{\Delta t} = \sigma \dfrac{(\delta^2 u)_{l,j} + (\delta^2 u)_{l,j+1}}{2(\Delta x)^2}$ $e = \mathrm{O}[(\Delta t)^2] + \mathrm{O}[(\Delta x)^2]$. Implicit, always stable.
3. Laasonen's method 	3. $\dfrac{u_{l,j+1} - u_{l,j}}{\Delta t} = \dfrac{(\delta^2 u)_{l,j+1}}{\Delta x}$ $e = \mathrm{O}[\Delta t] + \mathrm{O}[(\Delta x)^2]$. Implicit, always stable.

Table 6.5. (continued)

4. A special explicit method 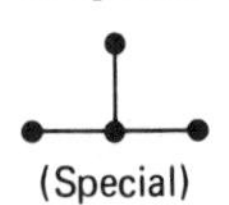	4. Same as 1, but with $\sigma\,\Delta t/(\Delta x)^2 = \frac{1}{6}$. $e = O[(\Delta t)^2] = O[(\Delta x)]$. Special case of 1, stable.
5. θ $1-\theta$	5. $\dfrac{u_{l,j+1}-u_{l,j}}{\Delta t} = \dfrac{\theta(\delta^2 u)_{l,j+1} + (1-\theta)(\delta^2 u)_{l,j}}{(\Delta x)^2}$ where $\theta = \text{const.}, 0 \leq \theta \leq 1$. $e = O(\Delta t) + O[(\Delta x)^2]$. For $0 \leq \theta \leq \frac{1}{2}$, stable if $\sigma\,\Delta t/(\Delta x)^2 = \text{const.} \leq 1/(2-4\theta)$; for $\frac{1}{2} \leq \theta \leq 1$, always stable. Includes 1, . . . , 4 as special cases.
6. $\frac{1}{2} - (\Delta x)^2/12\sigma\,\Delta t$ $\frac{1}{2} + (\Delta x)^2/12\sigma\,\Delta t$	6. Same as 5, but with $\theta = \frac{1}{2} - (\Delta x)^2/12\sigma\,\Delta t$. $e = O[(\Delta t)^2] + O[(\Delta x)^4]$. Stable.
7.	7. $\dfrac{u_{l,j+1}-u_{l,j-1}}{2\,\Delta t} = \sigma\dfrac{(\delta^2 u)_{l,j}}{(\Delta x)}$ Always unstable.
8. Dufort-Frankel method 	8. $\dfrac{u_{l,j+1}-u_{l,j-1}}{2\,\Delta t} = \sigma\dfrac{u_{l+1,j} - u_{l,j+1} - u_{l,j-1} + u_{l-1,j}}{(\Delta x)^2}$ where $\Delta t/\Delta x \longrightarrow 0$ as $\Delta t, \Delta x \longrightarrow 0$. $e = O[(\Delta t)^2] + O[(\Delta x)^2] + O\left[\left(\dfrac{\Delta t}{\Delta u}\right)^2\right]$. Explicit, always stable.
9. $\frac{3}{2}$ $-\frac{1}{2}$	9. $\dfrac{3}{2}\dfrac{u_{l,j+1}-u_{l,j}}{\Delta t} - \dfrac{1}{2}\dfrac{u_{l,j}-u_{l,j-1}}{\Delta t} = \sigma\dfrac{(\delta^2 u)_{l,j+1}}{(\Delta x)}$ $e = O[(\Delta t)^2] + O[(\Delta x)^2]$. Always stable.
10. $1+\theta$ $-\theta$	10. $(1+\theta)\dfrac{u_{l,j+1}-u_{l,j}}{\Delta t} - \theta\dfrac{u_{l,j}-u_{l,j-1}}{\Delta t} = \sigma\dfrac{(\delta^2 u)_{l,j+1}}{(\Delta x)^2}$ where $\theta = \text{const.} \geq 0$, $\sigma\,\Delta t/(\Delta x)^2 = \text{const.}$ $e = O[\Delta t] + O[(\Delta x)^2]$. Always stable. Contains 3, 9 as special cases.
11. $\frac{3}{2} - (\Delta x)^2/12\sigma\,\Delta t$ $-\frac{1}{2} + (\Delta x)^2/12\sigma\,\Delta t$	11. Same as 10 but with $\theta = \dfrac{1}{2} + \dfrac{(\Delta x)^2}{12\sigma\,\Delta t}$ $e = O[(\Delta t)^2] = O[(\Delta x)^2]$. Always stable.

Table 6.5. (continued)

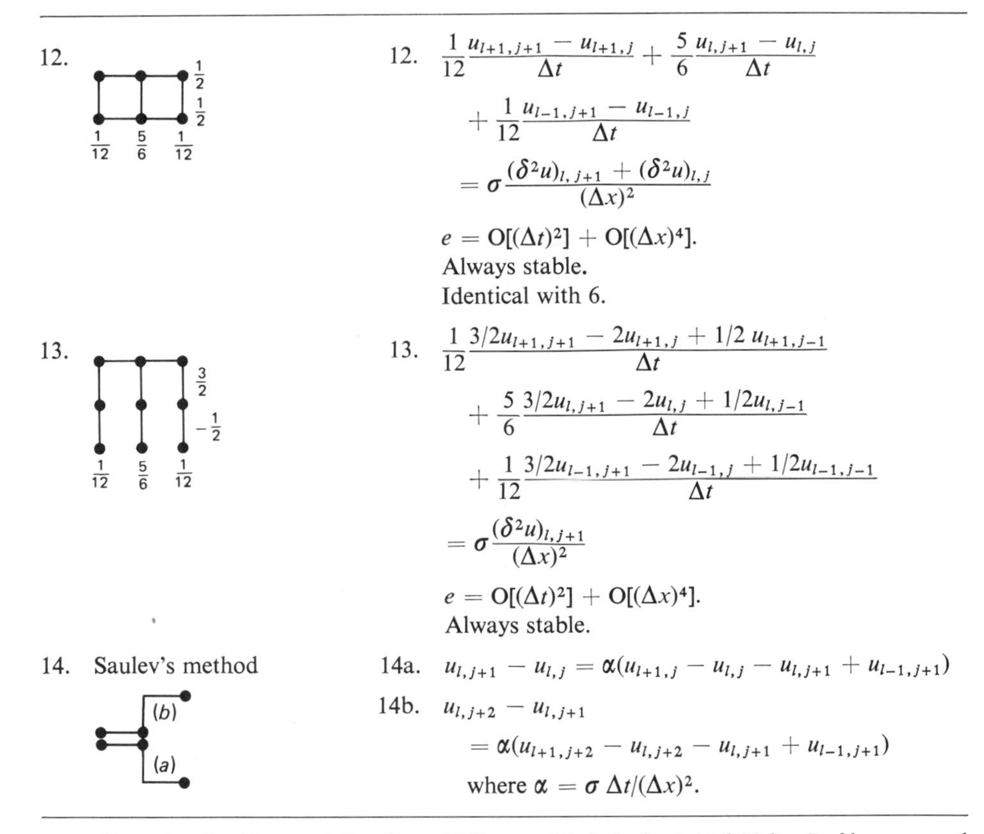

12. (stencil weights: $\frac{1}{12}$, $\frac{5}{6}$, $\frac{1}{12}$; $\frac{1}{2}$, $\frac{1}{2}$)	12. $\frac{1}{12}\frac{u_{l+1,j+1}-u_{l+1,j}}{\Delta t}+\frac{5}{6}\frac{u_{l,j+1}-u_{l,j}}{\Delta t}+\frac{1}{12}\frac{u_{l-1,j+1}-u_{l-1,j}}{\Delta t}=\sigma\frac{(\delta^2 u)_{l,j+1}+(\delta^2 u)_{l,j}}{(\Delta x)^2}$ $e = O[(\Delta t)^2] + O[(\Delta x)^4]$. Always stable. Identical with 6.
13. (stencil weights: $\frac{1}{12}$, $\frac{5}{6}$, $\frac{1}{12}$; $\frac{3}{2}$, $-\frac{1}{2}$)	13. $\frac{1}{12}\frac{3/2u_{l+1,j+1}-2u_{l+1,j}+1/2\,u_{l+1,j-1}}{\Delta t}+\frac{5}{6}\frac{3/2u_{l,j+1}-2u_{l,j}+1/2u_{l,j-1}}{\Delta t}+\frac{1}{12}\frac{3/2u_{l-1,j+1}-2u_{l-1,j}+1/2u_{l-1,j-1}}{\Delta t}=\sigma\frac{(\delta^2 u)_{l,j+1}}{(\Delta x)^2}$ $e = O[(\Delta t)^2] + O[(\Delta x)^4]$. Always stable.
14. Saulev's method ((b), (a))	14a. $u_{l,j+1}-u_{l,j}=\alpha(u_{l+1,j}-u_{l,j}-u_{l,j+1}+u_{l-1,j+1})$ 14b. $u_{l,j+2}-u_{l,j+1}=\alpha(u_{l+1,j+2}-u_{l,j+2}-u_{l,j+1}+u_{l-1,j+1})$ where $\alpha = \sigma\,\Delta t/(\Delta x)^2$.

HYPERBOLIC EQUATIONS

In the solution of hyperbolic systems in two independent variables, say x and t, the ***characteristic curves*** representing characteristic directions play an important role. A finite difference approach that utilizes characteristics consists of integration over a computational grid composed of arcs that are approximations to the characteristic curves of the system. Since the characteristic curves are not generally known in advance but must be obtained as a part of the solution, an approach can be used in which the computational grid is built up as the computation progresses.

The solution of hyperbolic systems in three or more independent variables by the method of characteristics is quite complicated as no general theory exists in this case. The major difficulty in using the method of characteristics in higher dimensions is the geometrical problem of locating, a priori, unknown characteristic surfaces on which the solution may be discontinuous. However, when successful, the characteristic method can be made quite accurate with relatively few grid points since the discontinuities are treated properly.

If the characteristic (discontinuity) surfaces of a hyperbolic system are not handled properly in a finite difference calculation, numerical errors can propagate in such a way as to make the calculation worthless, or in fact to cause numerical instability. An alternative approach to the characteristic methods is to recognize that the surfaces of discontinuity of the hyperbolic system represent, in fact, physical phenomena, such as shocks. These shocks can be viewed as regions of rapid but continuous change in the solution. The equations of the problem are then slightly modified so that the new system of equations has continuous solutions with large gradients in the vicinity of the characteristic (or discontinuity) surfaces. One version of this approach is the "pseudoviscocity" method. Another version of this approach, but somewhat less physical in nature, is to write the difference equations such that a "numerical" viscosity is introduced. This idea is indeed the source of stability in the Lax method, Lax-Wendroff method, donor-cell method, particle-in-cell (PIC) method, and so forth. As some of these methods will be discussed at length, no further discussion of the characteristic method appears hereafter.

As a majority of hyperbolic equations of significant practical interest occur while describing fluid flow phenomena, discussion here is primarily in terms of fluid flow terminology. When fluid flow is described, two distinctly different formulations are popular: the Eulerian and the Lagrangian. In an ***Eulerian formulation***, the spatial grid remains fixed with respect to time and the fluid moves relative to it. The finite difference equation for such a formulation requires an equation accounting for material flow into and out of each zone. In a ***Lagrangian formulation***, the coordinate system moves with the mass particles; thus, the mass in each zone remains constant, and we then describe the movement of the Lagrangian coordinate system. The Lagrangian approach has the advantage of providing higher-order accuracy at the least cost in computer time. However, since the finite difference grid follows the flow of the material, the grid can become hopelessly distorted for turbulent flow. The calculations can yield meaningless results under such circumstances, or the grid can even crash. The Eulerian approach can handle turbulent flow more readily, since the finite difference grid remains fixed, and material is transported through it. However, it is the more expensive approach. Thus, the advantage of the Lagrangian grid, which is clearly evident in one-space dimensional problems, quickly dwindles in higher-dimensional problems. Therefore, only Eulerian grids are considered, unless otherwise stated explicitly.

6.5. EXPLICIT METHODS FOR THE ADVECTIVE EQUATION

The basic computational ideas involved in solving initial-value problems associated with hyperbolic partial differential equations are introduced in this section. The major ideas are introduced using a simple, one-dimensional advective equation

$$v\frac{\partial u}{\partial x} + \frac{\partial u}{\partial t} = 0 \tag{6.54}$$

The advective equation is related to the wave equation and arises when properties of a fluid are advected (or convected) by the fluid. Equations (2.63) and (2.64), characterizing electrical transmission lines, are analogous to Eq. (6.54).

A finite difference analog to Eq. (6.54) can be obtained by replacing the spatial derivative by a central difference quotient and time derivative by a forward difference. Then Eq. (6.54) reduces to

$$v\left(\frac{u_{l+1,j} - u_{l-1,j}}{2(\Delta x)}\right) + \frac{u_{l,j+1} - u_{l,j}}{(\Delta t)} = 0 \tag{6.55}$$

or

$$u_{l,j+1} = u_{l,j} - \frac{v(\Delta t)}{2(\Delta x)}(u_{l+1,j} - u_{l-1,j}) \tag{6.56}$$

A mnemonic for this scheme is shown in Figure 6.7. It can be readily shown that the

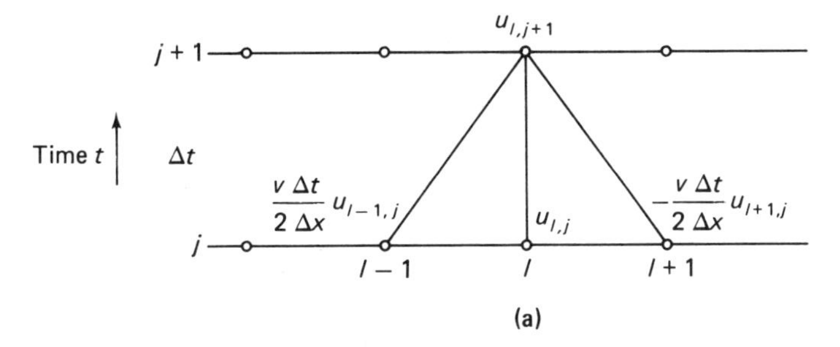

(a)

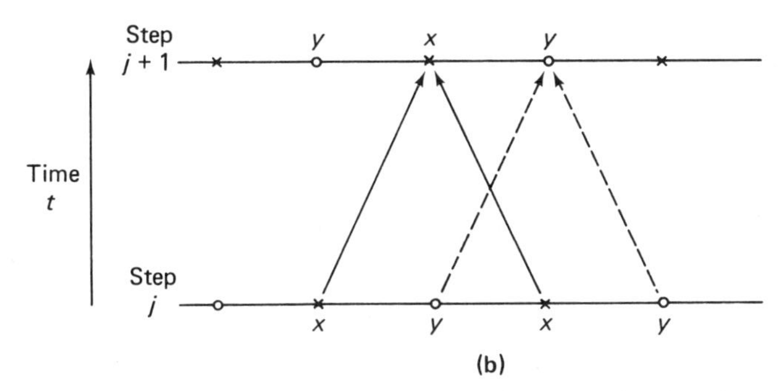

(b)

Figure 6.7 An unstable first-order scheme, Eq. (6.56), for the advective equation: (a) the time derivative is evaluated between the points $(l, j+1)$ and (l, j), while the space derivative is evaluated between the points $(l+1, j)$ and $(l-1, j)$; (b) the scheme is unstable since there exist two interlocking uncoupled meshes, in the sense that advection on the mesh X occurs independently of advection on the mesh Y. (Reproduced with permission from *Computational Physics* by D. Potter, Copyright © 1967, John Wiley & Sons, London.)

amplification factor g for this scheme is

$$g = 1 - i\frac{v(\Delta t)}{(\Delta x)} \sin (k(\Delta x)) = 1 - i\alpha \tag{6.57}$$

which is complex. The magnitude of this amplification factor is

$$g = \sqrt{gg^*} = \sqrt{1 + \alpha^2} \tag{6.58}$$

which is always greater than unity. Therefore, the Von Neumann stability condition cannot be satisfied for any nonzero value of α and for any step size. Thus, the difference scheme represented by Eq. (6.55) or (6.56) is completely useless.

LAX SCHEME

A slight modification of Eq. (6.56) yields a stable, useful, and first-order scheme for the advective equation. This useful scheme is obtained by replacing $u_{l,j}$ in Eq. (6.56) by $\frac{1}{2}(u_{l+1,j} + u_{l-1,j})$:

$$u_{l,j+1} = \tfrac{1}{2}(u_{l+1,j} + u_{l-1,j}) - \frac{v(\Delta t)}{2(\Delta x)}(u_{l+1,j} - u_{l-1,j}) \tag{6.59}$$

This is one of the widely used schemes to solve hyperbolic equations. Stability of this scheme can be studied once again by determining the amplification factor for a Fourier mode. Setting $u = \hat{u}(t) \exp(ikx)$, Eq. (6.59) becomes

$$\hat{u}_{j+1} = \left\{\cos k(\Delta x) - i\frac{v(\Delta t)}{\Delta x} \sin k(\Delta x)\right\} \hat{u}_j \tag{6.60}$$

Therefore, the amplification factor is

$$g = \cos k(\Delta x) - i\frac{v(\Delta t)}{\Delta x} \sin k(\Delta x) \tag{6.61}$$

which is complex. Therefore,

$$|g|^2 = gg^* = 1 - \sin k(\Delta x)\left\{1 - \left[\frac{v(\Delta t)}{(\Delta x)}\right]^2\right\} \tag{6.62}$$

Thus, the stability condition is satisfied if, for all wave numbers k,

$$\frac{|v\,\Delta t|}{(\Delta x)} \leq 1$$

or

$$\Delta t \leq \frac{\Delta x}{|v|} \tag{6.63}$$

This condition is generally known as the ***Courant-Friedrichs-Lewy*** condition for hyperbolic equations. The preceding result can be given a physical interpretation. Here the time step is required to be smaller than the smallest characteristic physical time in the problem. For the advective equation, the time scale of interest is simply the time required for a point in the fluid to move over a specified distance, say (Δx).

Thus, the Courant-Friedrichs-Lewy condition ensures that the physical velocity is less than the lattice speed or mesh speed defined as $(\Delta x)/(\Delta t)$.

6.6. DISPERSION AND DIFFUSION ON A DIFFERENCE GRID

At this point it is useful to pose a few questions to gain better insight into the stability problem. Why is Eq. (6.56) unconditionally unstable? Why did a minor change, resulting in Eq. (6.59), render it conditionally stable? To understand what happened in going from Eq. (6.56) to Eq. (6.59), two different approaches will be considered.

The first approach is based on an inspection of the mnemonic scheme of the difference equation. In Figure 6.7a, the mnemonic scheme of the unstable Eq. (6.56) indicates that in order to calculate u_l value at time level $(j+1)$, three u values at time level j are required. As one steps to calculate u_{l+1}, as the same time level $j+1$, two of the three values, at the j-level, that participated in calculating u_l are still required. Such a grid is said to be ***coupled***. The case with Eq. (6.59) is different. The conditional stability of this equation is achieved at the expense of uncoupling this grid. This is shown in Figure 6.7b, where the grid points are labeled with two letters x and y. Inspection of this figure in conjunction with Eq. (6.59) reveals that the set of grid points labeled x that participate in the evaluation of Eq. (6.59) for a given l is completely disjoint from the set of points labeled y that paritcipate in evaluating the equation for $l+1$. That is, there are two uncoupled interlocking grids, and advection on the x-grid occurs independently of advection on the y-grid. Further implication of this uncoupling of the grid will become evident later.

The second approach is based on an inspection of the difference scheme more closely. The only difference between Eqs. (6.56) and (6.59) lies in the first term; $u_{l,j}$ was replaced by a spatial average. Is this procedure admissible? In other words, is Eq. (6.59) consistent with the original differential equation? To answer this question, Eq. (6.59) is rewritten by adding some terms on both sides of the equality sign and rearranging, as

$$\tfrac{1}{2}(u_{l,j+1} - u_{l,j-1}) + \tfrac{1}{2}(u_{l,j+1} - 2u_{l,j} + u_{l,j-1}) = \tfrac{1}{2}(u_{l+1,j} - 2u_{l,j} + u_{l-1,j}) - \frac{v(\Delta t)}{2(\Delta x)}(u_{l+1,j} - u_{l-1,j}) \tag{6.64}$$

Inspection of this equation reveals that it is indeed a difference approximation to

$$\frac{\partial u}{\partial t} + \frac{(\Delta t)}{2}\frac{\partial^2 u}{\partial t^2} - \frac{(\Delta x)^2}{2(\Delta t)}\frac{\partial^2 u}{\partial x^2} + v\frac{\partial u}{\partial x} = 0 \tag{6.65}$$

but not to the original advection equation. That is, Eq. (6.59) is consistent with Eq. (6.65) but not with Eq. (6.54). That is, the conditionally stable Lax method is really attempting to simulate the advective equation by solving Eq. (6.65). As Eq. (6.65) differs from the advective equation in the two middle terms, it is reasonable to assume that they are stabilizing the process.

6.7. CONSERVATION ON A DIFFERENCE GRID

It was seen in Chapter 2 that many of the partial differential equations of engineering and applied physics are derived by using the principle of conservation and are therefore called conservative. When the problem is discretized, there is no reason to believe that the resulting finite difference equations also satisfy conservation laws. To investigate this problem in depth, it is useful to first write the original partial differential equations in a conservative form.

For the Euler formulation, let $\mathbf{x} = (x, y, z)$ denote the position vector of a particle of a fluid in Cartesian coordinate system. The properties of the fluid are characterized by the density $\rho = \rho(\mathbf{x}, t)$, the pressure $p = p(\mathbf{x}, t)$, the internal energy per gram $\mathcal{E} = \mathcal{E}(\mathbf{x}, t)$, and fluid velocity $\mathbf{v} = \mathbf{v}(\mathbf{x}, t) = (v_x, v_y, v_z)^T$. Given an equation of state

$$\mathcal{E} = f(p, \rho) \tag{6.66}$$

The equation for the conservation of matter (sometimes called the continuity equation) is

$$\left(\frac{\partial}{\partial t} + \mathbf{v}\nabla\right)\rho = -\rho\nabla \cdot \mathbf{v} \tag{6.67a}$$

The equation of motion is

$$\rho\left(\frac{\partial}{\partial t} + \mathbf{v} \cdot \nabla\right)\mathbf{v} = -\nabla p \tag{6.67b}$$

The equation of energy is

$$\rho\left(\frac{\partial}{\partial t} + \mathbf{v} \cdot \nabla\right)\mathcal{E} = -p\nabla \cdot \mathbf{v} \tag{6.67c}$$

For simplicity, only one-dimensional forms of these equations are considered. There is no loss of generality, because in a problem with slab symmetry nothing depends on y or z and Eq. (6.67) reduces to

$$\left(\frac{\partial}{\partial t} + v_x\frac{\partial}{\partial x}\right)\rho = -\rho\frac{\partial v_x}{\partial x} \tag{6.68a}$$

$$\rho\left(\frac{\partial}{\partial t} + v_x\frac{\partial}{\partial x}\right)v_x = -\frac{\partial p}{\partial x} \tag{6.68b}$$

$$\rho\left(\frac{\partial}{\partial t} + v_x\frac{\partial}{\partial x}\right)\mathcal{E} = -p\frac{\partial v_x}{\partial x} \tag{6.68c}$$

To put these Eulerian equations in the conservative form, new dependent variables m and e are defined as

$$m = \rho v_x, \qquad e = \rho(\mathcal{E} + \tfrac{1}{2}v_x^2) \tag{6.69}$$

where ρ, m, and e are mass, momentum, and energy, respectively, per unit volume. Now the set of Eqs. (6.68) can be written, using vector notation, as

$$\frac{\partial \mathbf{u}}{\partial t} + \frac{\partial \mathbf{f}(\mathbf{u})}{\partial x} = 0 \tag{6.70}$$

where **u** and **f** are vectors defined as

$$\mathbf{u} = \begin{bmatrix} \rho \\ m \\ e \end{bmatrix}, \qquad \mathbf{f}(\mathbf{u}) = \begin{bmatrix} m \\ (m^2/\rho) + p \\ (e + p)m/\rho \end{bmatrix} \tag{6.71}$$

Equation (6.70) is said to be the ***conservation-law form*** of the equations of fluid dynamics (in one-space dimension). To present the discrete version of the conservative law, it is more descriptive to consider at least a two-space dimensional problem. For this purpose, Eq. (6.70) is rewritten in a more general form as

$$\frac{\partial u}{\partial t} + \nabla \cdot \mathbf{f} = 0 \tag{6.72}$$

which is defined over a region Ω with a boundary Γ. A finite difference grid is superimposed on this region as shown in Figure 6.8. Equation (6.72) can now be integrated over each space-time box of volume $\Delta A\, \Delta t$. For example, integrating over the "space-time volume" associated with the center cell C,

$$\int_{t_j}^{t_{j+1}} dt \iint_C \frac{\partial \mathbf{u}}{\partial t}\, dA = \int_{t_j}^{t_{j+1}} dt \iint_C \nabla \cdot \mathbf{f}\, dA \tag{6.73}$$

Performing the time-integration on the left side and using the divergence theorem on the right side, one gets

$$\iint_C \mathbf{u}_{j+1}\, dA - \iint_C \mathbf{u}_j\, dA = -\int_{t_j}^{t_{j+1}} dt \oint_C \mathbf{f} \cdot d\mathbf{l} \tag{6.74}$$

which is valid for the center cell. The center of this center cell C can be thought of as

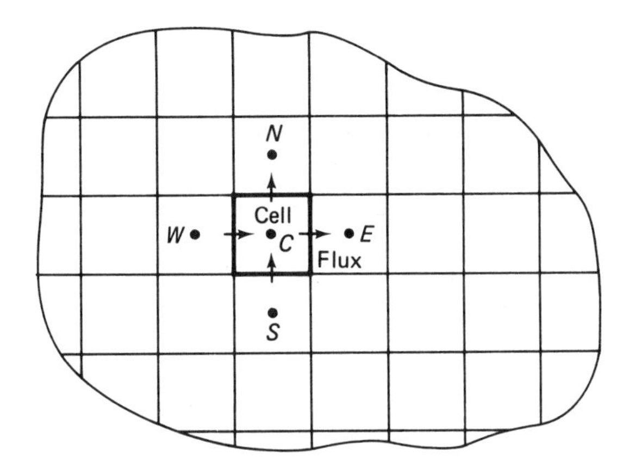

Figure 6.8 Conservation in two space dimensions on a difference mesh. The flux from cell C to cell E is exactly equal and opposite to the flux from cell E to cell C. (Reproduced with permission from *Computational Physics* by D. Potter, © 1967, John Wiley & Sons, London.)

representing the grid point (l, m). The following definitions with respect to cell C are self-explanatory.

$$\Delta A \mathbf{u}_{l,m,j} = \iint_C u_j \, dA \tag{6.75}$$

and the integral on the right side is

$$\sum_{\alpha} \mathbf{f}_{\alpha,l,m} = \oint_C \mathbf{f} \cdot d\mathbf{l} \tag{6.76}$$

where $\alpha = E, N, S, W$. Equation (6.75) essentially defines the total $\mathbf{u}$ (i.e., total density, total momentum, and total energy) associated with cell C. Equation (6.76) defines the total flux $\mathbf{f}$ associated with the cell C in terms of the fluxes of the surrounding four cells. Substituting Eqs. (6.75) and (6.76) in Eq. (6.74), one gets

$$\mathbf{u}_{l,m,j+1} = \mathbf{u}_{l,m,j} - \int_{t_j}^{t_{j+1}} dt \frac{1}{\Delta A} \sum_{\alpha} f_{\alpha,l,m} \tag{6.77}$$

or, in one space dimension,

$$\mathbf{u}_{l,j+1} = \mathbf{u}_{l,j} - \int_{t_j}^{t_{j+1}} dt \frac{1}{(\Delta x)} \sum f_{\alpha,l} \tag{6.78}$$

Equations (6.77) and (6.78) are said to be conservative.

6.8. CONSERVATIVE METHODS FOR HYPERBOLIC EQUATIONS

Unlike the case with parabolic equations, explicit methods, though conditionally stable, are more popular in solving hyperbolic equations than implicit methods. For hyperbolic equations, the suitability of implicit methods is not always clear. The success of implicit methods depends somewhat on the properties of the particular solution and how the equations are formulated. Therefore, only explicit methods will be described in the sequel.

CONSERVATIVE LAX METHOD

The Lax scheme, defined by Eq. (6.59) for solving the advective equation, in conservative form can be written as

$$\mathbf{u}_{l,j+1} = \tfrac{1}{2}(\mathbf{u}_{l+1,j} + \mathbf{u}_{l-1,j}) - \frac{\Delta t}{2(\Delta x)}(\mathbf{f}_{l+1,j} - \mathbf{f}_{l-1,j}) \tag{6.79}$$

It is useful to pause here and compare this equation with Eqs. (6.59) and (6.78). If $\mathbf{f}$ is replaced by $v\mathbf{u}$, then Eq. (6.79) is identical to the Lax's scheme in Eq. (6.59). However, Eq. (6.79) differs from Eq. (6.78) in the first term. The first term of Eq. (6.78) is identical to the first term in the unstable scheme described in Eq. (6.56). As before, Eq. (6.79) is only first order correct, and the stability condition is

$$\Delta t \leq \frac{(\Delta x)}{v}$$

CONSERVATIVE LEAPFROG METHOD

To get higher-order accuracy in (Δt), a three-level formula can be used in the t-direction. To go from the $(j-1)$ level to the $(j+1)$ level, the fluxes $\mathbf{f}$ are first defined at the intermediate level as

$$\mathbf{f}_{l,j} = \mathbf{f}(\mathbf{u}_{l,j}) \tag{6.80a}$$

These flux values can be used to calculate $\mathbf{u}$ at the $(j+1)$ level using

$$\mathbf{u}_{l,j+1} = \mathbf{u}_{l,j-1} - \frac{\Delta t}{(\Delta x)}(\mathbf{f}_{l+1,j} - \mathbf{f}_{l-1,j}) \tag{6.80b}$$

These $\mathbf{u}$ values at the $(j+1)$ level will now be used to calculate fluxes at the $(j+2)$ level, which will in turn be used to evaluate $\mathbf{u}$ values at the $(j+3)$ level. A mnemonic scheme for this "leapfrog" scheme is shown in Figure 6.9. An expression for the amplification factor for this leapfrog scheme can be easily derived to be

$$g = i\alpha \pm \sqrt{1-\alpha^2} \tag{6.81}$$

where $\alpha = (\Delta t/\Delta x)v \sin k(\Delta x)$. Therefore,

$$|g| = 1 \qquad \text{if } |\alpha| \leq 1 \tag{6.82}$$

Thus, for stability the condition on the size of (Δt) is

$$\Delta t \leq \frac{\Delta x}{v} \tag{6.83}$$

for all wave numbers.

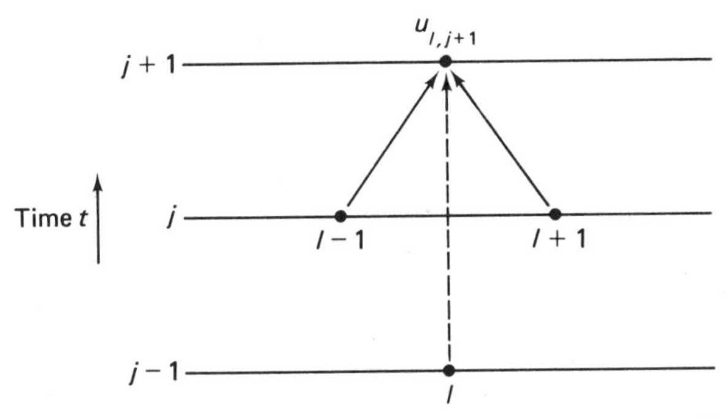

Figure 6.9 Computational stencil for the leapfrog scheme. Variables at the intermediate point (l, j) are not defined

It is of significant interest to note that the two roots obtained for g correspond to the two uncoupled grids in the leapfrog scheme. Although both roots can provide stable solutions under certain conditions, the two solutions may drift apart. The usual remedial measure to prevent this drifting is either to define the dependent variables only on one grid or to employ a small diffusion term to couple the two grids.

TWO-STEP LAX-WENDROFF SCHEME

An extremely useful difference scheme that is second-order accurate in time can be obtained by centering the time step. This is normally done in two steps. First, temporary or intermediate values of the dependent variables are defined at the half time steps, that is, $t_{j+1/2}$. An auxiliary calculation is performed at this step using Lax's method. This auxiliary step is

$$\mathbf{u}_{l+1/2,j+1/2} = \tfrac{1}{2}(\mathbf{u}_{l,j} + \mathbf{u}_{l+1,j}) - \frac{\Delta t}{2(\Delta x)}(\mathbf{f}_{l+1,j} - \mathbf{f}_{l,j}) \tag{6.84}$$

The **u**'s obtained from this step are used to define fluxes at the intermediate grid points,

$$\mathbf{f}_{l+1/2,j+1/2} = \mathbf{f}(\mathbf{u}_{l+1/2,\, j+1/2}) \tag{6.85}$$

Now, the main calculations can be performed using

$$\mathbf{u}_{l,j+1} = \mathbf{u}_{l,j} - \frac{(\Delta t)}{(\Delta x)}(\mathbf{f}_{l+1/2,\, j+1/2} - \mathbf{f}_{l-1/2,\, j+1/2}) \tag{6.86}$$

A mnemonic for this procedure is shown in Figure 6.10. Notice that the intermediate values of the dependent variables are discarded and play no part in the solution process after each step is advanced. The magnitude of the amplification factor for this scheme can be verified to be

$$gg^* = 1 - \alpha^2(1 - \alpha^2)\{1 - \cos{(k\,\Delta x)}\}^2$$

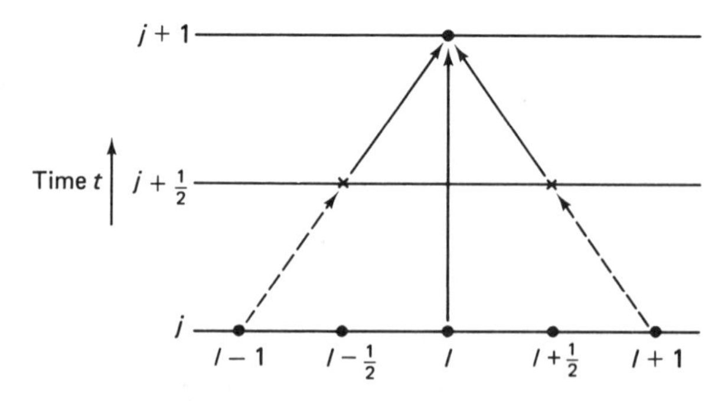

Figure 6.10 Computational stencil for the two-step Lax-Wendroff scheme. Notice the use of temporary points at the half time step $j + \frac{1}{2}$. These points are used to time center the integration for second-order accuracy

Therefore, $|g| \leq 1$ if $\alpha^2 \leq 1$ or if

$$\Delta t \leq \frac{\Delta x}{|v|} \tag{6.87}$$

which is the Courant-Friedrichs-Lewy condition.

QUASI-SECOND-ORDER METHOD

A particularly simple, second-order accurate method can be obtained by using Taylor series expansion and writing $\mathbf{u}_{j+1}$ as

$$\mathbf{u}_{j+1} = \mathbf{u}_j + \left(\frac{\partial \mathbf{u}}{\partial t}\right)_j (\Delta t) + \left(\frac{\partial^2 \mathbf{u}}{\partial t^2}\right)\frac{(\Delta t)^2}{2!} + \cdots$$

which can be rewritten as

$$\mathbf{u}_{j+1} \simeq \mathbf{u}_j + \left(\frac{\partial \mathbf{u}}{\partial t}\right)_j (\Delta t) + \left[\left(\frac{\partial \mathbf{u}}{\partial t}\right)_j - \left(\frac{\partial \mathbf{u}}{\partial t}\right)_{j-1}\right]\frac{(\Delta t)}{2}$$

Applying this approximation to Eq. (6.70)

$$\begin{aligned}\mathbf{u}_{l,j+1} = \mathbf{u}_{l,j} &- \left(\frac{3}{2} + \epsilon\right)\frac{\Delta t}{2(\Delta x)}(\mathbf{f}_{l+1,j} - \mathbf{f}_{l-1,j}) \\ &+ \left(\frac{1}{2} + \epsilon\right)\frac{\Delta t}{2(\Delta x)}(\mathbf{f}_{l+1,j-1} - \mathbf{f}_{l-1,j-1})\end{aligned} \tag{6.88}$$

where ϵ is a small number. Just as in the Lax-Wendroff scheme, a three-level formula has been used here to obtain a second-order accuracy. However, the fluxes, rather than $\mathbf{u}$'s, are used from the lowest time level. At each step in the calculation, the fluxes are calculated only once and stored directly for one time level instead of storing $\mathbf{u}$ at two time levels, as was done with the Lax-Wendroff scheme.

For the advective equation $F = vu$, the preceding scheme leads to an amplification factor

$$g^2 = g - (\tfrac{3}{2} + \epsilon)i\alpha g + (\tfrac{1}{2} + \epsilon)i\alpha \tag{6.89}$$

where $\alpha = \dfrac{\Delta t}{\Delta x}(v) \sin (k \Delta x)$

It can be shown that

$$|g| < 1 \qquad \text{if } \epsilon > \tfrac{1}{4}\alpha^2 + \tfrac{1}{2}\alpha^4 \quad \text{and} \quad |\alpha| \leq \tfrac{1}{2} \tag{6.90}$$

Thus, for small α, this method will be stable if ϵ is sufficiently large. For instance, if one chooses $\Delta t = (\Delta x)/2|v|$, then for stability one must choose $\epsilon \geq 3/32$. When $\epsilon = 0$, the method becomes second-order accurate.

6.9. METHOD OF LINES FOR INITIAL VALUE PROBLEMS

The method of lines, which was briefly introduced in Section 5.8, is a convenient technique to approximate the solution of initial value problems for systems of linear and nonlinear partial differential equations. The method, as described in this section, is capable of solving a wide variety of different types of linear as well as nonlinear PDEs. Although the following discussion is focused on parabolic and hyperbolic equations, the method is applicable also to elliptic systems provided one poses the elliptic boundary value problem as the stable time independent solution of a parabolic equation.

The idea behind the method of lines can be grasped by considering a general initial value problem in one space dimension and time dimension, such as

$$\frac{\partial u}{\partial t} = g(u), \qquad u(x, 0) = u(0) \qquad u(0, t) = u(1, t) \tag{6.91}$$

where g is, in general, a nonlinear parabolic or hyperbolic operator. The first step is to discretize the spatial variable and leave time continuous. This discrete-space, continuous-time (DSCT) approximation transforms Eq. (6.91) into

$$\frac{d\mathbf{U}}{dt} = \mathbf{G}(\mathbf{U}), \qquad \mathbf{U}(0) = \mathbf{U}_0 \tag{6.92}$$

where $\boldsymbol{U}$ represents a discrete approximation of u. In fact,

$$\mathbf{U} = \begin{bmatrix} u(x_0, t) \\ u(x_1, t) \\ \cdot \\ \cdot \\ \cdot \\ u(x_1, t) \\ \cdot \\ \cdot \\ \cdot \\ u(x_L, t) \end{bmatrix} = \begin{bmatrix} u_0 \\ u_1 \\ \cdot \\ \cdot \\ \cdot \\ u_1 \\ \cdot \\ \cdot \\ \cdot \\ u_L \end{bmatrix} \tag{6.93}$$

Now Eq. (6.92) is a system of ordinary differential equations (ODEs) posed as initial value problems (compare this to Section 5.8, where the method of lines resulted in a system of TPBVPs). These ordinary differential equations can be solved by using any standard integration method (e.g., Adams-Moulton method, Gear's method). Thus, the method of lines, among other things, allows one to use the powerful numerical techniques that are already available to solve ODEs.

Although the method at first sight appears simple, some practical difficulties do arise in its implementation. For example, the matrix $\mathbf{G}$, representing the discrete equivalent of g, tends to have both small and large eigenvalues. This makes Eq. (6.92) stiff, thus forcing one to use methods that are capable of handling stiff systems of ODEs. According to J. M. Hyman, there are two crucial aspects of the method of lines that determine the success of the method. The first involves a choice as to the method used to approximate the spatial derivatives and the manner of handling the boundary conditions. The second involves a choice as to the method used to solve the resulting ODEs. Regarding the choice of discretization scheme, Hyman recommends that the discrete model retain as closely as possible all the salient properties of the original PDE. That is, if the operator g in Eq. (6.91) is antisymmetric (as in the case with hyperbolic systems), then an attempt should be made to approximate it by an antisymmetric matrix $\mathbf{G}$ in Eq. (6.92). If g is dissipative (as is the case with parabolic systems), then an attempt should be made to approximate g by a dissipative matrix $\mathbf{G}$. Finally, when g is in conservation form, it is recommended that $\mathbf{G}$ also be in conservation form. In the process of the transformation from g to $\mathbf{G}$, it is recommended that one include the boundary conditions as restrictions on the domain of g, and therefore of $\mathbf{G}$. These ideas are now illustrated by considering an initial value problem for the hyperbolic equation

$$\begin{aligned} \frac{\partial u}{\partial t} &= c\frac{\partial u}{\partial x}, \quad 0 \leq x \leq 1 \\ u(x, 0) &= f(x) \end{aligned} \tag{6.94}$$

Case 1. Periodic Boundary Conditions. Suppose the boundary conditions are periodic, that is, $u(x = 0) = u(x = 1)$. To approximate Eq. (6.94) by the method of lines, we begin by replacing the spatial derivatives with, say, second-order centered differences; that is,

$$\left(\frac{\partial u}{\partial x}\right)_{x=x_l} \simeq \frac{u_{l+1}(x) - u_{l-1}(x)}{2h} \tag{6.95}$$

with $l = 0, 1, \ldots, L$ and $h = 1/(L + 1)$. This results in the system of ODEs

$$\frac{d\mathbf{U}}{dt} = \mathbf{AU} \tag{6.96}$$

where

$$\mathbf{U} = \begin{bmatrix} u_0 \\ u_1 \\ u_2 \\ \cdot \\ \cdot \\ \cdot \\ u_L \end{bmatrix} \quad \text{and} \quad \mathbf{A} = \frac{c}{2h}\begin{bmatrix} 0 & 1 & 0 & 0 \ldots\ldots\ldots 0 & -1 \\ -1 & 0 & 1 & 0 \ldots\ldots\ldots\ldots & 0 \\ 0 & -1 & 0 & 1 \quad 0 \ldots\ldots\ldots & 0 \\ \ldots & \ldots & \ldots & \ldots & \ldots \\ 1 & 0 \ldots\ldots\ldots\ldots & & -1 & 0 \end{bmatrix} \tag{6.97}$$

Notice how the periodic nature of the boundary conditions is utilized to incorporate the boundary conditions into the first and last rows of $\mathbf{A}$. The system in Eq. (6.96) can now be solved using a method such as the Runge-Kutta or the Adams-Moulton integration scheme.

Case 2. Dirichlet Boundary Conditions. Suppose the boundary conditions are specified as

$$\begin{aligned} u(0, t) &= g_0(t) \\ u(1, t) &= g_1(t) \end{aligned} \tag{6.98}$$

Let us also suppose that we are interested in a fourth-order approximation in approximating $\partial u/\partial x$ (rather than a second-order approximation as was done in Case 1).

Using five points, we can get a fourth-order difference approximation to $\partial u/\partial x$. If the five points are distributed symmetrically (see Exercises 4.9, through 4.11) then,

$$\left(\frac{\partial u}{\partial x}\right)_{x=x_l} = \frac{(u_{l-1} - 8u_l + 8u_{l+1} - u_{l+2})}{12h} \tag{6.99}$$

This formula is applicable at all points except those that are one mesh away from the boundary, and those on the boundary. This problem can be rectified using two fictitious grid points beyond the boundary and using a cubic extrapolation formula. Designating the fictitious grid points beyond the left boundary by subscripts -1 and -2 those beyond the right boundary by subscripts $L + 1$ and $L + 2$, we can write cubic extrapolation formulas (see Exericse 4.10) as

$$\begin{aligned} u_{-1} &= 4u_0 - 6u_1 + 4u_2 - u_3 \\ u_{-2} &= 4u_{-1} - 6u_0 + 4u_1 - u_2 \\ u_{L+1} &= 4u_L - 6u_{L-1} + 4u_{L-2} - u_{L-3} \\ u_{L+2} &= 4u_{L+1} - 6u_L + 4u_{L-1} - u_{L-2} \end{aligned} \tag{6.100}$$

Now a semidiscrete approximation to Eq. (6.91) can be written as

$$\frac{d\mathbf{U}}{dt} = \mathbf{A}\mathbf{U} \tag{6.101}$$

where $\mathbf{U} = [u_0, u_1, \ldots, u_N]^T$ and

$$\mathbf{A} = \frac{c}{12h}\begin{bmatrix} -22 & 36 & -18 & 4 & 0 & 0\ldots\ldots\ldots\ldots\ldots.0 & 0 \\ -4 & -6 & 12 & -2 & 0 & 0\ldots\ldots\ldots\ldots\ldots.0 & 0 \\ 1 & -8 & 0 & 8 & -1 & 0\ldots\ldots\ldots\ldots\ldots.0 & 0 \\ 0 & 1 & -8 & 0 & 8 & -1\ldots\ldots\ldots\ldots\ldots.0 & 0 \\ \cdots & \cdots & \cdots & \cdots & \cdots & \cdots & \cdots \\ \cdots & \cdots & \cdots & \cdots & \cdots & \cdots & \cdots \\ 0 & 0\ldots\ldots\ldots\ldots\ldots.1 & -8 & 0 & 8 & -1 \\ 0 & 0\ldots\ldots\ldots\ldots\ldots.0 & 2 & -12 & 6 & 4 \\ 0 & 0\ldots\ldots\ldots\ldots\ldots.0 & -4 & 18 & -36 & 22 \end{bmatrix} \tag{6.102}$$

Notice that $u(0, t) = g_0(t)$ and $u(1, t) = g_1(t)$ are defined in Eq. (6.98) as boundary conditions. They are included in the definition of $\mathbf{U}$ to simplify notation. If the specified boundary conditions are used to define the first and last entries of $\mathbf{U}$, the first and last rows of $\mathbf{A}$ play no role in the calculation; they are included here just to demonstrate the calculation. However, if Dirichlet conditions are not specified at the $x = 0$ boundary, say, then the first row of $\mathbf{A}$ in Eq. (6.102) would remain as shown.

Case 3. Mixed Boundary Conditions. If the boundary conditions contain spatial derivatives, as in

$$\alpha(t)u + \beta(t)\frac{\partial u}{\partial x} = \gamma(t) \qquad \text{at } x = x_0 \tag{6.103}$$

then we can replace $(\partial u/\partial x)$ by the third-order unsymmetric difference approximation

$$\left(\frac{\partial u}{dx}\right)_{x=x_0} = \frac{1}{6h}(-2u_{-1} - 3u_0 + 6u_1 - u_2) \tag{6.104}$$

and solve the resulting equation for u_{-1},

$$\frac{3h}{\beta(t)}[\alpha(t)u_0 - \gamma(t)] - \frac{1}{2}[3u_0 - 6u_1 + u_2]$$

SUGGESTIONS FOR FURTHER READING

1. *An excellent book on initial value problems with a wealth of material on theory and practice is*

   R. D. Richtmyer and K. W. Morton, *Difference Methods for Initial Value Problems*, 2nd ed., John Wiley & Sons, Inc. (Interscience Division), New York, 1967.

2. *The method of stability analysis using Fourier modes is presented at an elementary level in*

   D. Potter, *Computational Physics*, John Wiley & Sons, Inc., New York, 1973.

3. *For a book devoted exclusively to diffusion phenomena with a lot of practical insights, but not much on finite difference methods, see*

   J. Crank, *The Mathematics of Diffusion*, 2nd ed., Oxford University Press, Inc., New York, 1975.

4. *An elementary treatment of parabolic and hyperbolic equations with some computer programs can be found in*

   C. F. Gerald, *Applied Numerical Analysis*, 2nd ed., Addison-Wesley Publishing Co., Reading, Mass., 1978.

5. *The method of lines, although quite popular in the Soviet Union for a long time, is gaining ground in the West only in recent times. Useful references to this area include*

   O. A. Liskovets, "The Method of Lines," (review), English translation appeared in *Differential Equations*, Vol. 1, pp. 1308–1323, 1965.

   R. A. Willoughby (ed.), *Stiff Differential Systems*, Plenum Publishing Corp., New York, 1974.

   N. K. Madsen and R. F. Sincovec, *The Numerical Method of Lines for the Solution of Nonlinear Partial Differential Equations*, Lawrence Livermore Laboratory, UCRL-75142, September 1973; and

   J. M. Hyman, *The Method of Lines Solution of Partial Differential Equations*, Courant Institute of Mathematical Sciences, Mathematical and Computing Laboratory Report, COO-3077-139, 1976.

EXERCISES

1. Show that the effect of a discontinuity in the initial values decreases as t increases by reproducing Table 6.2 at the grid point $l = 5$. Pay special attention to the percentage error and discuss why it is behaving the way it is. (*Hint:* Notice the discontinuity in the initial value of $\partial u/\partial x$ at $l = 5$.)

2. It can be shown that the solution of the finite difference equations is very close to the solution of the partial differential equation if the initial condition and all its derivatives are continuous. Demonstrate the validity of this statement by solving

$$\frac{\partial^2 u}{\partial x^2} = \frac{\partial u}{\partial t}, \qquad 0 \leq x \leq 1$$

subject to the initial condition

$$u(x, 0) = \sin \pi x, \qquad 0 \leq x \leq 1$$

and the boundary conditions

$$\left.\begin{aligned} u(0, t) &= 0 \\ u(1, t) &= 0 \end{aligned}\right\} \quad \text{for } t > 0$$

using an explicit method with $\Delta x = 0.1$ and $r = 0.1$, and comparing this finite difference solution with the exact solution

$$u(x, t) = e^{-\pi^2 t} \sin \pi x$$

3. Consider

$$\sigma \frac{\partial^2 u}{\partial t^2} = \frac{\partial u}{\partial t}, \qquad 0 \leq x \leq \pi, \qquad t \geq 0$$

where $\sigma > 0$ is assumed to be a constant. The initial and boundary conditions are specified as

$$u(x, 0) = f(x), \qquad u(0, t) = u(\pi, t) = 0, \qquad \text{for } t > 0$$

Now consider the explicit difference scheme

$$\frac{u_{l,j+1} - u_{l,j}}{(\Delta t)} = \sigma \frac{(u_{l,j+1} - 2u_{l,j} + u_{l,j-1})}{(\Delta x)^2}$$

where $u_{0,j} = f(x_j)$ and $u_{l,0} = u_{l,j\,\max} = 0$. We know that the general solution of the differential equation can be written as

$$u(x, t) = \sum_{k=-\infty}^{+\infty} A_k \exp(ikx - k^2 \sigma t)$$

where

$$A_k = \frac{1}{2\pi} \int_{-\pi}^{\pi} f(x) \exp(-ikx)\, dx$$

The technique of separation of variables can be used to solve both the differential equation and the difference equation. For instance, for the difference equation, one can try the form

$$u_{l,j} = \sum_{k=-\infty}^{+\infty} A_k \exp(ikj(\Delta x))(g_k)^l$$

where g_k is the amplification factor. Substitute the preceding into the difference equation and derive an expression for g_k and show that the solution of the difference scheme converges to that of the differential equation as $\Delta x \longrightarrow 0$ and $\Delta t \longrightarrow 0$. Show also that the condition for stability is $2\sigma(\Delta t)/(\Delta x)^2 < 1$.

4. Consider the problem of approximating

$$\frac{du}{dt} + \Lambda u = f, \qquad u = g \text{ for } t = 0$$

where Λ is a partial differential operator that does not depend upon t. Show that a variety of difference approximations to the preceding can be written in the very general form

$$u_{j+1} = Tu_j + (\Delta t)Sf_j$$

where the transition operator T and the source operator S are defined as follows.

(i) $T = (I - (\Delta t)\Lambda)$, $S = I$ for the simplest first-order explicit scheme.

(ii) $T = (I + (\Delta t)\Lambda)^{-1}$, $S = T$ for the simplest first-order implicit scheme.

(iii) $T = \left(I + \frac{\Delta t}{2}\Lambda\right)^{-1}\left(I - \frac{\Delta t}{2}\Lambda\right)$, $S = \left(I + \frac{\Delta t}{2}\Lambda\right)^{-1}$ for the Crank-Nicolson scheme.

5. An important role in analyzing algorithms is played by the Fourier expansions with respect to the eigenfunctions of operators and their adjoints. For instance, consider the eigenvalue problems, for $L \geq 0$,

$$L\phi = \lambda\phi, \qquad L^*\phi^* = \lambda^*\phi^*$$

where L^* is the adjoint of L. Let U be the domain of L and U^* the domain of L^*. Let us further assume that the preceding eigenvalue problems generate a complete set of eigenfunctions $\{\phi_n\}$ and $\{\phi_n^*\}$. Then, it is well known that arbitrary functions $f \in U$ and $f^* \in U^*$ can be represented in the form of Fourier series as

$$f = \sum_n f_n\phi_n, \qquad f^* = \sum_n f_n^*\phi_n^*$$

where

$$f_n = \langle f, \phi_n^*\rangle, \qquad f_n^* = \langle f^*, \phi_n\rangle$$

Now consider Problem 3 once again with the difference scheme

$$u_{j+1} = (I - (\Delta t)\Lambda)u_j + (\Delta t)f_j, \qquad u_0 = g$$

Assume that the operator $\Lambda > 0$ induces a complete set of eigenfunctions $\{\phi_n\}$ along with the corresponding eigenvalues $\lambda_n > 0$ with respect to the eigenvalue problem $\Lambda\phi = \lambda\phi$. Introduce the Fourier series as follows:

$$u_j = \sum_n u_{nj}\phi_n, \qquad f_j = \sum_n f_{nj}\phi_n, \qquad g = \sum_n g_n\phi_n$$

$$u_{nj} = \langle u_j, \phi_n^*\rangle, \qquad f_{nj} = \langle f_j, \phi_n^*\rangle, \qquad g_n = \langle g, \phi_n^*\rangle$$

where ϕ_n^* are the eigenfunctions of the adjoint eigenvalue problem.

(i) Now show that

$$u_{n,j+1} = (1 - (\Delta t)\lambda_n)u_{n,j} + (\Delta t)f_{n,j}, \qquad u_{n,0} = g_n$$

(ii) Solve (i) by successive elimination of the unknowns and show

$$u_{n,j} = r_n^j g_n + (\Delta t)\sum_{i=1}^{j} r_n^{j-i} f_{n,i-1}$$

where

$$r_n = 1 - (\Delta t)\lambda_n$$

(iii) For $(\Delta t) > 0$, show that

$$|u_{n,j}| \leq |r_n|^j|g_n| + \frac{1 - |r_n|^j}{1 - |r_n|}(\Delta t)|f_n|$$

(iv) According to Von Neumann, the difference scheme under study is stable if for every Fourier coefficient $u_{n,j}$, one has

$$|u_{n,j}| \leq C_{1n}|g_n| + C_{2n}|f_n|, \qquad n = 1, 2, \ldots$$

Show that this criterion is satisfied if $|r_n| < 1$ for $n = 1, 2, \ldots$.

(v) If the eigenvalues of Λ are contained in the interval

$$0 < \lambda_{\min} \leq \lambda_n \leq \lambda_{\max}$$

then a constructive condition for stability is

$$(\Delta t) < 2/\lambda_{\max}$$

6. Repeat all steps in Problem 5 for the implicit scheme described in Problem 3. In particular show that

$$r_n = \frac{1}{1 + (\Delta t)\lambda_n}$$

and that the difference scheme is absolutely stable provided all the eigenvalues of Λ are positive.

7. Repeat all the steps in Problem 4 for the Crank-Nicolson scheme in Problem 3.

7

Special Topics

7.1. INTRODUCTION

So far attention has been focused on relatively simple problems. While solving problems of practical interest, however, one encounters complications such as singularities, shocks, free boundaries, and nonlinearities. A thorough treatment of all these topics would occupy too much of space. The purpose of this chapter is to briefly present some practical aspects of these problems. Some topics, such as the treatment of shock phenomena, are not discussed at all.

7.2. SINGULARITIES

There is no convenient way to define the word "singularity"; therefore, a pragmatic approach is taken here. For example, there exists a class of problems in which one of the coefficients of the partial differential equation becomes singular at one or more points s of Ω. In such problems, the solution u will ordinarily have a ***singularity*** at s, and the standard methods will hardly be applicable in their original form.

In rectangular coordinates the most common type of singularity is the ***boundary singularity***. This type of singularity can be illustrated by considering the domain of integration of a parabolic problem. For example, consider the simple diffusion equation

$$\frac{\partial^2 u}{\partial x^2} = \frac{\partial u}{\partial t}, \qquad 0 \leq x \leq 1, \qquad t \geq 0 \tag{7.1a}$$

subject to the auxiliary conditions

$$u(x, 0) = f(x) = 0, \qquad 0 < x < 1 \tag{7.1b}$$

$$u(0, t) = g(t) = 1, \qquad t \geq 0 \tag{7.1c}$$

Notice that this set of auxiliary conditions is incompatible at the corners ($x = 0$, $t = 0$) and ($x = 1, t = 0$) of the domain of integration. That is, the solution $u(x, t)$ is discontinuous at the corners (0, 0) and (1, 0). Stated formally,

$$\lim_{x \to 0} u(x, t) \neq \lim_{t \to 0} u(x, t) \tag{7.2a}$$

$$\lim_{x \to 1} u(x, t) \neq \lim_{t \to 0} u(x, t) \tag{7.2b}$$

A weaker form of discontinuity can also occur. For example, if

$$u(x, 0) = f(x) = 1, \qquad 0 < x < 1 \tag{7.3a}$$

$$u(0, t) = g(t) = e^{-t}, \qquad t \geq 0 \tag{7.3b}$$

then there is no discontinuity at the corner (0, 0). However, a more subtle type of singularity exists at (0, 0) since $\partial u/\partial t \neq \partial^2 u/\partial x^2$ at $x = 0, t = 0$.

Finite difference techniques generally produce erroneous results in the vicinity of singular points. However, fortunately, these errors do not penetrate deep into the field of integration provided the methods used are stable. Nevertheless, the solution near the singularity is generally useless. In the general case, one generally calculates the solution around a circular neighborhood of the singularity via analytical techniques. A less drastic way to deal with this problem is to reduce the sensitivity of the finite difference approximation to the singularity by using the value $\frac{1}{2}(f(0) + g(0))$ at (0, 0).

In contrast to the boundary singularities, some problems arise with interior singularities. ***Interior singularities*** occur when one, or more than one, coefficient of a partial differential equation becomes singular. Physical problems where interior singularities occur are generally characterized by the presence of sources, sinks, or concenterated point loads. A typical example of this category is the Poisson equation

$$\nabla^2 u = f(x, y)$$

where $f(x, y)$ is not smooth, such as the case when $f(x, y)$ represents a point source or line source. Some of the more commonly used methods of handling interior singularities are now described.

SUBTRACTING OUT THE SINGULARITY

The accepted technique in interior singularity problems is to subtract the singularity, where possible, thereby generating a new problem with different boundary conditions but with a well-behaved solution. In practice, this procedure works well with linear equations. Nonlinear equations require individual treatment.

GRID REFINEMENT

A brute-force method of dealing with singularities is to ignore them and attempt to diminish their effect by refining the grid in the vicinity of a singularity.

REMOVAL OF SINGULARITY

Some singularities can be removed by using a new independent variable. The transformation generally expands the singular point into a line or curve. For instance, the singularity at (0, 0) in Eq. (7.1) can be removed by using the transformation

$$\xi = x/\sqrt{t}, \qquad \tau = \sqrt{t} \tag{7.4}$$

The effect of this transformation is to expand the point (0, 0) into the positive half of the ξ-axis and to remove the whole of the positive half of the x-axis to $\xi = \infty$. Thus, the discontinuity in u at (0, 0) is transformed into a smooth change along the positive ξ-axis. Specifically, under this transformation, Eq. (7.1) becomes

$$2\frac{\partial^2 u}{\partial \xi^2} = \tau \frac{\partial u}{\partial \tau} - \xi \frac{\partial u}{\partial \xi} \tag{7.5a}$$

$$u(\infty, 0) = 0 \tag{7.5b}$$

$$u(0, \tau) = 1, \qquad \tau \geq 0$$

The numerical process starts with the solution of the ordinary differential equation for $u(\xi)$ obtained by setting $\tau = 0$ in Eq. (7.5a). There will come a time when the boundary condition at $x = 1$ will need to be taken into account. At this point one can return to the original xt plane.

METHOD OF MOTZ AND WOODS

Another interesting type of singularity occurs at the ***reentrant corner*** of an L-shaped region. A reentrant corner is a point where the boundary changes direction through an angle exceeding π. For concreteness, consider an example treated by J. Crank:[1]

$$\frac{\partial u}{\partial t} = D\left(\frac{\partial^2 u}{\partial x^2} + \frac{\partial^2 u}{\partial y^2}\right) \tag{7.6}$$

defined over an L-shaped region shown in Figure 7.1. In this figure P is a reentrant corner. Expressing Eq. (7.6) in polar coordinates r, θ centered at P, one gets

$$D\left(\frac{\partial^2 u}{\partial r^2} + \frac{1}{r}\frac{\partial u}{\partial r} + \frac{1}{r^2}\frac{\partial^2 u}{\partial \theta^2}\right) = \frac{\partial u}{\partial t} \tag{7.7a}$$

Let the boundary conditions in the neighborhood of the point P be

$$\frac{\partial u}{\partial \theta} = 0 \qquad \text{on } \theta = (0, \theta_0) \tag{7.7b}$$

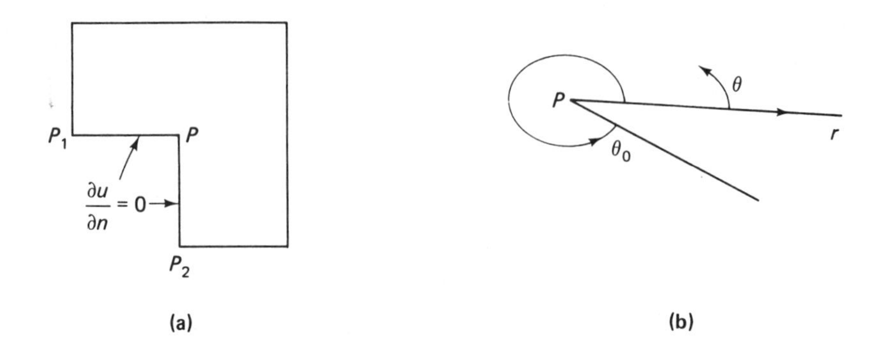

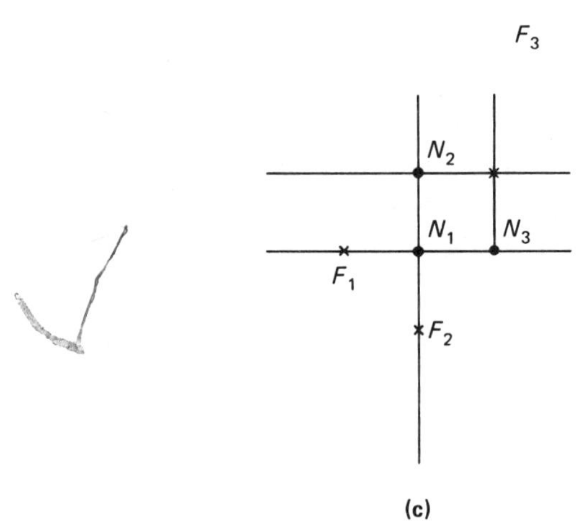

Figure 7.1 Treatment of a reentrant corner: (a) reentrant corner at P; (b) corner in polar coordinates; (c) portion of the grid near the reentrant corner. The x's indicate far points and $\cdot$'s indicate near points. (From *The Mathematics of Diffusion* by J. Crank, published by Oxford University Press, New York, © Oxford University Press, 1975)

Equation (7.7) can be solved by considering a separation of variables solution such as

$$u = \exp(-\alpha^2 Dt) R(r) \psi(\theta) + W(r, \theta) \tag{7.8}$$

where $W(r, \theta)$ is a solution of Laplace's equation representing the steady-state form of the singularity. Substituting Eq. (7.8) in Eq. (7.7a) and using Eq. (7.7b), we get

$$\psi'' = -\omega^2 \psi \tag{7.9a}$$

$$R'' + \frac{1}{r} R' + R\left(\alpha^2 - \frac{\omega^2}{r^2}\right) = 0 \tag{7.9b}$$

where ω is a new variable introduced in the separation process. Now the solutions of Eqs. (7.9a) and (7.9b), respectively, are

$$\psi = a \cos \omega\theta + b \sin \omega\theta \tag{7.10a}$$

$$R = \beta J_\omega(\alpha r), \qquad \omega \geq 0 \tag{7.10b}$$

where J_ω are the Bessel functions. Now Eq. (7.8) becomes

$$u(r, \theta, t) = \exp(-\alpha^2 Dt) J_\omega(\alpha r)\{A \cos \omega\theta + B \sin \omega\theta\} + W(r, \theta) \tag{7.11}$$

The boundary conditions in Eq. (7.7b) require that

$$B = 0, \qquad \omega = k\pi/\theta_0, \qquad k = 0, 1, 2, \ldots$$

Now an expression for $W(r, \theta)$ can be obtained in the standard fashion as

$$W(r, \theta) = \sum_{k=0}^{\infty} c_k r^{k\lambda} \cos k\lambda\theta$$

with $\lambda = \pi/\theta_0$.

For concreteness, if $\theta_0 = 3\pi/2$, as shown in Figure 7.1b, then $\lambda = \pi/\theta_0 = \frac{2}{3}$ and $\partial u/\partial y = 0$ on PP_1 and $\partial u/\partial x = 0$ on PP_2. Then Eq. (7.11) becomes

$$u(r, \theta, t) = \sum_{j=0}^{\infty} \sum_{k=0}^{\infty} A_{kj} \exp(-\alpha_j^2 Dt) J_{2k/3}(\alpha_j r) \cos \frac{2k\theta}{3} + \sum_{k=0}^{\infty} c_k r^{2k/3} \cos \frac{2k\theta}{3} \tag{7.12}$$

Notice that in Eq. (7.12) all the derivatives of u with respect to r contain singular terms at P even though u itself is finite at P. Hence the need to avoid finite differences at P. Instead, an approximation, for small r, based on Eq. (7.12) will be used.

Recognizing that $J_{2k/3}(z)$ is defined by

$$J_{2k/3}(z) = \sum_{m=0}^{\infty} \frac{(-1)^m \left(\frac{z}{2}\right)^{\frac{2}{3}k + 2m}}{m!\,\Gamma(\frac{2}{3}k + m + 1)}$$

and using this in Eq. (7.12) and collecting the like terms in r, one gets, after some algebra

$$u = a_0(t) + a_1(t) \cos \tfrac{2}{3}\theta r^{2/3} + a_2(t) \cos \tfrac{4}{3}\theta r^{4/3} + r^2\{a_2(t) \cos 2\theta - b_0(t)\} + O(r^{8/3}) \tag{7.13}$$

Here the a_i's are functions of t and represent the coefficients of the leading term in each Bessel function expansion together with the corresponding coefficient of the term $W(r, \theta)$.

Equation (7.13) can now be used to obtain the solution u at points near P in terms of the solution u at points far away from P. One far point is needed to determine each unknown coefficient in the series. Suppose our interest is to determine the first three coefficients; then Eq. (7.13) is approximated at a given time t_1 by

$$u(r, \theta, t_1) = a_0(t_1) + a_1(t_1) \cos \frac{2\theta}{3} r^{2/3} + a_2 \cos \frac{4\theta}{3} r^{4/3} \tag{7.14}$$

Notice that $a_0(t_1)$, $a_1(t_1)$, and $a_2(t_1)$ are now just three unknown numbers. Using the known values of u at the far points F_1, F_2, and F_3, one can determine a_0, a_1, and a_2 with the help of Eq. (7.14). Now Eq. (7.14) can be used once again to determine the values of u at the near points N_1, N_2, and N_3 (see Figure 7.1c).

SINGULARITIES IN NONLINEAR HYPERBOLIC EQUATIONS

The presence of nonlinearities often complicates the analysis in several ways. For example, initial value problems of nonlinear hyperbolic equations may develop singularities in their solution even though no such singularities exist at initial time. At these singularities the derivatives become unbounded and the solution ceases to exist in its differentiable form. Singularities of this type usually represent the appearance of important physical phenomena. Examples involving systems of hyperbolic first-order partial differential equations include shock formation in gas flow, breaking of water waves, formation of transverse shock waves, and velocity jump phenomena in traveling threadlines.

While dealing with singularities in nonlinear systems, one has to be careful in interpreting results obtained from numerical calculations. For example, it was observed that a convergent finite difference approximation for a nonlinear problem may not exhibit the singularities that are really present in the exact solution. Contrast this with the situation arising in the treatment of shocks by the use of artificial viscosity.

7.3. INTERFACES, MOVING BOUNDARIES, AND FREE SURFACES

Sometimes the solution of a partial differential equation is characterized by curves in the *xt* plane across which certain of the dependent variables are discontinuous but possess one-sided limits on both sides. This general problem can be subdivided into two categories: interfaces and shocks. The first category is treated here. Shocks are not treated in this book.

INTERFACES

By interfaces, one generally means boundaries separating media whose properties (physical, chemical, thermodynamical, etc.) differ. For example, interfaces separate two different regions, elastic and plastic domains, a liquid and a solid, or two adjacent portions of the same fluid whose thermodynamic states are different. Moving and unknown interfaces come under the category of moving boundaries (see the discussion which occurs later).

If the position of the interface is known, then one can adjust the grid of a finite-difference method so that some grid points fall on these boundaries. This requirement almost always makes it difficult to use a uniform square grid throughout the domain of interest. If some system property changes rather rapidly across the interface, then one may have to refine the grid in the vicinity of an interface. In either case, the computations become both difficult and inaccurate.

Treatment of interfaces in the finite element method is natural. It is easy to adjust the mesh size near an interface, and mesh refinement is not detrimental to computational efficiency.

MOVING BOUNDARIES

An important general class of problems, which is often referred to as the ***moving boundary*** or ***Stefan problems***, occurs in a variety of diffusion phenomena. In one class of problems some of the diffusing molecules are immobilized and prevented from taking further part in the diffusion process. For example, in chemical reactions, diffusing molecules may either get precipitated or may form a new immobile chemical compound. Very often the immobilization processes are irreversible and so rapid when compared to the rate of diffusion that they are considered instantaneous. An essential feature of an idealized version of these phenomena is the presence of a sharp boundary surface that moves through the medium. From this viewpoint, a moving boundary problem can also be regarded as a moving interface problem. Specific examples of such processes include the diffusion of oxygen into muscle where oxygen combines with lactic acid, the reaction of Cu^{+2} ions with CS_2 groups as they diffuse in cellulose xanthate, and the diffusion of periodate ions into cellulose fibers and their removal by combination with the glucose groups of the cellulose. Analogous mathematical problems arise in the study of heat flow in a medium that undergoes a phase change. An example that is easy to visualize is afforded by the melting of ice in contact with water; as heat flows from water to ice, ice disappears and water appears at the interface, so that both ice and water are in bodily movement with respect to the interface. A closely related problem is the ablation problem in which the molten part of a slab is continuously and immediately removed from the melting surface, as is the case with space vehicles reentering the atmosphere.

To fix the physical ideas, consider an idealized problem of diffusion accompanied by the instantaneous and irreversible immobilization of a limited number of the diffusing molecules; that is, a sharp boundary moves through the medium. Figure 7.2 illustrates the situation. Let u be the total concentration of molecules, that is, both freely diffusing and immobilized. Let u be the concentration of the immobilized component. For the front to advance a distance ΔX, we need to supply an amount $u_X \, \Delta X$ of diffusant. The amount arriving at X during an interval Δt is $-D(\Delta t)\partial u/\partial x$. Conser-

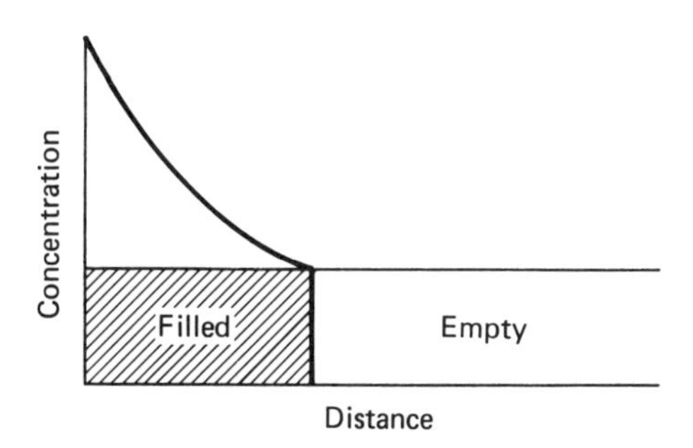

Figure 7.2 Treatment of a moving boundary (From *The Mathematics of Diffusion* by J. Crank, published by Oxford University Press, New York, © Oxford University Press, 1975)

vation at the moving boundary therefore requires

$$-D\frac{\partial u}{\partial x} = u_X \frac{dX}{dt} \tag{7.15}$$

In heat diffusion, $u_X = Lp$, where L is the latent heat and p is the density of the medium. In solute diffusion, $u_X = s$, where s is the sites per unit volume, on each of which one diffusing molecule can be instantaneously and irreversibly immobilized. Thus, for solute diffusion, the mathematical equation can be formulated as that of solving

$$\frac{\partial u}{\partial t} = D\frac{\partial^2 u}{\partial t^2} \tag{7.16a}$$

subject to

$$u = 0, \qquad t = 0, \qquad x > 0 \tag{7.16b}$$

$$\frac{\partial u}{\partial t} = D\frac{\partial u}{\partial x}, \qquad t \geq 0, \ x = 0 \tag{7.16c}$$

$$s\frac{dX}{dt} = -D\frac{\partial u}{\partial x}, \qquad t > 0, \qquad x = X \tag{7.16d}$$

where $X(t)$ denotes the position of the boundary. Note that condition in Eq. (7.16c) represents conservation of solute at the boundary.

Lagrangian Interpolation. One method of handling a moving boundary is to develop difference approximations on either side of the boundary using Lagrange interpolation polynomials. The procedure is not too unlike the method discussed briefly in Section 4.10. As a first step toward solving Eq. (7.16), let the moving boundary be located at $x = X$. This boundary need not coincide with the grid lines. Our goal is to track the boundary as it moves along. The location of a few grid points and that of the boundary are shown in Figure 7.3.

To develop difference formulas, let us start with the Lagrangian interpolation polynomial to approximate $u(x)$. This formula is

$$u(x) = \sum_{j=0}^{n} l_j(x)u_j \tag{7.17}$$

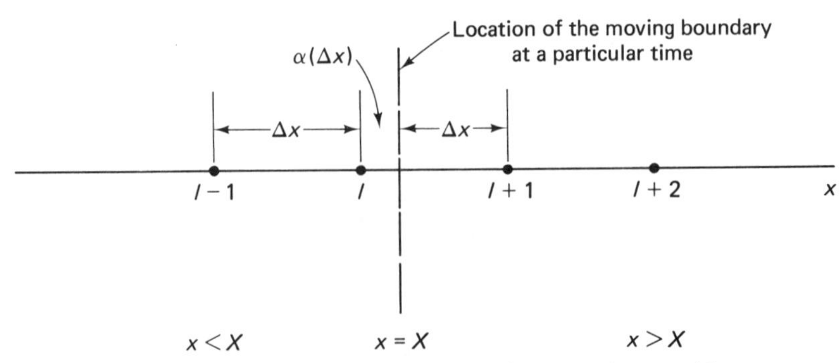

Figure 7.3 Difference grid for the moving boundary problem

where $u_j = u(x = x_j)$ and

$$l_j(x) = \frac{p_n(x)}{(x - x_j)p'_n(x_j)} \tag{7.18}$$

in which

$$p_n(x) = (x - x_0)(x - x_1) \cdots (x - x_{n-1})(x - x_n) \tag{7.19}$$

and $p'_n(x_j)$ is the derivative of $p_n(x)$ evaluated at $x = x_j$. Suppose we are interested only in three-point formulas. Then $n = 3$ and it is simple to write

$$\frac{du(x)}{dx} = l'_0(x)u_0 + l'_1(x)u_1 + l'_2(x)u_2 \tag{7.20}$$

$$\frac{1}{2}\frac{d^2u(x)}{dx^2} = \frac{u_0}{(x_0 - x_1)(x_0 - x_2)} + \frac{u_1}{(x_1 - x_0)(x_1 - x_2)} + \frac{u_2}{(x_2 - x_0)(x_2 - x_1)} \tag{7.21}$$

where

$$l'_0(x) = \frac{(x - x_1) + (x - x_2)}{(x_0 - x_1)(x_0 - x_2)}, \qquad l'_1(x) = \frac{(x - x_2) + (x - x_0)}{(x_1 - x_0)(x_1 - x_2)} \tag{7.22}$$

and so on. Now consider the case $x < X$. For this case, d^2u/dx^2 evaluated at $x = l(\Delta x)$ becomes

$$\frac{d^2u(x)}{dx^2} = \frac{2}{(\Delta x)^2}\left\{\frac{u_{l-1}}{\alpha + 1} - \frac{u_l}{\alpha} + \frac{u_X}{\alpha(\alpha + 1)}\right\} \tag{7.23}$$

Similarly, du/dx evaluated at $x = X$ becomes

$$\frac{du(x)}{dx} = \frac{1}{(\Delta x)}\left\{\frac{\alpha}{\alpha + 1}u_{l-1} - \frac{\alpha + 1}{\alpha}u_l + \frac{2\alpha + 1}{\alpha(\alpha + 1)}u_X\right\} \tag{7.24}$$

Similarly, for the case $x > X$, we get at $x = (l + 1)\,\Delta x$

$$\frac{d^2u(x)}{dx^2} = \frac{2}{(\Delta x)^2}\left\{\frac{u_X}{(1 - \alpha)(2 - \alpha)} - \frac{u_{l+1}}{1 - \alpha} + \frac{u_{l+2}}{2 - \alpha}\right\} \tag{7.25}$$

and at $x = X$

$$\frac{du}{dx} = \frac{1}{(\Delta x)}\left\{\frac{2\alpha - 3}{(1 - \alpha)(2 - \alpha)}u_X + \frac{2 - \alpha}{1 - \alpha}u_{l+1} - \frac{1 - \alpha}{2 - \alpha}u_{l+2}\right\} \tag{7.26}$$

Now we use these formulas to approximate space derivatives and the usual implicit or explicit formulas to approximate the time derivatives in the equation as well as the derivatives appearing on the moving boundary.

There are other methods of handling moving boundaries. One idea is to fix the boundary via a mathematical transformation. Alternatively, one can consider the use of a moving grid. Another possibility is based on the application of Green's functions to develop an integro-differential equation for the position of the moving boundary. A method popular with analog computer users is to discretize all independent variables except the one, say x, that specifies the location of the moving boundary. Typically, this results in a system of ordinary differential equations where the independent variable is x. By properly performing the "time scaling," it is possible to adjust the speed of the solution process (called "computer time") to match the movement of the boundary (called the "problem time" or real time). A discussion of these topics would take us too far afield. Interested readers should consult the literature.

FREE SURFACES

A somewhat different type of moving boundary problem is the free-surface problem. In a number of fluid flow problems, parts of the flow boundaries (free surfaces) are unknown a priori. These problems include sluice gate flows, flows over weirs, and the like. The free surface is a streamline, and the problem is to determine the location of this streamline as well as the streamline pattern for the remainder of the flow.

The procedure for solving free-surface potential flow problems involves solving Laplace's equation as usual, plus satisfying the conditions that the velocity normal to the free surface be zero and the pressure along the free surface be constant. One way is to select a trial free surface, solve $\nabla^2\psi = 0$, and calculate the velocity components along the assumed free-surface profile. If the velocity conditions on the free surface are not met, the procedure is repeated with a modified free surface until some error criterion is satisfied.

7.4. NONLINEARITIES

The substantial progress of the eighteenth and nineteenth centuries in the construction of effective theories for physical phenomena was, in large measure, due to a linear principle, that of superposition. Generally, these theories are only a first approximation; many problems in all areas of science and engineering lead to nonlinear partial differential equations. Formulations that omit or suppress nonlinear terms often lead to inadequate or faulty results. Considerable mathematical effort is being expended in this general area with little progress. Before substantial progress can occur in this difficult area, guiding general principles must be developed. The linear theory has its Fourier analysis, Strum-Liouville theory, and the like. The nonlinear theory is badly fragmented. The only significant progress of general utility has been in the numerical analyses. In this era of high-speed computers, some schools of thought are even questioning the wisdom of the preoccupation with general theories when workable solutions of adequate accuracy suffice.

Two kinds of nonlinearities are of concern to us: material nonlinearities and geometric nonlinearities. Material nonlinearities, which are often referred to simply as nonlinearities, are common in field problems of the type discussed in this book. These material nonlinearities can be further subdivided into two categories: (1) Nonlinearities can arise because the physical properties of the region under consideration are not constant. Saturation phenomena are typical of this class. Change in permeability induced by the wetting of a porous medium is another example. (2) Nonlinearities can also arise if the complex constitutive relationships of the phenomena supersede the simple linearity assumptions. Interaction between a charged particle and an electromagnetic field belongs to this class. Other examples can be easily cited. In solid mechanics, phenomena such as plasticity and creep belong to this class. Indeed, it is impossible to list and categorize the nonlinearities that occur in practice.

A different type of nonlinearity that is quite common in structural problems is the geometrical nonlinearity. In structural problems, one normally makes the assumption that the displacements and strains are small. This assumption fails frequently in practice. An example is a situation where very large displacements may occur without causing large strains, as is the case with a watch spring.

TWO-LEVEL DIFFERENCE METHODS

One of the most gratifying features of the finite difference methods is that many of the methods and proofs based on linear equations with constant coefficients carry over directly to nonlinear equations. Thus many of the simple implicit and explicit methods described in Chapters 5 and 6 can also be used to solve nonlinear equations. For example, consider a rather general nonlinear parabolic equation

$$\frac{\partial u}{\partial t} = f\left(x, t, u, \frac{\partial u}{\partial x}, \frac{\partial^2 u}{\partial u^2}\right) \tag{7.27}$$

over the region $0 \leq x \leq 1$ and $0 \leq t \leq T$. A finite difference analog to Eq. (7.27) can be written as

$$\frac{u_{l,j+1} - u_{l,j}}{(\Delta t)} = f\left(x_l, t_j, u_{l,j}, \frac{1}{2(\Delta x)}(u_{l+1,j} - u_{l-1,j}), \frac{1}{(\Delta x)^2}\delta_x^2 u_{l,j}\right) \tag{7.28}$$

This explicit method, though easy to solve, suffers from stability problems unless the size of (Δt) is strictly limited. In the linear case (i.e., when $f \equiv \partial^2/\partial x^2$), this limitation was expressed as $(\Delta t)/(\Delta x)^2 \leq 1/2$. In the nonlinear case, this limitation depends on f, which in turn depends on the solution u. That is, in nonlinear problems, stability depends not only on the difference scheme used but also on the solution being sought. Therefore, if one wishes to use explicit methods to solve nonlinear equations, it is prudent to include in the computer program a constant check on stability.

The stability limitation can be removed by using an implicit difference method of the Crank-Nicolson type. By this approach, Eq. (7.28) can be discretized as

$$\begin{aligned}\frac{1}{(\Delta t)}(u_{l,j+1} - u_{l,j}) = f\Bigg[& x_l, t_{j+1/2}, \frac{1}{2}(u_{l,j+1} + u_{l,j}), \\ & \frac{1}{2}\left\{\frac{1}{2(\Delta x)}(u_{l+1,j+1} - u_{l-1,j+1}) + \frac{1}{2(\Delta x)}(u_{l+1,j} - u_{l-1,j})\right\}, \\ & \frac{1}{2(\Delta x)^2}\delta_x^2(u_{l,j+1} + u_{l,j})\Bigg]\end{aligned} \tag{7.29}$$

Unfortunately, the algebraic problem of solving Eq. (7.29) for $u_{l,j+1}$ is generally very complicated because Eq. (7.29) represents a system of nonlinear algebraic equations. Iterative methods, such as the Newton-Raphson method, are available to solve such nonlinear systems. A brief description of the Newton-Raphson algorithm appears in this book under the heading of quasilinearization. For the purpose of present discussion, it is sufficient to note that the solution to a system of nonlinear algebraic equations is obtained via (1) an outer iteration (quasilinearization, say) that linearizes, (2) followed by an inner iteration using a method such as SOR.

THREE-LEVEL DIFFERENCE METHODS

Another approach to discretize nonlinear equations would be to try three-level difference schemes rather than two-level schemes such as the one discussed previously. For this purpose, consider an example treated by A. R. Mitchell,[2]

$$b(u)\frac{\partial u}{\partial t} = \frac{\partial}{\partial x}\left[a(u)\frac{\partial u}{\partial x}\right] \tag{7.30}$$

with $a(u) > 0$ and $b(u) > 0$. A simple difference approximation to Eq. (7.30) is

$$b(u_{l,j})\frac{1}{2(\Delta t)}(u_{l,j+1} - u_{l,j-1}) = \frac{1}{(\Delta x)^2}\delta_x(a(u_{l,j})\delta_x)u_{l,j} \tag{7.31}$$

where

$$\delta_x(a(u_{l,j})\delta_x)u_{l,j} = \delta_x(a(u_{l,j})(u_{l+1/2,j} - u_{l-1/2,j})) \tag{7.32a}$$

$$= a(u_{l+1/2,j})u_{l+1,j} - a(u_{l+1/2,j})u_{l,j} - a(u_{l-1/2,j})u_{l,j} + a(u_{l-1/2,j})u_{l-1,j} \tag{7.32b}$$

$$= a(u_{l+1/2,j})[u_{l+1,j} - u_{l,j}] - a(u_{l-1/2,j})[u_{l,j} - u_{l-1,j}] \tag{7.32c}$$

However, a scheme such as this would be unconditionally unstable if $a(u) = b(u) = 1$. However, unconditional stability can be obtained if $\delta_x^2 u_{l,j}$ is replaced [in the case of $a(u) = b(u) = 1$] by $\frac{1}{3}\delta_x^2(u_{l,j+1} + u_{l,j} + u_{l,j-1})$. (A proof of this assertion is left to the reader as an exercise.)

Following the lead from the linear case, the terms $u_{l+1,j}$, $u_{l,j}$ and $u_{l-1,j}$ in Eq. (7.32c) are replaced, respectively, as shown:

$$\begin{aligned} u_{l+1,j} &\longleftarrow \tfrac{1}{3}(u_{l+1,j+1} + u_{l+1,j} + u_{l+1,j-1}) \\ u_{l,j} &\longleftarrow \tfrac{1}{3}(u_{l,j+1} + u_{l,j} + u_{l,j-1}) \\ u_{l-1,j} &\longleftarrow \tfrac{1}{3}(u_{l-1,j+1} + u_{l-1,j} + u_{l-1,j-1}) \end{aligned} \tag{7.33}$$

Substitution of Eqs. (7.33) in Eq. (7.32c) and the latter in Eq. (7.31) leads to

$$\begin{aligned} &b(u_{l,j})[u_{l,j+1} - u_{l,j-1}] \\ &\quad = \tfrac{2}{3}r[\alpha^+\{(u_{l+1,j+1} - u_{l,j+1}) + (u_{l+1,j} - u_{l,j}) + (u_{l+1,j} - u_{l,j-1})\} \\ &\qquad - \alpha^-\{(u_{l,j+1} - u_{l-1,j+1}) + (u_{l,j} - u_{l-1,j}) + (u_{l,j-1} - u_{l-1,j-1})\}] \end{aligned} \tag{7.34}$$

where

$$\begin{aligned} r &= \frac{(\Delta t)}{(\Delta x)^2} \\ \alpha^+ &= a\left(\frac{u_{l+1,j} + u_{l,j}}{2}\right) \\ \alpha^- &= a\left(\frac{u_{l,j} + u_{l-1,j}}{2}\right) \end{aligned} \tag{7.35}$$

By the preceding calculation we have essentially replaced the problem of evaluating $a(u)$ at the grid points $(l + \frac{1}{2}, j)$ and $(l - \frac{1}{2}, j)$ by the problem of evaluating $a(u)$ at the grid points $(l + 1, j)$, (l, j) and (l, j), $(l - 1, j)$. Furthermore, Eq. (7.34) is now *linear*, and so the complication of solving sets of nonlinear equations is avoided.

NONLINEAR OVERRELAXATION

The strategy of this method is to bypass the complexity of inner and outer iterations by what is called nonlinear overrelaxation (NLOR). This method is illustrated by considering K nonlinear algebraic equations

$$f_p(x_1, x_2, \dots, x_k) = 0, \qquad p = 1, 2, \dots, k \tag{7.36}$$

Assume that each f_p has continuous first derivatives. For convenience set $f_{pq} = \partial f_p / \partial x_q$. The basic idea is the introduction of a relaxation factor ω so that

$$\begin{aligned} x_1^{(i+1)} &= x_1^{(i)} - \omega \frac{f_1(x_1^{(i)}, x_2^{(i)}, \dots, x_k^{(i)})}{f_{11}(x_1^{(i)}, x_2^{(i)}, \dots, x_k^{(i)})} \\ x_2^{(i+1)} &= x_2^{(i)} - \omega \frac{f_2(x_1^{(i+1)}, x_2^{(i)}, \dots, x_k^{(i)})}{f_{22}(x_1^{(i+1)}, x_2^{(i)}, \dots, x_k^{(i)})} \\ x_3^{(i+1)} &= x_3^{(i)} - \omega \frac{f_3(x_1^{(i+1)}, x_2^{(i+1)}, x_3^{(i)}, \dots, x_k^{(i)})}{f_{33}(x_1^{(i+1)}, x_2^{(i+1)}, x_3^{(i)}, \dots x_k^{(i)})} \end{aligned} \tag{7.37}$$

and so on.

The convergence criteria for NLOR can be shown to be the same as those for SOR with the coefficient matrix **A** replaced by the Jacobian $\mathbf{J}(f_{pq}^{(i)})$ of Eq. (7.36). For example, if the Jacobian **J** can be split into lower triangular, diagonal, and upper triangular parts as

$$\mathbf{J}^{(i)} = \mathbf{L}^{(i)} + \mathbf{D}^{(i)} + \mathbf{U}^{(i)} \tag{7.38}$$

then an iterative scheme can be defined as

$$\mathbf{x}^{(i+1)} = \mathbf{x}^{(i)} - \omega[\mathbf{D}^{(i)^{-1}}\mathbf{L}^{(i)}\mathbf{e}^{(i+1)} + (\mathbf{I} + \mathbf{D}^{(i)^{-1}}\mathbf{U}^{(i)})\mathbf{e}^{(i)}] \tag{7.39}$$

where $\mathbf{e}^{(i)}$ stands for the error vector $\mathbf{e}^{(i)} = \mathbf{x}^{(i)} - \mathbf{x}$. For convergence, the Jacobian at each stage of the iteration must satisfy all the properties required of **A** in SOR.

QUASILINEARIZATION

The quasilinearization method can be visualized as an application of Newton-Raphson-Kantorovich approximation technique in function space. For motivational reasons, let us begin with the problem of finding a sequence of approximations to the root of the nonlinear, scalar algebraic equation

$$f(x) = 0 \tag{7.40}$$

If we assume that $f(x)$ is monotone decreasing for all x and strictly convex, that is, $f''(x) > 0$, then the root r is simple and $f'(r) \neq 0$.

Let $x^{(0)}$ be an initial approximation to the root r with $x^{(0)} < r$ and $f(x^{(0)}) > 0$. Now suppose that we approximate $f(x)$ by a linear function of x determined by the slope of the function $f(x)$ at $x = x^{(0)}$. Then

$$f(x) \simeq f(x^{(0)}) + (x - x^{(0)})f'(x^{(0)}) \tag{7.41}$$

A further approximation to r is then obtained by solving the *linear* equation in x,

$$f(x^{(0)}) + (x - x^{(0)})f'(x^{(0)}) = 0 \tag{7.42}$$

This yields the next approximation

$$x^{(1)} = x^{(0)} - \frac{f(x^{(0)})}{f'(x^{(0)})} \tag{7.43}$$

This process can be repeated. That is, in general, we have the recurrence relation

$$x^{(i+1)} = x^{(i)} - \frac{f(x^{(i)})}{f'(x^{(i)})} \tag{7.44}$$

It is clear from Figure 7.4 that

$$x^{(0)} < x^{(1)} < \cdots < x^{(i)} < x^{(i+1)} < \cdots < r$$

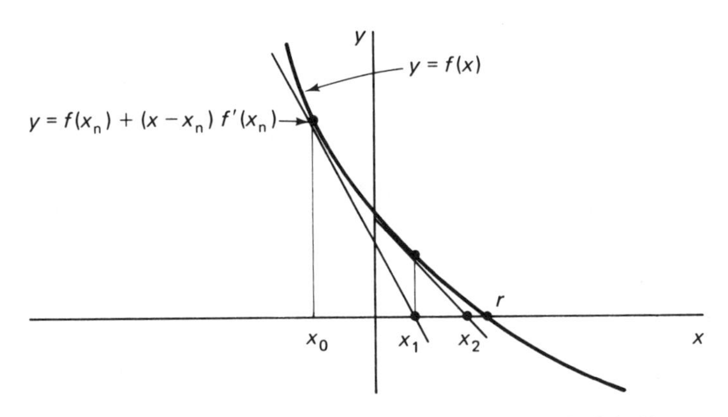

Figure 7.4 Graphical interpretation of Newton's method of finding a root of $f(x) = 0$

The method described here is the well-known Newton's method for finding the root of a scalar equation. This method readily generalizes to the vector case. For instance, if we want to solve the system of nonlinear equations described by

$$\mathbf{f}(\mathbf{x}) = \mathbf{0} \tag{7.45}$$

then we start as usual with an initial approximation $\mathbf{x}^{(0)}$ and repeatedly apply the recurrence relation

$$\mathbf{x}^{(i+1)} = \mathbf{x}^{(i)} - \mathbf{J}(x^{(i)})^{-1}\mathbf{f}(x^{(i)}) \tag{7.46}$$

where $\mathbf{J}(x^{(i)})$ is the Jacobian matrix whose (l, m)th element is

$$[J(x^{(i)})]_{l,m} = \left(\frac{\partial f_l}{\partial x_m}\right)\Bigg|_{\mathbf{x}=\mathbf{x}^{(i)}} \tag{7.47}$$

Extension of this basic idea to infinite dimensional spaces (function spaces) results in the method of quasilinearization or the Newton-Raphson-Kantorovich method. Without further ado, let us see how this method can be applied to solve nonlinear ordinary and partial differential equations.

Consider the nonlinear ordinary differential equation

$$\frac{d^2y}{dx^2} = y'' = f(y), \qquad y(0) = y(1) = 0 \tag{7.48}$$

If we expand $f(y)$ about a nominal solution $y^{(i)}(x)$, using Taylor series expansion, and keep only the linear terms, we get

$$y'' = f(y^{(i)}) + (y - y^{(i)})f'(y^{(i)}) \tag{7.49a}$$

with the boundary conditions

$$y^{(i)}(0) = y^{(i)}(1) = 0 \tag{7.49b}$$

In Eq. (7.49), $y^{(i)}$ are known and the equation itself is *linear* in the unknown $y(x)$. Therefore, Eq. (7.49) can be solved using standard techniques available for solving linear boundary value problems. The resulting solution $y(x)$ can now be labeled as $y^{(i+1)}(x)$ and used in place of $y^{(i)}$ to get an improved approximation. That is, starting from a nominal solution $y^{(0)}(x)$, we solve the recurrence relation

$$y^{(i+1)''} = f(y^{(i)}) + (y^{(i+1)} - y^{(i)})f'(y^{(i)}) \tag{7.50a}$$

$$y^{(i)}(0) = y^{(i)}(1) = 0 \tag{7.50b}$$

until convergence is achieved.

This idea can be applied to solve partial differential equations also. For this purpose, consider the nonlinear parabolic equation

$$\frac{\partial u}{\partial t} = \frac{\partial^2 u}{\partial x^2} + g\left(u, \frac{\partial u}{\partial x}\right) \tag{7.51}$$

with specified initial and boundary conditions. To get the recurrence relation, we expand $g(u, \partial u/\partial x)$ about a nominal solution $u^{(i)}(x, t)$ and write the recurrence relation as follows:

$$\begin{aligned}\frac{\partial u^{(i+1)}}{\partial t} = {} & \frac{\partial^2 u^{(i+1)}}{\partial x^2} + g\left(u^{(i)}, \frac{\partial u^{(i)}}{\partial x}\right) \\ & + (u^{(i+1)} - u^{(i)})\frac{\partial g}{\partial u}\left(u^{(i)}, \frac{\partial u^{(i)}}{\partial x}\right) \\ & + \left(\frac{\partial u^{(i+1)}}{\partial x} - \frac{\partial u^{(i)}}{\partial x}\right)\frac{\partial g}{\partial u_x}\left(u^{(i)}, \frac{\partial u^{(i)}}{\partial x}\right)\end{aligned} \tag{7.52}$$

where $\partial g/\partial u_x$ stands for the partial derivative of g with respect to $\partial u/\partial x$. In Eq. (7.52), $u^{(i)} = u^{(i)}(x, t)$ is known at any stage. Therefore g, $\partial g/\partial u$, and $\partial g/\partial u_x$ can all be evaluated. Thus, Eq. (7.52) is linear in $u^{(i+1)}$, and therefore can be solved using one of the techniques described in earlier chapters.

PREDICTOR-CORRECTOR METHODS

The idea here is to extend some of the well-developed concepts from the field of ordinary differential equations toward the solution of partial differential equations. Consider an ordinary differential equation such as

$$y' = f(y, t), \qquad y(0) = y_0$$

The predictor-corrector method of solving this equation proceeds in two steps. First, one "predicts" the value of y at time t_{n+1} by using several of the previous function values up to and including the value of y at t_n. These predicted values are relatively inaccurate. They are then improved by using the more accurate "corrector" formulas, which require information at t_{n+1}.

This general idea can be extended to partial differential equations. For illustrative purposes, let us consider

$$\frac{\partial^2 u}{\partial x^2} = \psi\left(x, t, u, \frac{\partial u}{\partial x}, \frac{\partial u}{\partial t}\right), \qquad 0 < x < 1, \qquad 0 < t \leq T \tag{7.53a}$$

subject to the auxiliary conditions

$$u(x, 0) = u_0(x) \tag{7.53b}$$

$$u(0, t) = u_0(t), \qquad u_1(t) = u_1(t) \tag{7.53c}$$

If the function ψ is restricted to be of the form

$$\psi = f_1(x, t, u)\frac{\partial u}{\partial t} + f_2(x, t, u)\frac{\partial u}{\partial x} + f_3(x, t, u) \tag{7.54}$$

or of the form

$$\psi = g_1\left[x, t, u, \frac{\partial u}{\partial x}\right]\frac{\partial u}{\partial t} + g_2\left[x, t, u, \frac{\partial u}{\partial x}\right] \tag{7.55}$$

then the classical Crank-Nicolson procedure can be modified and recast as a predictor-corrector method. The restrictions imposed in Eqs. (7.54) and (7.55) are not too severe, because the admissible class includes such classical nonlinear problems as the Burger's equation

$$\frac{\partial^2 u}{\partial x^2} = u\frac{\partial u}{\partial x} + \frac{\partial u}{\partial t} \tag{7.56}$$

and the nonlinear diffusion equation

$$\frac{\partial}{\partial x}\left[k(u)\frac{\partial u}{\partial x}\right] = \alpha(u)\frac{\partial u}{\partial t} \tag{7.57}$$

Assuming a square grid of size $\Delta x = \Delta y = h$ and a time step of size $\Delta t = k$, a predictor-corrector type formula to solve Eqs. (7.53) and (7.54) can be written as follows:

For the Predictor

$$(h^2)^{-1}\delta_x^2 u_{l,j+1/2} = \psi\left[lh, \left(j + \frac{1}{2}\right)k, u_{l,j}, (2h)^{-1}\mu\delta_x u_{l,j}, \frac{2}{k}(u_{l,j+1/2} - u_{l,j})\right] \tag{7.58}$$

for $l = 1, 2, \ldots, M - 1$.

For the Corrector

$$\tfrac{1}{2}\delta_x^2[u_{l,j+1} + u_{l,j}] = \psi[lh, (j + \tfrac{1}{2})k, u_{l,j+1/2}, (4h)^{-1}\mu\delta_x(u_{l,j+1} + u_{l,j}), k^{-1}(u_{l,j+1} - u_{l,j})] \tag{7.59}$$

where δ_x^2 and μ are defined by

$$\delta_x^2 u_{l,j} = u_{l+1,j} - 2u_{l,j} + u_{l-1,j} \tag{7.60}$$

$$\mu u_{l,j} = \tfrac{1}{2}[u_{l+1/2,j} + u_{l-1/2,j}] \tag{7.61}$$

Note that both the predictor formula and the corrector formula are linear algebraic equations. The predictor is a backward difference equation utilizing the intermediate time points $(j + \frac{1}{2})\,\Delta t$. Since Eq. (7.54) is linear in $\partial u/\partial t$, the calculation of $u_{l,j+1/2}$ is a linear algebraic problem. Similarly, since Eq. (7.54) is linear in $\partial u/\partial x$, the progression from $(j + \frac{1}{2})\ \Delta t$ to $(j + 1)\ \Delta t$ also involves a linear algebraic problem.

REFERENCES

1. J. Crank, *The Mathematics of Diffusion.* New York: Oxford University Press, Inc., 1975.
2. A. R. Mitchell, *Computational Methods for Partial Differential Equations.* New York: John Wiley & Sons, Inc., 1969.

SUGGESTIONS FOR FURTHER READING

1. *A treatment of singularities, reentrant corners, and moving boundaries can be found in* J. Crank, *The Mathematics of Diffusion,* Oxford University Press, Inc., New York, 1975.
2. *For a thorough coverage of various aspects of solving nonlinear partial differential equations, see* W. F. Ames, *Nonlinear Partial Differential Equations in Engineering,* Academic Press, Inc., New York, 1965.
3. *For a concise treatment of nonlinear partial differential equations with a number of useful worked out examples and Fortran programs, see*
   D. U. Von Rosenberg, *Methods for the Numerical Solution of Partial Differential Equations,* American Elsevier, New York, 1969.
3. *For an authoritative work on quasilinearization, consult*
   R. E. Bellman and R. E. Kalaba, *Quasilinearization and Nonlinear Boundary Value Problems,* American Elsevier, New York, 1965.
4. *A good discussion on iterative methods to solve systems of nonlinear algebriac equations can be found in Chapter 3 of*
   E. Issacson and H. B. Keller, *Analysis of Numerical Methods,* John Wiley & Sons, Inc., New York, 1967.
   Also consult
   J. M. Ortega and W. C. Rheinboldt, *Iterative Solution of Nonlinear Equations in Several Variables,* Academic Press, Inc., New York, 1970.

5. *Additional information on nonlinear overrelaxation can be obtained from*

H. M. LIBERSTEIN, "A Numerical Test Case for the Nonlinear Overrelaxation Algorithm," Tech. Rept. No. MRC-TR-122, University of Wisconsin Mathematics Research Center, Madison, Wis., 1960.

EXERCISES

1. Find a root of $e^x - 3x^2 = 0$ near $x = -0.5$ using Newton's method.
2. Solve the set of nonlinear equations

$$x^2 + y^2 = 4$$
$$e^x + y = 1$$

(a) by a graphical method, and (b) by an iterative method. (c) Solve the set by first rearranging it as

(i) $x = \pm\sqrt{4 - y^2}$, $y = 1 - e^x$ (ii) $x = \ln(1 - y)$, $y = \pm\sqrt{4 - x^2}$

Use $y = -1.7$ as an initial guess for (b) and (c).

3. Solve by using Newton's method:

$$x^3 + 3y^2 = 21$$
$$x^2 + 2y + 2 = 0$$

4. Consider the PDE

$$u_t - u_{xx} = (1 + u^2)(1 - 2u)$$

over the region $0 \leq t \leq (1 - x)$, $0 \leq x \leq 1$.

(a) Linearize this equation by writing a set of recurrence relations obtained by using quasilinearization.

(b) Solve the resulting sequence of linear equations using a Crank-Nicolson scheme with $\Delta x = \Delta t = 0.001$. Stop your iterations when the following condition is satisfied:

$$\max_{l} \left| \frac{u_{l,j}^{(i)} - u_{l,j}^{(i-1)}}{u_{l,j}^{(i)}} \right| \leq 10^{-6}.$$

(c) Compare the efficiency of the method in (b) by comparing the number of iterations required with those needed by the usual Picard iteration, which is defined by

$$(u^{(i+1)})_t - (u^{(i+1)})_{xx} = (1 + (u^{(i)})^2)(1 - 2u^{(i)})$$

5. Derive the recurrence relation by applying quasilinearization to

$$u_{xx} + u_{yy} = e^u$$

on a square with Dirichlet boundary conditions. Compare the convergence with that obtained by Picard iteration.

8

Interpolation and Approximation

8.1. INTRODUCTION

The purpose of this chapter is to lay the groundwork for the finite element method, which is the subject matter of Chapter 9. There are similarities and differences between finite difference and finite element methods. The finite difference method has a local character, whereas the finite element method has a global character. To bring forth the similarities and differences between these two methods, it is necessary to stop and look back at some of the more elementary ideas in mathematics. A more formal comparison of finite difference and finite element methods appears in Section 9.6

8.2. THE APPROXIMATION PROBLEM

Consider the problem of solving a simple, linear, two-point boundary value problem

$$-\frac{d^2u(x)}{dx^2} + \sigma(x)u(x) = g(x), \qquad a \leq x \leq b \tag{8.1}$$

subject to the boundary conditions

$$u(a) = \alpha, \qquad u(b) = \beta \tag{8.2}$$

This problem was already discussed in Section 5.4. If $\sigma(x)$ is not a constant, as it is here, then it is not possible, in general, to get an exact solution $u(x)$ to this problem.

Then, one generally settles for an approximate solution $\tilde{u}(x)$, which is, in some sense, close to $u(x)$. Thus, the problem of approximating a function $u(x)$ by another function $\tilde{u}(x)$ is very important.

The idea of approximating one function by other functions is not new. It dates back at least to Fourier in 1822 and to Euler and Lagrange in the eighteenth century. One of the oldest approximating techniques is that of approximating a given function $u(x)$ by $\tilde{u}(x)$, which can be written as a finite sum. That is,

$$u(x) \simeq \tilde{u}(x) = c_1 N_1(x) + c_2 N_2(x) + \cdots + c_n N_n(x) \tag{8.3}$$

This is done, of course, with the hope that the functions $N_i(x)$, chosen by the analyst, are simpler to handle. The coefficients c_i are constants to be determined using constraints that one has to impose on $u(x)$. This simple idea was the basis of Fourier analysis in which one attempts to write $u(x)$ as an infinite sum

$$\sum_{k=0}^{\infty} (c_k \sin(akx) + b_k \cos(akx))$$

where a is a given constant. When using Fourier series, one usually truncates the series and writes

$$u(x) = c_0 + c_1 \sin(ax) + b_1 \cos(ax) + \cdots + c_n \sin(anx) + b_n \cos(anx) \tag{8.4}$$

and the constants b_k and c_k are determined by imposing constraints on $\tilde{u}(x)$. That is, in Fourier analysis, $N_i(x)$ are chosen to be the trigonometric functions.

The trend of thought in the preceding can be continued. There is nothing sacred about trigonometric functions. For example, one can choose the $N_i(x)$'s to be polynomials or piecewise polynomials. In particular, if $N_k(x) = x^{k-1}$, $1 \leq k \leq n+1$, then

$$\tilde{u}(x) = c_n x^n + c_{n-1} x^{n-1} + \cdots + c_1 x + c_0 \tag{8.5}$$

which is an nth-degree polynomial. This simple idea will be specialized in the subsequent sections.

The next question to be faced is: How good is $\tilde{u}(x)$ in approximating $u(x)$? There are many ways of measuring the goodness of fit between $u(x)$ and $\tilde{u}(x)$. Relegating the details to a later stage, let us represent the distance between u and $\tilde{u}$ by $\|\tilde{u} - u\|$; the smaller this quantity, the better the desired fit. Now the approximation problem can be stated as follows. Given the N_i's, find the coefficients c_i's such that $\|u - \tilde{u}\|$, that is,

$$\|u - (c_1 N_1 + c_2 N_2 + \cdots + c_n N_n)\|$$

is as small as possible. The set of coefficients $\{c_i\}$ is hopefully unique. Since there are n unknown constants c_i, we must impose at least n additional constraints.

Interpolatory Constraints. For instance, one can demand

$$u(x_i) = \tilde{u}(x_i) \tag{8.6}$$

at some distinct values x_i, $i = 1, 2, \ldots, n$ in the interval $[a, b]$.

Interpolatory and Smoothness Constraints. For instance, one can demand

1. $u(x_i) = \tilde{u}(x_i)$ at some distinct values x_i, $i = 1, 2, \ldots, k$ in $[a, b]$
2. $u'(x_1) = \tilde{u}'(x_1)$ and $u'(x_k) = \tilde{u}'(x_k)$.
3. $\tilde{u}(x)$ is twice continuously differentiable.

The third requirement essentially is the smoothness constraint.

Pure Variational Constraints. For instance, one can demand that $\tilde{u}$ be chosen from, say, $P_n[a, b]$, the set of all polynomials of degree n or less, defined on the interval $[a, b]$. Stated formally, choose c_i such that

$$\{\|u - \tilde{u}\| : \tilde{u} \in P_n[a, b]\}$$

is minimized.

Orthogonality Constraints. For instance, one can demand that $(u - \tilde{u})$ be orthogonal to a set of n given functions $w_1(x), \ldots, w_n(x)$. Stated formally, choose c_i such that the inner product $\langle (u - \tilde{u}), w_i \rangle$ vanishes. That is, for $i = 1, 2, \ldots, n$

$$\langle (u - \tilde{u}), w_i \rangle = \int_a^b [(u(x) - \tilde{u}(x))] w_i(x)\, dx = 0$$

It is useful to comment, in passing, that variational-type constraint leads to the finite element method and orthogonality constraint leads to the Galerkin method. Having thus set the stage for the finite element method, the next few sections will be devoted to some of the routine details involved in approximation.

8.3. PIECEWISE LAGRANGIAN INTERPOLATION

In Section 8.2 it was stated that, in approximating $u(x)$ by $\tilde{u}(x) = \sum c_i N_i(x)$, one hopes that the functions $N_i(x)$ are easier to handle. A widely popular choice for $N_i(x)$ are polynomials because they exhibit remarkable flexibility. Sums and differences of polynomials are once again polynomials. One can differentiate or integrate polynomials and still work with polynomials. A number of other properties make polynomials very attractive as candidates for approximation purposes. For instance, the Weierstrass approximation theorem states: Given any interval $[a, b]$, any real number $\epsilon > 0$, and any real-valued continuous function on $[a, b]$, there exists a polynomial $p(x)$ such that $\|u - p\| < \epsilon$.

Now the approximating problem described in Section 8.2 can also be stated as follows: Find a polynomial $u(x) = p(x) = a_n x^n + a_{n-1} x^{n-1} + \cdots + a_1 x + a_0$ that satisfies the interpolatory constraints, that is,

$$u(x_i) = p(x_i), \qquad i = 0, \ldots, n$$

where $x_0 < x_1 < \cdots < x_n$ are $(n + 1)$ distinct points in $[a, b]$. That is, we demand that the values of the approximating polynomial $p(x)$ be identical to the values of the function $u(x)$ at the points x_i, $i = 0, \ldots, n$. These points are called the ***grid points*** or ***knots*** of $p(x)$, and the polynomial $p(x)$ is said to be an ***interpolating polynomial.*** Notice that the set $\{1, x, x^2, \ldots, x^n\}$ is a linearly independent subset of $P_n[a, b]$, the set of all polynomials in $[a, b]$, and is therefore called a ***basis*** for $P_n[a, b]$.

INTERPOLATION WITH LAGRANGIAN POLYNOMIALS

It is well known that, given a pair of points (x_0, y_0) and (x_1, y_1), there exists a unique straight line passing through these points, and the equation of the straight line is a polynomial of degree 1. Can we generalize this result? That is, given three points, does there exist a unique quadratic? In general, given $(n + 1)$ points, does there exist a unique nth-degree polynomial passing through them? The answer is yes. The resulting polynomial $p_n(x)$ is the ***Lagrange interpolating polynomial of degree n,*** which assumes prescribed real values at the $(n + 1)$ distinct knots $x_0 < x_1 < x_2 < \cdots < x_n$. But the unanswered question is how to compute $p_n(x)$?

Matrix Method. Computation of Lagrange interpolating polynomials is fairly simple. Consider a real-valued function $u(x)$. Consider n distinct real numbers $x_0 < x_1 < \cdots < x_n$, and let $p_n(x)$ be an interpolating polynomial of degree n interpolating $u(x)$ at $x_0 < x_1 < \cdots < x_n$. If we choose the set $\{1, x, x^2, \ldots, x^n\}$ as a basis, then we can write

$$p_n(x) = c_0 + c_1 x + c_2 x^2 + \cdots + c_n x^n \tag{8.7}$$

The condition this has to satisfy is

$$p_n(x_i) = u(x_i), \qquad i = 0, 1, 2, \ldots, n \tag{8.8}$$

where $u(x_i)$ are the values assumed by the given function $u(x)$. Thus Eqs. (8.7) and (8.8) represent a set of $(n + 1)$ linear equations and have a unique solution $(c_0, c_1, \ldots, c_n)$, which can be obtained by inverting the coefficient matrix **A** defined by

$$\mathbf{A} = \begin{bmatrix} x_0^n & x_0^{n-1} & \cdots & x_0 & 1 \\ x_1^n & x_1^{n-1} & \cdots & x_1 & 1 \\ \cdot & \cdot & \cdots & \cdot & \cdot \\ \cdot & \cdot & \cdots & \cdot & \cdot \\ \cdot & \cdot & \cdots & \cdot & \cdot \\ x_n^n & x_n^{n-1} & \cdots & x_n & 1 \end{bmatrix}$$

Because a unique solution, that is, $(c_0, c_1, \ldots, c_n)$, exists, theoretically matrix **A** is nonsingular; however, for large n, the matrix tends to be ill conditioned. Therefore, this method is computationally not attractive.

Canonical Basis Method. The set $[1, x, x^2, \ldots, x^n]$ is not the only basis for $P_n[a, b]$ from which the Lagrange interpolating polynomials $p_n(x)$ are picked. For example, the set $\{L_0^n(x), L_1^n(x), \ldots, L_n^n(x)\}$, where

$$L_i^n(x) = \frac{(x - x_0)\cdots(x - x_{i-1})(x - x_{i+1})\cdots(x - x_n)}{(x_i - x_0)\cdots(x_i - x_{i-1})(x_i - x_{i+1})\cdots(x_i - x_n)} \tag{8.9}$$

is another basis for $P_n[a, b]$. Therefore, $p_n(x)$ can also be written as

$$p_n(x) = c_0 L_0^n(x) + c_1 L_1^n(x) + \cdots + c_n L_n^n(x) \tag{8.10}$$

Imposing the constraint $p_n(x_i) = u(x_i)$, Eqs. (8.9) and (8.10) readily yield $c_i = u(x_i)$. Therefore,

$$p_n(x) = u(x_0)L_0^n(x) + u(x_1)L_1^n(x) + \cdots + u(x_n)L_n^n(x) \tag{8.11}$$

For this reason, the coefficients c_i's are sometimes called the ***generalized displacement amplitudes*** or, simply, the ***generalized coordinates***. The preceding procedure requires no matrix inversion and is therefore computationally simple.

The purpose of describing these two methods is twofold. First, note that every polynomial of degree n can be expressed in the form of Eq. (8.11). Second, computing $p_n(x)$ in terms of the basis $\{L_0^n(x), L_1^n(x), \ldots, L_n^n(x)\}$ is much simpler than in terms of the basis $\{1, x, x^2, \ldots, x^n\}$. Thus proper choice of basis is of crucial importance if ease of computation is the goal.

Example. Suppose we wish to approximate $u = (1 + x)^{-1}$ in the interval $(0, 2)$ by choosing equispaced knots as prescribed: for Case 1, $x_0 = 0, x_1 = 2$; for Case 2, $x_0 = 0, x_1 = 1, x_2 = 2$; for Case 3, $x_0 = 0, x_1 = \frac{2}{3}, x_2 = \frac{4}{3}, x_3 = 2$.

Case 1. From the text, we know that $c_i = u(x_i)$. In this case, therefore,

$$c_0 = u(x_0) = \left.\frac{1}{1+x}\right|_{x=0} = 1, \qquad c_1 = u(x_1) = \left.\frac{1}{1+x}\right|_{x=2} = \frac{1}{3}$$

$$\begin{aligned} p_1(x) &= u(x_0)L_0^n(x) + u(x_1)L_1^n(x) \\ &= 1L_0^n(x) + \tfrac{1}{3}L_1^n(x) \end{aligned}$$

Because there are only two knots, a first-degree ($n = 1$) polynomial can be passed through them. Therefore, from Eq. (8.9)

$$\begin{aligned} L_0^1(x) &= \left.\frac{(x - x_1)(x - x_2)\cdots(x - x_n)}{(x_0 - x_1)(x_0 - x_2)\cdots(x_0 - x_n)}\right|_{n=1} = \frac{(x - x_1)}{(x_0 - x_1)} \\ &= \frac{x - 2}{0 - 2} = -\frac{1}{2}(x - 2) \end{aligned}$$

$$\begin{aligned} L_1^1(x) &= \left.\frac{(x - x_0)(x - x_2)\cdots(x - x_n)}{(x_1 - x_0)(x_1 - x_2)\cdots(x_1 - x_n)}\right|_{n=1} = \frac{(x - x_0)}{(x_1 - x_0)} \\ &= \frac{x - 0}{2 - 0} = \frac{x}{2} \end{aligned}$$

Therefore,

$$u_1(x) = p_1(x) = -\frac{1}{2}(x - 2) + \frac{1}{3}\left(\frac{x}{2}\right) = 1 - \frac{1}{3}x$$

Case 2. Following the same procedure,

$$c_0 = \frac{1}{1+x}\bigg|_{x=0} = 1, \qquad c_1 = \frac{1}{1+x}\bigg|_{x=1} = \frac{1}{2}, \qquad c_2 = \frac{1}{1+x}\bigg|_{x=2} = \frac{1}{3}$$

$$p_2(x) = L_0^n(x) + \tfrac{1}{2}L_1^n(x) + \tfrac{1}{3}L_2^n(x)$$

Now, we are looking for a second-degree polynomial ($n = 2$)

$$L_0^2(x) = \frac{(x - x_1)(x - x_2)}{(x_0 - x_1)(x_0 - x_2)} = \frac{(x-1)(x-2)}{(-1)(-2)} = \frac{1}{2}(x-1)(x-2)$$

$$L_1^2(x) = \frac{(x - x_0)(x - x_2)}{(x_1 - x_0)(x_1 - x_2)} = \frac{(x-0)(x-2)}{(1)(-1)} = -x(x-2)$$

$$L_2^2(x) = \frac{(x - x_0)(x - x_1)}{(x_2 - x_0)(x_2 - x_1)} = \frac{x(x-1)}{(2)(1)} = \frac{1}{2}x(x-1)$$

Therefore,

$$\begin{aligned} u_2(x) = p_2(x) &= \tfrac{1}{2}(x-1)(x-2) - \tfrac{1}{2}x(x-2) + (\tfrac{1}{3})\tfrac{1}{2}x(x-1) \\ &= 1 - \tfrac{2}{3}x + \tfrac{1}{6}x^2 \end{aligned}$$

Case 3. Analogous calculations lead to

$$u_3(x) = p_3(x) = 1 - \tfrac{89}{105}x + \tfrac{3}{7}x^2 - \tfrac{3}{35}x$$

A graph of these approximations is shown in Figure 8.1. Note that $p_3(x)$ is too close to $u(x)$ to be clearly exhibited.

Notice that the interpolation polynomial of degree 1 replaced the given curves by a straight line. An interpolation polynomial of degree 2, that is, a quadratic, gave an apparently better approximation. In general, however, one should not expect that as n goes to infinity (i.e., as the knots are placed closer and closer), $\|u - p_n(x)\|$ goes to zero. Indeed, a crude estimate of the error can be derived to be

$$\|u - p_n(x)\| \le \frac{\|u^{(n+1)}\|\, h^{n+1}}{4(n+1)} \tag{8.12}$$

where $u^{(n+1)}$ is the $(n + 1)$st derivative of u and $h = (x_{i+1} - x_i)$, the grid size. If u is not sufficiently differentiable (i.e., u is not smooth enough), then $\|u^{(n+1)}\|$ grows too fast as n becomes large. Then increasing n may prove to be a matter of diminishing returns. That is, if $u(x)$ is a rapidly fluctuating function, then one has to look for ways of improving the accuracy of approximation in a more subtle way rather than increasing n in a brute-force manner. An often-used remedy is to use Lagrangian polynomials in a piecewise fashion.

PIECEWISE LAGRANGIAN INTERPOLATION

Consider a function $u(x)$ over an interval $[a, b]$. Select $(n + 1)$ distinct points $x_0, x_1, \ldots, x_n$ in $[a, b]$ as usual. Now these knots can be grouped with $(m + 1)$ knots per group, and one can pass an mth-degree interpolation polynomial through each group of points. For $m = 1$, this results in a piecewise linear Lagrange polynomial

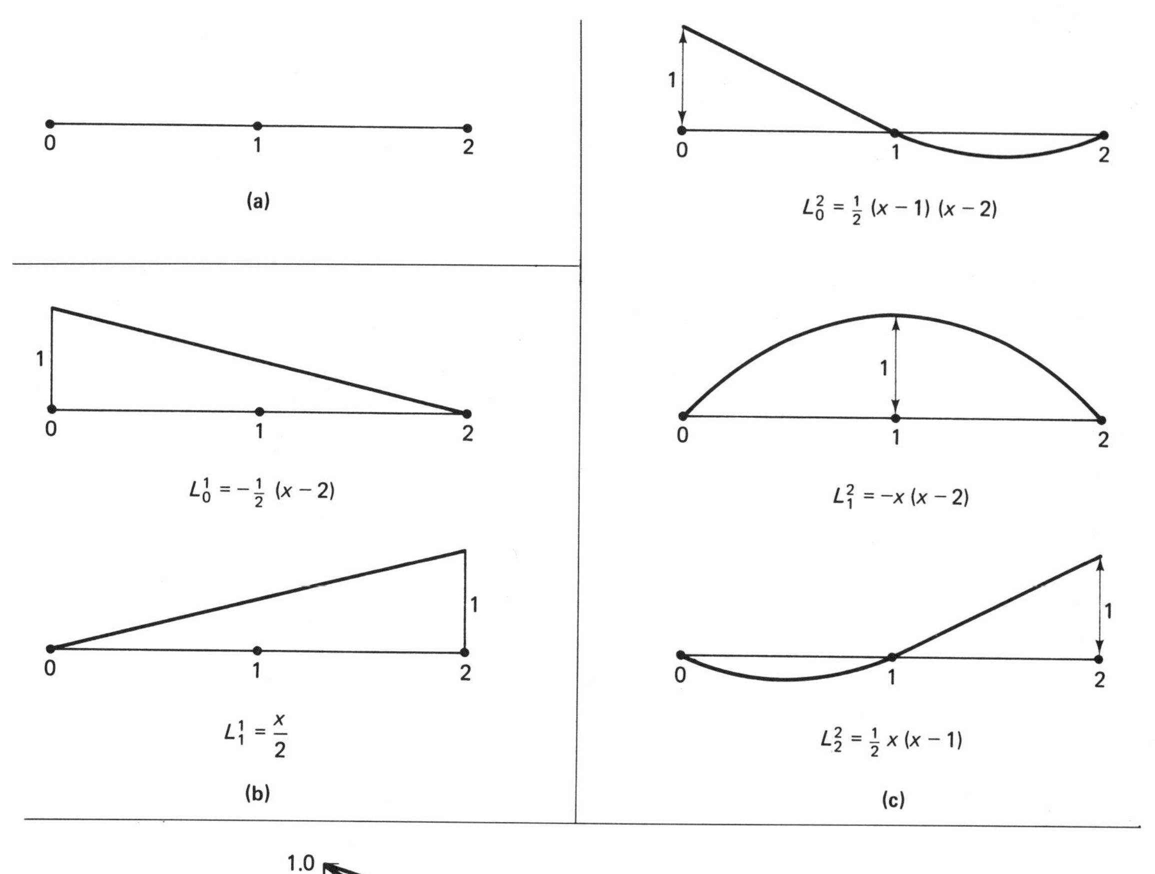

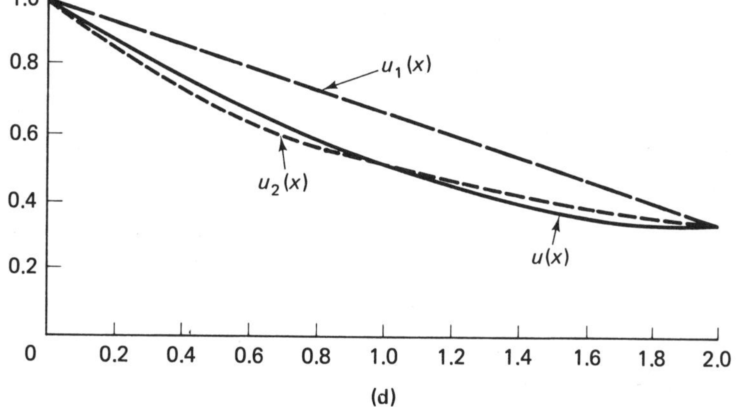

Figure 8.1 Various stages in the approximation of $u(x) = (1 + x)^{-1}$: (a) the interval (0, 2); (b) linear interpolation functions; (c) quadratic interpolation functions; (d) the function $u(x) = (1 + x)^{-1}$ and its successive approximations by Lagrangian polynomials when the approximations are made perfect at specified points in the interval $0 \leq x \leq 2$. Note that $u_1(x)$ is the result of using linear interpolation polynomials, and $u_2(x)$ is the result of quadratic interpolation

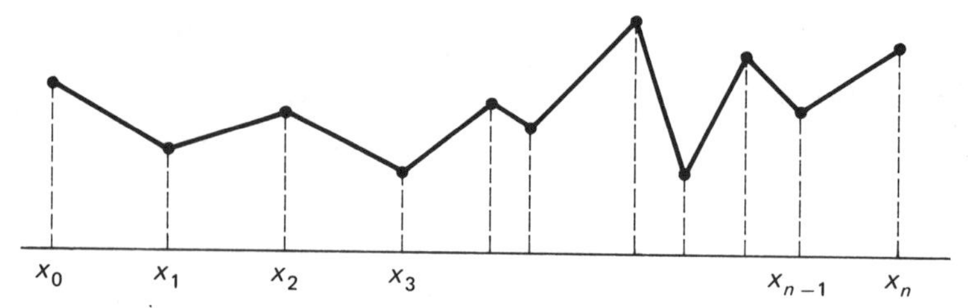

Figure 8.2 Piecewise linear Lagrange polynomial (Reproduced with permission from *Splines and Variational Methods* by P. M. Prenter, © 1975, John Wiley & Sons, Inc., New York)

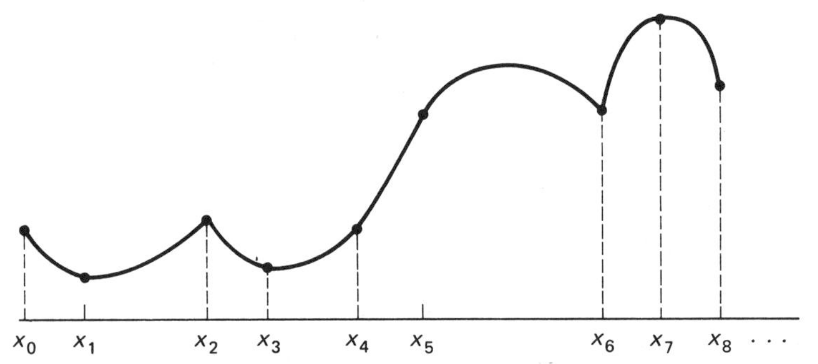

Figure 8.3 Piecewise quadratic Lagrange polynomial (Reproduced with permission from *Splines and Variational Methods* by P. M. Prenter, © 1975, John Wiley & Sons, Inc., New York)

(see Figure 8.2). For $m = 2$, one gets a piecewise quadratic Lagrange polynomial (see Figure 8.3). One can go higher for piecewise cubics, quartics, quintics, and so on. However, the pieced-together function may fail to be differentiable (i.e., possess sharp corners as in Figures 8.2 and 8.3). This idea of piecewise interpolation is not at all new; this is precisely what one does in deriving numerical integration formulas such as the trapezoidal rule, Simpson's rule, and so on.

The actual computation of piecewise Lagrange polynomials is relatively straightfoward. For simplicity, let us consider the problem of using piecewise linear (i.e., $m = 1$) functions. Now $u(x)$ can be approximated by

$$u(x) = u(x_0)L_0^1(x) + u(x_1)L_1^1(x) + \cdots + u(x_n)L_n^1(x) \tag{8.13}$$

where $L_i^1(x)$ has the form shown in Figure 8.4. Such functions are called "hat functions" or ***cardinal basis functions***. Mathematically, $L_i^1(x)$ can be written as

$$L_i^1(x) = \begin{cases} 0, & x \leq x_{i-1} \\ \dfrac{x - x_{i-1}}{x_i - x_{i-1}}, & x_{i-1} \leq x \leq x_i \\ \dfrac{x_{i+1} - x}{x_{i+1} - x_i}, & x_i \leq x \leq x_{i+1} \\ 0, & x \geq x_{i+1} \end{cases} \tag{8.14a}$$

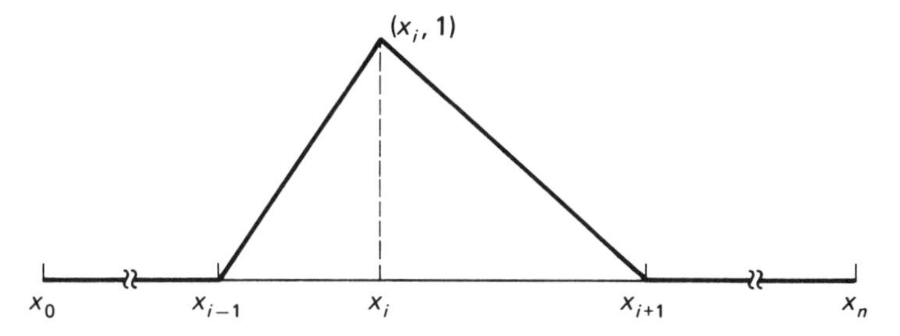

Figure 8.4 Graph of $L_i^1(x)$, $1 \leq i \leq n - 1$ (Reproduced with permission from *Splines and Variational Methods* by P. M. Prenter, © 1975, John Wiley & Sons, Inc., New York)

Note that Eq. (8.14a) merely represents two Lagrange interpolation formulas of the first degree in each of the two intervals $[x_{i-1}, x_i]$ and $[x_i, x_{i+1}]$. Care should be taken in writing $L_i^1(x)$ at both ends of the interval $[a, b]$. For completeness, the graphs of $L_0^1(x)$ and $L_0^1(x)$ are shown in Figure 8.5; the formulas are

$$L_n^1(x) = \begin{cases} \dfrac{x - x_{n-1}}{x_n - x_{n-1}}, & x_{n-1} \leq x \leq x_n \\ 0, & x_0 \leq x \leq x_{n-1} \end{cases} \tag{8.14b}$$

and

$$L_0^1(x) = \begin{cases} 0, & x \geq x_1 \\ \dfrac{x_1 - x}{x_1 - x_0}, & x_0 \leq x \leq x_1 \end{cases} \tag{8.14c}$$

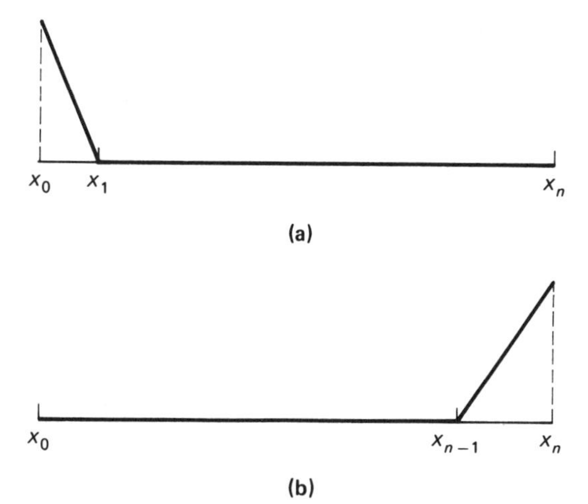

Figure 8.5 (a) Graph of $L_0^1(x)$; (b) graph of $L_n^1(x)$ (Reproduced with permission from *Splines and Variational Methods* by P. M. Prenter, © 1975, John Wiley & Sons, Inc., New York)

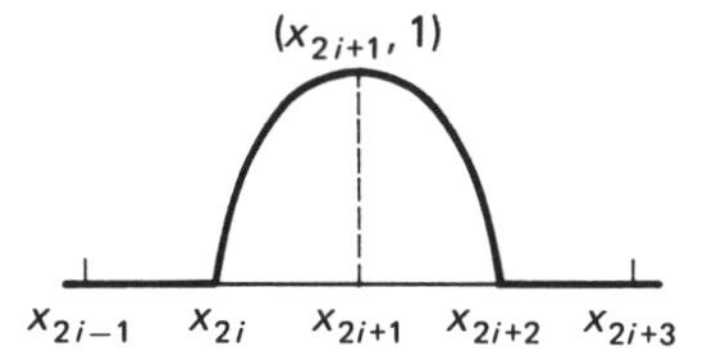

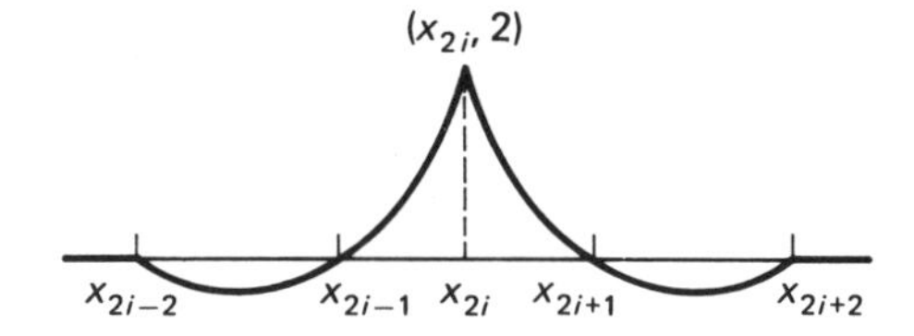

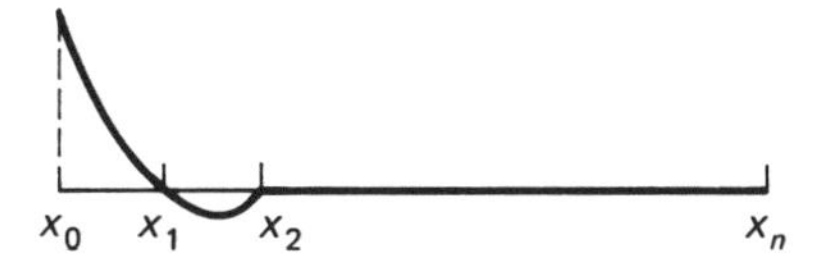

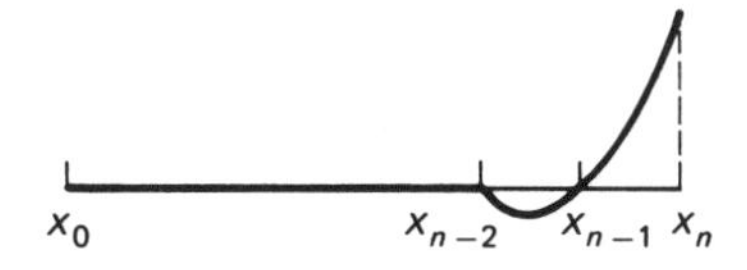

Figure 8.6 Basis piecewise quadratic Lagrange polynomials (Reproduced with permission from *Splines and Variational Methods* by P. M. Prenter, © 1975, John Wiley & Sons, Inc., New York)

Extension of these ideas to higher-order polynomials is straightforward. Figure 8.6 shows typical shapes of cardinal basis functions ($L_i^2(x)$) used in piecewise quadratic Lagrange interpolation. Actual derivation of the algebraic expressions for these basis functions is left as an exercise.

A NOTE ON NOTATION

To avoid confusion, some of the notation being used is restated here: The functions $N_i(x)$, as they are used in Eq. (8.3), are variously called ***trial functions***, ***interpolation functions***, or ***shape functions***. Throughout this book N_i stands for a very general situation. In Lagrangian interpolation, $N_i = L_i^n$, where the superscript n stands for the degree of the polynomial. The polynomial $p_n(x)$, $a \leq x \leq b$, is obtained from a linear combination of the functions N_i. See Eq. (8.7). Therefore, the functions $N_i(x) = L_i^n(x)$ are also called ***basis functions***. In piecewise Lagrange interpolation, the polynomial $p_n(x)$, $x_i \leq x \leq x_{i+1}$, is however defined over the subinterval $[x_i, x_{i+1}] \subseteq [a, b]$. In such cases, the notation $s(x)$ will be used to describe a concatenation of the polynomials $p_n(x)$ defined in each subinterval.

For notational simplicity, the cardinal basis function is often denoted by

$$L_i^n(x_j) = \delta_{ij} = \begin{cases} 1 & \text{if } i = j \\ 0 & \text{if } i \neq j \end{cases} \tag{8.15}$$

This notation only conveys the information that the cardinal basis function $L_i^n(x)$ attains the value unity at the knot x_i and zero at all other knots. Each of the functions L_i^n, $i = 0, 1, \ldots, n$, is a polynomial of the nth degree between the knots. This is an important point to remember. In Fourier series analysis, for example, the basis functions sine and cosine assume values throughout $[a, b]$; the cardinal basis functions, though defined throughout $[a, b]$, are zero outside a small subinterval of $[a, b]$.

8.4. PIECEWISE HERMITIAN INTERPOLATION

One major drawback of the piecewise Lagrangian interpolation is the presence of sharp corners on the graph. One method of overcoming this lack of smoothness is to glue, or piece together, Hermite polynomials rather than Lagrange polynomials. In interpolations using Hermite polynomials, one not only interpolates $u(x)$ at each knot but also interpolates a specified number of consecutive higher derivatives of u at each knot. The simplest situation in this context is to demand only that

$$u(x_i) = p(x_i) \qquad \text{and} \qquad u'(x_i) = p'(x_i) \tag{8.16}$$

for $0 \leq i \leq n$, and the resulting polynomial would be a cubic Hermite polynomial. If one demands

$$u(x_i) = p(x_i), \qquad u'(x_i) = p'(x_i), \qquad u''(x_i) = p''(x_i) \tag{8.17}$$

for $0 \leq i \leq n$, the result would be a quintic Hermite polynomial. This process can be carried on and on. Thus one method of eliminating sharp corners occurring in piecewise Lagrange interpolation is to simply stop using them and use instead piecewise Hermite polynomials, the simplest of them being piecewise cubic Hermites. Of course, other options are available, but, for a moment, let us concentrate on the construction of piecewise cubic Hermites.

Formally, given $u(x)$ and the knots $a = x_0 < x_1 < \cdots < x_n = b$, how would one go about constructing piecewise cubic Hermite polynomials? One considers any typical interval $[x_{i-1}, x_i]$ and constructs a cubic Hermite polynomial $p_i(x)$ in that interval, and the procedure is repeated for all intervals. To construct a cubic Hermite in $[x_{i-1}, x_i]$, the necessary constraints are the interpolatory constraints, that is,

$$\begin{aligned} u(x_{i-1}) &= p_i(x_{i-1}), \qquad & u'(x_{i-1}) &= p_i'(x_{i-1}) \\ u(x_i) &= p_i(x_i), \qquad & u'(x_i) &= p_i'(x_i) \end{aligned} \tag{8.18}$$

The concatenation of all these polynomials $p_i(x)$ for $i = 1, 2, \ldots, n-1$ is denoted by $s(x)$. That is,

$$s(x) = p_i(x), \qquad x_{i-1} \leq x \leq x_i, \qquad i = 1, 2, \ldots, n-1 \tag{8.19}$$

This is to say that $s(x)$ is a cubic polynomial inside each of the intervals (x_{i-1}, x_i) and matches the function values and first derivative values [i.e., satisfies Eq. (8.16)] at each of the interior knots x_i, $1 \leq i \leq n-1$. Because $s(x)$ is a polynomial inside (x_{i-1}, x_i), it is infinitely differentiable in that open interval. However, only the first derivatives are guaranteed to exist at the knots. Therefore, $s \in C^1[a, b]$. Because no attempt was made to match higher derivatives at the knots, $s''(x)$ is not likely to exist at the interior knots. To get greater smoothness, one must turn either to higher-order Hermite polynomials, as suggested already, or turn to the use of cubic polynomial splines, which is the subject matter of Section 8.5. For the moment, attention is focused on the actual computation of cubic Hermitian polynomials.

Matrix Method. Let

$$s(x) = \begin{cases} a_{10} + a_{11}x + a_{12}x^2 + a_{13}x^3 = p_1(x), & x_0 \le x \le x_1 \\ a_{20} + a_{21}x + a_{22}x^2 + a_{23}x^3 = p_2(x), & x_1 \le x \le x_2 \\ \cdot \qquad\qquad \cdot \quad \cdot & \cdot \\ \cdot \qquad\qquad \cdot \quad \cdot & \cdot \\ \cdot \qquad\qquad \cdot \quad \cdot & \cdot \\ a_{n0} + a_{n1}x + a_{n2}x^2 + a_{n3}x^3 = p_n(x), & x_{n-1} \le x \le x_n \end{cases} \tag{8.20}$$

and force the conditions in Eq. (8.18) for $i = 1, 2, \ldots, n$. For instance, at $i = 1$,

$$p_1(x) = a_{10} + a_{11}x + a_{12}x^2 + a_{13}x^3$$

Using Eq. (8.18),

$$\begin{aligned} p_1(x_0) &= a_{10} + a_{11}x_0 + a_{12}x_0^2 + a_{13}x_0^3 = y(x_0) \\ p_1(x_1) &= a_{10} + a_{11}x_1 + a_{12}x_1^2 + a_{13}x_1^3 = y(x_1) \\ p_1'(x_0) &= 0 + a_{11} + 2a_{12}x_0 + 3a_{13}x_0^2 = y'(x_0) \\ p_1'(x_1) &= 0 + a_{11} + 2a_{12}x_1 + 3a_{13}x_1^2 = y'(x_0) \end{aligned} \tag{8.21}$$

The right sides are known quantities, and x_0 and x_1 are the known locations of the knots. Therefore, a_{10}, a_{11}, a_{12}, and a_{13} can be computed by inverting the matrix

$$\begin{bmatrix} 1 & x_0 & x_0^2 & x_0^3 \\ 1 & x_1 & x_1^2 & x_1^3 \\ 0 & 1 & 2x_0 & 3x_0^2 \\ 0 & 1 & 2x_1 & 3x_1^2 \end{bmatrix}$$

This process is repeated for $i = 2, 3, \ldots, n$.

Cardinal Basis Method. The preceding method tends to become cumbersome for large n. An appropriate choice of a basis simplifies the computation considerably. Once the basis functions are chosen, one can write

$$s(x) = \sum_{i=0}^{n} u(x_i)H_{0i}(x) + \sum_{i=0}^{n} u'(x_i)H_{1i}(x) \tag{8.22}$$

The first part of the right side is our attempt to interpolate the function values, and the second part represents our attempt to interpolate the first derivative values. The first subscript on H mnemonically conveys this information: a 0 (zero) reminding function values, a 1 (one) reminding first derivatives. To construct H_{0i}, one sets

$$H_{0i} = \begin{cases} a_{i3}x^3 + a_{i2}x^2 + a_{i1}x + a_{i0}, & x_{i-1} \le x \le x_i \\ a_{i+1,3}x^3 + a_{i+1,2}x^2 + a_{i+1,1}x + a_{i+1,0}, & x_i \le x \le x_{i+1} \\ 0, & \text{elsewhere} \end{cases} \tag{8.23}$$

The coefficients can now be calculated using the required constraints as described next.

First, $H_{0i}(x)$ should satisfy

$$H_{0i}(x_j) = \delta_{ij}, \qquad H'_{0i}(x_j) = 0, \qquad 0 \leq j \leq n \tag{8.24}$$

In particular, notice that $H_{0i}(x) \equiv 1$ when $x = x_i$, and is zero at all other knots. Further, $H_{0i}(x)$ has zero slope at all knots and is a cubic polynomial between the knots. Also $H_{0i}(x) \equiv 0$ for all $x \geq x_{i+1}$ and $x \leq x_{i-1}$. Now notice that $H_{0i}(x)$ has a double zero (root) at $x = x_{i+1}$. Therefore, one can write

$$H_{0i}(x) = (x - x_{i+1})^2(\alpha x + \beta), \qquad x_i \leq x \leq x_{i+1} \tag{8.25}$$

Now forcing $H_{0i}(x_i) = 1$ and $H'_{0i}(x_i) = 0$, Eq. (8.25) becomes, for $x_i \leq x \leq x_{i+1}$,

$$H_{0i}(x) = \frac{(x_{i+1} - x)^2}{(x_{i+1} - x_i)^3}[2(x_{i+1} - x) + (x_{i+1} - x_i)] \tag{8.26}$$

An analogous argument can be made for the interval $x_{i-1} \leq x \leq x_i$, and one can conclude

$$H_{0i}(x) = \begin{cases} \dfrac{(x - x_{i-1})^2}{(x_i - x_{i-1})^3}[2(x_i - x) + (x_i - x_{i-1})], & x_{i-1} \leq x \leq x_i \\ \dfrac{(x_{i+1} - x)^2}{(x_{i+1} - x_i)^3}[2(x_{i+1} - x) + (x_{i+1} - x_i)], & x_i \leq x \leq x_{i+1} \\ 0, \quad \text{elsewhere} \end{cases} \tag{8.27}$$

The procedure can be repeated for $H_{1i}(x)$ with the constraints

$$H_{1i}(x_j) = 0, \qquad H'_{1i}(x_j) = \delta_{ij}, \qquad 0 \leq j \leq n \tag{8.28}$$

and conclude

$$H_{1i}(x) = \begin{cases} \dfrac{(x - x_{i-1})^2(x - x_i)}{(x_i - x_{i-1})^2}, & x_{i-1} \leq x \leq x_i \\ \dfrac{(x - x_{i+1})^2(x - x_i)}{(x_{i+1} - x_i)^2}, & x_i \leq x \leq x_{i+1} \\ 0, \quad \text{elsewhere} \end{cases} \tag{8.29}$$

Graphs of H_{0i}, H_{1i}, H_{00}, H_{n0}, H_{01}, and H_{n1} are shown in Figure 8.7.

8.5. INTERPOLATION WITH POLYNOMIAL SPLINES

The next logical step in interpolation is to approximate $u(x)$ by

$$\tilde{u}(x) = c_1N_1(x) + c_2N_2(x) + \cdots + c_nN_n(x) \tag{8.30}$$

which satisfies the interpolation and the smoothness constraints. Recall that the characteristic of the smoothness constraint is the requirement that $\tilde{u}(x)$ be twice continuously differentiable. This can be done, for instance, by going to piecewise quintic Hermite polynomials. If one prefers to work with lower-order polynomials, then the desired smoothness can be built into the basis functions $\mathcal{B} = \{B_1, B_2, \ldots, B_n\}$, and one is thus left with purely interpolatory constraints to worry about. This is

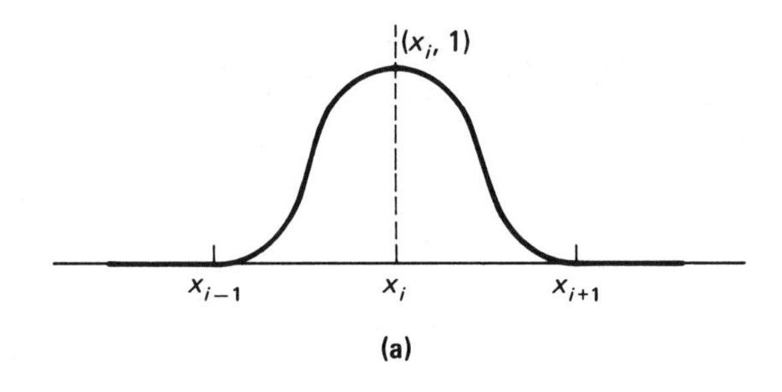

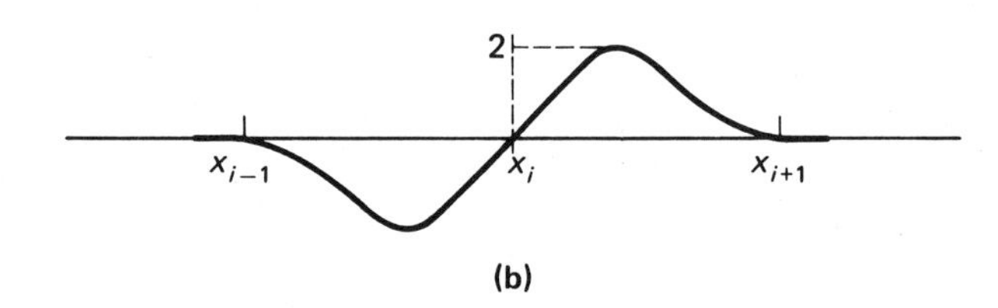

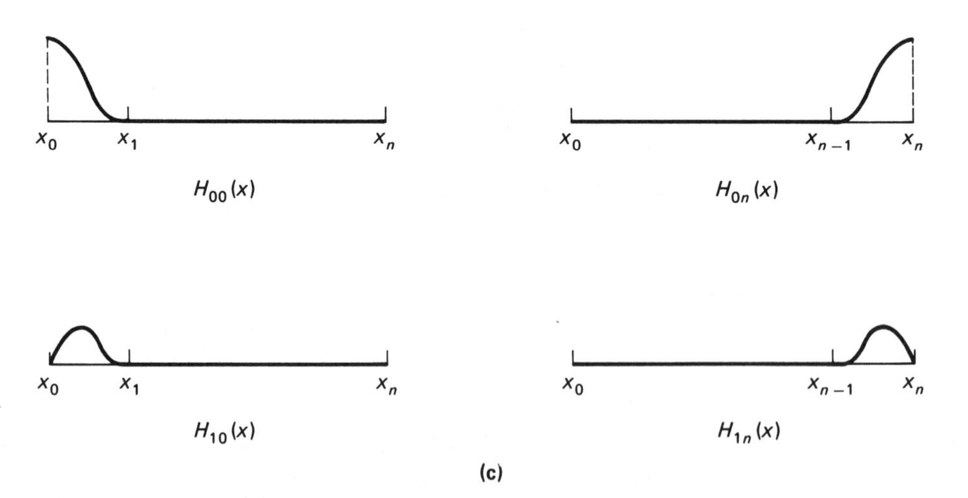

Figure 8.7 Graphs of basis Hermitian polynomials: (a) graph of $H_{0i}(x)$, $1 \leq i \leq n-1$; (b) graph of $H_{1i}(x)$, $1 \leq i \leq n-1$; (c) graphs of $H_{00}(x)$, $H_{0n}(x)$, $H_{10}(x)$, and $H_{1n}(x)$ (Reproduced with permission from *Splines and Variational Methods* by P. M. Prenter, © 1975, John Wiley & Sons, Inc., New York)

the central idea behind the development of spline functions. Although the concept of spline function can be greatly generalized, particular attention will be paid here to cubic splines or simply "splines."

A cubic spline function is a piecewise cubic polynomial that is twice continuously differentiable. On a given interval $[a, b]$ with a given set of knots, denoted by the "partition" π, there are several such cubic splines, and the set of all such cubic spline functions is generally denoted by $S_3(\pi)$. Therefore, the space $S_3(\pi)$ is nothing but the set of all functions $s(t) \in C^2[a, b]$ that reduce to cubic polynomials on each subinterval (x_i, x_{i+1}) of $[a, b]$. It can be proved that there exists a unique function $s(x) \in S_3(\pi)$ satisfying the pure interpolatory constraints

$$s(x_i) = u(x_i), \qquad 0 \leq i \leq n \tag{8.31}$$

$$s'(x_0) = u'(x_0), \qquad s'(x_n) = u'(x_n) \tag{8.32}$$

Notice that, by stating the problem as here, we have already replaced a mixed constraint problem with a pure interpolatory constraint problem by seeking $s \in C^2[a, b]$.

The cubic spline interpolate $s(x)$ can now be written, once again as before, in terms of an appropriately chosen set of basis functions $B_i(x)$. Particularly, $s(x)$ can be written as

$$s(x) = c_{-1}B_{-1}(x) + c_0 B_0(x) + \cdots + c_n B_n(x) + c_{n+1}B_{n+1}(x) \tag{8.33}$$

The $B_i(x)$ in Eq. (8.33), called the B-splines, are defined by (see Figure 8.8 also)

$$B_i(x) = \begin{cases} \dfrac{1}{h^3}[(x - x_{i-2})^3], & \text{if } x \in [x_{i-2}, x_{i-1}] \\ \dfrac{1}{h^3}[h^3 + 3h^2(x - x_{i-1}) + 3h(x - x_{i-1})^2 - 3(x - x_{i-1})^3], & \\ \qquad\qquad\qquad\qquad\qquad\qquad\qquad\qquad \text{if } x \in [x_{i-1}, x_i] & \\ \dfrac{1}{h^3}[h^3 + 3h^2(x_{i+1} - x) + 3h(x_{i+1} - x)^2 - 3(x_{i+1} - x)^3], & \\ \qquad\qquad\qquad\qquad\qquad\qquad\qquad\qquad \text{if } x \in [x_i, x_{i+1}] & \\ \dfrac{1}{h^3}[(x_{i+2} - x)^3], & \text{if } x \in [x_{i+1}, x_{i+2}] \\ 0, & \text{otherwise} \end{cases} \tag{8.34}$$

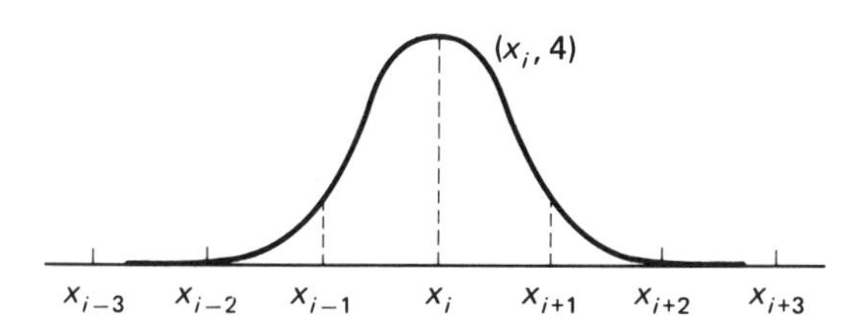

Figure 8.8 Graph of $B_i(x)$ (Reproduced with permission from *Splines and Variational Methods* by P. M. Prenter, © 1975, John Wiley & Sons, Inc., New York)

Notice that the definition of B-splines involves the introduction of four additional knots x_{-2}, x_{-1}, x_{n+1}, and x_{n+2}. The following facts about each $B_i(x)$ can be quickly verified: (1) $B_i(x)$ is twice continuously differentiable on the entire real line; (2) $B_i(x) \equiv 0$ for $x \geq x_{i+1}$ and $x \leq x_{i-1}$. Furthermore,

$$B_i(x_j) = \begin{cases} 4, & \text{if } j = i \\ 1, & \text{if } j = i-1 \text{ or } j = i+1 \\ 0, & \text{if } j = i-2 \text{ or } j = i+2 \end{cases} \tag{8.35}$$

Imposing the constraints in Eqs. (8.31) and (8.32) on Eq. (8.33), one gets

$$\begin{aligned} s'(x_0) &= c_{-1}B'_{-1}(x_0) + c_0 B'_0(x_0) + \cdots + c_{n+1}B'_{n+1}(x_0) \\ &= y'(x_0) \\ s(x_i) &= c_{-1}B_{-1}(x_i) + c_0 B_0(x_i) + \cdots + c_{n+1}B'_{n+1}(x_i) \\ &= y(x_i) \\ s'(x_n) &= c_{-1}B'_{-1}(x_n) + c_0 B'_0(x_n) + \cdots + c_{n+1}B'_{n+1}(x_n) \\ &= y'(x_n) \end{aligned} \tag{8.36}$$

with $i = 0, 1, 2, \ldots, n$. Thus, Eq. (8.36) represents a system of $n + 3$ equations $\mathbf{Ac} = \mathbf{b}$, where $\mathbf{c} = (c_{-1}, c_0, c_1, \ldots, c_n, c_{n+1})^T$ and $\mathbf{b} = (y'(x_0), y(x_0), y(x_1), \ldots, y(x_n), y'(x_n))^T$, and the coefficient matrix $\mathbf{A}$ is given by

$$\mathbf{A} = \begin{bmatrix} -\frac{3}{h} & 0 & \frac{3}{h} & 0 & 0 & \cdots & 0 \\ 1 & 4 & 1 & 0 & 0 & \cdots & 0 \\ 0 & 1 & 4 & 1 & 0 & \cdots & 0 \\ \cdot & \cdot & \cdot & \cdot & \cdot & \cdot & \cdot \\ \cdot & \cdot & \cdot & \cdot & \cdot & \cdot & \cdot \\ \cdot & \cdot & \cdot & \cdot & \cdot & \cdot & \cdot \\ 0 & 0 & \cdots & & 1 & 4 & 1 \\ 0 & 0 & \cdots & & -\frac{3}{h} & 0 & \frac{3}{h} \end{bmatrix}$$

Example. Approximate $u = (1 + x)^{-1}$ in the interval $[0, 2]$ by choosing equispaced knots at $0, \frac{1}{2}, 1, \frac{3}{2}$, and 2.

Choose the four extra knots at $x_{-2} = -1, x_{-1} = -\frac{1}{2}, x_5 = \frac{5}{2}$, and $x_6 = \frac{6}{2}$. Now set

$$s(x) = c_{-1}B_{-1}(x) + c_0 B_0(x) + c_1 B_1(x) + \cdots + c_4 B_4(x) + c_5 B_5(x)$$

The value of grid spacing is $h = \frac{1}{2}$. So $1/h = 2$. Also

$$s'(0) = y'(0) = -(1 + 0)^{-2} = -1$$

$$s(x_i) = s\left(\frac{i}{2}\right) = y\left(\frac{i}{2}\right) = \left(1 + \frac{i}{2}\right)^{-1}$$

$$s'(2) = y'(2) = -(1 + 2)^{-2} = -\frac{1}{9}$$

Thus the system $\mathbf{Ac} = \mathbf{b}$ is simply

$$\begin{bmatrix} -6 & 0 & 6 & 0 & 0 & 0 & 0 \\ 1 & 4 & 1 & 0 & 0 & 0 & 0 \\ 0 & 1 & 4 & 1 & 0 & 0 & 0 \\ 0 & 0 & 1 & 4 & 1 & 0 & 0 \\ 0 & 0 & 0 & 1 & 4 & 1 & 0 \\ 0 & 0 & 0 & 0 & 1 & 4 & 1 \\ 0 & 0 & 0 & 0 & -6 & 0 & 6 \end{bmatrix} \begin{bmatrix} c_{-1} \\ c_0 \\ c_1 \\ c_2 \\ c_3 \\ c_4 \\ c_5 \end{bmatrix} = \begin{bmatrix} -1 \\ 1 \\ \frac{2}{3} \\ \frac{1}{2} \\ \frac{2}{5} \\ \frac{1}{3} \\ -\frac{1}{9} \end{bmatrix}$$

The solution, rounded to five decimal places, is

$$\begin{aligned} c_{-1} &= 0.27429 & c_3 &= 0.06583 \\ c_0 &= 0.15452 & c_4 &= 0.05505 \\ c_1 &= 0.10763 & c_5 &= 0.04731 \\ c_2 &= 0.08163 \end{aligned}$$

Using these values in the preceding equation for $s(x)$, the values of $s(x)$ for various of x were evaluated by writing a simple computer program. The results are

x	0.2	0.4	0.6	0.8	1.2	1.4	1.8
$y(x)$	0.83333	0.71429	0.62500	0.55556	0.45456	0.41667	0.35714
$s(x)$	0.83166	0.71332	0.62536	0.55573	0.45440	0.41660	0.35714

8.6. INTERPOLATING FUNCTIONS OF SEVERAL VARIABLES

The basic ideas described in the preceding sections can be generalized to higher dimensions. The crux of the finite element method lies in these generalizations. Consider the problem of approximating a function $u(x, y)$, of two independent variables, which is defined over a region Ω. A set of randomly chosen points $\{(x_1, y_1), (x_2, y_2), \ldots, (x_n, y_n)\}$, called the knots, or random grid points, are chosen in Ω, and the problem is to find a polynomial $P(x, y)$ of degree n in two variables that solves the interpolation problem

$$P(x_i, y_i) = u(x_i, y_i), \qquad 0 \leq i \leq n \tag{8.37}$$

By $P(x, y)$ of degree n, we mean any function such as

$$P(x, y) = c_0 + c_1 x + c_2 y + c_3 x^2 + c_4 xy + c_5 y^2 + \cdots + c_k y^n \tag{8.38}$$

One should exercise care in visualizing higher-dimensional analogs to the simple piecewise interpolation schemes discussed earlier. This problem will be considerably simplified if the discussion is confined to regular grids (such as triangular and rectangular) instead of a purely random grid. That is, the points (x_1, y_1), (x_2, y_2), . . . , (x_n, y_n) in Ω are connected to form elemental shapes such as rectangles and triangles. Now one attempts piecewise interpolation by developing interpolation polynomials on each of these elemental shapes. This procedure has one drawback: If Ω has a curved boundary, one cannot possibly cover it with rectangles or triangles. If Ω indeed has a curved boundary, then one has to use the ***isoparametric*** transformations. This topic will be briefly described in Section 9.8.

APPROXIMATION ON TRIANGULAR GRIDS

Consider a region Ω. This region is covered by a triangular grid. The vertexes of the triangles are denoted by $v_0, v_1, \ldots, v_n$. A typical triangulation of a region Ω and a typical elemental triangle are shown in Figure 8.9. (As will be seen subsequently, the triangular shapes possess a number of properties that will simplify computation.) One can now develop piecewise interpolates on these triangles. For instance, one can

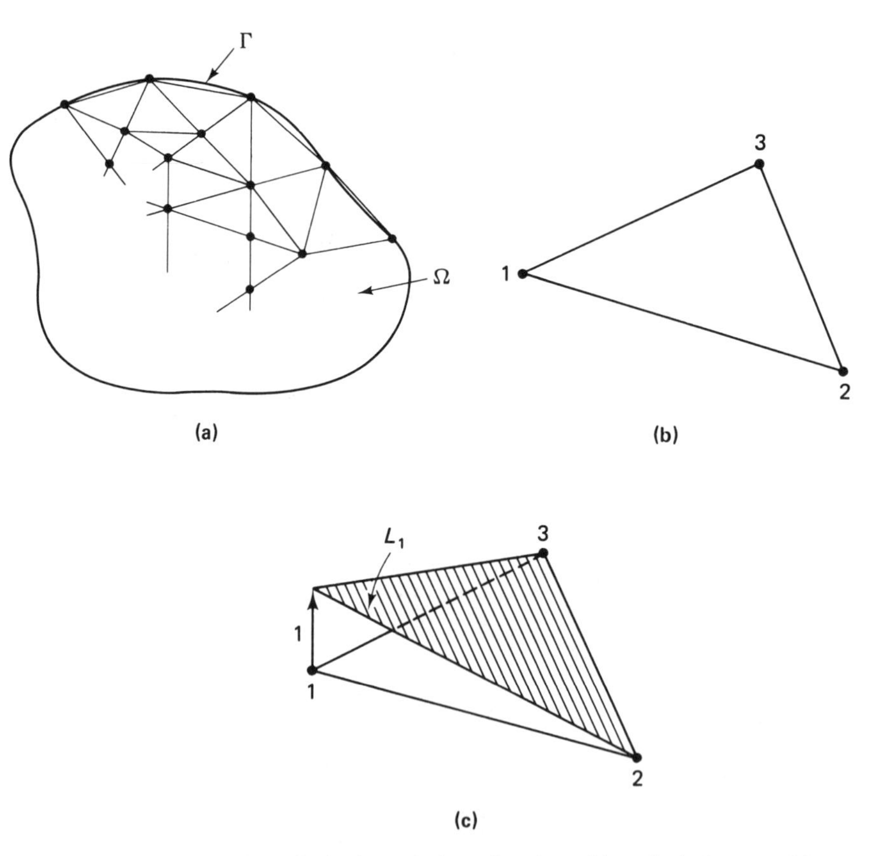

Figure 8.9 (a) Typical triangulation of region; (b) typical element; (c) isometric view of a typical linear shape function

develop a piecewise linear Lagrange polynomial fit. That is, on each triangle T_i, we construct a plane

$$P_i(x, y) = a_i x + b_i y + c_i \tag{8.39}$$

interpolating $u(x, y)$ at the vertexes of the triangle T_i. Then the piecewise interpolate $s(x, y)$ to u is

$$s(x, y) = P_i(x, y) \qquad \text{if } (x, y) \in T_i \tag{8.40}$$

The canonical basis for computing $s(x, y)$ is the set $\{L_1(x, y), L_2(x, y), \ldots, L_n(x, y)\}$, which solves the interpolation problem

$$L_i(v_j) = \delta_{ij}, \qquad 0 \leq i, j \leq n \tag{8.41}$$

One can also develop a piecewise quadratic polynomial fit on each triangle T_i by constructing a quadratic surface

$$P_i(x, y) = c_0 + c_1 x + c_2 y + c_3 x^2 + c_4 xy + c_5 y^2 \tag{8.42}$$

This procedure obviously requires three points on each side of the triangle. This extra point is obtained by placing an additional knot at the midpoint of each side of the triangle (see Figure 8.9). Then $u(x, y)$ is interpolated not only at the vertexes of the triangles but also at midpoints.

Natural Coordinate Systems. The interpolations indicated in the preceding become computationally cumbersome because the sides of a typical triangle can no longer be parallel to the Cartesian coordinate axes. That is, the Cartesian coordinate system is not a ***natural coordinate system*** for triangular grids. However, a natural coordinate system that is local to a given triangle can be developed as described next.

The development of natural coordinate systems is best understood by tracing the idea from a one-dimensional case. Consider a line segment of length l as shown in Figure 8.10(a). Consider a point P whose "natural coordinates" are designated by (L_1, L_2) [for the moment, do not confuse these L's and the $L_i^n(x)$, etc., used for Lagrange polynomials]. Now a relationship between the natural coordinates (L_1, L_2) and the Cartesian coordinates, x_1, x_2, and x, can be written, from inspection, as

$$\begin{bmatrix} 1 \\ x \end{bmatrix} = \begin{bmatrix} 1 & 1 \\ x_1 & x_2 \end{bmatrix} \begin{bmatrix} L_1 \\ L_2 \end{bmatrix} \tag{8.43}$$

where

$$L_1 = \frac{l_1}{l} \qquad \text{and} \qquad L_2 = \frac{l_2}{l} \tag{8.44}$$

Because $L_1 + L_2 = 1$, only one of the natural coordinates is independent, as it should be in a one-dimensional problem. The inverse of Eq. (8.43) is

$$\begin{bmatrix} L_1 \\ L_2 \end{bmatrix} = \frac{1}{l} \begin{bmatrix} x_2 & -1 \\ -x_1 & 1 \end{bmatrix} \begin{bmatrix} 1 \\ x \end{bmatrix} \tag{8.45}$$

Notice that $L_1 = (x_2 - x)/l = (x_2 - x)/(x_2 - x_1)$ and $L_2 = (x - x_1)/(x_2 - x_1)$, which are nothing but the linear Lagrange interpolates of the first degree. Notice also that $L_1 = 1$ and $L_2 = 0$ at node 1. Similarly, $L_2 = 1$ and $L_1 = 0$ at node 2. Therefore,

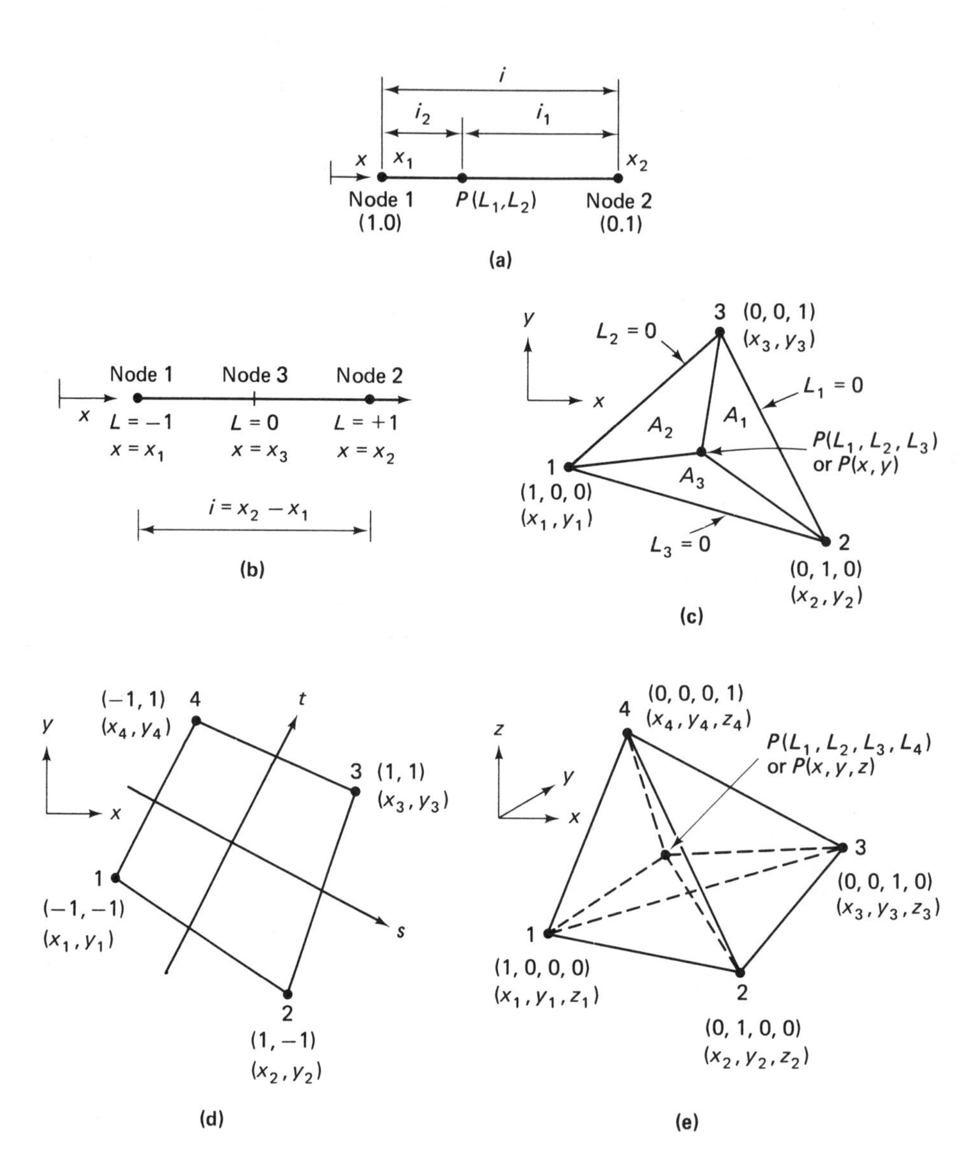

Figure 8.10 Natural coordinate systems: (a) natural coordinates for line element; (b) simple natural coordinate for line element; (c) triangular coordinates; (d) quadrilateral coordinates; (e) tetrahedral coordinates (From *Introduction to the Finite Element Method* by C. S. Desai and John F. Abel, © 1972 by Litton Educational Publishing, Inc. Reprinted by permission of Van Nostrand Reinhold Company)

linear interpolation formulas can be written in terms of the natural coordinates (here, they can also be referred to as ***length coordinates***) as

$$u(x) = u_1 L_1 + u_2 L_2$$

Sometimes it becomes necessary to differentiate and integrate $u(x)$. Using the chain rule,

$$\frac{du}{dx} = \frac{\partial u}{\partial L_1}\frac{\partial L_1}{\partial x} + \frac{\partial u}{\partial L_2}\frac{\partial L_2}{\partial x}$$

where

$$\frac{\partial L_1}{\partial x} = \frac{-1}{x_2 - x_1}, \qquad \frac{\partial L_2}{\partial x} = \frac{1}{x_2 - x_1}$$

It is also often necessary to perform integration of $u(x)$ over the length of an element. A convenient formula for this purpose is

$$\int_{x_1}^{x_2} L_1^{\alpha} L_2^{\beta}\, dx = \frac{\alpha!\,\beta!(x_2 - x_1)}{(\alpha + \beta + 1)!} \tag{8.46}$$

Table 8.1 gives the values of this integral for various values of α and β.

The idea of natural coordinates is now extended to triangles. Consider a triangle with vertexes at (x_1, y_1), (x_2, y_2), and (x_3, y_3) in the Cartesian coordinate system. Now consider a point P inside the triangle whose natural coordinates are designated by (L_1, L_2, L_3). A relationship between (L_1, L_2, L_3) and (x, y) can be written, from

Table 8.1. Integrals of length coordinates

$$\frac{1}{x_2 - x_1}\int_{x_1}^{x_2} L_1^{\alpha} L_2^{\beta}\, dx = \frac{x}{y}$$

$\alpha + \beta$	α	β	x	y
0	0	0	1	1
1	1	0	1	2
2	2	0	2	6
2	1	1	1	6
3	3	0	3	12
3	2	1	1	12
4	4	0	12	60
4	3	1	3	60
4	2	2	2	60
5	5	0	10	60
5	4	1	2	60
5	3	2	1	60
6	6	0	60	420
6	5	1	10	420
6	4	2	4	420
6	3	3	3	420

inspection, as

$$\begin{bmatrix} 1 \\ x \\ y \end{bmatrix} = \begin{bmatrix} 1 & 1 & 1 \\ x_1 & x_2 & x_3 \\ y_1 & y_2 & y_3 \end{bmatrix} \begin{bmatrix} L_1 \\ L_2 \\ L_3 \end{bmatrix} \tag{8.47}$$

If one lets

$$L_1 = \frac{A_1}{A}, \qquad L_2 = \frac{A_2}{A}, \qquad L_3 = \frac{A_3}{A} \tag{8.48}$$

where A_1, A_2, and A_3 are the areas, respectively, of the smaller triangles opposite vertexes V_1, V_2, and V_3 (denoted in Figure 8.10 respectively by the numbers 1, 2, and 3), then $L_1 + L_2 + L_3 = 1$. Because of this, L_1, L_2, and L_3 are called ***area coordinates.*** It is instructive to note once again that only two of these area coordinates are independent. Solving for L_1, L_2, and L_3,

$$\begin{bmatrix} L_1 \\ L_2 \\ L_3 \end{bmatrix} = \frac{1}{2A} \begin{bmatrix} 2A_{23} & b_1 & a_1 \\ 2A_{31} & b_2 & a_2 \\ 2A_{12} & b_3 & a_3 \end{bmatrix} \begin{bmatrix} 1 \\ x \\ y \end{bmatrix} \tag{8.49}$$

where $2A$, twice the area of triangle $V_1V_2V_3$, is given by

$$2A = a_3b_2 - a_2b_3 = a_1b_3 - a_3b_1 = a_2b_1 - a_1b_2 \tag{8.50}$$

and A_{23}, for instance, defines the area of the triangle for which nodes V_2, V_3 and the origin of the xy coordinate system are the vertexes. Also

$$\begin{aligned} a_1 &= x_3 - x_2, \qquad & b_1 &= y_2 - y_3 \\ a_2 &= x_1 - x_3, \qquad & b_2 &= y_3 - y_1 \\ a_3 &= x_2 - x_1, \qquad & b_3 &= y_1 - y_2 \end{aligned} \tag{8.51}$$

These formulas can be written in a general notation. For a triangle whose vertexes V_i, V_j, and V_k are located, respectively, at (x_i, y_i), (x_j, y_j), and (x_k, y_k), one can write compactly

$$2(\text{Area of triangle } V_iV_jV_k) = \det \begin{vmatrix} 1 & x_i & y_i \\ 1 & x_j & y_j \\ 1 & x_k & y_k \end{vmatrix} \tag{8.52}$$

$$\begin{aligned} A_{ij} &= x_iy_j - x_jy_i \\ a_i &= x_k - x_j \\ b_i &= y_j - y_k \end{aligned} \tag{8.53}$$

The other formulas are obtained by changing the subscripts in a cyclic permutation as

$$\begin{matrix} i & j & k \\ \downarrow & \downarrow & \downarrow \\ j & k & i \\ \downarrow & \downarrow & \downarrow \\ k & i & j \end{matrix}$$

Now let us look at some of the properties of the area coordinates. For instance, at vertex V_i, the value $L_i = 1$ and the values $L_j = 0$ for $j \neq i$. That is, each area coordinate is unity at one node and zero at the other nodes. Thus, in the case of a *linear triangle* (such as the one shown in Figure 8.9b), the shape functions N_i are the same as the area coordinates. That is,

$$N_1 = L_1, \qquad N_2 = L_2, \qquad N_3 = L_3 \tag{8.54}$$

Some of the other properties of area coordinates are shown in Figure 8.11.

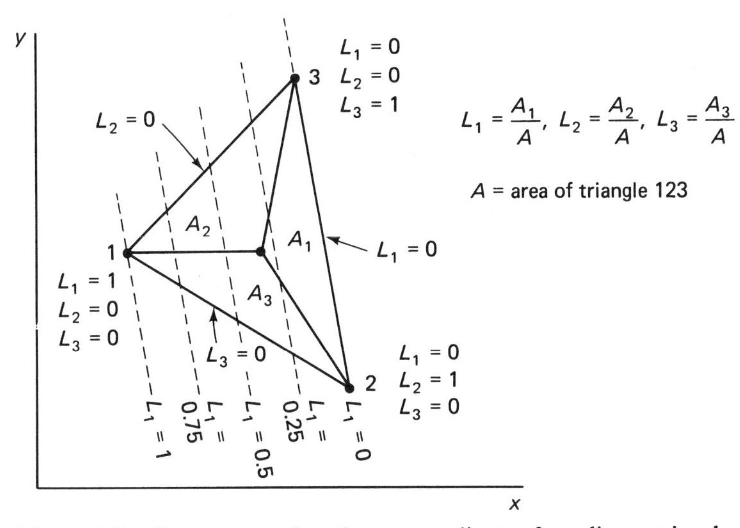

Figure 8.11 Some properties of area coordinates for a linear triangle (Reproduced with permission from *The Finite Element Method for Engineers* by K. H. Huebner, © 1975, John Wiley & Sons, Inc., New York)

As in the linear case, the field variable u can be interpreted as a function of L_1, L_2, and L_3 instead of x and y. Then differentiation and integration of u can be performed as follows:

$$\frac{\partial u}{\partial x} = \frac{\partial u}{\partial L_1}\frac{\partial L_1}{\partial x} + \frac{\partial u}{\partial L_2}\frac{\partial L_2}{\partial x} + \frac{\partial u}{\partial L_3}\frac{\partial L_3}{\partial x} = \sum_{i=1}^{3} \frac{b_i}{2A}\frac{\partial u}{\partial L_i} \tag{8.55a}$$

$$\frac{\partial u}{\partial y} = \frac{\partial u}{\partial L_1}\frac{\partial L_1}{\partial y} + \frac{\partial u}{\partial L_2}\frac{\partial L_2}{\partial y} + \frac{\partial u}{\partial L_3}\frac{\partial L_3}{\partial y} = \sum_{i=1}^{3} \frac{a_i}{2A}\frac{\partial u}{\partial L_i} \tag{8.55b}$$

where

$$\frac{\partial L_i}{\partial x} = \frac{b_i}{2A}, \qquad \frac{\partial L_i}{\partial y} = \frac{c_i}{2A}, \qquad i = 1, 2, 3 \tag{8.55c}$$

Similarly,

$$\int_A L_1^{\alpha} L_2^{\beta} L_3^{\gamma}\, dA = \frac{\alpha!\beta!\gamma!}{(\alpha + \beta + \gamma + 2)!} 2A \tag{8.56}$$

Table 8.2 gives the values of this integral for various values of α, β, and γ.

Table 8.2. Integrals of area coordinates

$$\frac{1}{A}\int_A L_1^{\alpha}L_2^{\beta}L_3^{\gamma}\,dA = \frac{x}{y}$$

$\alpha+\beta+\gamma$	α	β	γ	x	y
0	0	0	0	1	1
1	1	0	0	1	3
2	2	0	0	2	12
2	1	1	0	1	12
3	3	0	0	6	60
3	2	1	0	2	60
3	1	1	1	1	60
4	4	0	0	12	180
4	3	1	0	3	180
4	2	2	0	2	180
4	2	1	1	1	180
5	5	0	0	60	1,260
5	4	1	0	12	1,260
5	3	2	0	6	1,260
5	3	1	1	3	1,260
5	2	2	1	2	1,260
6	6	0	0	180	5,040
6	5	1	0	30	5,040
6	4	2	0	12	5,040
6	4	1	1	6	5,040
6	3	3	0	9	5,040
6	3	2	1	3	5,040
6	2	2	2	1	5,040

HIGHER-ORDER TRIANGLES

The three-node linear triangle discussed previously is only the first in an infinite series of triangular elements that can be specified. Figure 8.12 shows a portion of the family of higher-order elements obtained by assigning additional nodes to the triangles. Note that the quadratic triangle has three nodes in each direction, and these three nodes are sufficient to specify a quadratic variation of u within the element and along the element boundaries. Stated more formally, for triangles with nodes arrayed as in Figure 8.12, a complete polynomial of order n (i.e., of an order necessary to give C^0 continuity) requires $\frac{1}{2}(n+1)(n+2)$ nodes for its specification.

Proceeding in a fashion analogous to the development for a linear triangle, one can easily establish the relations

$$N_1 = L_1(2L_1 - 1), \qquad N_2 = L_2(2L_2 - 1) \qquad N_3 = L_3(2L_3 - 1) \qquad (8.57a)$$

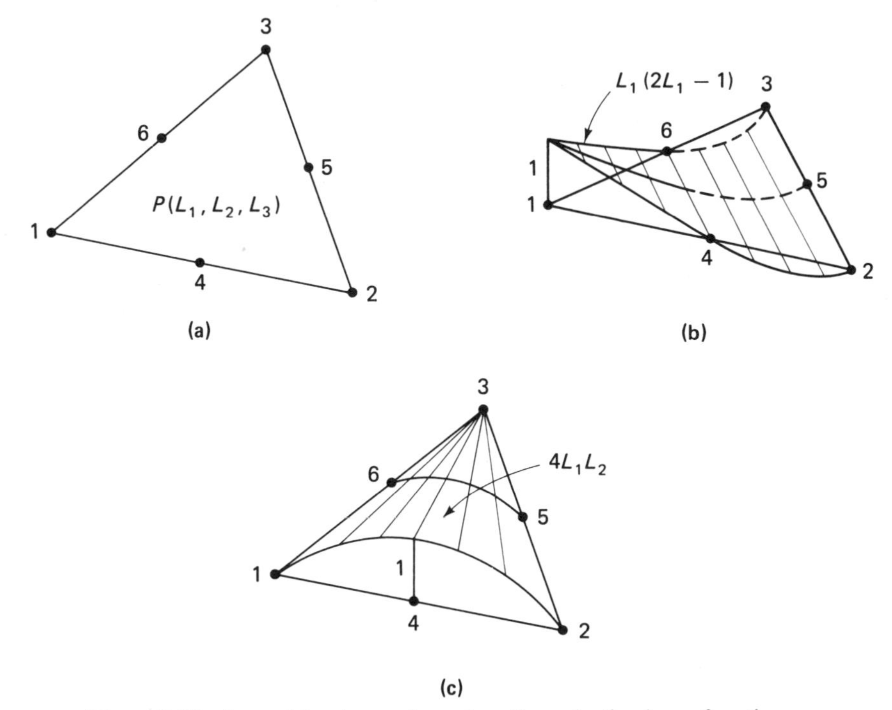

Figure 8.12 Isometric views of graphs of quadratic shape functions. Note that the magnitude of the functions are plotted normal to the planar element; however, the functions may represent displacements that occur either in the plane or normal to it (From *Introduction to the Finite Element Method* by C. S. Desai and John F. Abel, © 1972 by Litton Educational Publishing, Inc. Reprinted by permission of Van Nostrand Reinhold Company)

for the corner (or primary) nodes of the quadratic triangle, and the relations

$$N_4 = 4L_1L_2, \qquad N_5 = 4L_2L_3, \qquad N_6 = 4L_3L_1 \tag{8.57b}$$

for the midside (or secondary) nodes.

For a ***cubic triangle***, the corresponding formulas are

$$N_1 = \tfrac{1}{2}L_1(3L_1 - 1)(3L_1 - 2), \quad \text{etc., for corner nodes} \tag{8.58a}$$

$$N_4 = \tfrac{9}{2}L_1L_2(3L_1 - 1), \quad \text{etc., for the midside nodes} \tag{8.58b}$$

$$N_{10} = 27L_1L_2L_3, \quad \text{for the interval node} \tag{8.58c}$$

APPROXIMATION ON RECTANGULAR GRIDS

Consider now a typical rectangle $R = \{(x, y)\colon a \leq x \leq b, c \leq y \leq d\}$ obtained after superimposing a rectangular grid on the region Ω. There is no loss of generality in assuming that the sides of the rectangle are parallel to the coordinate axes. Now one can attempt to construct higher-dimensional analogs to the one-dimensional piecewise Lagrange, and other, polynomial interpolates that were discussed earlier. Toward this end, let $\Pi_x: a = x_0 < x_1 < \cdots < x_n = b$ and $\Pi_y: c = y_0 < y_1 < \cdots < y_m = d$ be the partitions of the intervals $[a, b]$ and $[c, d]$, respectively.

For concreteness, let $\{L_i^n(x): 0 \leq i \leq n\}$ and $\{L_j^m(y): 0 \leq j \leq m\}$ be, respectively, the Lagrange polynomials of degree m and n solving the interpolation problem

$$\begin{aligned} L_i^n(x_k) &= \delta_{ik}, \qquad 0 \leq i, k \leq n \\ L_j^m(y_\mu) &= \delta_{j\mu}, \qquad 0 \leq j, \mu \leq m \end{aligned} \tag{8.59}$$

Now, if

$$L_{ij}(x, y) = L_i^n(x) \cdot L_j^m(y) \tag{8.60}$$

then it can be shown that the required interpolating polynomial

$$P(x, y) = \sum_{\substack{0 \leq i \leq n \\ 0 \leq j \leq m}} u(x_i, y_j) L_{ij}(x, y) \tag{8.61}$$

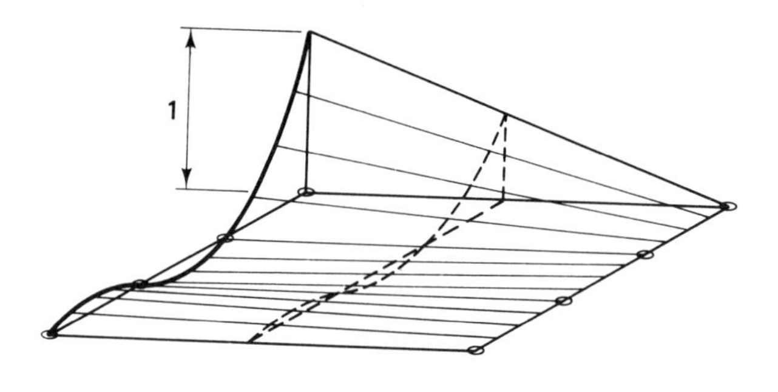

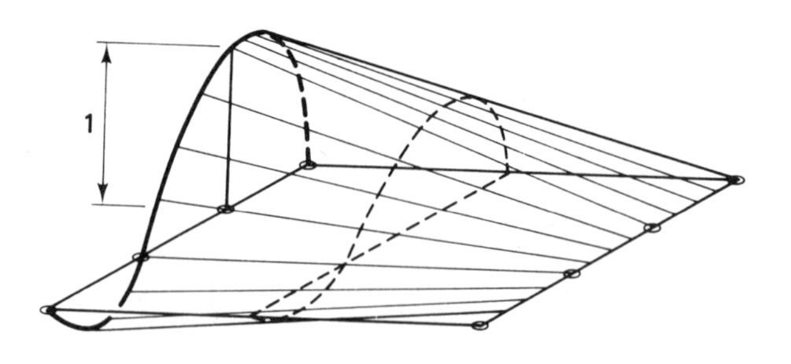

(a)

Figure 8.13 Typical basis Lagrangian polynomials on a quadrilateral element: (a) Degree 1 in x and degree 3 in y; (b) degree 2 in both x and y; (c) degree 5 in x and degree 4 in y (Reproduced with permission from *The Finite Element Method*, 3rd ed., by O. C. Zienkiewicz, © 1977, McGraw-Hill Book Company (U.K.) Limited)

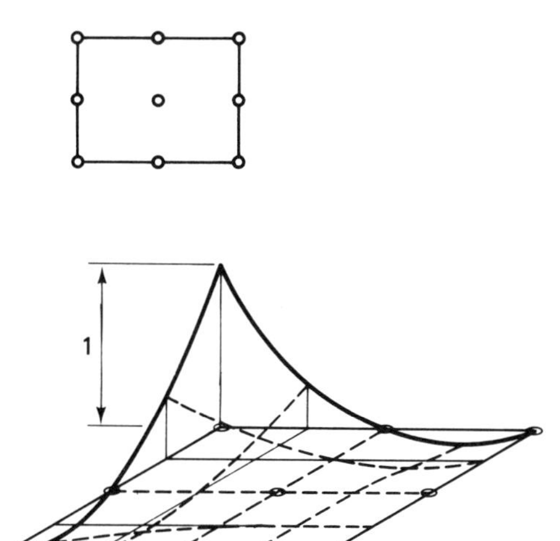

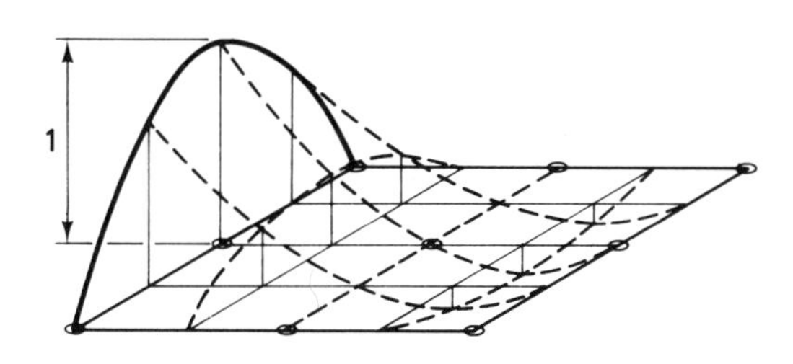

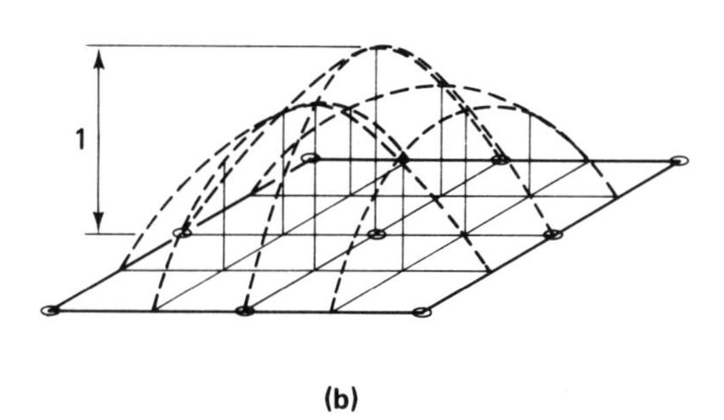

(b)

Figure 8.13 (Continued)

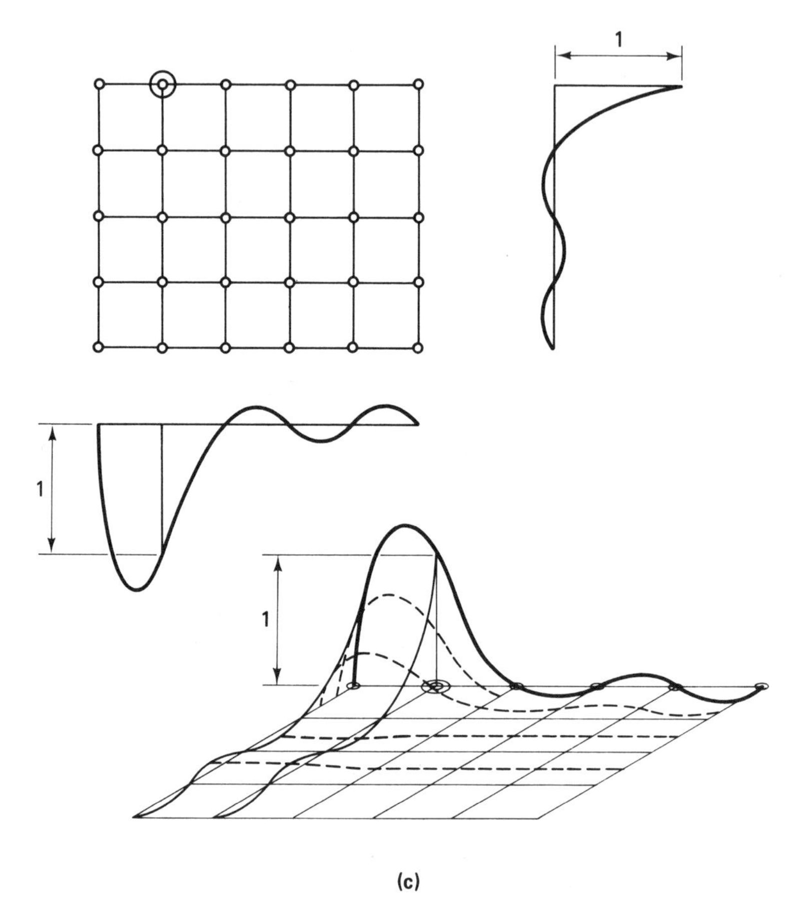

(c)

Figure 8.13 (Continued)

is a polynomial of degree n in x and m in y that solves the interpolation problem posed in Eq. (8.37) at the beginning of this section. The choice of $L_{ij}(x, y)$, which is tantamount to choosing n and m in Eq. (8.41) is usually made either to simplify computations or to satisfy some physical requirements in a given problem. Graphs of some of these shape functions are displayed in Figure 8.13. Such figures help in visualizing the fairly complex shape of the Lagrange polynomials in two dimensions. Notice that Figure 8.13b, describing the quadratic Lagrange function, has a knot in the center of the rectangle R. For cubic Lagrange polynomials, four such internal knots will be required.

Piecewise Interpolating Polynomials. Among all types of polynomials, the piecewise Lagrange polynomials are very popular with engineers. These piecewise Lagrange polynomials are formed by taking sums of products of piecewise basis functions in the x and y directions. If the basis functions are linear in both directions, then they are called piecewise bilinear. The names biquadratic, bicubic, and so on, carry similar meaning. Figure 8.14 shows the graph of a basis piecewise bilinear Lagrange polynomial (i.e., a shape function, in engineering parlance) $L_{ij}(x, y)$. The four regions around a grid point (x_i, y_j) are shown in Figure 8.14a and the four parts of the shape function in Figure 8.14b. It is significant to note that each shape function vanishes outside the four quadrants shown. This feature makes the shape functions shown in Figure 8.14 very attractive in solving partial differential equations. The

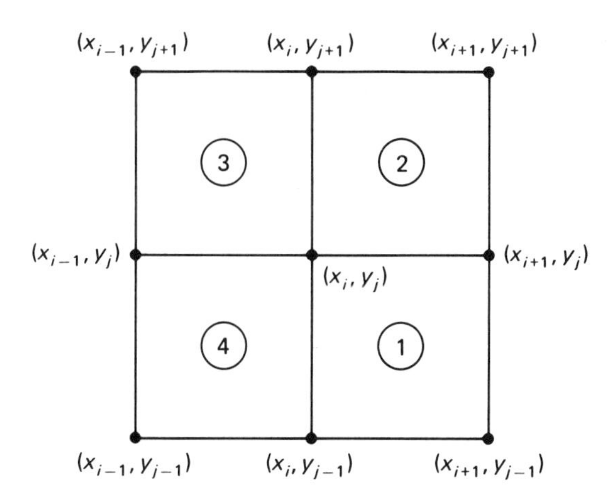

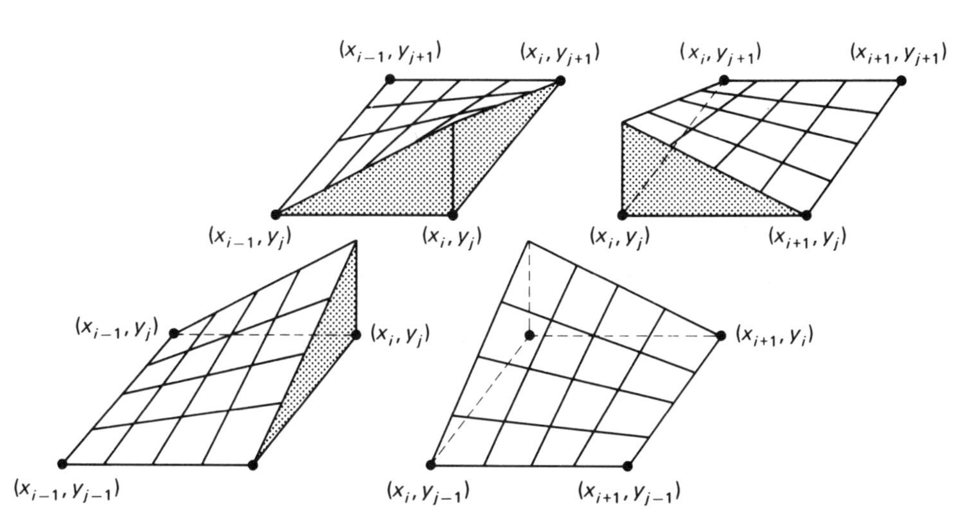

Figure 8.14 A canonical basis bilinear Lagrange polynomial over its region of support (Reproduced with permission from *Splines and Variational Methods* by P. M. Prenter, © 1975, John Wiley & Sons, Inc., New York)

earlier versions of the finite element method used these piecewise linear functions extensively. A mathematical expression for these bilinear piecewise basis functions can be written, by inspection of Figure 8.14, as

$$L_{ij}(x, y) = \begin{cases} \dfrac{(x_{i+1} - x)(y - y_{j-1})}{(x_{i+1} - x_i)(y_j - y_{j-1})} & \text{on region 1} \\ \dfrac{(x_{i+1} - x)(y_{j+1} - y)}{(x_{i+1} - x_i)(y_{j+1} - y_j)} & \text{on region 2} \\ \dfrac{(x - x_{i-1})(y_{j+1} - y)}{(x_i - x_{i-1})(y_{j+1} - y_j)} & \text{on region 3} \\ \dfrac{(x - x_{i-1})(y - y_{j-1})}{(x_i - x_{i-1})(y_j - y_{j-1})} & \text{on region 4} \\ 0, & \text{otherwise} \end{cases} \tag{8.62}$$

Now the piecewise bilinear interpolate $s(x, y)$ can be written as

$$s(x, y) = \sum_{\substack{0 \le i \le n \\ 1 \le j \le m}} u(x_i, y_j) L_{ij}(x, y) \tag{8.63}$$

Other higher-order rectangular elements can be formulated in precisely the same way. Figure 8.15 shows a small sample out of the infinite number of elements that can be constructed.

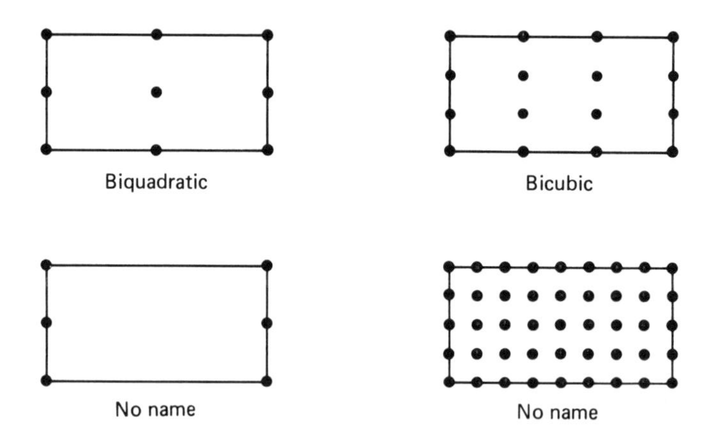

Figure 8.15 A sample of elements from the infinite series of Lagrange rectangles (Reproduced with permission from *The Finite Element Method for Engineers* by K. H. Huebner, © 1975, John Wiley & Sons, Inc., New York)

The usefulness of rectangular elements is somewhat limited because higher-order Lagrangian elements require internal nodes. These may be eliminated by the condensation process when desired, but this requires extra manipulation. Another important aspect of Lagrangian elements is that interpolation functions are never complete polynomials, and they possess geometric isotropy only when equal numbers of nodes are used in both x and y directions. The appeal of rectangular elements stems from the ease with which interpolation functions can be found for them.

A more useful set of rectangular elements is known as the serendipity family. These elements contain only exterior nodes. In terms of the natural coordinate system shown in Figure 8.16, the interpolation functions for the serendipity family are as follows.

1. Linear element:

$$N_i(s, t) = \tfrac{1}{4}(1 + ss_i)(1 + tt_i) \tag{8.64}$$

2. Quadratic element:

$$N_i(s, t) = \begin{cases} \tfrac{1}{4}(1 + ss_i)(1 + tt_i)(ss_i + tt_{i-1}), & \text{for nodes at } s = \pm 1, t = \pm 1 \\ \tfrac{1}{2}(1 - s^2)(1 + tt_i), & \text{for nodes at } s = 0,\ t = \pm 1 \\ \tfrac{1}{2}(1 + ss_i)(1 - t^2), & \text{for nodes at } s = \pm 1,\ t = 0 \end{cases} \tag{8.65}$$

3. Cubic element:

$$N_i(s, t) = \begin{cases} \tfrac{1}{32}(1 + ss_i)(1 + tt_i)\,[9(s^2 + t^2) - 10], & \text{for nodes at } s = \pm 1, t = \pm 1 \\ \tfrac{9}{32}(1 + ss_i)(1 - t^2)(1 + 9tt_i), & \text{for nodes at } s = \pm 1, t = \pm \tfrac{1}{3}, \text{ etc.} \end{cases} \tag{8.66}$$

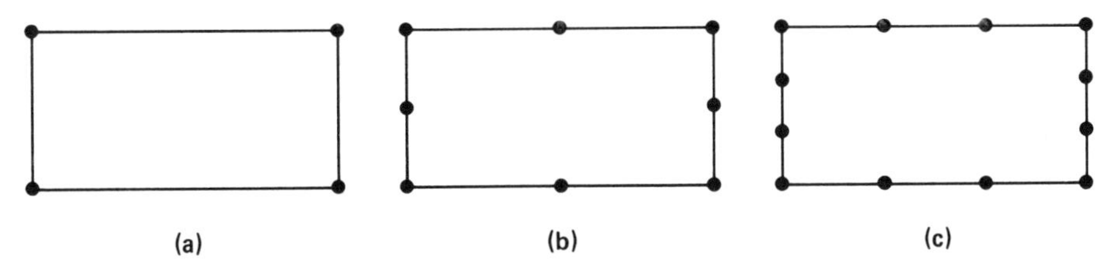

Figure 8.16 Useful elements of the serendipity family: (a) linear; (b) quadratic; (c) cubic (Reproduced with permission form *The Finite Element Method for Engineers* by K. H. Huebner, © 1975, John Wiley & Sons, Inc., New York)

SUGGESTIONS FOR FURTHER READING

1. *Many books covering the material in this chapter are available. For an excellent mathematical coverage of interpolation and approximation, refer to*

   P. M. PRENTER, *Splines and Variational Methods*, John Wiley & Sons, Inc., New York, 1975.

2. *For books devoted to splines, consult*

   M. SCHULTZ, *Spline Analysis*, Prentice-Hall, Inc., Englewood Cliffs, N.J., 1973.

3. *For a description of area coordinates, integration procedures, and isoparametric elements, refer to*

O. C. ZIENKIEWICZ, *The Finite Element Method in Engineering Science*, McGraw-Hill Book Co., New York, 1971, and

C. S. DESAI and J. F. ABEL, *Introduction to the Finite Element Method: A Numerical Method for Engineerng Analysis*, Van Nostrand Reinhold Co., New York, 1972.

4. *Another standard book covering various aspects of finite element method and approximation is by*

G. STRANG and G. J. FIX, *An Analysis of the Finite Element Method*, Prentice-Hall, Inc., Englewood Cliffs, N.J., 1973.

EXERCISES

1. Consider a segment of a rod of length $L = x_j - x_i$. The temperature of the rod at x_i is u_i and at x_j it is u_j. The temperature at any point x in (x_i, x_j) is $u(x)$, where $u(x) = c_0 + c_1 x$. Show that an expression for $u(x)$ can be written as $u(x) = u(x_i)N_i + u(x_j)N_j$, where

$$N_i = \frac{x_j - x}{L} \quad \text{and} \quad N_j = \frac{x - x_i}{L}$$

(The line segment representing the rod is also called a one-dimensional simplex.)

2. Consider a triangular plate of area A in the xy plane with vertexes i, j, and k located, respectively, at (x_i, y_i), (x_j, y_j), and (x_k, y_k). The temperature of the plate at these vertexes is known to be u_i, u_j, and u_k. We wish to determine the temperature $u = u(x, y)$ at any point within the plate. If u can be written as $u = c_0 + c_1 x + c_2 y = u_i N_i + u_j N_j + u_k N_k$, then

(a) show that

$$N_i = \frac{1}{2A}(a_i + b_i x + c_i y)$$

where $a_i = x_j y_k - x_k y_j$, $b_i = y_j - y_k$, and $c_i = x_k - x_j$. Derive similar expressions for N_j and N_k.

(b) Show also that N_i evaluated at node i is unity.

(c) Evaluate $\partial u/\partial x$ and show that it is a constant.

3. Extend the idea presented in Problems 1 and 2 to three dimensions. For this purpose, consider a three-dimensional simplex element (a tetrahedron) located in the xyz-plane with nodes at i, j, k, and l. Assume that u can be written as $u = c_0 + c_1 x + c_2 y + c_3 z$. Specifically, show that

$$u = [1 \quad x \quad y \quad z]\mathbf{C}^{-1}\mathbf{u}$$

where $\mathbf{C}$ is a 4×4 array, analogous to the one used in Eq. (8.52), and the determinant of $\mathbf{C}$ is equal to six times the volume of the tetrahedral element and $\mathbf{u} = (u_i, u_j, u_k, u_l)^T$.

4. Repeat Problem 3 if the coordinates of the tetrahedral element are given to be i: (1, 2, 1), j: (0, 0, 0), k: (2, 0, 0), and l: (1, 0, 3). Verify that, for this case, $N_i = y/2$, $N_j = (6 - 3x - y - z)/6$, $N_k = (3x - y - z)/6$, and $N_l = (-y + 2z)/6$.

5. So far we considered only one unknown (the temperature) at each node. That is, the unknown is a scalar quantity. Consider a two-dimensional problem. Let the unknown quantity at node i be a vector quantity $\mathbf{u}_i = (u_{1i}, u_{2i})$. For easy visualization, assume that $\mathbf{u}_i$ is the displacement at node i. Then u_{1i} is the component of displacement in the x-direction, and so on. Now write expressions for the displacements along x and y directions at a typical point in a triangular element.

6. For the one-, two- and three-dimensional simplex elements considered previously show that the sum of the shape function values always adds up to unity at any point within the element in question. (Indeed, $\sum_{\beta=1}^{r} N_\beta = 1$ serves as a convergence criterion for the finite element method.)

7. What can you conclude about the rows of $\mathbf{C}^{-1}$ in Problem 4 if the convergence criterion $N_i + N_j + N_k + N_l = 1$ is to be satisfied?

8. Show that

$$\int_{l_{ij}} \begin{bmatrix} N_i \\ N_j \\ N_k \end{bmatrix} [N_i \quad N_j \quad N_k]\, dl = \frac{l_{ij}}{6} \begin{bmatrix} 2 & 1 & 0 \\ 1 & 2 & 0 \\ 0 & 0 & 0 \end{bmatrix}$$

where l_{ij} is the length of a side of a simplex triangle between nodes i and j, and N_i, N_j, and N_k are linear shape functions.

9. Consider an element of length l with three nodes i, j, and k located at 0, $l/2$, and l. The interpolation polynomial $u = c_0 + c_1x + c_2x^2$ is used to approximate the value of temperature $u(x)$ along the element. If u_i, u_j, and u_k are known values of temperature at nodes i, j, and k, then

(a) Show that u can be written as $u = N_iu_i + N_ju_j + N_ku_k$, where

$$N_i = \left(1 - \frac{2x}{l}\right)\left(1 - \frac{x}{l}\right), \qquad N_j = \frac{4x}{l}\left(1 - \frac{x}{l}\right), \qquad N_k = -\frac{x}{l}\left(1 - \frac{2x}{l}\right)$$

(b) Verify that $\sum_{\beta=i,j,k} N_\beta = 1$.

10. The general form of the approximating polynomial for a one-dimensional element is

$$u = c_0 + c_1x + c_2x^2 + \cdots + c_{r-1}x^{r-1}$$

where r is the number of nodes. Show that the shape functions N_β can always be rearranged to look like the approximating function. That is, show that N_β can be rearranged to look like

$$N_\beta = a_0 + a_1x + a_2x^2 + \cdots + a_{r-1}x^{r-1}$$

11. Consider the problem of approximating a function $y(x)$ using cubic splines over the interval [0, 1]. For this purpose, [0, 1] is divided into $0 = x_0 < x_1 < \cdots < x_i < x_{i+1} < \cdots < x_n = 1$. The function $f(x)$ is now approximated over the ith interval (x_i, x_{i+1}) by the cubic

$$y = a_i(x - x_i)^3 + b_i(x - x_i)^2 + c_i(x - x_i) + d_i$$

(a) Show that the following relations are valid:

$$a_i = \frac{(S_{i+1} - S_i)}{6h_i}, \qquad b_i = \frac{S_i}{2},$$

$$c_i = \frac{y_{i+1} - y_i}{h_i} - \frac{2h_iS_i + h_iS_{i+1}}{6}, \qquad d_i = y_i$$

where $h_i = x_{i+1} - x_i$ and $S_i = y''$ evaluated at x_i. In this interpolation, we not only want the function values to match, but also require the slopes and curvatures to match. Consider two consecutive intervals (x_{i-1}, x_i) and (x_i, x_{i+1}) and invoke the condition that the slopes of y be the same at x_i. Now show that the following relation is valid:

$$h_{i-1}S_{i-1} + (2h_{i-1} + 2h_i)S_i + h_iS_{i+1} = 6\frac{y_{i+1} - y_i}{h_i} - \frac{y_i - y_{i-1}}{h_{i-1}}$$

(b) Write the preceding equation in full for $i = 2$ to $n - 1$ and discuss how you would solve the system for S_i, $i = 1$ to n. Explicitly consider three cases: (i) $S_1 = 0$ and $S_n = 0$, (ii) $S_1 = S_2$ and $S_{n-1} = S_n$, and (iii) S_1 is a linear extrapolation of S_3 and S_2, and S_n is a linear extrapolation of S_{n-2} and S_{n-1}.

(c) Fit the following data (for $y = x^3 - 8$) to a cubic spline curve.

x	0	1	2	3	4
y	-8	-7	0	19	56

Calculate the second-derivative values of y and compare with the S_i values obtained previously.

12. A general equation for calculating the shape functions on triangular elements is

$$N_\beta = \prod_{\alpha=1}^{n} \frac{F_\alpha}{F_{\alpha\,|\,L_1, L_2, L_3}}$$

where n is the order of the triangle. The functions F_α are derived from equations (written in terms of area coordinates) of n lines that pass through all the nodes except the one of interest. If the equation of the straight line is of the form $L_1 = c$, then $F_\alpha = L_1 - c$. The denominator is the value of F_α evaluated using the coordinates of node β.

(a) Usng the preceding rule, show that the shape functions given in Eq. (8.57) are valid.

(b) Repeat and verify Eq. (8.58).

(c) Show that N_{15} for a quintic triangular element is

$$N_{15} = 32L_1L_2L_3(4L_3 - 1)$$

where the location of node 15 for the quintic case is shown by an x in the following figure:

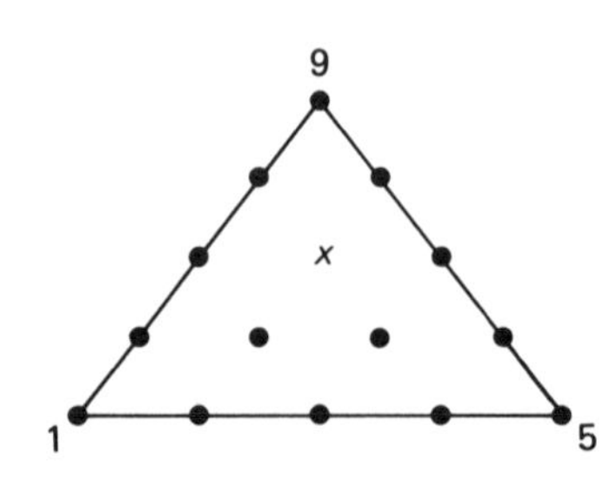

13. Consider the following rectangular element:

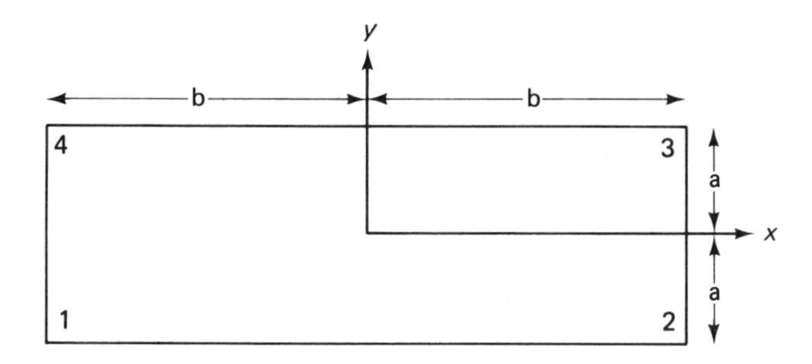

An interpolation polynomial for this four-node element can be written as $u = c_0 + c_1x + c_2y + c_3xy$. The field variable attains the values u_1, u_2, u_3, and u_4 at the four corner nodes.

(a) Show that the expression for u can be rearranged as $u = N_1u_1 + N_2u_2 + N_3u_3 + N_4u_4$ and determine the values of N_β, $\beta = 1, 2, 3, 4$.

(b) Let $\xi = x/b$ and $\eta = y/a$. Now write expressions for N_1, N_2, N_3, and N_4 in terms of ξ and η.

(c) Suppose the interpolation polynomial is written as $u = c_0 + c_1x + c_2y + c_3x^2$. What will be the difficulty, if any? (Study the gradients of u.)

14. The focus of all the preceding exercises has been on an individual element. The purpose of this exercise is to lay the groundwork for the development of a set of equations for an entire region. Specifically, we would like to embed each element into the region and express the interpolation polynomial in terms of the global coordinates. Toward this end, consider a general interpolation polynomial

$$u^{(e)}(x, y) = [N_i^{(e)}, N_j^{(e)}, N_k^{(e)}, \ldots, N_r^{(e)}] \begin{bmatrix} u_i \\ u_j \\ u_k \\ \cdot \\ \cdot \\ \cdot \\ u_r \end{bmatrix} \qquad \text{(A)}$$

where each element (e) in the region has r nodes. Note that u_i stands for the value of u_i at node number i (in the global context) and $u^{(e)}(x, y)$ stands for the value of u at any point (x, y) in the element (e).

Now consider a five-sided region. Divide this region into five triangular elements by drawing straight lines from each vertex to an arbitrarily chosen point in the region. Label the central node by the number 3 and peripheral vertexes (or nodes) by the numbers 1, 2, 4, 5, and 6 in a counterclockwise direction. Starting from triangle 123, label the elements also in a like manner by the labels (1), (2), (3), (4) and (5). Assume that the coordinates of any node point i are known to be (x_i, y_i). Our objective is to relate the local coordinate indexes i, j, k of a typical element (e) to the global indexes 1, 2, 3, 4, 5, and 6. Toward this objective, first write down the five equations by specializing Eq. (A) given above to the five elements and show that the collection of equations can be arranged to look like

$$\mathbf{u}^{(e)}(x, y) = \mathfrak{N}\mathbf{u}$$

or, in expanded form

$$\begin{bmatrix} u^{(1)}(x, y) \\ u^{(2)}(x, y) \\ u^{(3)}(x, y) \\ u^{(4)}(x, y) \\ u^{(5)}(x, y) \end{bmatrix} = \begin{bmatrix} N_1^{(1)} & N_2^{(1)} & N_3^{(1)} & 0 & 0 & 0 \\ 0 & N_2^{(2)} & N_3^{(2)} & N_4^{(2)} & 0 & 0 \\ 0 & 0 & N_3^{(3)} & N_4^{(3)} & N_5^{(3)} & 0 \\ 0 & 0 & N_3^{(4)} & 0 & N_5^{(4)} & N_6^{(4)} \\ N_1^{(5)} & 0 & N_3^{(5)} & 0 & 0 & N_6^{(5)} \end{bmatrix} \begin{bmatrix} u_1 \\ u_2 \\ u_3 \\ u_4 \\ u_5 \\ u_6 \end{bmatrix}$$

[*Hint*. Take an imaginary element (e) with vertexes i, j, k and overlay it on each of the five elements and identify which of the global node numbers correspond to the local vertexes i, j, k.]

15. Repeat the process described in Problem 14 when we are interested (not in the scalar quantity u) in the vector quantity (u, v) at each node.

9

The Finite Element Method

9.1. INTRODUCTION

The history of the evolution of the finite element method is an interesting one. Since the early stages of its inception, the method had been nurtured and developed by practicing engineers on an ad hoc basis. Particular credit goes to structural engineers; the complex configuration of the structures they had to deal with almost always defied the application of traditional mathematical analysis. One recourse available was to use approximate numerical methods. One of the most popular numerical methods available, even before the era of modern digital computers, was ***finite difference method*** (see Chapters 5 and 6). It is significant to note that the starting point for the application of finite difference method is a partial differential equation with specified initial and boundary conditions. By this procedure one is implicitly assuming that the PDE being solved truly represents the physical system. This assumption made the structural engineers very uncomfortable. What, for example, is the partial differential equation that governs the distribution of stress and strain in a complex structure such as a bridge or a ship hull? Even though one can conceivably derive the required equations for a single structural element, like a beam or a plate, the problem of assembling the solutions of all such elements to get the total picture is truly formidable. So the structural engineers asked themselves: why not forget about the partial differential equation formulation, go back to the physical problem, and seek reasonable solutions to reasonable questions using engineering insight and first principles of physics? An answer to this question is the ***finite element method.***

The finite element method thus developed by structural engineers was later generalized and formalized. A first step of this formalization is the recognition that the finite element method was really a variation of the well known Raleigh-Ritz procedure. This led to the recognition that the finite element process can be visualized as an approximate minimization of a functional (such as the potential energy of the system), which in turn led to the discovery that the said minimization in fact was an application of the Hellinger-Reissner variational principle. Once the variational approach was recognized, the suitability of the finite-element approach to problems beyond the structural framework also became apparent. Indeed, any mathematical, or physical situation in which a stationary value of a definable functional provides the solution could immediately be approximated by the finite element process. In addition, the classical variational formulation was found to have many advantages over the differential equation formulation. First, the functional, which may actually represent some physical quantity in the problem, contains lower-order derivatives than the differential operator, and consequently an approximate solution can be sought in a larger class of functions. Second, the problem may possess a reciprocal variational formulation, that is, one functional to be minimized and another functional of a different form to be maximized. In such cases we have a means for finding upper and lower bounds on the functional, and this capability may have significant engineering value. Third, the variational formulation allows one to treat very complicated boundary conditions as natural boundary conditions. Finally, from a mathematical viewpoint, the variational formulation allows one to provide proofs for the existence of a solution. The variational formulation also has a computational advantage; this approach always leads to a symmetric system of algebraic equations.

The variational formulation, however, is restrictive; it is not always possible to find an equivalent variational formulation to a given PDE. The range of applicability of the finite element method was considerably enlarged when it was shown that the required equilibrium equations could also be derived by using a weighted residual procedure, such as Galerkin's method or the least squares approach. This knowledge played an important role in extending the finite element method to problems that lie beyond the realm of structural mechanics. Thus the finite element method has advanced from an ad hoc numerical procedure to solve structural problems to a general numerical procedure for solving systems of differential equations.

9.2. VARIATIONAL FORMULATION OF THE FINITE ELEMENT METHOD

To appreciate the significance and limitations of this method, the student should review Sections 3.5 and 5.6, where some of the related issues are discussed. Briefly, a variational principle specifies a scalar quantity J, which is defined by an integral form such as

$$J = \int_{\Omega} F\left(u, \frac{\partial u}{\partial x}, \ldots\right) d\Omega + \int_{\Gamma} E\left(u, \frac{\partial u}{\partial x}, \ldots\right) d\Gamma$$

where F and E are specified operators and u is the unknown function to be determined. The solution u is found by making J stationary with respect to small changes δu. That is, the variation δJ is set equal to zero.

In this finite element method, we approximate u by

$$u \simeq \hat{u} = \mathbf{N}^T\mathbf{u}$$

where $\mathbf{u} = (u_1, u_2, \ldots, u_r)^T$. Therefore, the condition $\delta J = 0$ becomes

$$\delta J = \frac{\partial J}{\partial \mathbf{u}} \delta \mathbf{u} = 0$$

As this must be true for any $\delta \mathbf{u}$, the required condition is

$$\frac{\partial J}{\partial \mathbf{u}} = \begin{bmatrix} \frac{\partial J}{\partial u_1} \\ \cdot \\ \cdot \\ \cdot \\ \cdot \\ \frac{\partial J}{\partial u_r} \end{bmatrix} = \mathbf{0}$$

If the functional J is "quadratic," that is, if the function $\mathbf{u}$ and its derivatives occur in powers not exceeding 2, then the preceding equation reduces to the standard linear form:

$$\frac{\partial J}{\partial \mathbf{u}} = \mathbf{A}\mathbf{u} - \mathbf{f} = \mathbf{0}$$

Of course, all this depends on our ability to write the variational principle from the given differential equation and the boundary conditions. The rules for deriving variational principles from nonlinear differential equations are complicated. A brief discussion of this topic can be found at the end of this section.

A SIMPLE ONE-DIMENSIONAL PROBLEM

Consider the one-dimensional elliptic equation

$$k\frac{d^2u}{dx^2} = 0, \qquad 0 \leq x \leq l \tag{9.1a}$$

subject to the boundary condition

$$k\frac{du}{dx} + q = 0 \qquad \text{at } x = 0 \tag{9.1b}$$

and

$$k\frac{du}{dx} + h(u - u_\infty) = 0 \qquad \text{at } x = l \tag{9.1c}$$

Physically, these equations can be used to characterize the flow of heat along an insulated rod attached to a wall, which supplies a specified heat input q. The other end of the rod is free and has a convection coefficient h and a surrounding media temperature u_∞. Note that in any portion of Γ, either the term containing h or the term containing q are present, but not both.

The first step of the variational method is to recognize that solving the preceding is equivalent to minimizing

$$J = \int_\Omega \frac{k}{2}\left(\frac{du}{dx}\right)^2 d\Omega + \int_\Gamma \left[qu + \frac{1}{2}h(u - u_\infty)^2\right] d\Gamma \tag{9.2}$$

This minimization is done by setting $\partial J/\partial u = 0$.

The second step of the finite element method is to divide the rod into several segments, called finite elements. For convenience, the rod is divided into two elements of length l_1 and l_2. The rod, the two elements, the three node points, and the temperature at these points are labeled as shown in Figure 9.1. Note that the element sizes need not be equal.

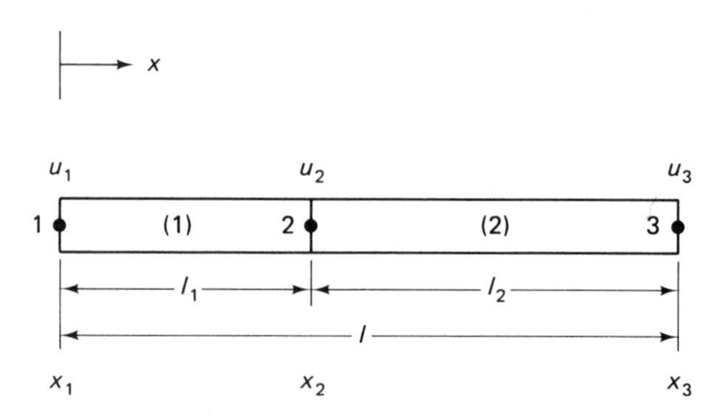

Figure 9.1 Rod of length l divided into two elements of length l_1 and l_2

Using the interpolation techniques discussed in Chapter 8, we can write expressions for the temperature u at any point x in each of the two elements by

$$u^{(1)}(x) \triangleq u^{(1)} = N_1^{(1)}(x)u_1 + N_2^{(1)}(x)u_2 \quad \text{in element 1} \tag{9.3a}$$

$$u^{(2)}(x) \triangleq u^{(2)} = N_2^{(2)}(x)u_2 + N_3^{(2)}(x)u_3 \quad \text{in element 2} \tag{9.3b}$$

where u_1, u_2, u_3 are the temperatures, respectively, at nodes 1, 2, and 3, and the $N_\beta(x)$'s are appropriately chosen shape functions. The superscripts in parentheses refer to element numbers. If linear Lagrange interpolation functions are chosen as the shape functions, then

$$N_1^{(1)} = \frac{x_2 - x}{l_1}, \qquad N_2^{(1)} = \frac{x - x_1}{l_1} \tag{9.4a}$$

$$N_2^{(2)} = \frac{x_3 - x}{l_2}, \qquad N_3^{(2)} = \frac{x - x_3}{l_2} \tag{9.4b}$$

The third step of the finite element method is to rewrite J as the summation of two components evaluated over the two elements. That is,

$$J = J^{(1)} + J^{(2)} \tag{9.5}$$

where

$$J^{(\alpha)} = \int_{\Omega^{(\alpha)}} (\ )\, d\Omega^{(\alpha)} + \int_{\Gamma^{(\alpha)}} (\ )\, d\Gamma^{(\alpha)}, \qquad \alpha = 1, 2 \tag{9.6}$$

The integrals over Γ are easily evaluated because the required boundary is located at the nodal points. For example, for element 1, $\alpha = 1$ and

$$\int_{\Gamma^{(1)}} (\ )\, d\Gamma^{(1)} = \int_{\Gamma^{(1)}} qu(x)\, d\Gamma^{(1)} = qu_1 \int_{\Gamma^{(1)}} d\Gamma = qu_1 A_1 \tag{9.7}$$

where A_1 is the cross-sectional area of the rod at node 1. Note that at node 1, the term containing h plays no role while dealing with element 1 and is therefore dropped from consideration. Similarly, for element 2, $\alpha = 2$, and

$$\int_{\Gamma^{(2)}} (\ )\, d\Gamma^{(2)} = \int_{\Gamma^{(2)}} \frac{h}{2}[u(x) - u_\infty]^2\, d\Gamma^{(2)} = \frac{h}{2}(u_3 - u_\infty)^2 \int_{\Gamma^{(2)}} d\Gamma^{(2)}$$

$$= \frac{hA_3(u_3 - u_\infty)^2}{2} \tag{9.8}$$

where A_3 is the cross-sectional area of the rod at node 3.

To evaluate the integrals over Ω, it is necessary to evaluate the derivative of the temperature. Differentiating Eq. (9.3),

$$\frac{du^{(1)}}{dx} = u_1 \frac{dN_1^{(1)}}{dx} + u_2 \frac{dN_2^{(1)}}{dx} = -\frac{1}{l^{(1)}}(-u_1 + u_2) \tag{9.9a}$$

$$\frac{du^{(2)}}{dx} = u_2 \frac{dN_2^{(2)}}{dx} + u_3 \frac{dN_3^{(2)}}{dx} = \frac{1}{l^{(2)}}(-u_2 + u_3) \tag{9.9b}$$

Note that it is necessary to evaluate the volume integral separately over each element because du/dx is not continuous at node 2. Thus,

$$\int_{\Omega} (\ )\, d\Omega = \int_{\Omega^{(1)}} (\ )\, d\Omega^{(1)} + \int_{\Omega^{(2)}} (\ )\, d\Omega^{(2)} \tag{9.10a}$$

$$= \frac{k^{(1)}A^{(1)}}{2l_1}(-u_1 + u_2)^2 + \frac{k^{(2)}A^{(2)}}{2l_2}(-u_2 + u_3)^2 \tag{9.10b}$$

$$= \frac{c^{(1)}}{2}(-u_1 + u_2)^2 + \frac{c^{(2)}}{2}(-u_2 + u_3)^2 \tag{9.10c}$$

where $k^{(1)}$ and $k^{(2)}$ are the value of k in each element and $A^{(1)}$, $A^{(2)}$ are the cross-sectional areas of the rod in each element. A constant cross sectional area over each element was assumed, and $d\Omega = A^{(e)}\, dx$ was utilized when evaluating the integral. Notice how the need to separate the integral, as done in Eq. (9.10a), helps in realistic problems; it permits variation of the properties of the rod from element to element.

Now the values of $J^{(1)}$ and $J^{(2)}$ can be written as

$$J^{(1)} = \frac{c^{(1)}}{2}(-u_1 + u_2)^2 + qu_1 A_1 \tag{9.11a}$$

$$J^{(2)} = \frac{c^{(2)}}{2}(-u_2 + u_3)^2 + \frac{hA_3}{2}(u_3 - u_\infty)^2 \tag{9.11b}$$

Differentiating $J^{(1)}$ with respect to u_1, u_2, and u_3,

$$\frac{\partial J^{(1)}}{\partial u_1} = c^{(1)}(u_1 - u_2) + q_1 A_1 \tag{9.12a}$$

$$\frac{\partial J^{(1)}}{\partial u_2} = c^{(1)}(-u_1 + u_2) \tag{9.12b}$$

$$\frac{\partial J^{(1)}}{\partial u_3} = 0 \tag{9.12c}$$

These three equations can be written compactly as

$$\frac{\partial J^{(1)}}{\partial \mathbf{u}} = \begin{bmatrix} c^{(1)} & -c^{(1)} & 0 \\ -c^{(1)} & c^{(1)} & 0 \\ 0 & 0 & 0 \end{bmatrix} \begin{bmatrix} u_1 \\ u_2 \\ u_3 \end{bmatrix} + \begin{bmatrix} qA_1 \\ 0 \\ 0 \end{bmatrix} \tag{9.13a}$$

Similarly,

$$\frac{\partial J^{(2)}}{\partial \mathbf{u}} = \begin{bmatrix} 0 & 0 & 0 \\ 0 & c^{(2)} & -c^{(2)} \\ 0 & -c^{(2)} & (c^{(2)} + hA_3) \end{bmatrix} \begin{bmatrix} u_1 \\ u_2 \\ u_3 \end{bmatrix} + \begin{bmatrix} 0 \\ 0 \\ -hA_3 u_\infty \end{bmatrix} \tag{9.13b}$$

However,

$$\frac{\partial J}{\partial \mathbf{u}} = \frac{\partial J^{(1)}}{\partial \mathbf{u}} + \frac{\partial J^{(2)}}{\partial \mathbf{u}} = 0 \tag{9.14}$$

Therefore, we have

$$\begin{bmatrix} c^{(1)} & -c^{(1)} & 0 \\ -c^{(1)} & [c^{(1)} + c^{(2)}] & -c^{(2)} \\ 0 & -c^{(2)} & [c^{(2)} + + hA_3] \end{bmatrix} \begin{bmatrix} u_1 \\ u_2 \\ u_3 \end{bmatrix} = \begin{bmatrix} -qA_1 \\ 0 \\ hA_3 u_\infty \end{bmatrix} \tag{9.15}$$

This is of the form $\mathbf{Ax} = \mathbf{f}$, and any of the methods described in Chapter 10 can be used to solve for u_1, u_2, and u_3.

A SIMPLE TWO-DIMENSIONAL PROBLEM

The example considered earlier is too simple to bring forth all the important ideas associated with problems of practical interest. Now let us consider the problem of two-dimensional heat conduction

$$\frac{\partial}{\partial x}\left(k_x \frac{\partial}{\partial x}\right) + \frac{\partial}{\partial y}\left(k_y \frac{\partial u}{\partial y}\right) + Q = 0 \tag{9.16a}$$

defined on a region Ω, and subjected to the general boundary conditions

$$u = u_\beta \qquad \text{on } \Gamma_1 \tag{9.16b}$$

$$k_x \frac{\partial u}{\partial x} l_x + k_y \frac{\partial u}{\partial y} l_y + q + h(u - u_\infty) = 0 \qquad \text{on } \Gamma_2 \tag{9.16c}$$

The union of Γ_1 and Γ_2 forms the complete boundary. The quantity Q represents a source (or sink) in Ω, and q represents flow across the boundary. In a heat conduction problem, q represents heat loss at the boundary Γ_2 due to conduction. The term $h(u - u_\infty)$ represents heat loss at the boundary Γ_2 due to convection, where h represents convection heat transfer coefficient and u_∞ represents the ambient temperature. Notice that both q and $h(u - u_\infty)$ cannot appear on the same segment of the boundary. In a given problem, one of the two will appear. The coefficients k_x and k_y represent thermal conductivities along the x- and y-axes, while l_x and l_y are the direction cosines of the outward drawn normal from the boundary of Ω.

The first step, as before, is to recognize the functional formulation that is equivalent to Eq. (9.16). For the present case,

$$J = \int_\Omega \left[\frac{1}{2}k_x\left(\frac{\partial u}{\partial x}\right)^2 + \frac{1}{2}k_y\left(\frac{\partial u}{\partial y}\right)^2 - 2Qu\right] d\Omega + \int_{\Gamma_2}\left[qu + \frac{1}{2}h(u - u_\infty)^2\right] d\Gamma_2 \tag{9.17}$$

Our goal is to minimize J with respect to $\mathbf{u}$, whose components represent the required solution.

Before embarking on the second step, a useful notation is introduced. Define

$$\mathbf{g} = \begin{bmatrix}\frac{\partial u}{\partial x} & \frac{\partial u}{\partial y}\end{bmatrix}^T \tag{9.18}$$

and

$$\mathbf{D} = \begin{bmatrix} k_x & 0 \\ 0 & k_y \end{bmatrix} \tag{9.19}$$

Equation (9.17) can now be written as

$$J = \int_\Omega \frac{1}{2}[\mathbf{g}\,\mathbf{D}\,\mathbf{g}^T - 2Qu]\, d\Omega + \int_{\Gamma_2} qu\, d\Gamma_2 + \int_{\Gamma_2}\frac{h}{2}[u^2 - 2uu_\infty + u_\infty^2]\, d\Gamma_2 \tag{9.20}$$

The next step is to superimpose a grid on Ω. It is customary to number the elements as well as the grid points. The element numbers are usually enclosed within parentheses, and the node numbers are shown beside the node, as in Figure 9.2. Recalling that u is not continuous over the entire region Ω but, instead, is defined over individual elements, the preceding integrals are separated into integrals defined over individual elements. This operation yields

$$J = \sum_{e=1}^{E}\left[\int_{\Omega^{(e)}} \frac{1}{2}\mathbf{g}^{(e)T}\mathbf{D}^{(e)}\mathbf{g}^{(e)}\, d\Omega^{(e)} - \int_{\Omega^{(e)}} Q^{(e)}u^{(e)}\, d\Omega^{(e)} + \int_{\Gamma_2^{(e)}} q^{(e)}u^{(e)}\, d\Gamma_2^{(e)} + \int_{\Gamma_2^{(e)}} \frac{h^{(e)}}{2}[u^{(e)}u^{(e)} - 2u^{(e)}u_\infty + u_\infty^2]\, d\Gamma_2^{(e)}\right] \tag{9.21}$$

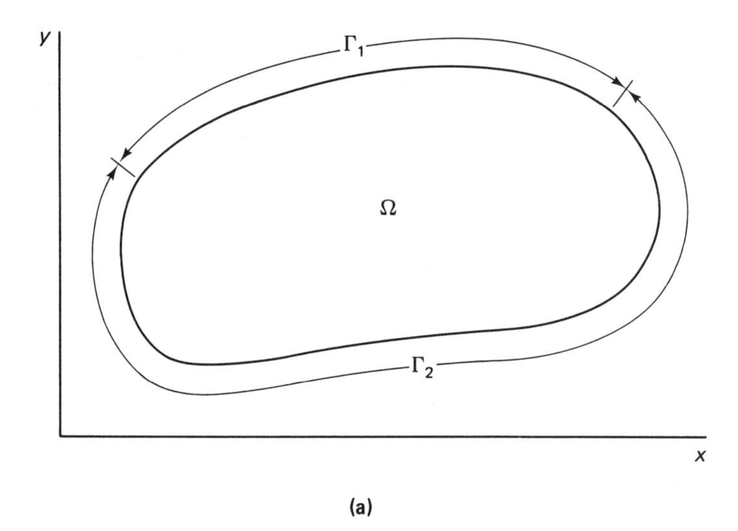

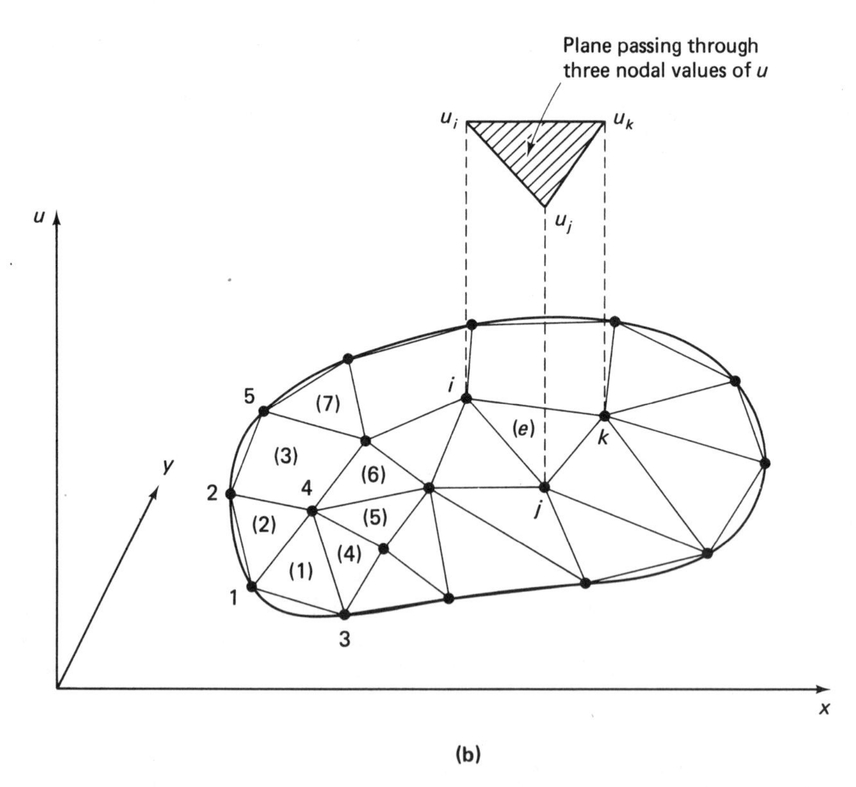

Figure 9.2 (a) A region Ω; (b) subdivision of Ω into finite elements. The figure shows a sample numbering of the elements and nodes. Also shown are a typical element (e) and a piecewise linear solution over this element

where E is the total number of elements in Ω. It is important to note that the last two integrals vanish if the element in question does not contribute anything to the definition of the boundary Γ_2. Equation (9.21) can also be written as

$$J = J^{(1)} + J^{(2)} + \cdots + J^{(E)} = \sum_{e=1}^{E} J^{(e)} \tag{9.22}$$

where $J^{(e)}$ is a contribution of element e to the total value of J. Now J is minimized by setting

$$\frac{\partial J}{\partial \mathbf{u}} = \frac{\partial}{\partial \mathbf{u}}\left[\sum_{e=1}^{E} J^{(e)}\right] = \sum_{e=1}^{E} \frac{\partial J^{(e)}}{\partial \mathbf{u}} = 0 \tag{9.23}$$

To the uninitiated, a note of caution about the notation is appropriate here. The superscript (e) is used whenever the discussion refers to an element $\Omega^{(e)}$ of Ω. Thus $u^{(e)} = u^{(e)}(x, y)$ refers to the solution at a point (x, y) in $\Omega^{(e)}$. However, the vector $\mathbf{u}^{(e)}$, for example, refers to a set of values attained by $u^{(e)}(x, y)$ at the nodal points of $\Omega^{(e)}$.

The derivatives $\partial J^{(e)}/\partial \mathbf{u}$ in Eq. (9.23) cannot be evaluated until the integrals in Eq. (9.21) have been written in terms of the nodal values $\mathbf{u}$. To get these nodal values, it is necessary to approximate $u^{(e)} = u^{(e)}(x, y)$ in terms of an interpolating polynomial. Recalling the discussion in Chapter 8, the solution $u^{(e)}(x, y)$ is written as

$$u^{(e)} = u^{(e)}(x, y) = N_1 u_1^{(e)} + N_2 u_2^{(e)} + \cdots + N_n u_n^{(e)} = \mathbf{N}^T \mathbf{u}^{(e)} \tag{9.24}$$

where

$$\mathbf{N} = \begin{bmatrix} N_1(x, y) \\ N_2(x, y) \\ \cdot \\ \cdot \\ \cdot \\ N_n(x, y) \end{bmatrix}, \qquad \mathbf{u}^{(e)} = \begin{bmatrix} u_1^{(e)} \\ u_2^{(e)} \\ \cdot \\ \cdot \\ \cdot \\ u_n^{(e)} \end{bmatrix} \tag{9.25}$$

Note that $N_1, N_2, \ldots, N_n$ are suitably chosen shape functions. Substituting Eq. (9.24) in Eq. (9.18), we get for element e,

$$\mathbf{g}^{(e)} = \begin{bmatrix} \dfrac{\partial u^{(e)}}{\partial x} \\ \dfrac{\partial u^{(e)}}{\partial y} \end{bmatrix} = \begin{bmatrix} \dfrac{\partial N_1^{(e)}}{\partial x} & \cdots & \dfrac{\partial N_n^{(e)}}{\partial x} \\ \dfrac{\partial N_1^{(e)}}{\partial y} & \cdots & \dfrac{\partial N_n^{(e)}}{\partial y} \end{bmatrix} \begin{bmatrix} u_1 \\ u_2 \\ \cdot \\ \cdot \\ \cdot \\ u_n \end{bmatrix} \tag{9.26}$$

or

$$\mathbf{g}^{(e)} = \mathbf{B}^{(e)} \mathbf{u}^{(e)} \tag{9.27}$$

The matrix $\mathbf{B}^{(e)}$ contains information concerning the derivatives of the shape functions. The evaluation of the entries of $\mathbf{B}^{(e)}$ can be done once the nature of the shape functions is decided upon. Utilizing Eqs. (9.24) and (9.27), we can now write the element integrals as

$$
\begin{aligned}
J^{(e)} = & \int_{\Omega^{(e)}} \frac{1}{2}\mathbf{u}^T\mathbf{B}^{(e)T}\mathbf{D}^{(e)}\mathbf{B}^{(e)}\mathbf{u}\,d\Omega^{(e)} \\
& - \int_{\Omega^{(e)}} Q^{(e)}\mathbf{N}^{(e)T}\mathbf{u}\,d\Omega^{(e)} + \int_{\Gamma_2^{(e)}} q\mathbf{N}^{(e)T}\mathbf{u}\,d\Gamma_2^{(e)} \\
& + \int_{\Gamma_2^{(e)}} \frac{h}{2}\mathbf{N}^{(e)T}\mathbf{u}\mathbf{u}^T\mathbf{N}^{(e)}\,d\Gamma_2^{(e)} \\
& - \int_{\Gamma_2^{(e)}} hu_\infty\mathbf{N}^{(e)T}\mathbf{u}\,d\Gamma_2^{(e)} + \int_{\Gamma_2^{(e)}} \frac{h}{2}u_\infty^2\,d\Gamma_2^{(e)}
\end{aligned}
\tag{9.28}
$$

Now Eq. (9.28) is differentiated with respect to $\mathbf{u}$ and the derivatives are written here term by term. (Verify these results!)

$$\frac{\partial}{\partial\mathbf{u}}(\text{first term of Eq. (9.28)}) = \int_{\Omega^{(e)}} \mathbf{B}^{(e)T}\mathbf{D}^{(e)}\mathbf{B}^{(e)}\mathbf{u}\,d\Omega^{(e)} \tag{9.29a}$$

$$\frac{\partial}{\partial\mathbf{u}}(\text{second term}) = \int_{\Omega^{(e)}} Q\mathbf{N}^{(e)}\,d\Omega^{(e)} \tag{9.29b}$$

$$\frac{\partial}{\partial\mathbf{u}}(\text{third term}) = \int_{\Gamma_2^{(e)}} q\mathbf{N}^{(e)}\,d\Gamma_2^{(e)} \tag{9.29c}$$

$$\frac{d}{du}(\text{fourth term}) = \int_{\Gamma_2^{(e)}} h\mathbf{N}^{(e)}\mathbf{N}^{(e)T}\mathbf{u}\,d\Gamma_2^{(e)} \tag{9.29d}$$

$$\frac{\partial}{\partial\mathbf{u}}(\text{fifth term}) = \int_{\Gamma_2^{(e)}} hu_\infty\mathbf{N}^{(e)}\,d\Gamma_2^{(e)} \tag{9.29e}$$

$$\frac{\partial}{\partial\mathbf{u}}(\text{sixth term}) = \int_{\Gamma_2^{(e)}} \frac{h}{2}u_\infty^2\,d\Gamma_2^{(e)} \tag{9.29f}$$

Therefore,

$$
\begin{aligned}
\frac{\partial J^{(e)}}{\partial\mathbf{u}} = & \left(\int_{\Omega^{(e)}} \mathbf{B}^{(e)T}\mathbf{D}^{(e)}\mathbf{B}^{(e)}\,d\Omega^{(e)} + \int_{\Gamma_2^{(e)}} h\mathbf{N}^{(e)}\mathbf{N}^{(e)T}\,d\Gamma_2^{(e)}\right)\mathbf{u} \\
& - \int_{\Omega^{(e)}} Q\mathbf{N}^{(e)}\,d\Omega^{(e)} + \int_{\Gamma_2^{(e)}} q\mathbf{N}^{(e)}\,d\Gamma_2^{(e)} \\
& - \int_{\Gamma_2^{(e)}} hu_\infty\mathbf{N}^{(e)}\,d\Gamma_2^{(e)} = \mathbf{0}
\end{aligned}
\tag{9.30}
$$

where

$$\frac{\partial J}{\partial\mathbf{u}} = \left(\frac{\partial J}{\partial u_1}\,\frac{\partial J}{\partial u_2}\cdots\frac{\partial J}{\partial u_n}\right)^T$$

Equation (9.30) can be compactly written as

$$\frac{\partial J^{(e)}}{\partial\mathbf{u}} = \mathbf{A}^{(e)}\mathbf{u}^{(e)} - \mathbf{f}^{(e)} = \mathbf{0} \tag{9.31}$$

where the matrix **A** is

$$\mathbf{A}^{(e)} = \int_{\Omega^{(e)}} \mathbf{B}^{(e)T}\mathbf{D}^{(e)}\mathbf{B}^{(e)}\,d\Omega^{(e)} + \int_{\Gamma_2^{(e)}} h\mathbf{N}^{(e)}\mathbf{N}^{(e)T}\,d\Gamma_2^{(e)} \tag{9.32a}$$

or, equivalently, the (i, j)th entry of $\mathbf{A}^{(e)}$ is

$$a_{ij}^{(e)} = \int_{\Omega^{(e)}} \left[k_x \frac{\partial N_i}{\partial x}\frac{\partial N_j}{\partial x} + k_y \frac{\partial N_i}{\partial y}\frac{\partial N_j}{\partial y}\right] dx\,dy + \int_{\Gamma_2^{(e)}} hN_iN_j\,d\Gamma_2^{(e)} \tag{9.32b}$$

Similarly, the quantity $\mathbf{f}^{(e)}$ in Eq. (9.31) is

$$-\mathbf{f}^{(e)} = -\int_{\Omega^{(e)}} Q\mathbf{N}^{(e)}\,d\Omega^{(e)} + \int_{\Gamma_2^{(e)}} q\mathbf{N}^{(e)}\,d\Gamma_2^{(e)} - \int_{\Gamma_2^{(e)}} hu_\infty\mathbf{N}^{(e)}\,d\Gamma_2^{(e)} \tag{9.33a}$$

or, equivalently, the ith entry of $\mathbf{f}^{(e)}$ is

$$-f_i^{(e)} = -\int_{\Omega^{(e)}} QN_i\,dx\,dy + \int_{\Gamma_2^{(e)}} qN_i\,d\Gamma_2^{(e)} - \int_{\Gamma_2^{(e)}} hu_\infty N_i\,d\Gamma_2^{(e)} \tag{9.33b}$$

Now Eq. (9.23) can be written as

$$\frac{\partial J}{\partial \mathbf{u}} = \sum_{e=1}^{E} [\mathbf{A}^{(e)}\mathbf{u}^{(e)} - \mathbf{f}^{(e)}] = \mathbf{0} \tag{9.34}$$

or

$$\mathbf{A}\mathbf{u} = \mathbf{f} \tag{9.35}$$

where

$$\mathbf{A} = \sum_{e=1}^{E} \mathbf{A}^{(e)} \quad \text{and} \quad \mathbf{f} = \sum_{e=1}^{E} \mathbf{f}^{(e)} \tag{9.36}$$

The process involved in implementing Eq. (9.36) goes by the name ***assembly***. Assuming that $\mathbf{A}^{(e)}$ and $\mathbf{f}^{(e)}$ are known for each element, the next step is to perform the "summation" indicated in Eq. (9.36). The procedure for doing this is the same regardless of the type of problem being considered or the complexity of the elements. The assembly procedure is based on our insistence of ***compatibility*** at the element nodes. By this we mean that at nodes where elements are connected the value of the unknown nodal variable (here, the value of u) is the same for all the elements connecting at that node. The consequence of this rule is the basis for the assembly process. The actual assembly process is best illustrated by an example.

Example. Consider two parallel flat plates of identical size and shape in the form of equilateral triangles. Assume that the plates are separated by a thin fluid film. If the plates move together with a velocity V while the fluid film is squeezed out, the pressure field generated within the space between the plates satisfies the equation

$$\frac{\partial^2 u}{\partial x^2} + \frac{\partial^2 u}{\partial y^2} = \frac{12\mu V}{h^3} = -Q \tag{9.37a}$$

subject to the boundary conditions

$$u = 0 \text{ on all three sides.} \tag{9.37b}$$

This problem was first discussed by K. H. Huebner [1], and we are indebted to him for his permission to use this here.

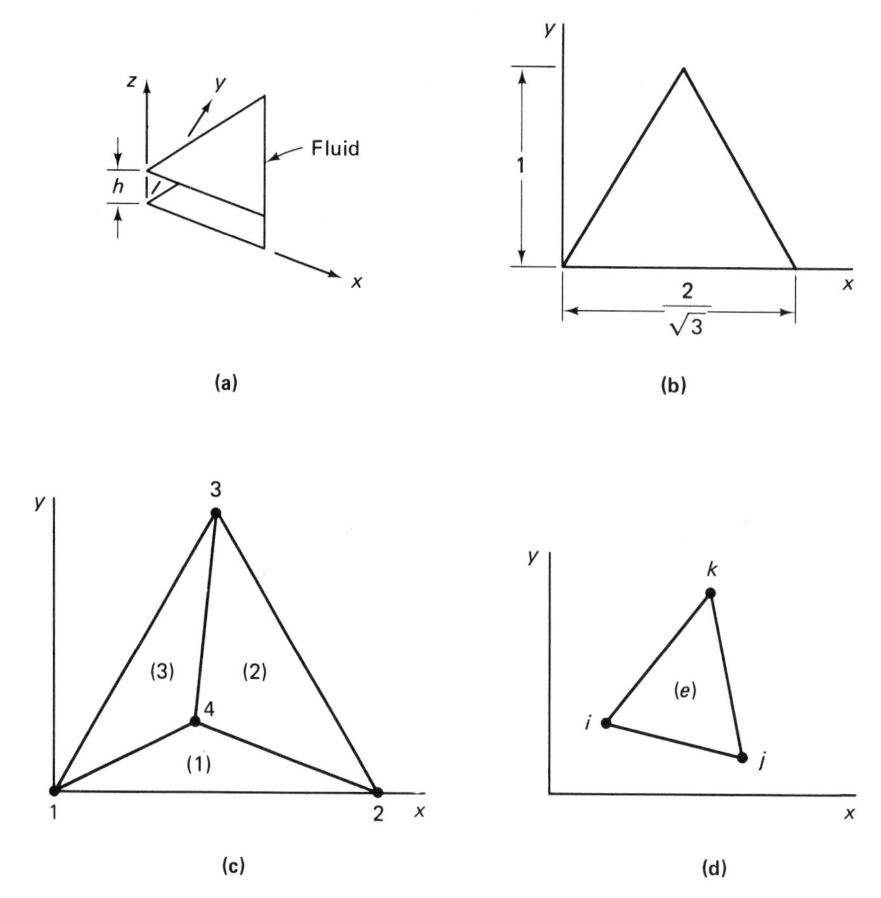

Figure 9.3 Two-dimensional fluid film problem discussed in the text: (a) triangular plates generating the film; (b) geometry of the solution domain; (c) discretization of the solution domain and the numbering of the nodes and elements; (d) notation used for a typical element (Reproduced with permission from *The Finite Element Method for Engineers* by K. H. Huebner, © 1975, John Wiley & Sons, Inc., New York)

Let the triangle be located on the coordinate axes as shown in Figure 9.3a. For this configuration the boundary conditions are specified on the lines

$$y = 0, \qquad y = 2 - \sqrt{3}\,x, \qquad y = \sqrt{3}\,x \tag{9.38}$$

and the equation can be solved exactly to yield

$$u(x, y) = \frac{Q}{4}(y - 2 + \sqrt{3}\,x)(y - \sqrt{3}\,x)y \tag{9.39}$$

This problem is sufficiently simple enough to demonstrate the various steps in the finite element implementation. Let the region Ω (i.e., the plate) be divided into three triangular elements as shown in Figure 9.3b. This situation is simple enough and further complications will be avoided as much as possible. For instance, the simplicity can be maintained by deriving all equations in the global coordinate system, thus avoiding transformations from local to global. For analogous reasons, linear interpolation functions will be used to approximate u over each element. Therefore, Eq. (9.24) for an element (e) can be written as

$$u^{(e)}(x, y) = u = N_i u_i^{(e)} + N_j u_j^{(e)} + N_k u_k^{(e)} \tag{9.40a}$$

$$= [N_i \quad N_j \quad N_k] \begin{bmatrix} u_i^{(e)} \\ u_j^{(e)} \\ u_k^{(e)} \end{bmatrix} \tag{9.40b}$$

where the shape functions for the linear triangular element are (see also Eq. (8.49))

$$N_\beta = \frac{1}{2\Omega^{(e)}}(c_\beta + b_\beta x + a_\beta y), \qquad \beta = i, j, k \tag{9.41}$$

where $c_i = 2A_{jk} = 2(x_j y_k - x_k y_j)$, $b_i = y_j - y_k$, and $a_i = x_k - x_j$. Now, the matrix $\mathbf{B}^{(e)}$ can be written as

$$\mathbf{B}^{(e)} = \frac{1}{2\Omega^{(e)}} \begin{bmatrix} b_i & b_j & b_k \\ a_i & a_j & a_k \end{bmatrix} \tag{9.42}$$

Similarly, the material property matrix $\mathbf{D}^{(e)}$ becomes

$$\mathbf{D}^{(e)} = \begin{bmatrix} 1 & 0 \\ 0 & 1 \end{bmatrix} \tag{9.43}$$

Now, the first term of Eq. (9.32) becomes

$$\int_{\Omega^{(e)}} \mathbf{B}^{(e)T}\mathbf{D}^{(e)}\mathbf{B}^{(e)}\, d\Omega^{(e)} = \int_{\Omega^{(e)}} \frac{1}{4\Omega^{(e)2}} \begin{bmatrix} b_i & a_i \\ b_j & a_j \\ b_k & a_k \end{bmatrix} \begin{bmatrix} b_i & b_j & b_k \\ a_i & a_j & a_k \end{bmatrix} d\Omega^{(e)} \tag{9.44}$$

Because all the terms under the integral sign are constant, they can be pulled out of the integral sign. Therefore,

$$\int_{\Omega^{(e)}} \mathbf{B}^{(e)T}\mathbf{D}^{(e)}\mathbf{B}^{(e)}\, d\Omega^{(e)} = \mathbf{B}^{(e)T}\mathbf{D}^{(e)}\mathbf{B}^{(e)} \int_{\Omega^{(e)}} d\Omega^{(e)} = \Omega^{(e)}\mathbf{B}^{(e)T}\mathbf{D}^{(e)}\mathbf{B}^{(e)} \tag{9.45}$$

The second term of Eq. (9.32) must be evaluated on Γ_2, that portion of the boundary over which normal derivative boundary conditions are specified. In this example, this term contributes nothing to $\mathbf{A}^{(e)}$ because of Eq. (9.37b). Therefore, from Eq. (9.44),

$$\mathbf{A}^{(e)} = \Omega^{(e)}\mathbf{B}^{(e)T}\mathbf{D}^{(e)}\mathbf{B}^{(e)}$$

$$= \frac{1}{4\Omega^{(e)}} \begin{bmatrix} (b_i b_i + a_i a_i) & (b_i b_j + a_i a_j) & (b_i b_k + a_i a_k) \\ (b_j b_i + a_j a_i) & (b_j b_j + a_j a_j) & (b_j b_k + a_j a_k) \\ (b_k b_i + a_k a_i) & (b_k b_j + a_k a_j) & (b_k b_k + a_k a_k) \end{bmatrix} \tag{9.46}$$

Evaluation of $\mathbf{f}$ on the right side of Eq. (9.35) is accomplished by using the integration formulas developed in Chapter 8. Starting from Eq. (9.33),

$$\int_{\Omega^{(e)}} Q\mathbf{N}^{(e)}\,d\Omega^{(e)} = Q\int_{\Omega^{(e)}} \mathbf{N}^{(e)}\,d\Omega^{(e)} = Q\int_{\Omega^{(e)}} \begin{bmatrix} N_i \\ N_j \\ N_k \end{bmatrix} d\Omega^{(e)}$$

$$= \frac{Q\Omega^{(e)}}{3}$$

The second equality is obtained by evaluating the area integral using the second row of Table 8.2. The remaining two terms in Eq. (9.33) are zero for this example.

Now matrix equations, analogous to Eq. (9.31), can be written for each element as follows. Note $a_{ij}^{(e)}$ stands for the (i, j)th entry of $\mathbf{A}^{(e)}$.

For element (1), whose vertexes are 1, 2, and 4:

$$\begin{bmatrix} a_{11}^{(1)} & a_{12}^{(1)} & a_{14}^{(1)} \\ a_{21}^{(1)} & a_{22}^{(1)} & a_{24}^{(1)} \\ a_{41}^{(1)} & a_{42}^{(1)} & a_{44}^{(1)} \end{bmatrix} \begin{bmatrix} u_1^{(1)} \\ u_2^{(1)} \\ u_4^{(1)} \end{bmatrix} = -\frac{Q}{3} \begin{bmatrix} \Omega^{(1)} \\ \Omega^{(1)} \\ \Omega^{(1)} \end{bmatrix} \tag{9.47}$$

For element (2), whose vertexes are 2, 3, and 4:

$$\begin{bmatrix} a_{22}^{(2)} & a_{23}^{(2)} & a_{24}^{(2)} \\ a_{32}^{(2)} & a_{33}^{(2)} & a_{34}^{(2)} \\ a_{42}^{(2)} & a_{43}^{(2)} & a_{44}^{(2)} \end{bmatrix} \begin{bmatrix} u_2^{(2)} \\ u_3^{(2)} \\ u_4^{(2)} \end{bmatrix} = -\frac{Q}{3} \begin{bmatrix} \Omega^{(2)} \\ \Omega^{(2)} \\ \Omega^{(2)} \end{bmatrix} \tag{9.48}$$

For element (3), whose vertexes are 3, 1 and 4:

$$\begin{bmatrix} a_{33}^{(3)} & a_{31}^{(3)} & a_{34}^{(3)} \\ a_{13}^{(3)} & a_{11}^{(3)} & a_{14}^{(3)} \\ a_{43}^{(3)} & a_{41}^{(3)} & a_{44}^{(3)} \end{bmatrix} \begin{bmatrix} u_3^{(3)} \\ u_1^{(3)} \\ u_4^{(3)} \end{bmatrix} = -\frac{Q}{3} \begin{bmatrix} \Omega^{(3)} \\ \Omega^{(3)} \\ \Omega^{(3)} \end{bmatrix} \tag{9.49}$$

These individual element equations are now assembled as follows. As there are four nodes, the assembled matrix contains four rows and four columns. Now we must find a systematic procedure to fill the entries of this 4×4 matrix.

Step 1. For each element, a correspondence between local and global numbering schemes is established as shown in Table 9.1. It is important to understand how the numbering is done as this step is an essential part of the solution process. Toward this end, an arbitrary global numbering scheme, as indicated in Figure 9.3b, is established to identify the elements and nodes. Once the numbering scheme has been established for a finite element mesh, one has to make a record of which nodes belong to each of the elements. For example, Table 9.1 tells that element 3 has nodes 3, 1, and 4, and that node 1 of the typical element (see Figure 9.3c) actually is node 3 of the system, while node 2 of the typical element actually is node 1 of the system, and node 3 of the element coincides with node 4 of the system.

Table 9.1. Correspondence between local and global numbering schemes

Element Number	*Local Node Number*	*Global Node Number*
1	$i = 1$	1
	$j = 2$	2
	$k = 3$	4
2	$i = 1$	2
	$j = 2$	3
	$k = 3$	4
3	$i = 1$	3
	$j = 2$	1
	$k = 3$	4

Step 2. The individual element equations, Eqs. (9.47 to 9.49), are now assembled. Since the local and global coordinate systems are chosen to be the same, in this particular example, no complicated coordinate transformations are required. Now the assembly process can be explained as follows.

Since the total system has 4 nodes, the assembled matrix is a 4×4 matrix. Therefore, a 4×4 matrix with blank entries is constructed first, as shown:

$$\text{Assembled matrix or system matrix} = \begin{bmatrix} & & & \\ & & & \\ & & & \\ & & & \end{bmatrix} \tag{9.50}$$

Now the 3×3 element matrixes shown in Eqs. (9.47 to 9.49) can be thought of as submatrixes of the above 4×4 matrix. As there is only one 4×4 matrix with 16 entries and three 3×3 matrixes with a total of 27 entries, it is evident that some entries overlap each other when the entries of the element matrixes are entered into their proper locations in the assembled matrix. Then the resultant entries of the system matrix are obtained by simple addition. This addition, of course, is not done in a random fashion. Because matrix addition is defined only for matrixes of the same size, the element matrixes are first "inflated" so as to conform them with the dimension of the system matrix. These inflated element matrix equations are obtained by slightly modifying Eqs. (9.47 to 9.49) as follows.

For element (1),

$$\begin{bmatrix} a_{11}^{(1)} & a_{12}^{(1)} & 0 & a_{14}^{(1)} \\ a_{21}^{(1)} & a_{22}^{(1)} & 0 & a_{24}^{(1)} \\ 0 & 0 & 0 & 0 \\ a_{41}^{(1)} & a_{42}^{(1)} & 0 & a_{44}^{(1)} \end{bmatrix} \begin{bmatrix} u_1^{(1)} \\ u_2^{(1)} \\ 0 \\ u_4^{(1)} \end{bmatrix} = -\frac{Q}{3} \begin{bmatrix} \Omega^{(1)} \\ \Omega^{(1)} \\ 0 \\ \Omega^{(1)} \end{bmatrix} \tag{9.51}$$

For element (2),

$$\begin{bmatrix} 0 & 0 & 0 & 0 \\ 0 & a_{22}^{(2)} & a_{23}^{(2)} & a_{24}^{(2)} \\ 0 & a_{32}^{(2)} & a_{33}^{(2)} & a_{34}^{(2)} \\ 0 & a_{42}^{(2)} & a_{43}^{(2)} & a_{44}^{(2)} \end{bmatrix} \begin{bmatrix} 0 \\ u_2^{(2)} \\ u_3^{(2)} \\ u_4^{(2)} \end{bmatrix} = -\frac{Q}{3} \begin{bmatrix} 0 \\ \Omega^{(2)} \\ \Omega^{(2)} \\ \Omega^{(2)} \end{bmatrix} \tag{9.52}$$

For element (3), after a slight rearrangement,

$$\begin{bmatrix} a_{11}^{(3)} & 0 & a_{13}^{(3)} & a_{14}^{(3)} \\ 0 & 0 & 0 & 0 \\ a_{31}^{(3)} & 0 & a_{33}^{(3)} & a_{34}^{(3)} \\ a_{41}^{(3)} & 0 & a_{43}^{(3)} & a_{44}^{(3)} \end{bmatrix} \begin{bmatrix} u_1^{(3)} \\ 0 \\ u_3^{(3)} \\ u_4^{(3)} \end{bmatrix} = -\frac{Q}{3} \begin{bmatrix} \Omega^{(3)} \\ 0 \\ \Omega^{(3)} \\ \Omega^{(3)} \end{bmatrix} \tag{9.53}$$

Now the matrixes in these three equations are overlayed on each other. The entries in the blank spaces in Eq. (9.50) are the sums of the individual entries on the overlays. Thus, the assembled matrix equation now becomes

$$\begin{bmatrix} a_{11}^{(1)} + a_{13}^{(3)} & a_{12}^{(1)} & a_{13}^{(1)} & a_{14}^{(1)} + a_{14}^{(3)} \\ a_{21}^{(1)} & a_{22}^{(1)} + a_{22}^{(2)} & a_{23}^{(2)} & a_{24}^{(1)} + a_{24}^{(2)} \\ a_{31}^{(3)} & a_{32}^{(2)} & a_{33}^{(2)} + a_{33}^{(3)} & a_{34}^{(2)} + a_{34}^{(3)} \\ a_{41}^{(1)} + a_{41}^{(3)} & a_{42}^{(1)} + a_{42}^{(2)} & a_{43}^{(2)} + a_{43}^{(3)} & a_{44}^{(1)} + a_{44}^{(2)} + a_{44}^{(3)} \end{bmatrix} \begin{bmatrix} u_1 \\ u_2 \\ u_3 \\ u_4 \end{bmatrix}$$

$$= -\frac{Q}{3} \begin{bmatrix} \Omega^{(1)} + \Omega^{(3)} \\ \Omega^{(1)} + \Omega^{(2)} \\ \Omega^{(2)} + \Omega^{(3)} \\ \Omega^{(1)} + \Omega^{(2)} + \Omega^{(3)} \end{bmatrix} \tag{9.54}$$

In Eq. (9.54), u_1, u_2, and u_3 are known quantities because they represent specified boundary conditions; that is, $u = 0$ on the boundary. The only unknown, u_4, can be determined by solving the last equation of the system in Eq. (9.54). That is,

$$(a_{44}^{(1)} + a_{44}^{(2)} + a_{44}^{(3)})u_4 = -\frac{Q}{3}(\Omega^{(1)} + \Omega^{(2)} + \Omega^{(3)}) \tag{9.55}$$

But from Eq. (9.46),

$$\begin{aligned} a_{44}^{(1)} &= \frac{1}{4\Omega^{(1)}}(b_4^2 + c_4^2)^{(1)} \\ &= \frac{1}{4\Omega^{(1)}}[(y_1 - y_2)^2 + (x_2 - x_1)^2] \\ &= \frac{1}{4\Omega^{(1)}} \cdot \left(\frac{4}{3}\right) = \frac{1}{3\Omega^{(1)}} \end{aligned}$$

Similarly,

$$a_{44}^{(2)} = \frac{1}{4\Omega^{(2)}}(b_4^2 + c_4^2)^{(2)} = \frac{1}{3\Omega^{(2)}}$$

$$a_{44}^{(3)} = \frac{1}{4\Omega^{(3)}}(b_4^2 + c_4^2)^{(3)} = \frac{1}{3\Omega^{(3)}}$$

Also,

$$\Omega^{(1)} = \frac{1}{2}\begin{vmatrix} 1 & x_1 & y_1 \\ 1 & x_2 & y_2 \\ 1 & x_3 & y_3 \end{vmatrix} = \frac{y_4}{\sqrt{3}}$$

$$\Omega^{(2)} = \frac{1}{2\sqrt{3}}(2 - y_4 - \sqrt{3}\,x_4)$$

$$\Omega^{(3)} = \frac{1}{2\sqrt{3}}(\sqrt{3}\,x_4 - y_4)$$

Substituting all these in Eq. (9.55) and solving for u_4, one gets

$$u_4 = -\frac{Q\sqrt{3}}{3}\frac{\Omega^{(1)}\Omega^{(2)}\Omega^{(3)}}{\Omega^{(2)}\Omega^{(3)} + \Omega^{(1)}\Omega^{(3)} + \Omega^{(1)}\Omega^{(2)}}$$

$$= -Q\frac{(y_4 - 2 + \sqrt{3}\,x_4)(y_4 - \sqrt{3}\,x_4)y_4}{3 \cdot [2(y_4 + \sqrt{3}\,x_4) - 3(x_4^2 + y_4^2)]}$$

Notice that if the centroid of the triangle 123 coincides with node 4, then the preceding solution becomes identically equal to the exact solution given in Eq. (9.39).

FINDING VARIATIONAL PRINCIPLES

From the preceding discussion it is clear that one major question still remains unanswered. Given a differential equation, how do we obtain its variational equivalent? The functional we seek is some integrated quantity that is characteristic of the problem. By looking at the physics of the problem, it is sometimes possible to identify the quantity to be maximized or minimized. For example, in classical thermodynamics we know that for an isolated system at equilibrium, the entropy is a maximum, whereas for a system at constant volume and temperature, the Helmholtz free energy is a minimum. Therefore, to find functionals for these cases, we express the entropy or the Helmholtz free energy in integral form. It is possible to identify similar variational principles in other disciplines. For example, in an incompressible, inviscid fluid experiencing an irrotational flow, the kinetic energy is a minimum. However, it is not always possible to identify the relevant variational principle. Then, one has to resort to one of many possible manipulations.

Method 1. Given a differential equation

$$L(u) - f = 0 \tag{9.56}$$

we multiply the equation by the first variation of u, that is, δu, and integrate over Ω to get

$$\int_\Omega \delta u[L(u) - f]d\Omega = 0 \tag{9.57}$$

Now the goal is to manipulate Eq. (9.57) so that the variational operator δ can be moved outside the integral sign. Through such a manipulation, suppose we obtain

$$\delta \int_\Omega [\tilde{L}(u) - fu]\, d\Omega = 0 \tag{9.58}$$

where $\tilde{L}$ is a new operator resulting from the manipulation. Because $\delta J(u) = 0$, the desired functional is

$$J(u) = \int_\Omega [\tilde{L}(u) - fu]\, d\Omega \tag{9.59}$$

Example. Consider the Dirichlet problem

$$\nabla^2 u = 0 \qquad \text{in } \Omega \tag{9.60a}$$

$$u = \alpha \qquad \text{on } \Gamma \tag{9.60b}$$

To derive the corresponding variational form, we simply substitute Eq. (9.60) in Eq. (9.57), and get

$$\int_\Omega \delta u \nabla^2 u \, d\Omega = 0 \tag{9.61}$$

By virtue of the identities

$$\nabla^2 u = \nabla \cdot \nabla u \tag{9.62}$$

and

$$\nabla \cdot (u\mathbf{v}) = \nabla u \cdot \mathbf{v} + u \nabla \cdot \mathbf{v} \tag{9.63}$$

we can write

$$\nabla \cdot (\delta u \nabla u) = \nabla(\delta u) \cdot \nabla u + \delta u \nabla \cdot \nabla u$$

$$= \nabla(\delta u) \cdot \nabla u + \delta u \nabla^2 u$$

or

$$\delta u \nabla^2 u = \nabla \cdot (\delta u \nabla u) - \nabla(\delta u) \cdot \nabla u$$

Substituting in Eq. (9.61),

$$\int_\Omega \nabla \cdot (\delta u \nabla u)\, d\Omega - \int_\Omega \nabla(\delta u) \cdot \nabla u \, d\Omega = 0 \tag{9.64}$$

But Green's theorem states that

$$\oint_\Gamma \hat{n} \cdot \mathbf{v} \, d\Gamma = \int_\Omega \nabla \cdot \mathbf{v} \, d\Omega$$

where $\hat{n}$ is a unit vector normal to Γ.
Therefore, Eq. (9.64) becomes

$$\int_\Gamma \hat{n} \cdot \delta u \nabla u \, d\Gamma - \int_\Omega \nabla(\delta u) \cdot \nabla u \, d\Omega = 0 \tag{9.65}$$

Since $u = \alpha$ on Γ, $\delta u = 0$ on Γ, and the first integral vanishes. Thus we have

$$\int_\Omega \nabla(\delta u) \cdot \nabla u \, d\Omega = 0 \tag{9.66}$$

But

$$\nabla(\delta u) \cdot \nabla u = \tfrac{1}{2}\, \delta(\nabla u \cdot \nabla u)$$

Therefore, Eq. (9.66) becomes

$$\frac{1}{2}\, \delta \int_\Omega (\nabla u \cdot \nabla u)\, d\Omega = 0$$

or, simply, the variational expression equivalent to Eq. (9.66) is

$$J(u) = \int_\Omega (\nabla u \cdot \nabla u)\, d\Omega = 0 \tag{9.67}$$

Method 2. Another convenient procedure for deriving the variational expression invokes the self-adjoint property of the differential operator. Suppose the given linear differential equation can be written as

$$\mathbf{Lu} + \mathbf{b} = \mathbf{0} \tag{9.68}$$

where $\mathbf{L}$ is self-adjoint; then $J(u)$ can be written as

$$J(u) = \int_\Omega \left[\frac{1}{2}\mathbf{u}^T\mathbf{Lu} + \mathbf{u}^T\mathbf{b}\right] d\Omega \tag{9.69}$$

Example. Consider

$$\frac{\partial^2 u}{\partial x^2} + \frac{\partial^2 u}{\partial y^2} + cu + Q = 0$$

This can be written as

$$\mathbf{Lu} + \mathbf{b} = \mathbf{0}$$

where

$$\mathbf{L} = \left[\frac{\partial^2}{\partial x^2} + \frac{\partial^2}{\partial y^2} + c\right], \qquad \mathbf{u} = u, \qquad \mathbf{b} = Q$$

Therefore,

$$\frac{1}{2}\mathbf{u}^T\mathbf{Lu} + \mathbf{u}^T\mathbf{b} = \frac{1}{2}u\left[\frac{\partial^2 u}{\partial x^2} + \frac{\partial^2 u}{\partial y^2} + cu\right] + Qu$$

Therefore,

$$J(u) = \int_\Omega \frac{1}{2}\left[u\left(\frac{\partial^2 u}{\partial x^2} + \frac{\partial^2 u}{\partial y^2} + cu\right) + Qu\right] d\Omega$$

Integration by parts results in

$$J(u) = -\int_\Omega \left[\frac{1}{2}\left(\frac{\partial u}{\partial x}\right)^2 + \frac{1}{2}\left(\frac{\partial u}{\partial y}\right)^2 - \frac{1}{2}cu^2 - Qu\right] d\Omega$$

which is the required result. The test for the self-adjointness of $\mathbf{L}$ is left as an exercise to the reader.

Method 3. A general way to determine whether a classical variational principle exists for a given differential equation and boundary conditions is to use Frechet derivatives. This approach is quite general and applies to nonlinear and non-self-adjoint operators. Interested readers should consult the procedures described in the books by Mikhlin [2] and Finlayson [3] cited at the end of this chapter.

9.3. METHOD OF WEIGHTED RESIDUALS

The purpose of this section is to briefly present the highlights of the method of weighted residuals (MWR) first in a context-free manner, and then show how it can be used to develop the Galerkin approach to the formulation of finite element equations. One should bear in mind that MWR is an important method in its own right; the present intent is only to see its relevance to the finite element method. To set the stage for MWR, consider a general boundary value problem whose governing differential equation is given by

$$Lu(\mathbf{x}) = 0, \qquad \mathbf{x} \in \Omega \tag{9.70a}$$

$$u(\mathbf{x}) = g(\mathbf{x}), \qquad \mathbf{x} \in \Gamma \tag{9.70b}$$

where L is a differential operator and $u = u(\mathbf{x})$ is the dependent variable defined in a region Ω with the boundary Γ, and $\mathbf{x}$ represents the space coordinates. In MWR, one attempts to approximate the solution $u(\mathbf{x})$ of Eq. (9.70) by a *trial solution* $u_n(\mathbf{x})$, which is chosen in some convenient fashion. This trial solution will not, in general, satisfy the governing differential equation. Thus substitution of the trial solution into the governing differential equation will result in a ***residual***, denoted by R. To obtain the "best" solution, one then attempts to distribute the residual throughout the region Ω by trying to minimize the integral of the residual throughout Ω. That is,

$$\text{Minimize} \int_{\Omega} R \, d\Omega$$

The range of opportunities in meeting this objective can be enlarged by requiring that a *weighted* value of the residual be minimum throughout the region of interest. The weighting permits one to achieve a minimum value of zero to the weighted integral. Denoting the weighting functions by w, the required objective of the MWR becomes

$$\int_{\Omega} wR \, d\Omega = 0 \tag{9.71}$$

Thus central to the idea of MWR is the selection of weights and trial solutions. These two topics are addressed next.

SELECTION OF TRIAL SOLUTION

The idea of approximating the solution $u(\mathbf{x})$ of a differential equation by a trial solution is not a new one. Nevertheless, proper choice of the trial solution is crucial to the success of MWR. This choice provides the power of the method in that known information about the problem can be incorporated into the trial solution. In lower-order approximations [i.e., for small n in $u_n(\mathbf{x})$] this choice may influence the accuracy of the results, and in higher-order approximation it may influence convergence.

Of all the trial solutions employed by several people, perhaps the polynomial series such as

$$u_n(\mathbf{x}) = \sum_{i=1}^{n} c_i N(\mathbf{x}) = \sum_{i=1}^{n} c_i x^i \tag{9.72}$$

is most popular. In Eq. (9.72), c_i are arbitrary constants to be determined during the minimization of Eq. (9.71), and $N(\mathbf{x})$ are preselected functions called ***trial functions*** or ***shape functions***. The overwhelming preference for polynomials stems largely because they are easy to manipulate.

Of course, there are several guidelines in choosing the trial solution $u_n(\mathbf{x})$. First, the shape functions N_i must be complete. A set of functions $\{N_i\}$ is said to be complete in a space if any function u in that space can be expanded in terms of the set of functions $\{N_i\}$. The polynomials, for example, are complete. The completeness property ensures that we can represent the exact solution, provided enough number of terms are used in Eq. (9.72). Second, the choice of trial solution can be guided by any symmetry in the problem and by the type of boundary conditions. For Dirichlet-type boundary conditions, that is, if $u(\mathbf{x}) = g(\mathbf{x})$ as in Eq. (9.70b), a convenient form of the trial solution is

$$u_n(\mathbf{x}) = g(\mathbf{x}) + \sum_{i=1}^{n} c_i N_i(\mathbf{x}) \tag{9.73}$$

where $N_i(\mathbf{x})$ is chosen in such a fashion that $N_i(\mathbf{x}) = 0$ on the boundary. The shape functions, $N_i(\mathbf{x})$, themselves can be found sometimes by starting with a general polynomial, such as Eq. (9.73), and applying the boundary conditions and symmetry conditions. Finally, the chosen shape functions should not unduly complicate the analysis.

CHOICE OF WEIGHTS

The weighting functions w can be chosen in many different ways, and each choice corresponds to a different *criterion* of MWR.

Subdomain Method. In this procedure, the domain Ω is subdivided into m smaller subdomains $\Omega_j, j = 1, 2, \ldots, m$, which are not necessarily disjoint, and the weights are chosen such that

$$w_j = \begin{cases} 1, & \mathbf{x} \in \Omega_j \\ 0, & \mathbf{x} \notin \Omega_j \end{cases} \tag{9.74}$$

Now Eq. (9.71) becomes

$$\int_{\Omega_j} R \, d\Omega_j = 0, \qquad j = 1, 2, \ldots, m \tag{9.75}$$

That is, the differential equation integrated over the subdomain is zero. As m increases, the differential equation is satisfied on the average in smaller and smaller subdomains, and presumably approaches zero everywhere.

Collocation Method. In this method, the weighting functions w_j are chosen to be the displaced Dirac delta functions

$$w_j = \delta(\mathbf{x} - \mathbf{x}_j) \triangleq \delta_j \tag{9.76}$$

Now Eq. (9.71) becomes

$$\int_{\Omega} w_j R \, d\Omega = \int \delta_j \, R \, d\Omega = R_j = 0, \qquad j = 1, 2, \ldots, m \tag{9.77}$$

where R_j stands for the value of R evaluated at the point $(\mathbf{x}_j)$. Thus the residual is made to vanish at m specified ***collocation points*** $x_j, j = 1, 2, \ldots, m$. As m increases, the residual vanishes at more and more points and presumably approaches zero everywhere.

Least Squares Method. In this method, the weighting functions w_j are chosen to be

$$w_j = \frac{\partial R}{\partial c_j} \tag{9.78}$$

Now Eq. (9.71) becomes

$$\frac{\partial}{\partial c_j}\int_\Omega R^2\, d\Omega = 2\int_\Omega \frac{\partial R}{\partial c_j} R\, d\Omega = 0, \qquad j = 1, 2, \ldots, n \tag{9.79}$$

That is, the integral of the square of the residual is minimized with respect to the undetermined parameters to provide N simultaneous equations for the c_j's.

Method of Moments. In this method, the weighting functions w_j are chosen to be

$$w_j = P_j(\mathbf{x}) \tag{9.80}$$

where $P_j(x)$ are orthogonal polynomials defined over the domain Ω. This procedure is most useful in one-dimensional problems where the theory of orthogonal polynomials is well understood. In one-dimensional problems, the use of weighting functions $w_j = x^j$, that is, $1, x, x^2, \ldots$, is widespread, and with this choice Eq. (9.73) becomes

$$\int_\Omega x^j R\, d\Omega = 0 \tag{9.81}$$

The structure of Eq. (9.81) led to the name "method of moments." Note, however, that the set $\{w_j\} = \{x^j\} = \{1, x, x^2, \ldots\}$ is not orthogonal over $0 \le x \le 1$, and generally one can expect better results if they are orthogonalized before use.

Galerkin Method. In this method, the weighting functions w_j are chosen to be identical to the shape functions N_j themselves. That is,

$$w_j = N_j(\mathbf{x}), \qquad j = 1, 2, \ldots, m \tag{9.82}$$

Therefore, Eq. (9.71) becomes

$$\int_\Omega N_j(\mathbf{x}) R\, d\Omega = 0, \qquad j = 1, 2, \ldots, m \tag{9.83}$$

or, in vector-matrix notation

$$\int_\Omega N R\, d\Omega = 0 \tag{9.84}$$

where $N = (N_1, N_2, \ldots, N_m)^T$. Using the well-known fact that a continuous function is zero if it is orthogonal to every member of a complete set, it can be seen that the Galerkin method forces the residual to be zero by making it to be orthogonal to each member of a complete set of basis functions.

9.4. GALERKIN FORMULATION OF THE FINITE ELEMENT METHOD

The Galerkin approach to the formulation of finite element equilibrium equations is illustrated in this section. As before, a one-dimensional problem is treated first. This simple example helps to fix the ideas as well as prepares the ground for the subsequent treatment of transient field problems.

A SIMPLE INITIAL VALUE PROBLEM

Consider the following example discussed by Segerlind [4]:

$$\frac{d^2y}{dt^2} + 4y = 0, \qquad 0 \le t \le 1 \tag{9.85a}$$

$$y(0) = 0, \qquad y'(0) = 4 \tag{9.85b}$$

Application of Eq. (9.84) results in

$$\int_0^1 \mathbf{N}\left(\frac{d^2y}{dt^2} + 4y\right) dt = 0 \tag{9.86}$$

To perform this integration, the interval [0, 1] is divided into several elements, thus: $0 = T_1 < T_2 < \cdots < T_E = 1$. Now Eq. (9.86) can be rewritten as

$$\sum_{e=1}^{E} \int_{L^{(e)}} \mathbf{N}^{(e)}\left[\frac{d^2y^{(e)}}{dt^2} + 4y^{(e)}\right] dt = 0 \tag{9.87}$$

where E is the number of elements and $L^{(e)}$ is the length of an individual element.

Before proceeding further, it is useful to digress and discuss a point of some theoretical interest. There is no limit on the order of the highest derivative that may be contained within the square brackets of Eq. (9.87). However, if $y^{(e)}$ is to be approximated by an interpolation polynomial, then the highest-order derivative that is allowed within the square brackets is one greater than the order of continuity in the interpolation polynomial. If the interpolation polynomial used is of order zero (i.e., continuity in u but is not in the first derivative), then derivatives greater than the first cannot appear in Eq. (9.87). This restriction can be bypassed by reducing the order of the derivatives using integration by parts.

Integrating the first term of Eq. (9.87) by parts, we get

$$\int_{L^{(e)}} \mathbf{N}\frac{d^2y}{dt^2}\, dt = \mathbf{N}\frac{dy}{dk}\bigg|_{T_i}^{T_j} - \int_{L^{(e)}} \frac{d\mathbf{N}}{dt}\frac{dy}{dt}\, dt \tag{9.88}$$

Therefore, Eq. (9.87) becomes

$$-\mathbf{N}^{(1)}\frac{dy}{dt}\bigg|_{t=0} - \sum_{e=0}^{E} \int_{T_i}^{T_j} \left(\frac{d\mathbf{N}^{(e)}}{dt}\frac{dy}{dt} - 4\mathbf{N}^{(e)}y\right) dt = 0 \tag{9.89}$$

where T_i and T_j are the time values at nodes i and j.

If linear interpolation polynomials are used, then

$$y = \mathbf{N}^{(e)T}\mathbf{y} = \begin{bmatrix}\left(1 - \frac{t}{T_e}\right) & \frac{t}{T_e}\end{bmatrix}\begin{bmatrix} y_i \\ y_j \end{bmatrix} \tag{9.90}$$

Equation (9.90) is defined relative to a local coordinate system located at node i. Therefore, integration limits are $T_i = 0$ and $T_j = L^{(e)}$. Substituting Eq. (9.90) in Eq. (9.89), we get

$$-\mathbf{N}^{(1)}\frac{dy}{dt}\bigg|_{t=0} - \sum_{e=1}^{E}\int_0^{L^{(e)}}\left(\frac{d\mathbf{N}^{(e)}}{dt}\frac{d\mathbf{N}^{(e)}}{dt}\mathbf{y} - 4\mathbf{N}^{(e)}\mathbf{N}^{(e)}\mathbf{y}\right)dt = 0 \tag{9.91}$$

For the first element, $e = 1$, and Eq. (9.91) reduces to

$$-\begin{bmatrix} 4 \\ 0 \end{bmatrix} - \frac{1}{L^{(1)}}\begin{bmatrix} 1 & -1 \\ -1 & 1 \end{bmatrix}\begin{bmatrix} y_1 \\ y_2 \end{bmatrix} + \frac{4L^{(1)}}{6}\begin{bmatrix} 2 & 1 \\ 1 & 2 \end{bmatrix}\begin{bmatrix} y_1 \\ y_2 \end{bmatrix} = \begin{bmatrix} 0 \\ 0 \end{bmatrix} \tag{9.92}$$

For every other element, Eq. (9.91) becomes

$$-\frac{1}{L^{(1)}}\begin{bmatrix} 1 & -1 \\ -1 & 1 \end{bmatrix}\begin{bmatrix} y_i \\ y_j \end{bmatrix} + \frac{4L^{(e)}}{6}\begin{bmatrix} 2 & 1 \\ 1 & 2 \end{bmatrix}\begin{bmatrix} y_i \\ y_j \end{bmatrix} = \begin{bmatrix} 0 \\ 0 \end{bmatrix} \tag{9.93}$$

Assembling these element equations,

$$\begin{bmatrix} 4 \\ 0 \\ 0 \\ \cdot \\ \cdot \\ \cdot \end{bmatrix} - \frac{1}{L^{(e)}}\begin{bmatrix} 1 & -1 & & & \\ -1 & 2 & -1 & & \\ & -1 & 2 & -1 & \\ & & \cdot & \cdot & \\ & & \cdot & \cdot & \\ & & \cdot & \cdot & \end{bmatrix}\begin{bmatrix} y_1 \\ y_2 \\ \cdot \\ \cdot \\ \cdot \\ \cdot \end{bmatrix} + \frac{4L^{(e)}}{6}\begin{bmatrix} 2 & 1 & & & \\ 1 & 4 & 1 & & \\ & 1 & 4 & 1 & \\ & & \cdot & \cdot & \\ & & \cdot & \cdot & \\ & & \cdot & \cdot & \end{bmatrix}\begin{bmatrix} y_1 \\ y_2 \\ \cdot \\ \cdot \\ \cdot \\ \cdot \end{bmatrix} = \begin{bmatrix} 0 \\ 0 \\ \cdot \\ \cdot \\ \cdot \\ \cdot \end{bmatrix} \tag{9.94}$$

This system of equations can also be rewritten as

$$-4 - \frac{1}{L^{(e)}}(y_1 - y_2) + \frac{4L^{(e)}}{6}(2y_1 + y_2) = 0 \tag{9.95}$$

$$-\frac{1}{L^{(e)}}(-y_{n-1} + 2y_n - y_{n+1}) + \frac{4L^{(e)}}{6}(y_{n-1} + 4y_n + y_{n+1}) = 0, \qquad n \geq 2 \tag{9.96}$$

Once the length of the time step $L^{(e)}$ is selected, the preceding equations can be solved. Actual computation of the numerical solution is left as an exercise. It is constructive to compare this numerical solution with the exact solution $y = 4 \sin 2t$.

A SIMPLE TWO-DIMENSIONAL PROBLEM

Consider once again

$$\frac{\partial}{\partial x}\left(k_x \frac{\partial u}{\partial x}\right) + \frac{\partial}{\partial y}\left(k_y \frac{\partial u}{\partial y}\right) + Q = 0 \tag{9.97a}$$

with the boundary conditions

$$u = u_B \qquad \text{on } \Gamma_1 \tag{9.97b}$$

$$k_x \frac{\partial u}{\partial x} l_x + k_y \frac{\partial u}{\partial y} l_y + q + h(u - u_\infty) = 0 \qquad \text{on } \Gamma_2 \tag{9.97c}$$

To bring forth some of the features of the Galerkin method more clearly, we shall start with the MWR and then specialize the case for the Galerkin criterion. Toward the end, a grid is superimposed on Ω and attention is focused on a typical element $\Omega^{(e)}$. Let the trial solution be of the form $\mathbf{N}^T\mathbf{u}^{(e)}$. That is, on element (e),

$$u(x, y) \cong \hat{u} = \mathbf{N}^T\mathbf{u}^{(e)} \tag{9.98}$$

Now Eq. (9.71) can be written, for element e, as

$$\int_{\Omega^{(e)}} w_i \left[\frac{\partial}{\partial x}\left(k_x \frac{\partial \hat{u}}{\partial x}\right) + \frac{\partial}{\partial y}\left(k_y \frac{\partial \hat{u}}{\partial y}\right) + Q\right] dx\, dy = 0 \tag{9.99a}$$

or in vector-matrix notation as

$$\int_{\Omega} \mathbf{w} \left[\frac{\partial}{\partial x}\left(k_x \frac{\partial \hat{u}}{\partial x}\right) + \frac{\partial}{\partial}\left(k_y \frac{\partial \hat{u}}{\partial y}\right) + Q\right] dx\, dy = 0 \tag{9.99b}$$

Notice that in Eq. (9.99b) the trial solution $\hat{u} = \mathbf{N}^T\mathbf{u}^e$ is required to be differentiated twice because it appears under $\partial^2/\partial x^2$ and $\partial^2/\partial y^2$. Also recall that if we propose to use, say, piecewise interpolation polynomial for u then the first derivatives will not be continuous and second derivatives will be infinitely large. Stated differently, to perform the integration in Eq. (9.99) we would like to avoid infinities in second derivatives, which is equivalent to saying that we want a continuous first derivative. However, if we wish to use piecewise interpolation polynomials, this restriction cannot be met. The alternative is to avoid this restriction by a transformation, as described next.

Using integration by parts (which is called Green's theorem in two dimensions and Gauss's theorem in three dimensions), we can write

$$\int_{\Omega^{(e)}} \mathbf{w} \left[\frac{\partial}{\partial x}\left(k_x \frac{\partial \hat{u}}{\partial x}\right)\right] dx\, dy = \int_{\Gamma^{(e)}} \mathbf{w} k_x \frac{\partial \hat{u}}{\partial x}\bigg|_A^B dy - \int_{\Omega^{(e)}} k_x \frac{\partial \mathbf{w}}{\partial x}\frac{\partial \hat{u}}{\partial x} dx\, dy \tag{9.100}$$

where A and B refer to the limits of integration to be imposed in the x-direction. Because $\Omega^{(e)}$ here is a plane region, the first term on the right of Eq. (9.100) can be written as a contour integral. Therefore, Eq. (9.100) becomes

$$\int_{\Omega^{(e)}} \mathbf{w} \left[\frac{\partial}{\partial x}\left(k_x \frac{\partial \hat{u}}{\partial x}\right)\right] dx\, dy = \int_{\Gamma^{(e)}} k_x \mathbf{w} \frac{\partial \hat{u}}{\partial x} l_x \, d\Gamma^{(e)} - \int_{\Omega^{(e)}} k_x \frac{\partial \mathbf{w}}{\partial x}\frac{\partial \hat{u}}{\partial x} dx\, dy \tag{9.101}$$

where l_x is the direction cosine of the outward normal and the x-axis, and the integral is evaluated along the contour $\Gamma^{(e)}$. Now Eq. (9.99) can be written as

$$\int_{\Gamma^{(e)}} \mathbf{w}\left(k_x \frac{\partial \hat{u}}{\partial x} l_x + k_y \frac{\partial \hat{u}}{\partial y} l_y\right) d\Gamma^{(e)}$$
$$- \int_{\Omega^{(e)}} \left[k_x \frac{\partial \mathbf{w}}{\partial x}\frac{\partial \hat{u}}{\partial x} + k_y \frac{\partial \mathbf{w}}{\partial y}\frac{\partial \hat{u}}{\partial y} - \mathbf{w}Q\right] dx\, dy = 0 \tag{9.102}$$

Now the following observation can be made. The quantity inside the parentheses of the first integral is readily recognized as $\partial u/\partial n$. Therefore, by virtue of the boundary condition in Eq. (9.97c), the first integral need be evaluated only along Γ_2. Therefore, Eq. (9.102) can be rewritten as

$$\int_{\Omega^{(e)}} \left[k_x \frac{\partial \mathbf{w}}{\partial x}\frac{\partial \hat{u}}{\partial x} + k_y \frac{\partial \mathbf{w}}{\partial y}\frac{\partial \hat{u}}{\partial y}\right] dx\, dy - \int_{\Omega^{(e)}} \mathbf{w} Q\, dx\, dy - \int_{\Gamma_2^{(e)}} \mathbf{w} \frac{\partial \hat{u}}{\partial n}\, d\Gamma_2^{(e)} = 0 \qquad (9.103)$$

Note that in Eq. (9.103) the derivatives of the weighting functions appear. This forces us to make a restriction on the weighting function, that the w_i's be continuous. This restriction, if imposed, excludes the subdomain and collocation methods from further consideration. However, any other WRM in which the weights are continuous functions is admissible. In the Galerkin procedure, one chooses $w_i = N_i$. Thus, substituting

$$\mathbf{w} = \mathbf{N}, \qquad \hat{u} = \mathbf{N}^T\mathbf{u}^{(e)} \qquad (9.104)$$

in Eq. (9.103), we get

$$\int_{\Omega^{(e)}} \left[k_x \frac{\partial \mathbf{N}}{\partial x}\frac{\partial \mathbf{N}^T}{\partial x} + k_y \frac{\partial \mathbf{N}}{\partial y}\frac{\partial \mathbf{N}^T}{\partial y}\right] dx\, dy\, \mathbf{u}^{(e)} - \int_{\Omega^{(e)}} \mathbf{N} Q\, dx\, dy$$

$$- \int_{\Gamma_2^{(e)}} \mathbf{N} \frac{\partial \hat{u}}{\partial n}\, d\Gamma_2^{(e)} = 0 \qquad (9.105a)$$

which can also be written as

$$\int_{\Omega^{(e)}} [\mathbf{B}^{(e)T}\mathbf{D}^{(e)}\mathbf{B}^{(e)}\, dx\, dy]\mathbf{u}^{(e)} - \int_{\Omega^{(e)}} \mathbf{N} Q\, dx\, dy - \int_{\Gamma_2^{(e)}} \mathbf{N} \frac{\partial \hat{u}}{\partial n}\, d\Gamma_2^{(e)} = 0 \qquad (9.105b)$$

where the matrixes $\mathbf{B}^{(e)}$ and $\mathbf{D}^{(e)}$ are the same as those defined respectively in Eqs. (9.27) and (9.19). Notice that Eq. (9.105b) can also be written as

$$\int_{\Omega^{(e)}} [\mathbf{B}^{(e)T}\mathbf{D}^{(e)}\mathbf{B}^{(e)}\, dx\, dy]\mathbf{u}^{(e)} - \int_{\Omega^{(e)}} \mathbf{N} Q\, dx\, dy$$

$$+ \int_{\Gamma_2^{(e)}} \mathbf{N}[q + h(\hat{u} - u_\infty)]\, d\Gamma_2^{(e)} = 0 \qquad (9.105c)$$

Now the matrix $\mathbf{A}^{(e)}$ can be recognized as

$$\mathbf{A}^{(e)} = \int_{\Omega^{(e)}} \mathbf{B}^{(e)T}\mathbf{D}\mathbf{B}^{(e)}\, dx\, dy + \int_{\Gamma_2^{(e)}} h\mathbf{N}\,\mathbf{N}^T\, d\Gamma_2^{(e)} \qquad (9.106)$$

Similarly, the vector $\mathbf{f}^{(e)}$ can be identified to be

$$-\mathbf{f}^{(e)} = -\int_{\Omega^{(e)}} Q\mathbf{N}\, dx\, dy + \int_{\Gamma_2^{(e)}} q\mathbf{N}\, d\Gamma_2^{(e)} - \int_{\Gamma_2^{(e)}} hu_\infty \mathbf{N}\, d\Gamma_2^{(e)} \qquad (9.107)$$

Notice that the equation pairs (9.106) and (9.32a) and (9.107) and (9.33a) are indeed identical.

Note that the first integral in Eq. (9.105b) carries information about $\mathbf{A}^{(e)}$, the second integral carries information about the forcing function, and the third integral carries information about boundary conditions. Evidently, this third integral does not contribute anything to the equation at the internal grid points. When a grid point lies on a boundary and if $\partial u/\partial n$ is specified on that boundary, then the integral can be evaluated as shown next.

For convenience, consider a portion of a straight boundary, as shown in Figure 9.4, on which $\partial u/\partial n$ is specified. We wish to evaluate the last integral of Eq. 9.106 on this boundary. Recall that $\hat{u}$ on this element is given by (for a linear triangle)

$$\hat{u} = N_i u_i + N_j u_j + N_k u_k \tag{9.108}$$

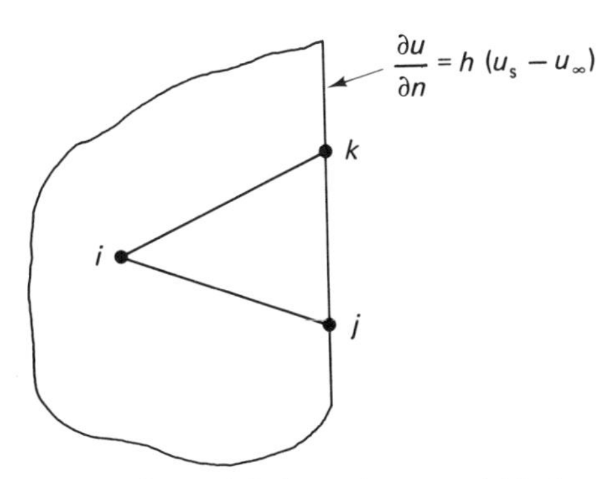

Figure 9.4 Portion of a straight boundary on which the normal derivative is specified

Now the value of u on the vertical surface can be written as

$$u_s = 0u_i + L_2 u_j + L_3 u_k \tag{9.109}$$

because $L_1 = 0$ on the vertical boundary (see Figure 8.11, for example). Therefore,

$$\frac{\partial \hat{u}}{\partial n} = h(u_s - u_\infty) = h[0 \quad L_2 \quad L_3]\begin{bmatrix} u_i \\ u_j \\ u_k \end{bmatrix} - hu_\infty \tag{9.110}$$

Substitution of Eq. (9.110) into the last integral of Eq. (9.106) gives

$$\int_{\Gamma^{(e)}} \mathbf{N}\frac{\partial \hat{u}}{\partial n}\,d\Gamma^{(e)} = h\int_{\Gamma^{(e)}} \mathbf{N}\mathbf{N}^T\mathbf{u}\,d\Gamma^{(e)} - \int_{\Gamma^{(e)}} \mathbf{N}hu_\infty\,d\Gamma^{(e)} \tag{9.111a}$$

$$= h\int_{\Gamma^{(e)}} \begin{bmatrix} 0 & 0 & 0 \\ 0 & L_2^2 & L_2L_3 \\ 0 & L_3L_2 & L_3^2 \end{bmatrix}\begin{bmatrix} u_i \\ u_j \\ u_k \end{bmatrix} d\Gamma^{(e)} - \int_{\Gamma^{(e)}} \mathbf{N}hu_\infty\,d\Gamma^{(e)} \tag{9.111b}$$

Using Eq. (8.46), we get for example

$$\int L_2^2\,d\Gamma^{(e)} = \frac{2!0!}{(2+0+1)!}l = \frac{l}{3}$$

$$\int L_2L_3\,d\Gamma^{(e)} = \frac{1!1!}{(1+1+1)!}l = \frac{l}{6}$$

where l is the length of the side of the triangular element. Therefore, Eq. (9.111) becomes

$$\int_{\Gamma^{(e)}} \mathbf{N}\frac{\partial \hat{u}}{\partial n}\,\partial\Gamma^{(e)} = \frac{hl}{6}\begin{bmatrix} 0 & 0 & 0 \\ 0 & 2 & 1 \\ 0 & 1 & 2 \end{bmatrix}\begin{bmatrix} u_i \\ u_j \\ u_k \end{bmatrix} - \int_{\Gamma^{(e)}} \mathbf{N}hu_\infty\, d\Gamma^{(e)} \tag{9.112}$$

Evaluation of the last integral is done as follows:

$$\int_{\Gamma^{(e)}} \mathbf{N}hu_\infty\, d\Gamma^{(e)} = hu_\infty \int_{\Gamma^{(e)}} \mathbf{N}\, d\Gamma^{(e)} = hu_\infty \int_{\Gamma^{(e)}} \begin{bmatrix} N_i \\ N_j \\ N_k \end{bmatrix} d\Gamma^{(e)}$$

Because $N_i = 0$, $N_j = L_j$, and $N_k = L_k$, this equation becomes

$$\int_{\Gamma^{(e)}} \mathbf{N}hu\, d\Gamma^{(e)} = \frac{hu_\infty l}{2}\begin{bmatrix} 0 \\ 1 \\ 1 \end{bmatrix} \tag{9.113}$$

Once the element matrixes are obtained, they can be assembled to get the global matrixes as was done earlier.

9.5. TIME-DEPENDENT FIELD PROBLEMS

Suppose Eq. (9.16a) is modified to include the transient nature of the problem, thus:

$$\frac{\partial}{\partial x}\left(k_x\frac{\partial u}{\partial x}\right) + \frac{\partial}{\partial y}\left(k_y\frac{\partial u}{\partial y}\right) + Q = \frac{\partial u}{dt} \tag{9.114a}$$

subject to the initial and boundary conditions

$$u(x, y, 0) = u_0(x, y) \qquad \text{in } \Omega \tag{9.114b}$$

$$u = u_B \qquad \text{on } \Gamma_1,\ t > 0 \tag{9.114c}$$

$$k_x\frac{\partial u}{\partial x}l_x + k_y\frac{\partial u}{\partial y}l_y + q + h(u - u_\infty) = 0 \qquad \text{on } \Gamma_2,\ t > 0 \tag{9.114d}$$

VARIATIONAL APPROACH

One method of handling this problem is to replace Q by $(Q - \partial u/\partial t)$ in Eq. (9.17). Now the functional equivalent of Eq. (9.114) can be written as

$$J = \int_\Omega \left[\frac{1}{2}k_x\left(\frac{\partial u}{\partial x}\right)^2 + \frac{1}{2}k_y\left(\frac{\partial u}{\partial y}\right)^2 - 2\left(Q - \frac{\partial u}{\partial t}\right)u\right] d\Omega$$

$$+ \int_{\Gamma_2}\left[qu + \frac{h}{2}(u - u_\infty)^2\right] d\Gamma_2 \tag{9.115}$$

The only difference between this and Eq. (9.17), for example, is in the treatment of the Q term. Therefore, it is sufficient to consider the contribution of the modified term to J. This can be written as

$$J_Q = -\int_\Omega u\left(Q - \frac{\partial u}{\partial t}\right) d\Omega \tag{9.116a}$$

$$= \sum_{e=1}^{E} \int_\Omega u^{(e)}\left[\frac{\partial u^{(e)}}{dt} - Q^{(e)}\right] d\Omega \tag{9.116b}$$

where the field variable $u^{(e)} = u^{(e)}(x, y)$ is defined throughout each element. Indeed, if the collection of all $u^{(e)}(x, y)$'s, for $e = 1, 2, \ldots$, E, is written as a vector $\mathbf{u}^{(e)}(x, y)$, then we have

$$\mathbf{u}^{(e)}(x, y) = \mathfrak{N}\mathbf{u} \tag{9.116c}$$

where $\mathfrak{N}$ is a rectangular matrix with as many rows as there are elements and as many columns as there are nodes in the region. The components of $\mathfrak{N}$ are the shape functions N_i, evaluated at the nodes. See Exercise 8.14 for details. Note also that the entries of $\mathfrak{N}$ are only functions of x and y. Therefore,

$$\frac{\partial \mathbf{u}^{(e)}}{\partial t}(x, y) = \mathfrak{N}\frac{\partial \mathbf{u}}{\partial t} \tag{9.117}$$

Now we can write

$$J_Q = \sum_{e=1}^{E} \int_\Omega \left[\mathfrak{N}\mathbf{u}\mathfrak{N}\frac{\partial \mathbf{u}}{\partial t} - \mathfrak{N}\mathbf{u}Q\right] d\Omega \tag{9.118}$$

Differentiation with respect to $\mathbf{u}$ yields (do this and verify!)

$$\frac{dJ_Q}{d\mathbf{u}} = \sum_{e=1}^{E} -\int_\Omega \mathfrak{N}^T Q\, d\Omega + \sum_{e=1}^{E}\left(\int_\Omega \mathfrak{N}^T\mathfrak{N}\, d\Omega\right)\frac{d\mathbf{u}}{dt} \tag{9.119}$$

This equation replaces Eq. (9.29b). Now the modified form of Eq. (9.35) becomes

$$\mathbf{Au} = \mathbf{f} + \mathbf{C}\frac{\partial \mathbf{u}}{\partial t} \tag{9.120}$$

where $\mathbf{A}$ and $\mathbf{f}$ are the same as those defined in Eqs. (9.32), (9.33), and (9.36), and $\mathbf{C}$ is given by

$$\mathbf{C}^{(e)} = \int_\Omega \mathfrak{N}^T\mathfrak{N}\, d\Omega \tag{9.121}$$

Equation (9.120) represents a system of first-order equations and can be solved using standard finite difference methods.

GALERKIN APPROACH

The procedure here is very similar to the one used in Section 9.4. Now Eq. (9.99a) for this time-dependent problem becomes

$$\int_{\Omega^{(e)}} w_i\left[\frac{\partial}{\partial x}\left(k_x\frac{\partial \hat{u}}{\partial x}\right) + \frac{\partial}{\partial y}\left(k_y\frac{\partial \hat{u}}{\partial y}\right) + Q - \frac{\partial \hat{u}}{\partial t}\right] dx\, dy = 0 \tag{9.122}$$

Using integration by parts, as before, we get

$$\int_{\Gamma^{(e)}} \mathbf{w}\left(k_x \frac{\partial \hat{u}}{\partial x} l_x + k_y \frac{\partial \hat{u}}{\partial y} l_y\right) d\Gamma^{(e)}$$

$$- \int_{\Omega^{(e)}} \left[k_x \frac{\partial \mathbf{w}}{\partial x}\frac{\partial \hat{u}}{\partial x} + k_y \frac{\partial \mathbf{w}}{\partial y}\frac{\partial \hat{u}}{\partial y} - \mathbf{w}Q\right] dx\, dy - \int_{\Omega^{(e)}} \mathbf{w} \frac{\partial \hat{u}}{\partial t} dx\, dy = 0 \quad (9.123)$$

instead of Eq. (9.102). The only difference between Eq. (9.123) and Eq. (9.102) is in the last term.

Now it is assumed that $u(x, y, t)$ within each element can be approximated by

$$\hat{u}^{(e)}(x, y, t) = \mathbf{N}^T\mathbf{u} = \sum_{i=1}^{r} N_i(x, y)u_i(t) \quad (9.124)$$

Using Galerkin criterion (i.e., $w_i = N_i$) and substituting Eq. (9.124) in Eq. (9.123), one gets an equation identical to Eq. (9.105a) except for an additional term, which is

$$- \int_{\Omega^{(e)}} \mathbf{N} \frac{\partial}{\partial t}(\mathbf{N}^T\mathbf{u})\, dx\, dy$$

Therefore, the coefficient matrix in this case is

$$\int_{\Omega^{(e)}} \mathbf{B}^{(e)}\mathbf{D}^{(e)}\mathbf{B}^{(e)T}\, dx\, dy + \int_{\Gamma_2^{(e)}} h\mathbf{N}\mathbf{N}^T\, d\Gamma_2^{(e)} - \int_{\Omega^{(e)}} \mathbf{N}\mathbf{N}^T \frac{\partial \mathbf{u}}{\partial t}\, dx\, dy \quad (9.125)$$

Therefore, instead of getting an $\mathbf{Au} = \mathbf{f}$ type of equation, we get

$$\mathbf{Au} = \mathbf{f} + \mathbf{C}\frac{\partial \mathbf{u}}{\partial t} \quad (9.126)$$

Now it is only necessary to solve the system of ordinary differential equations obtained here subject to the initial condition in Eq. (9.114a).

SOLUTION OF THE DIFFERENTIAL EQUATION

The ordinary differential equation $\mathbf{C}\dot{\mathbf{u}} = \mathbf{Au} - \mathbf{f}$ can be solved using one of many available standard techniques, such as the Runge-Kutta method or the predictor-corrector methods. It is also possible to solve the differential equation using finite element procedures. One such procedure is briefly described here.

The central idea of the finite element procedure described here lies in the discretization of the time domain. (Here we are deliberately avoiding the possibility of discretizing the space-time continuum using finite elements.) To fix the ideas, first let us assume that the entire domain of investigation corresponds with that of one element of length Δt. Proceeding in the usual manner of discretization, with time as the independent variable, we can write at any node

$$u^{(e)}(t) = N_i^{(e)}u_i + N_{i+1}^{(e)}u_{i+1} \quad (9.127)$$

where u_i stands for the value of $u^{(e)}(t)$ at time t_i at the node in question, and $t_{i+1} - t_i = \Delta t$ is the length of the time domain under investigation. If there are r nodes in the element, we can rewrite Eq. (9.127) as

$$\mathbf{u}^{(e)}(t) = N_i^{(e)}\mathbf{u}_i + N_{i+1}^{(e)}\mathbf{u}_{i+1} \tag{9.128}$$

where $\mathbf{u}(t) = (u_1(t), u_2(t), \ldots, u_r(t))^T$. The shape functions $N_i(t)$ are assumed to be the same for each component of $\mathbf{u}$, and therefore N_i is a scalar.

Now applying the WRM, an equation analogous to Eq. (9.87) can be written as

$$\int_{\Delta t} w_\beta[\mathbf{C}(\mathbf{u}_i\dot{N}_i + \mathbf{u}_{i+1}\dot{N}_{i+1}) - \mathbf{A}(\mathbf{u}_iN_i + \mathbf{u}_{i+1}N_{i+1}) + \mathbf{f}]\,dt = 0 \tag{9.129}$$

$$\text{for } \beta = i, i+1$$

As only the first derivatives are involved in Eq. (9.126), first-order polynomials are sufficient to represent the shape functions N_i. In terms of the local coordinate ξ, these shape functions and their derivatives are (see Figure 9.5)

$$\begin{aligned} &0 \le \xi \le 1, \qquad \xi = \frac{t}{\Delta t} \\ &N_i = 1 - \xi, \qquad \dot{N}_i = -\frac{1}{\Delta t} \\ &N_{i+1} = \xi, \qquad \dot{N}_{i+1} = \frac{1}{\Delta t} \end{aligned} \tag{9.130}$$

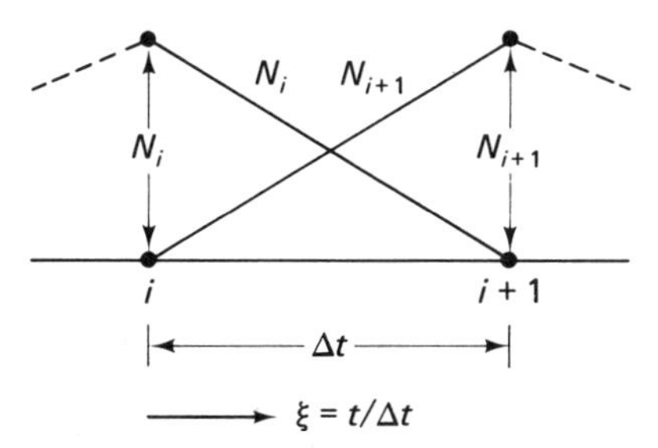

Figure 9.5 Shape functions in the time domain

Now the various integrals in Eq. (9.129) can be evaluated by using information in Eq. (9.130). A sample calculation for $\beta = i$ is shown here.

$$\begin{aligned} \int_{\Delta t} w_\beta \mathbf{C}\mathbf{u}_i\dot{N}_i\,dt &= \mathbf{C}\mathbf{u}_i\int_0^1 (1-\xi)\left(-\frac{1}{\Delta t}\right)(\Delta t)\,d\xi = -\frac{1}{2}\mathbf{C}\mathbf{u}_i \\ \int_{\Delta t} w_\beta \mathbf{C}\mathbf{u}_{i+1}\dot{N}_{i+1}\,dt &= \mathbf{C}\mathbf{u}_{i+1}\int_0^1 (1-\xi)\left(\frac{1}{\Delta t}\right)(\Delta t)\,d\xi = \frac{1}{2}\mathbf{C}\mathbf{u}_{i+1} \\ -\int_{\Delta t} w_\beta \mathbf{A}\mathbf{u}_iN_i\,dt &= -\mathbf{A}\mathbf{u}_i\int_0^1 (1-\xi)^2(\Delta t)\,d\xi = -\frac{1}{3}(\Delta t)\mathbf{A}\mathbf{u}_i \\ -\int_{\Delta t} w_\beta \mathbf{A}\mathbf{u}_{i+1}N_{i+1}\,dt &= -\mathbf{A}\mathbf{u}_{i+1}\int_0^1 (1-\xi)\xi(\Delta t)\,d\xi = -\frac{1}{6}(\Delta t)\mathbf{A}\mathbf{u}_{i+1} \\ \int_{\Delta t} w_\beta \mathbf{f}\,dt &= \int_0^1 \mathbf{f}(1-\xi)(\Delta t)\,d\xi = \frac{1}{2}(\Delta t)\mathbf{f}_i \end{aligned} \tag{9.131}$$

Thus for $\beta = i$, Eq. (9.129) becomes

$$\left(-\frac{1}{2}\mathbf{C} - \frac{\Delta t}{3}\mathbf{A}\right)\mathbf{u}_i + \left(\frac{1}{2}\mathbf{C} - \frac{\Delta t}{6}\mathbf{A}\right)\mathbf{u}_{i+1} + \frac{\Delta t}{2}\mathbf{f}_i = \mathbf{0} \tag{9.132a}$$

A similar calculation for $\beta = i + 1$ transforms Eq. (9.129) into

$$\left(-\frac{1}{2}\mathbf{C} - \frac{\Delta t}{6}\mathbf{A}\right)\mathbf{u}_i + \left(\frac{1}{2}\mathbf{C} - \frac{\Delta t}{3}\mathbf{A}\right)\mathbf{u}_{i+1} + \frac{\Delta t}{2}\mathbf{f}_{i+1} = \mathbf{0} \tag{9.132b}$$

Or, in matrix notation,

$$\begin{bmatrix} \left(-\frac{1}{2}\mathbf{C} - \frac{\Delta t}{3}\mathbf{A}\right) & \left(\frac{1}{2}\mathbf{C} - \frac{\Delta t}{6}\mathbf{A}\right) \\ \left(-\frac{1}{2}\mathbf{C} - \frac{\Delta t}{6}\mathbf{A}\right) & \left(\frac{1}{2}\mathbf{C} - \frac{\Delta t}{3}\mathbf{A}\right) \end{bmatrix} \begin{bmatrix} \mathbf{u}_i \\ \mathbf{u}_{i+1} \end{bmatrix} + \frac{\Delta t}{2} \begin{bmatrix} \mathbf{f}_i \\ \mathbf{f}_{i+1} \end{bmatrix} = \begin{bmatrix} \mathbf{0} \\ \mathbf{0} \end{bmatrix} \tag{9.133}$$

Equation (9.133) is valid for one time step of size Δt. The total time domain is traveled in several time steps. Then the nature of the shape functions over any successive pair of elements will be as shown in Figure 9.6. Therefore, it is necessary to assemble the equations for adjacent time elements. As an equation analogous to Eq. (9.133) holds good for time element $i - 1$, we can write

$$\begin{bmatrix} \left(-\frac{1}{2}\mathbf{C} - \frac{\Delta t}{3}\mathbf{A}\right) & \left(\frac{1}{2}\mathbf{C} - \frac{\Delta t}{6}\mathbf{A}\right) \\ \left(-\frac{1}{2}\mathbf{C} - \frac{\Delta t}{6}\mathbf{A}\right) & \left(\frac{1}{2}\mathbf{C} - \frac{\Delta t}{3}\mathbf{A}\right) \end{bmatrix} \begin{bmatrix} \mathbf{u}_{i-1} \\ \mathbf{u}_i \end{bmatrix} + \frac{\Delta t}{2} \begin{bmatrix} \mathbf{f}_{i-1} \\ \mathbf{f}_i \end{bmatrix} = \begin{bmatrix} \mathbf{0} \\ \mathbf{0} \end{bmatrix} \tag{9.134}$$

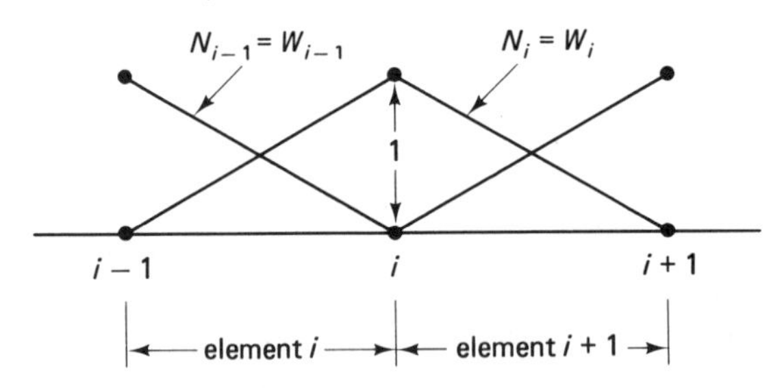

Figure 9.6 Shape functions over two successive elements in the time domain

Assembling Eqs. (9.133) and (9.134), we get

$$\begin{bmatrix} \left(-\frac{1}{2}\mathbf{C} - \frac{\Delta t}{3}\mathbf{A}\right) & \left(\frac{1}{2}\mathbf{C} - \frac{\Delta t}{6}\mathbf{A}\right) & 0 \\ \left(-\frac{1}{2}\mathbf{C} - \frac{\Delta t}{6}\mathbf{A}\right) & \left(\frac{-2(\Delta t)}{3}\mathbf{A}\right) & \left(\frac{1}{2}\mathbf{C} - \frac{\Delta t}{6}\mathbf{A}\right) \\ 0 & \left(-\frac{1}{2}\mathbf{C} - \frac{\Delta t}{6}\mathbf{A}\right) & \left(\frac{1}{2}\mathbf{C} - \frac{\Delta t}{3}\mathbf{A}\right) \end{bmatrix} \begin{bmatrix} \mathbf{u}_{i-1} \\ \mathbf{u}_i \\ \mathbf{u}_{i+1} \end{bmatrix} + \Delta t \begin{bmatrix} \frac{1}{2}\mathbf{f}_{i-1} \\ \mathbf{f}_i \\ \frac{1}{2}\mathbf{f}_{i+1} \end{bmatrix} = \begin{bmatrix} 0 \\ 0 \\ 0 \end{bmatrix} \tag{9.135}$$

Since this is an initial value problem, $\mathbf{u}_0$ is assumed to be known. Then Eq. (9.134) can be used to calculate $\mathbf{u}_1$. Then using $\mathbf{u}_0$ and $\mathbf{u}_1$, one can calculate $\mathbf{u}_i$ for other values of i.

9.6. A COMPARISON OF FINITE DIFFERENCE AND FINITE ELEMENT METHODS

There are many similarities and differences between the finite difference and finite element methods. The finite difference method starts with a differential equation and its boundary conditions and proceeds to replace the derivatives by function values at a discrete set of grid points. The finite element method starts with a variational statement of the problem and introduces piecewise definitions of the functions defined by a set of values at some grid points. If the processes are thus conceived, then they differ considerably from each other. First, in the finite difference approach, each equation appears immediately in its assembled form, whereas assembly is a distinct process in the finite element formulations. Second, unsymmetric coefficient matrixes often appear in finite difference methods owing to boundary condition approximation. Evidently, the first point appears to favor the finite difference approach, while the second point definitely favors the finite element approach.

In terms of similarities, both methods result in a set of algebraic equations relating a discrete set of variables in place of relations in the continuous variables. These algebraic equations are remarkably similar and provide the basis for identifying the two methods as essentially similar.

At a different level of comparison, it is an easy matter to observe that setting up the system of algebraic equations is relatively simple in the case of finite difference methods while it is somewhat involved in the finite element approach. The complexity of the finite element calculations in turn demands elaborate computer programs and computer facilities. The computations involved in the finite element method are too numerous for hand calculations even when solving very simple problems. This, from a pedagogical viewpoint, is a disadvantage. However, some of the more important and fundamental advantages of the finite element method appear to outweigh these disadvantages. Some of the advantages commonly attributed to the finite element method are as follows.

1. The material properties in adjacent elements do not have to be the same. This allows the method to be applied to nonhomogeneous and anisotropic regions, as well as regions with interfaces, with relative ease.
2. Irregularly shaped boundaries can be approximated using elements with straight sides or matched exactly using elements with curved boundaries. Thus, the finite element method is not limited to "nice" shapes with easily defined boundaries.
3. The size of the elements can be varied. This property allows the element grid to be expanded or refined as the need exists.

4. Boundary conditions such as discontinuous surface loadings and point and line sources do not present any special difficulties. Mixed boundary conditions can be easily handled.
5. The preceding properties can be incorporated into one general purpose computer code.

From a practical viewpoint, one question that can be posed is: Which process gives a better approximation for the same effort? While no general answer can be given to this question, it is useful to observe that, for a Dirichlet-type problem, a regular mesh of simplest triangular elements and the corresponding simplest finite difference expression give identical equations with no relative merit. However, the extension of the same formulation to more complicated boundary conditions immediately shows some differences that appear to favor the finite element formulation.

The similarities and differences between these two methods are brought forth more elegantly by working out some simple examples.

Example. Consider

$$\frac{d^2u}{dx^2} + f(x) = 0, \qquad 0 \leq x \leq L \tag{9.136}$$

Using a regular grid of size h, a finite difference approximation to Eq. (9.136) can be readily written, at a grid point i, as

$$u_{i+2} - 4u_{i+1} + 6u_i - 4u_{i-1} + u_{i-2} + f_i h^4 = 0 \tag{9.137}$$

Alternatively, we can start from the variational form

$$J = \int_L \left[\frac{1}{2}\left(\frac{d^2u}{dx^2}\right)^2 + fu\right] dx \tag{9.138}$$

Dividing the interval $[0, L]$ into several elements of length h, and writing $J = \sum_e J^{(e)}$, we can write

$$J^{(e)} = \int_0^h \left[\frac{1}{2}\left(\frac{d^2u}{dx^2}\right)^2 + fu\right]^{(e)} dx \tag{9.139}$$

If the element of length h is assumed to be centered about the grid point i, and approximating d^2u/dx^2 about the grid point, Eq. (9.137) becomes

$$\begin{aligned} J^{(e)} &= \int_0^h \left[\frac{\frac{1}{2}(u_{i+1} - 2u_i + u_{i-1})^2}{h^2} + fu_i\right] dx \\ &= \frac{1}{2h}(u_{i+1} - 2u_i + u_{i-1})^2 + fhu_i \end{aligned} \tag{9.140}$$

An analogous equation is valid for each element. Now differentiating $J^{(e)}$ and evaluating $\partial J^{(e)}/\partial u_i$, $\partial J^{(e)}/\partial u_{i-1}$, $\partial J^{(e)}/\partial u_{i+1}$, and assembling using standard assembly rules, one gets an equation identical to Eq. (9.137). It is significant to note that the grid points $i - 1$ and $i + 1$ are *external* to the finite element that was considered.

Example. For a somewhat more involved example, consider the example discussed by Key and Krieg [5],

$$\frac{\partial}{\partial x}\left(\sigma(x)\frac{\partial u}{\partial x}\right) = \rho\frac{\partial^2 u}{\partial t^2}, \qquad 0 \leq x \leq L, t \geq 0 \tag{9.141a}$$

subject to the initial conditions

$$u(x, 0) = \alpha(x), \qquad \frac{\partial u}{\partial t}(x, 0) = \beta(x) \tag{9.141b}$$

and the boundary conditions

$$\frac{\partial u}{\partial x}(0, t) = \frac{g(t)}{\sigma(0)}, \qquad \frac{\partial u}{\partial x}(L, t) = \frac{k(t)}{\sigma(L)} \tag{9.141c}$$

The variational formulation for this problem is

$$J = \int_0^t \left\{\int_0^L \left[\rho\left(\frac{\partial u}{\partial t}\right)^2 - \sigma(x)\left(\frac{\partial u}{\partial x}\right)^2\right] dx - 2u(0, t)g(t) - 2u(L, t)k(t)\right\} dt \tag{9.142}$$

The problem is to find the extremum of J from the class of functions that are continuous in the domain $(0, L) \times (0, t)$.

In the finite element method, we divide $(0, L)$ into several elements. Now attention is focused on one element whose coordinates are x_i and x_j. Let $u_i(t)$ and $u_j(t)$ represent the values of $u(x, y)$, respectively, at x_i and x_j. Assuming linear interpolation,

$$u(x, t) = u_i(t)\frac{(x_j - x)}{(x_j - x_i)} + u_j(t)\frac{x - x_i}{x_j - x_i}, \qquad x_i \leq x \leq x_j \tag{9.143}$$

Differentiating with respect to t and using matrix notation, we have

$$\frac{\partial u}{\partial t}(x, t) = \frac{1}{(x_j - x_i)}[(x_j - x)(x - x_i)]\begin{bmatrix}\dfrac{\partial u_i}{\partial t}\\[2ex] \dfrac{\partial u_j}{\partial t}\end{bmatrix} - [1 \quad -1]\begin{bmatrix} u_i \\[2ex] u_j\end{bmatrix} \tag{9.144}$$

Now the inner integral in Eq. (9.142) can be written, for the element (x_i, x_j), as

$$\int_{x_i}^{x_j} \frac{1}{(x_j - x_i)^2}\left\{\rho\begin{bmatrix}\dfrac{\partial u_i}{\partial t}\\[2ex] \dfrac{\partial u_j}{\partial t}\end{bmatrix}^T \begin{bmatrix}(x_j - x)^2 & (x_j - x)(x - x_i)\\[2ex] (x_j - x)(x - x_i) & (x - x_i)^2\end{bmatrix}\begin{bmatrix}\dfrac{\partial u_i}{\partial t}\\[2ex] \dfrac{\partial u_j}{\partial t}\end{bmatrix} - \sigma(x)\begin{bmatrix}u_i(t)\\[2ex] u_j(t)\end{bmatrix}^T\begin{bmatrix}1 & -1\\[2ex] -1 & 1\end{bmatrix}\begin{bmatrix}u_i(t)\\[2ex] u_j(t)\end{bmatrix}\right\} dx\, dt \tag{9.145}$$

For expediency, a constant value for ρ and σ will be used in each interval by using their values at the center of the interval. That is, we set

$$\sigma_e = \sigma \qquad \text{evaluated at } (x_i + x_j)/2$$

$$\rho_e = \rho \qquad \text{evaluated at } (x_i + x_j)/2$$

Now the expression in Eq. (9.145) can be evaluated for the interval (x_i, x_j) to yield

$$h_e\rho_e\begin{bmatrix}\dot{u}_i\\ \dot{u}_j\end{bmatrix}\begin{bmatrix}\frac{1}{3} & \frac{1}{6}\\ \frac{1}{6} & \frac{1}{3}\end{bmatrix}\begin{bmatrix}\dot{u}_i\\ \dot{u}_j\end{bmatrix} - \frac{\sigma_e}{h_e}\begin{bmatrix}u_i\\ u_j\end{bmatrix}^T\begin{bmatrix}1 & -1\\ -1 & 1\end{bmatrix}\begin{bmatrix}u_i\\ u_j\end{bmatrix} \tag{9.146}$$

where $h_e = x_j - x_i$ and $\dot{u}_i = \partial u_i/\partial t$.

Assembling all the elements, we get a discretized version of Eq. (9.142) as

$$J = \int_0^t \left\{\sum_{e=1}^{E}\left[h_e\rho_e\begin{bmatrix}\dot{u}_i\\ \dot{u}_j\end{bmatrix}\begin{bmatrix}\frac{1}{3} & \frac{1}{6}\\ \frac{1}{6} & \frac{1}{3}\end{bmatrix}\begin{bmatrix}\dot{u}_i\\ \dot{u}_j\end{bmatrix} - \frac{\sigma_e}{h_e}\begin{bmatrix}u_i\\ u_j\end{bmatrix}\begin{bmatrix}1 & -1\\ -1 & 1\end{bmatrix}\begin{bmatrix}u_i\\ u_j\end{bmatrix}\right.\right.$$
$$\left.- 2u(0, t)g(t) - 2u(L, t)k(t)\right\} dt \tag{9.147}$$

where E is the total number of elements. Equation (9.147) can be written more compactly as

$$J = \int_0^t \{\dot{\mathbf{u}}^T\mathbf{A}\dot{\mathbf{u}} - \mathbf{u}^T\mathbf{B}\mathbf{u} - 2u(0, t)g(t) - 2u(L, t)k(t)\}\, dt \tag{9.148}$$

Setting $\delta J = 0$ and simplifying, one gets

$$\mathbf{A}\ddot{\mathbf{u}} + \mathbf{B}\mathbf{u} = \mathbf{f}(t) \tag{9.149}$$

As our purpose here, however, is to show the relation between finite element and finite difference formulation, it is more instructive to go back to Eq. (9.147) and assemble the system for the first two elements with nodes numbered as 1, 2, and 3. This gives

$$J = \int\left\{\begin{bmatrix}\dot{u}_1\\ \dot{u}_2\\ \dot{u}_3\end{bmatrix}^T\begin{bmatrix}\frac{1}{3}h_1\rho_1 & \frac{1}{6}h_1\rho_1 & 0\\ \frac{1}{6}h_1\rho_1 & \frac{1}{3}(h_1\rho_1 + h_2\rho_2) & \frac{1}{6}h_2\rho_2\\ 0 & \frac{1}{6}h_2\rho_2 & \frac{1}{3}h_2\rho_2\end{bmatrix}\begin{bmatrix}\dot{u}_1\\ \dot{u}_2\\ \dot{u}_3\end{bmatrix}\right.$$
$$\left.-\begin{bmatrix}u_1\\ u_2\\ u_3\end{bmatrix}^T\begin{bmatrix}\frac{\sigma_1}{h_1} & -\frac{\sigma_1}{h_1} & 0\\ -\frac{\sigma_1}{h_1} & \frac{\sigma_1}{h_1} + \frac{\sigma_2}{h_2} & -\frac{\sigma_2}{h_2}\\ 0 & -\frac{\sigma_2}{h_2} & \frac{\sigma^2}{h_2}\end{bmatrix}\begin{bmatrix}u_1\\ u_2\\ u_3\end{bmatrix} - 2u_1g(t) - 2u_3k(t)\right\} dt \tag{9.150}$$

The dashed 2×2 submatrixes are the contributions of the two individual elements to the assembled matrix. Now setting $\delta J = 0$ and simplifying, one gets, in general, an equation such as

$$\frac{1}{6}(h_{i-1/2}\rho_{i-1/2})\ddot{u}_{i-1} + \frac{1}{3}(h_{i-1/2}\rho_{i-1/2} + h_{i+1/2}\rho_{i+1/2})\ddot{u}_i + \frac{1}{6}(h_{i+1/2}\rho_{i+1/2})\ddot{u}_{i+1} - \left(\frac{\sigma_{i-1/2}}{h_{i-1/2}}\right)u_{i-1} + \left(\frac{\sigma_{i-1/2}}{h_{i-1/2}} + \frac{\sigma_{i+1/2}}{h_{i+1/2}}\right)u_i - \left(\frac{\sigma_{i+1/2}}{h_{i+1/2}}\right)u_{i+1} = 0 \tag{9.151}$$

for any *internal node* i. If h is uniform and σ and ρ are constant, then Eq. (9.151) simplifies to

$$h\rho\ddot{u}_i + \left(\frac{\sigma}{h}\right)(-u_{i-1} + 2u_i - u_{i+1}) = 0 \tag{9.152}$$

To see how the preceding equations compare with those derived from a finite difference approach, we start once again from Eq. (9.141a). The interval $[0, L]$ is divided once again by the same set of grid points. Let h_1 and h_2 be the distances between the grid point pairs (1, 2) and (2, 3), respectively. Denote the right end and left end of the element in question, respectively, by the coordinates $x_L = \frac{1}{2}(x_1 + x_2)$ and $x_R = \frac{1}{2}(x_2 + x_3)$. Now integrating Eq. (9.141a) between the limits x_L and x_R,

$$\int_{x_L}^{x_R} \rho\frac{\partial^2 u}{\partial x^2}\,dx = \int_{x_L}^{x_R} \frac{\partial}{\partial x}\left(\sigma(x)\frac{\partial u}{\partial x}\right)dx = \sigma(x)\frac{\partial x}{\partial u}\bigg|_{x_R} - \sigma(x)\frac{\partial u}{\partial x}\bigg|_{x_L} = \sigma_2\left(\frac{u_3 - u_2}{h_2}\right) - \sigma_1\left(\frac{u_2 - u_1}{h_1}\right) \tag{9.153}$$

if u is assumed to be quadratic over each element.

Similarly, if $\rho(\partial^2 u/\partial x^2)$ is assumed to be quadratic over (x_1, x_3), then the left side integral can be written as

$$\int_{x_L}^{x_R} \rho\left(\frac{\partial^2 u}{\partial x^2}\right)dx = \int_{x_L}^{x_1} (\;)\,dx + \int_{x_1}^{x_R} (\;)\,dx \tag{9.154}$$

Performing the indicated integration (do them!), we get

$$\int_{x_L}^{x_R} \rho\left(\frac{\partial^2 u}{\partial x^2}\right)dx = (\rho\ddot{u})_2\frac{1}{2}(h_2 + h_1) + \frac{a}{2}\left(\frac{1}{4}\right)(h_2^2 - h_1^2) + \frac{b}{3}\left(\frac{1}{8}\right)(h_2^3 + h_1^3) \tag{9.155}$$

where

$$a = \frac{(\rho\ddot{u})_3 h_1^2 + (\rho\ddot{u})_2(h_2^2 - h_1^2) - (\rho\ddot{u})_1 h_2}{h_1 h_2(h_1 + h_2)} \tag{9.156}$$

and

$$b = \frac{(\rho\ddot{u})_3 h_1 - (\rho\ddot{u})_2(h_1 + h_2) + (\rho\ddot{u})_1 h_2}{h_1 h_2(h_1 + h_2)} \tag{9.157}$$

Combining Eqs. (9.153) and (9.155) and writing the combined equation for a general grid point i, we get

$$\frac{1}{24h_{i-1/2}}(-2h_{i+1/2}^2 + 2h_{i+1/2}h_{i-1/2} + h_{i-1/2}^2)(\rho_{i-1}\ddot{u}_{i-1})$$
$$+ \frac{1}{24h_{i-1/2}h_{i+1/2}}(2h_{i-1/2}^2 + 7h_{i-1/2}h_{i+1/2} + 2h_{i+1/2}^2)(h_{i-1/2} + h_{i+1/2})(\rho_i\ddot{u}_i)$$
$$+ \frac{1}{24h_{i+1/2}}(h_{i+1/2}^2 + 2h_{i+1/2}h_{i-1/2} - 2h_{i-1/2}^2)(\rho_{i+1}\ddot{u}_{i+1})$$
$$- \left(\frac{\sigma_{i-1/2}}{h_{i-1/2}}\right)u_{i-1} + \left(\frac{\sigma_{i-1/2}}{h_{i-1/2}} + \frac{\sigma_{i+1/2}}{h_{i+1/2}}\right)u_i - \left(\frac{\sigma_{i+1/2}}{h_{i+1/2}}\right)u_{i+1} = 0 \qquad (9.158)$$

If h is uniform and σ and ρ are constant, Eq. (9.158) becomes identical to Eq. (9.152), the one obtained from the finite element formulation.

Notice that both methods arrived at discrete analogs to the problem. However, the averaging involved differed in the two methods. In the finite difference derivation, the assumption regarding the quadratic nature of u and ρu was made for convenience. If, for example, ρu had been assumed to be piecewise linear then the results would have been different.

In spite of the apparant differences, many practitioners generally use both methods in an intertwined way. Both methods are adopting features of the other that prove to be attractive. In dynamic problems, finite element equations are invariably integrated forward in time with finite difference techniques, rather than using a time dimension on an element and using the integration scheme that results. The finite difference method has found the variational approach of the finite element method useful in producing symmetric difference equations.

9.7. FINITE ELEMENT TREATMENT OF SOURCES

Point and line sources (or sinks) occur in a wide variety of physical problems. If, for example, water is being pumped out of an aquifer, then one can regard the location where water leaves the aquifer as a point sink. In solid-state integrated circuits, a number of transistors are fabricated on a small piece of silicon wafer. The junction of each of these transistors generates heat as the transistor is switched. Thus each junction acts as a point source of heat. One of the thorny problems in electronic packaging is to dissipate this heat to an external heat sink. Physical examples of line sources include steam and/or hot-water pipes within the earth and conducting electrical wires embedded in a product.

Suppose a heat source of strength Q^* kW/m is located at (x_0, y_0) in a triangular element (see Figure 9.7). Then, at any point (x, y) in the neighborhood, the quantity of heat can be written as

$$Q(x, y) = Q^*\delta(x - x_0)\delta(y - y_0) \qquad (9.159)$$

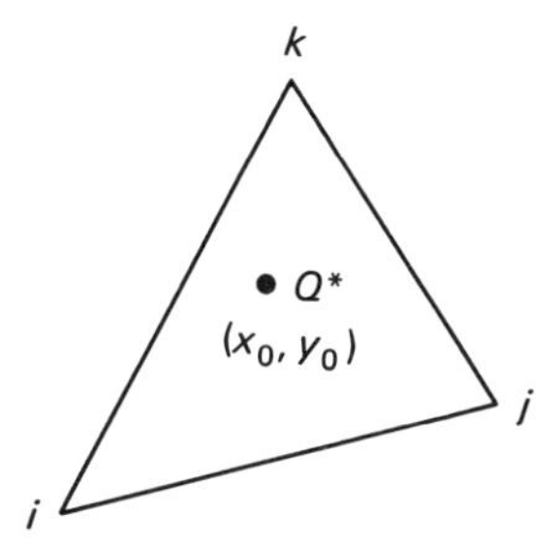

Figure 9.7 Source of strength Q^* located at (x_0, y_0) in an element

where $\delta(x - x_0)$ is the displaced unit impulse function. Now the integral [see Eq. (9.29b)]

$$\int_\Omega \mathbf{N}Q \, d\Omega$$

for a linear triangular element can be written as

$$\int_\Omega \mathbf{N}Q \, d\Omega = Q^* \int_\Omega \begin{bmatrix} N_i \\ N_j \\ N_k \end{bmatrix} \delta(x - x_0)\delta(y - y_0) \, dx \, dy$$

$$= Q^* \begin{bmatrix} N_i \\ N_j \\ N_k \end{bmatrix} \Bigg|_{\substack{x=x_0 \\ y=y_0}} \qquad (9.160)$$

The second equality follows from a property of the delta function.

Example. We shall illustrate this method by an example treated by Segerlind [4]. A line source $Q^* = 48$ W/cm is located at (5, 2) in the element shown in Figure 9.8. Determine the amount of Q^* allocated to each node.

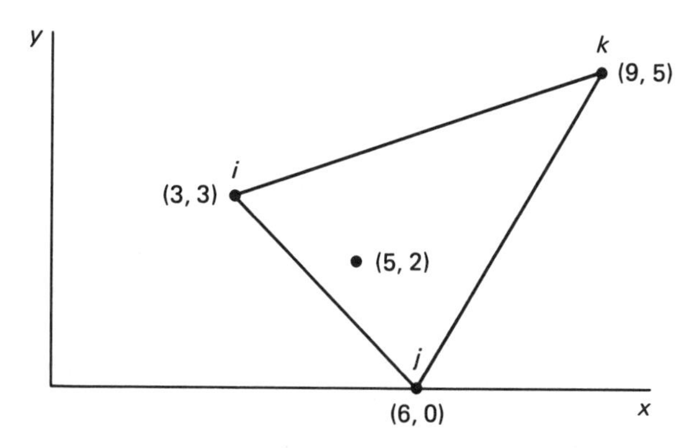

Figure 9.8 Treatment of the source discussed in the text

First we calculate the constants a, b, and c using the formulas (these are same as those discussed in Section 8.6)

$$a_i = x_j y_k - x_k y_j = (6)(5) - (9)(0) = 30$$
$$a_j = x_k y_i - x_j y_k = (9)(3) - (3)(5) = 12$$
$$a_k = x_i y_j - x_j y_i = (3)(0) - (6)(3) = -18$$
$$b_i = y_j - y_k = 0 - 5 = -5, \qquad c_i = x_k - x_j = 9 - 6 = 3$$
$$b_j = y_k - y_i = 5 - 3 = 2, \qquad c_j = x_i - x_k = 3 - 9 = -6$$
$$b_k = y_i - y_j = 3 - 0 = 3, \qquad c_k = x_j - x_i = 6 - 3 = 3$$

$$2A = \begin{vmatrix} 1 & x_i & y_i \\ 1 & x_j & y_j \\ 1 & x_k & y_k \end{vmatrix} = \begin{vmatrix} 1 & 3 & 3 \\ 1 & 6 & 0 \\ 1 & 9 & 5 \end{vmatrix} = 24$$

Now N_i, N_j, and N_k can be calculated from the formula for the shape functions

$$N_\beta = \frac{1}{2A}(a_\beta + b_\beta x + c_\beta y), \qquad \beta = i, j, k$$

as

$$N_i = \tfrac{1}{24}(30 - 5x + 3y)$$
$$N_j = \tfrac{1}{24}(12 + 2x - 6y)$$
$$N_k = \tfrac{1}{24}(-18 + 3x + 3y)$$

Substitution of $x_0 = 5$, $y_0 = 2$ produces

$$N_i = \tfrac{11}{24}, \qquad N_j = \tfrac{10}{24}, \qquad N_k = \tfrac{3}{24}$$

Now

$$\int_\Omega \mathbf{N}^T Q \, d\Omega = \frac{48}{24}\begin{bmatrix} 11 \\ 10 \\ 3 \end{bmatrix} = \begin{bmatrix} 22 \\ 20 \\ 6 \end{bmatrix} \text{W/cm}$$

Therefore, $Q^* = 48$ is allocated, respectively, to the nodes i, j, and k as 22, 20, and 6.

9.8. TREATMENT OF CURVED BOUNDARIES

An important topic that has not been treated so far is the case when the boundary Γ of the region Ω is curved in shape. In the finite difference methods, it was seen that the only viable course of action is to replace the original boundary by one with a polygonal shape. This procedure tends to destroy the symmetry of the resulting matrix, which in turn means poor convergence. An analogous procedure in the context of finite element method is far more fruitful. If the region Ω has curved boundaries, then in the finite element method a satisfactory approximation to the boundary can be obtained by replacing it with one of a suitable polygonal shape; then the use of elements with straight sides to cover the boundary will suffice. However, the approximation will be even better if elements with curved sides can be formulated to handle the special task. If elements with curved sides are available, then it would be possible to use a smaller number of larger elements and still achieve an accurate boundary representation. Use of fewer elements has the further advantage of reduced computational effort.

One method of developing elements with curved sides is to use the same interpolation function to define element shapes as are used to define displacements within an element. This approach is called the ***isoparametric*** formulation. Once the idea of an isoparametric element is grasped, it is easy to visualize other extensions. For example, one can define subparametric and superparametric elements for which element shape is defined respectively by lower-order and higher-order polynomials than used to define element displacements. However, the discussion here is confined only to the isoparametric family of elements.

The simplest member of the isoparametric family is the "linear" element and, by definition, this may not have curved sides. A more useful isoparametric element is the "quadratic" element because it may have curved sides and therefore provides a better fit to curved shape of a region Ω. The essential idea underlying the development of elements with curved sides centers on transforming simple geometric shapes in some local coordinate system into distorted shapes in the global system.

A. R. Mitchell [6] illustrates the nature of the computations involved by considering an example consisting of a triangular element with two straight sides and one curved side. To maintain generality, first a triangle with three curvilinear sides, as shown in Figure 9.9, is considered. Mitchell proceeds to transform this triangle into the standard triangle in the (s, t) plane by using the transformation formulas

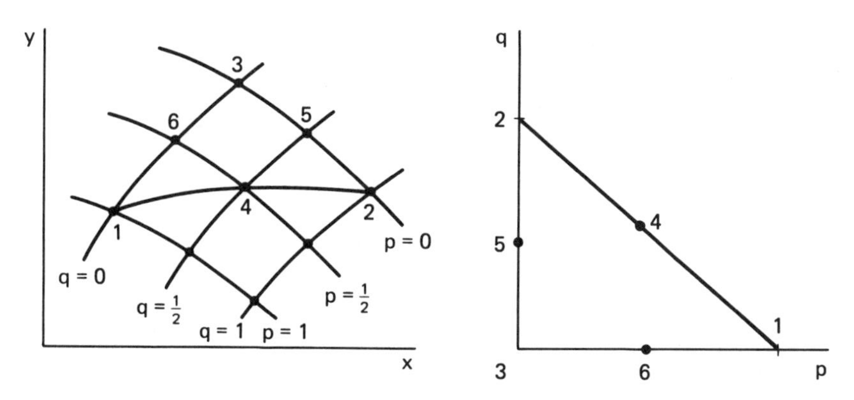

Figure 9.9 Treatment of curved boundaries via isoparametric transformations (Reproduced with permission from *The Mathematics of Finite Elements*, edited by J. R. Whiteman, © 1973, Academic Press, New York, 1973)

$$\begin{aligned} x = x_3 &+ 2(x_1 + x_3 - 2x_6)s^2 + 2(x_2 + x_3 - 2x_5)t^2 \\ &+ 4(x_3 + x_4 - x_5 - x_6)st + (4x_6 - x_1 - 3x_3)t \\ &+ (4x_5 - x_2 - 3x_3)t \end{aligned} \tag{9.161}$$

$$\begin{aligned} y = y_3 &+ 2(y_1 + y_3 - 2y_6)s^2 + 2(y_2 + y_3 - 2y_5)t^2 \\ &+ 4(y_3 + y_4 - y_5 - y_6)st + (4y_6 - y_1 - 3y_3)t \\ &+ (4y_5 - y_2 - 3y_3)t \end{aligned} \tag{9.162}$$

However, because the sides 253 and 361 are straight with points 5 and 6 as midpoints, these formulas simplify to

$$\begin{aligned} \bar{x} &= Xst + \bar{x}_1 s + \bar{x}_2 t \\ \bar{y} &= Yst + \bar{y}_1 s + \bar{y}_2 t \end{aligned} \tag{9.163}$$

where

$$\begin{aligned} \bar{x} = x - x_3, \qquad & \bar{y} = y - y_3 \\ X &= 2[2\bar{x}_4 - (\bar{x}_1 + \bar{x}_2)] \\ Y &= 2[2\bar{y}_4 - (\bar{y}_1 + \bar{y}_2)] \end{aligned} \tag{9.164}$$

After substantial manipulation, it can be shown that the line $s + t = 1$ in the (s, t) plane corresponds to the quadratic curve

$$[(\tilde{y}_1\tilde{x} - \tilde{x}_1\tilde{y}) + (\tilde{y}_2\tilde{x} - \tilde{x}_2\tilde{y})]^2 = (\tilde{y}_1\tilde{x}_2 - \tilde{x}_1\tilde{y}_2)[(\tilde{y}_1\tilde{x} - \tilde{x}_1\tilde{y}) - (\tilde{y}_2\tilde{x} - \tilde{x}_2\tilde{y})] \tag{9.165}$$

where $\tilde{x} = x - x_4$ and $\tilde{y} = y - y_4$. In the special case where the points are given by $1 \equiv (1, 0)$, $2 \equiv (0, 1)$, $3 \equiv (0, 0)$, and $4 = (l, l)$, Eq. (9.165) reduces to

$$(x - y)^2 = \frac{x + y - 2l}{1 - 2l} \tag{9.166}$$

The quadratic curve given by Eq. (9.165) is, of course, only an approximation to the original curvilinear side of the triangle in Figure 9.9.

This example illustrates a method of handling curvilinear sides. For a more thorough discussion on the methods of treating isoparametric elements, the reader is encouraged to consult the extensive literature available on this subject.

REFERENCES

1. K. H. Huebner, *The Finite Element Method for Engineers.* New York: John Wiley & Sons, Inc., 1975.

2. S. G. Mikhlin, *Variational Methods in Mathematical Physics.* New York: Macmillan, Inc., 1964.

3. B. A. Finlayson, *The Method of Weighted Residuals and Variational Principles.* New York: Academic Press, Inc., 1972.

4. L. J. Segerlind, *Applied Finite Element Analysis.* New York: John Wiley & Sons, Inc., 1976.

5. S. W. KEY and R. D. KRIEG, "Comparison of Finite-Element and Finite-Difference Methods," in S. J. Fenves and others, eds., *Numerical and Computer Methods in Structural Mechanics*. New York: Academic Press, Inc., 1973.

6. A. R. MITCHELL, "An Introduction to the Mathematics of the Finite Element Method," in J. R. Whiteman, ed. *The Mathematics of Finite Elements and Applications*. New York: Academic Press, Inc., 1973.

SUGGESTION FOR FURTHER READING

The preceding references and those cited at the end of Chapter 8 are a representative sample from a huge collection of references in this area. Additional material on specific topics can be found in the following.

1. *An elementary introduction to the finite element method as it applies to structural engineering problems can be found in*
   R. D. COOK, *Concepts and Applications of Finite Element Analysis*, John Wiley & Sons, Inc., New York, 1974.

2. *For an introductory level text on the application of the finite element method to fluid dynamical problems, consult*
   T. J. CHUNG, *Finite Element Analysis in Fluid Dynamics*, McGraw-Hill Book Co., New York, 1978.

3. *The finite element method is being widely used in the simulation of water resource systems. A good reference in this area is*
   G. F. PINDER and W. G. GRAY, *Finite Element Simulation in Surface and Subsurface Hydrology*, Academic Press, Inc., New York, 1977.

4. *For a discussion of electric and magnetic field problems by the finite element method, see*
   M. V. K. CHARI and P. SILVESTER (eds.), *Finite Elements in Electrical and Magnetic Field Problems*, John Wiley & Sons, Inc., New York, 1980.

5. *For a useful, but somewhat dated, bibliography in this area, refer to*
   J. R. WHITEMAN, *A Bibliography for Finite Elements*, Academic Press, Inc., New York, 1975.

EXERCISES

1. In Eq. (9.1), assume the following values: $k = 75$ W/cm-°C, $h = 10$ W/cm^2-°C, $A = \pi$ cm^2, $L = 7.5$ cm, $q = -150$ W/cm^2, and $u = 40$°C. Perform the necessary calculations as indicated in the text and show that $u_1 = 70$, $u_2 = 62.5$, and $u_3 = 55$.

2. Starting from Eq. (9.28), verify the results shown in Eq. (9.29).

3. "Mesh generation" refers to automatic generation of node locations and numbering of nodes and elements. Conduct a literature survey and write a term paper on this subject. For a survey on this subject up to 1973, consult W. R. Buell and B. A. Bush, "Mesh Generation: A Survey," *Transactions of ASME, Journal of Engineering for Industry*, Vol. 95, No. 1, pp. 332–338, 1973.

4. Write a term paper on the formulation, use, and relative merits and demerits of triangular and quadrilateral isoparametric elements.

5. Write a term paper discussing various strategies you would consider while solving (a) a diffusion equation and (b) a wave equation using the finite element method. Your discussion should explicitly consider, among other things, the various alternatives in handling the time dimension.

6. Consider

$$A(\phi) \triangleq \frac{\partial}{\partial x}\left(k\frac{\partial \phi}{\partial x}\right) + \frac{\partial}{\partial y}\left(k\frac{\partial \phi}{\partial y}\right) + Q = 0 \qquad \text{on } \Omega$$

$$B(\phi) \triangleq \begin{cases} \phi - \phi = 0 & \text{on } \Gamma_1 \\ k\dfrac{\partial \phi}{\partial n} - q = 0 & \text{on } \Gamma_2 \end{cases}$$

where $\Gamma = \Gamma_1 \cup \Gamma_2$ is the boundary of the region Ω. This problem can be reformulated using integral statements as

$$\int_\Omega WA(\phi)\, d\Omega + \int_\Gamma \bar{W}B(\phi)\, d\Gamma = 0 \tag{A}$$

(a) What conditions are required to be imposed on ϕ in formulation (A)?
(b) Reformulate (A) into a weaker form so that some of the restrictions imposed on ϕ can be lifted.
(c) What values should be assigned to W if one wants to get Galerkin method? Collocation method? Subdomain method? Least squares method? Method of moments?

(*Hint:* Refer to the book of Strang and Fix cited at the end of Chapter 8.)

7. Perform the necessary integrations and derive Eq. (9.155) from Eq. (9.144).

8. Make appropriate assumptions and show that Eq. (9.158) indeed is equivalent to Eq. (9.152) as claimed in the text.

COMPUTATIONS

10

Computational Linear Algebra

10.1. INTRODUCTION

The application of transformation methods, described in Part II, transforms a differential equation into a finite algebraic form. The purpose of this chapter is to focus attention on the practical and computational aspects of solving these algebraic equations on a digital computer. In some situations, such as the explicit formulation of the initial value problem, solutions are explicitly obtainable, and therefore the algebraic and computational manipulations required are trivial. However, boundary value problems lead to large systems of equations $\mathbf{Au} = \mathbf{b}$, and there is a need to solve these systems of equations in an efficient manner. In the execution of most computer programs, the solution of these equations constitutes a major time-consuming job. In linear problems 20% to 50% of the total execution time may be required to solve the set of equations depending on the size of the problem and the amount of peripheral processing involved. In nonlinear analysis, this figure may creep up to 80%. Consequently, the problem of solving large systems of linear algebraic equations elegantly and efficiently is of central importance.

It would be ideal if one could find a single method of solution that is acceptably fast and accurate regardless of the manner in which the particular system $\mathbf{Au} = \mathbf{b}$ was derived and the particular computer available. However, efficient and accurate methods are available when $\mathbf{A}$ exhibits certain properties such as sparsity, bandedness, symmetry, and so on. One reason for this dependence can be seen by recognizing that the entries of a matrix define the interactions between the components of a physical assemblage. Therefore, the nature of the matrix $\mathbf{A}$ is not only influenced by the discretization procedure used on the PDE but also by the nature of the problem itself.

Essentially, there are two different classes of methods that are widely used for the solution of systems of equations: ***direct*** solution techniques and ***iterative methods***. In a direct method, $\mathbf{Au} = \mathbf{b}$ is solved using a number of steps and operations that are predetermined in an exact manner, whereas iteration is used in the iterative methods. Either solution scheme will be seen to have certain advantages; however, direct methods appear to hold an advantage in solving systems arising out of the finite element method, and iterative methods appear to have the edge in solving systems arising from finite difference equations.

10.2. DIRECT METHODS OF SOLVING $\mathbf{Au} = \mathbf{b}$

In matrix terminology, the solution of the system of equations $\mathbf{Au} = \mathbf{b}$ involves the inversion of the matrix $\mathbf{A}$. Calculation of this inverse constitutes the crux of the problem in the solution of the given set of equations. Probably the best known of the direct methods of solving $\mathbf{Au} = \mathbf{b}$ involves the application of Cramer's rule to express each term of $\mathbf{A}^{-1}$ as a quotient of two determinants. This method is of little utility in the solution of large systems of equations because of the large number of multiplications that are required. If no refinements in technique are employed and if no advantage is taken of the many zero elements in the $\mathbf{A}$ matrix, the inversion of an $n \times n$ matrix would require $(n + 1)!$ multiplications. Note that n in the case of field computations is equal to the number of internal net points. Thus a two-dimensional rectangular field with 25 internal net points in each of the two dimensions, that is a very coarse net spacing, would require 26! or approximately 10^{28} multiplications. This would take even the best of the modern digital computers a good many years. Of course, an enormous saving in computing time can be effected by employing more refined direct methods and taking advantage of the sparsity of $\mathbf{A}$. Nonetheless, computing times rapidly become unduly large as n exceeds 100.

GAUSSIAN ELIMINATION ALGORITHM

The most effective and popular direct solution techniques currently used are basically applications of Gauss elimination. Consider a simply supported beam characterized by

$$EI\frac{d^4u}{dx^4} = q, \qquad 0 \leq x \leq 5$$

with $u = 0$ and $u'' = 0$ at $x = 0$ and $x = 5$. If applied load $q = 0$ everywhere except at $x = 2$, where it is $5/L$, and if $EI = 1$, then the finite difference equations with central difference approximations would be

$$\begin{bmatrix} 5 & -4 & 1 & 0 \\ -4 & 6 & -4 & 1 \\ 1 & -4 & 6 & -4 \\ 0 & 1 & -4 & 5 \end{bmatrix} \begin{bmatrix} u_1 \\ u_2 \\ u_3 \\ u_4 \end{bmatrix} = \begin{bmatrix} 0 \\ 1 \\ 0 \\ 0 \end{bmatrix} \tag{10.1a}$$

or

$$\mathbf{Au} = \mathbf{b}$$

or

$$\mathbf{A}^{(1)}\mathbf{u} = \mathbf{b}^{(1)}, \qquad \text{where } \mathbf{A} = \mathbf{A}^{(1)}, \qquad \mathbf{b} = \mathbf{b}^{(1)} \tag{10.1b}$$

Step 1. The essence of this method is to reduce the matrix in the preceding equation into a lower triangular form. Toward this end $(-\frac{4}{5})$ times the first row is subtracted from the second row, and $(\frac{1}{5})$ times the first row is subtracted from the third row. The resulting equations are

$$\begin{bmatrix} 5 & -4 & 1 & 0 \\ 0 & \frac{14}{5} & -\frac{16}{5} & 1 \\ 0 & -\frac{16}{5} & \frac{29}{5} & -4 \\ 0 & 1 & -4 & 5 \end{bmatrix} \begin{bmatrix} u_1 \\ u_2 \\ u_3 \\ u_4 \end{bmatrix} = \begin{bmatrix} 0 \\ 1 \\ 0 \\ 0 \end{bmatrix} \tag{10.2a}$$

or

$$\mathbf{A}^{(2)}\mathbf{u} = \mathbf{b}^{(2)} \tag{10.2b}$$

It is easy to recognize that the preceding step is equivalent to the operation defined as $\mathbf{A}^{(2)} = \mathbf{M}_1^{-1}\mathbf{A}^{(1)}$, where

$$\mathbf{M}_1^{-1} = \begin{bmatrix} 1 & & & \\ -m_{21} & 1 & & \\ -m_{31} & 0 & 1 & \\ -m_{41} & 0 & 0 & 1 \end{bmatrix}, \qquad m_{i,1} = \frac{a_{i,1}^{(1)}}{a_{1,1}^{(1)}} \tag{10.3}$$

Step 2. Now consider Eq. (10.2a). Subtracting $(-\frac{16}{14})$ times the second equation from the third equation, and $(\frac{5}{14})$ times the second equation from the fourth equation, one gets

$$\begin{bmatrix} 5 & -4 & 1 & 0 \\ 0 & \frac{14}{5} & -\frac{16}{5} & 1 \\ 0 & 0 & \frac{15}{7} & -\frac{20}{7} \\ 0 & 0 & -\frac{20}{7} & \frac{65}{14} \end{bmatrix} \begin{bmatrix} u_1 \\ u_2 \\ u_3 \\ u_4 \end{bmatrix} = \begin{bmatrix} 0 \\ 1 \\ \frac{8}{7} \\ -\frac{5}{14} \end{bmatrix} \tag{10.4a}$$

or

$$\mathbf{A}^{(3)}\mathbf{u} = \mathbf{b}^{(3)} \tag{10.4b}$$

It is easy to recognize that $\mathbf{A}^{(3)} = \mathbf{M}_2^{-1}\mathbf{A}^{(2)} = \mathbf{M}_2^{-1}\mathbf{M}_1^{-1}\mathbf{A}^{(1)}$, where $\mathbf{M}_2^{-1}$ is defined in a manner analogous to $\mathbf{M}_1^{-1}$.

Step 3. Subtracting $(-\frac{20}{15})$ times the third equation from the fourth, one gets

$$\begin{bmatrix} 5 & -4 & 1 & 0 \\ 0 & \frac{14}{5} & -\frac{16}{5} & 1 \\ 0 & 0 & \frac{15}{7} & -\frac{20}{7} \\ 0 & 0 & 0 & \frac{5}{6} \end{bmatrix} \begin{bmatrix} u_1 \\ u_2 \\ u_3 \\ u_4 \end{bmatrix} = \begin{bmatrix} 0 \\ 1 \\ \frac{8}{7} \\ \frac{7}{6} \end{bmatrix} \tag{10.5a}$$

or

$$\mathbf{A}^{(4)}\mathbf{u} = \mathbf{b}^{(4)} = \mathbf{S}\mathbf{u} \tag{10.5b}$$

where

$$\mathbf{A}^{(4)} = \mathbf{M}_3^{-1}\mathbf{M}_2^{-1}\mathbf{M}_1^{-1}\mathbf{A}^{(1)} \triangleq \mathbf{S} \tag{10.6}$$

Step 4. Now u_4, u_3, u_2, and u_1 are obtained in that order by backward substitutions.

$$u_4 = \frac{(7/6)}{(5/6)} = \frac{7}{5}, \qquad u_3 = \frac{(8/7) - (-20/7)u_4}{15/7} = \frac{12}{5}$$

$$u_2 = \frac{1 - (-16/5)u_3}{(14/5)} = \frac{13}{5}$$

$$u_1 = \frac{0 - (-4)(19/35) - 1(36/15) - 0(7/5)}{5} = \frac{8}{5}$$

It is useful to make two important observations at this point. At the end of Step i, the lower right submatrix of order $n - i$ (indicated by dashed lines) is symmetric. Therefore, during computer implementation it is sufficient to work with the upper triangular part of $\mathbf{A}^{(i)}$. Second, the backward substitution process was successfully carried out because the ith diagonal element of $\mathbf{A}^{(i)}$ is nonzero. This fact also enabled us to make all the entries below the main diagonal to be zero.

The operations performed in the preceding elimination procedure can be compactly written, for any $n \times n$ matrix $\mathbf{A}$, as follows. Starting from

$$\mathbf{A} = \begin{bmatrix} a_{11}^{(1)} & \cdots & a_{1n}^{(1)} \\ a_{21}^{(1)} & \cdots & a_{2n}^{(1)} \\ \cdots & \cdots & \cdots \\ a_{n1}^{(1)} & \cdots & a_{nn}^{(1)} \end{bmatrix} \tag{10.7}$$

we have

$$\mathbf{M}_1^{-1} = \begin{bmatrix} 1 & & & & \\ -m_{21} & 1 & & & \\ -m_{31} & 0 & & & \\ \cdots & \cdots & \cdots & \cdots & \cdots \\ -m_{n1} & 0 & \cdots & 0 & 1 \end{bmatrix}, \qquad m_{21} = \frac{a_{21}^{(1)}}{a_{11}^{(1)}}, \quad \text{etc.} \tag{10.8}$$

and

$$\mathbf{M}_2^{-1} = \begin{bmatrix} 1 & & & & & \\ 0 & 1 & & & & \\ 0 & -m_{32} & 1 & & & \\ 0 & -m_{42} & 0 & & & \\ \cdots & \cdots & \cdots & \cdots & \cdots & \cdots \\ 0 & -m_{n2} & 0 & \cdots & 0 & 1 \end{bmatrix}, \qquad m_{32} = \frac{a_{32}^{(2)}}{a_{22}^{(2)}}, \quad \text{etc.} \tag{10.9}$$

Now,

$$\mathbf{U} = \begin{bmatrix} a_{11}^{(1)} & a_{12}^{(1)} & \cdots\cdots & & a_{1n}^{(1)} \\ 0 & a_{22}^{(2)} & \cdots\cdots & & a_{2n}^{(2)} \\ 0 & 0 & a_{33}^{(3)} & \cdots & a_{3n}^{(3)} \\ \cdots & \cdots & \cdots & \cdots & \cdots \\ & & & & a_{nn}^{(n)} \end{bmatrix} = \mathbf{M}_{n-1}^{-1}\mathbf{M}_{n-2}^{-1}\ldots\mathbf{M}_1\mathbf{A}^{(1)} \tag{10.10}$$

Equation (10.10) can be equivalently written as

$$\mathbf{A}^{(1)} = \mathbf{LU} \tag{10.11}$$

where

$$\mathbf{L} = \mathbf{M}_1\mathbf{M}_2\ldots\mathbf{M}_{n-1} \tag{10.12}$$

$$= \begin{bmatrix} 1 & & & \\ m_{21} & 1 & & \\ m_{31} & m_{32} & 1 & \\ \cdots & \cdots & \cdots & \cdots \\ m_{n1} & m_{n2} & m_{n,n-1} & 1 \end{bmatrix} \tag{10.13}$$

Thus, the solution process can be described by the pair of equations

$$\mathbf{Ly} = \mathbf{b}, \qquad \mathbf{Uu} = \mathbf{y} \tag{10.14}$$

If $\mathbf{D}$ denotes the diagonal entries of $\mathbf{S}$, then it is evident that $\mathbf{U} = \mathbf{DL}^T$. Now Eq. (10.14) can be rewritten as

$$(\mathbf{LDL}^T)\mathbf{u} = \mathbf{b} \tag{10.15}$$

For this reason, the procedure described above is called "$\mathbf{LDL}^T$ decomposition" or "**LU** decomposition." Notice that $\mathbf{y}$ and $\mathbf{b}$ are related via

$$\mathbf{y} = \mathbf{L}^{-1}\mathbf{b} = \mathbf{M}_{n-1}^{-1}\mathbf{M}_{n-2}^{-1}\cdots\mathbf{M}_2^{-1}\mathbf{M}_1^{-1}\mathbf{b} \tag{10.16}$$

With reference to the problem worked earlier, it is easy to see that

$$\mathbf{M}_1^{-1} = \begin{bmatrix} 1 & & & \\ \frac{4}{5} & 1 & & \\ -\frac{1}{5} & 0 & 1 & \\ 0 & 0 & 0 & 1 \end{bmatrix}, \qquad \mathbf{M}_2^{-1} = \begin{bmatrix} 1 & & & \\ 0 & 1 & & \\ 0 & \frac{8}{7} & 1 & \\ 0 & -\frac{5}{14} & 0 & 1 \end{bmatrix}$$

$$\mathbf{M}_3^{-1} = \begin{bmatrix} 1 & & & \\ 0 & 1 & & \\ 0 & 0 & 1 & \\ 0 & 0 & \frac{4}{3} & 1 \end{bmatrix}, \qquad \mathbf{U} = \begin{bmatrix} 5 & -\frac{4}{5} & \frac{1}{5} & 0 \\ & \frac{14}{5} & -\frac{16}{5} & 1 \\ & & \frac{15}{7} & -\frac{20}{7} \\ & & & \frac{5}{6} \end{bmatrix}$$

$$\mathbf{L} = \begin{bmatrix} 1 & & & \\ -\frac{4}{5} & 1 & & \\ \frac{1}{5} & -\frac{8}{7} & 1 & \\ 0 & \frac{5}{14} & -\frac{4}{3} & 1 \end{bmatrix}, \qquad \mathbf{D} = \begin{bmatrix} 5 & & & \\ & \frac{14}{5} & & \\ & & \frac{15}{7} & \\ & & & \frac{5}{6} \end{bmatrix}$$

and

$$\mathbf{y} = [0 \quad 1 \quad \tfrac{8}{7} \quad \tfrac{7}{6}]^T$$

Thus Gaussian elimination is nothing but the factorization of **A** into the product, $\mathbf{A} = \mathbf{LU}$, of a lower triangular matrix **L** times an upper triangular matrix **U**. Thus $\mathbf{A}^{-1}\mathbf{b}$, which is the solution **u** we want to compute, is identical with $\mathbf{U}^{-1}\mathbf{L}^{-1}\mathbf{b}$, and the two triangular matrixes are easy to invert. If there are many systems $\mathbf{Au} = \mathbf{b}_n$ to be solved with different data $\mathbf{b}_n$, but the same coefficient matrix, then **L** and **U** must be stored.

The preceding example is too well behaved to expose some of the difficulties with Gaussian elimination. For example, it is important to guard against dividing by zero during the computations. Even if there are no zeros on the main diagonal at the start of the computation, zeros may get created in subsequent steps. A useful strategy in overcoming this problem is to rearrange the equations, before implementing each of the preceding steps, such that the coefficient with the largest magnitude appears at the left top corner of the array undergoing reduction. This idea is called ***pivoting***. If both row and column interchanges are made toward this goal, then it is called ***complete pivoting***. If only row interchanges are made, to implement the preceding strategy only partially, then we have ***partial pivoting***. In many cases partial pivoting is usually adequate.

A step-by-step procedure, to facilitate writing a computer program, is now presented to solve $\mathbf{Ax} = \mathbf{b}$.

Step 1. Augment the $n \times n$ matrix **A** with the right side **b** to form an $n \times (n + 1)$ matrix.

Step 2. Interchange the rows (of the augmented matrix), if necessary, to make the value in the (1, 1) position the largest in magnitude of any coefficient in the first column.

Step 3. Subtract a_{i1}/a_{11} times the first row from the ith row $(2 \leq i \leq n)$. This should leave a column of zeros below the pivot element in the first column.

Step 4. Repeat Steps 2 and 3 for the second through $(n - 1)$st rows as follows.
(a) Implement partial pivoting by considering only rows j to n.
(b) Subtract a_{ij}/a_{ji} times the jth row from the ith row so as to create zeros in all positions in the jth column below the main diagonal. At this stage, the matrix **A** is upper triangular.

Step 5. Solve for x_n by solving the nth equation of the upper triangular matrix **A**, by using

$$x_n = \frac{a_{n,n+1}}{a_{nn}}$$

Step 6. Solve for $x_{n-1}, x_{n-2}, \ldots, x_2, x_1$ from the $(n - 1)$st to the first equation, in turn, by using

$$x_i = \frac{a_{i,n+1} - \sum_{j=i+1}^{n} a_{ij}x_j}{a_{ii}}$$

It is important to reflect upon the elimination process if **A** is symmetric, positive definite, and tridiagonal. First, the elimination process succeeds because the factorization $\mathbf{A} = \mathbf{LU}$ exists. The condition for success is that each of the submatrixes in the upper left corner of **A**, that is,

$$\mathbf{A}_1 = [a_{11}], \qquad \mathbf{A}_2 = \begin{bmatrix} a_{11} & a_{12} \\ a_{21} & a_{22} \end{bmatrix}, \ldots$$

should have a nonzero determinant. For a positive-definite matrix, these determinants are all positive, and elimination can be carried out with no exchanges of rows. In fact, the determinant of $\mathbf{A}_j$ equals the product $a_{11}a_{22} \ldots a_{jj}$ so that the pivot elements a_{jj} lying on the main diagonal are all positive.

Another requirement for the successful completion of the Gaussian elimination algorithm is that the pivot elements be not only nonzero but also sufficiently large. Otherwise, roundoff errors will dominate. This type of intrinsic sensitivity of **A** to small perturbations is measured by what is called the ***condition number*** of **A**. This condition number, which is roughly the ratio of the largest to the smallest eigenvalue of **A**, will depend upon the grid size h and on the order of the differential equation.

Finally, the fact that **A** is tridiagonal means an enormous reduction in the number of computations. An algorithm that takes advantage of the property is described below.

In higher-dimensional problems, the matrix **A** will be sparse but it may not have a well-defined band structure. In fact, the band structure is influenced by the ordering of the grid points.

TRIDIAGONAL ALGORITHM

The special nature of the matrixes arising from the application of finite difference techniques to field problems has given rise to a number of computational techniques or algorithms. The most important of these, generally credited to L. H. Thomas, is derived from the Gaussian elimination method. This algorithm is applicable to ***tridiagonal matrixes***, matrixes having nonzero elements only on the main diagonal and on the two diagonals adjacent to the main diagonal. Such systems may be written in the form

$$\begin{bmatrix} b_1 & c_1 & & & & & \\ a_2 & b_2 & c_2 & & & & \\ & a_3 & b_3 & c_3 & & & \\ & & \ddots & \ddots & \ddots & & \\ & & & a_{M-2} & b_{M-2} & c_{M-2} & \\ & & & & a_{M-1} & b_{M-1} & c_{M-1} \\ & & & & & a_M & b_M \end{bmatrix} \begin{bmatrix} u_1 \\ u_2 \\ u_3 \\ \cdot \\ \cdot \\ \cdot \\ u_{M-2} \\ u_{M-1} \\ u_M \end{bmatrix} = \begin{bmatrix} d_1 \\ d_2 \\ d_3 \\ \cdot \\ \cdot \\ \cdot \\ d_{M-2} \\ d_{M-1} \\ d_M \end{bmatrix} \tag{10.17}$$

where all elements in the matrix other than those identified by a, b, and c are zero.

Generally, u_1 and u_M are given boundary conditions, and $d_1, d_2, \ldots, d_M$ are also given or known. This system of simultaneous equations therefore takes the form

$$\begin{aligned} b_1u_1 + c_1u_2 &= d_1 \\ a_iu_{i-1} + b_iu_i + c_iu_{i+1} &= d_i, \qquad i = 2, 3 \ldots (M-1) \\ a_Mu_{M-1} + b_Mu_M &= d_M \end{aligned} \tag{10.18}$$

where the subscript i identifies the unknowns. Two terms β_i and γ_i are now defined such that

$$\beta_i = b_i - \frac{a_ic_{i-1}}{\beta_{i-1}}, \qquad \text{with } \beta_1 = b_1 \tag{10.19}$$

$$\gamma_i = \frac{d_i - a_i\gamma_{i-1}}{\beta_i}, \qquad \text{with } \gamma_1 = \frac{d_1}{b_1} \tag{10.20}$$

The values of the dependent variables are now computed from

$$u_M = \gamma_M \qquad \text{and} \qquad u_i = \gamma_i - \frac{c_iu_{i+1}}{\beta_i} \tag{10.21}$$

To apply the algorithm, Eqs. (10.19) and (10.20) are employed to compute the terms β_i and γ_i beginning with $i = 2$ and proceeding in order of increasing i up to $i = M - 1$. Once all these terms have been calculated, Eq. (10.21) can be employed to determine the unknowns u_i starting with u_{M-1} and proceeding in the direction of decreasing i to u_1. This method is direct and generally involves far fewer computational steps than any available alternative technique. For this reason, in formulating approximate expressions and solution techniques, every effort is made to arrange the data in the form of Eq. (10.17) to permit the application of this algorithm.

QUIDIAGONAL ALGORITHM

In the treatment of problems characterized by the biharmonic equation, as well as in the application of higher-order difference techniques to other problems, there arise matrixes of the type $\mathbf{Au} = \mathbf{f}$; that is,

$$\left[\begin{array}{ccccc|cccccc} c_1 & d_1 & e_1 & & & & & & & & \\ b_2 & c_2 & d_2 & e_2 & & & & & & & \\ a_3 & b_3 & c_3 & d_3 & e_3 & & & & & & \\ & a_4 & b_4 & c_4 & d_4 & e_4 & & & & & \\ \hline & & & & & a_{M-3} & b_{M-3} & c_{M-3} & d_{M-3} & e_{M-3} & \\ & & & & & & a_{M-2} & b_{M-2} & c_{M-2} & d_{M-2} & e_{M-2} \\ & & & & & & & a_{M-1} & b_{M-1} & c_{M-1} & d_{M-1} \\ & & & & & & & & a_M & b_M & c_M \end{array}\right] \left[\begin{array}{c} u_1 \\ u_2 \\ u_3 \\ u_4 \\ \vdots \\ u_{M-3} \\ u_{M-2} \\ u_{M-1} \\ u_M \end{array}\right] = \left[\begin{array}{c} f_1 \\ f_2 \\ f_3 \\ f_4 \\ \vdots \\ f_{M-3} \\ f_{M-2} \\ f_{M-1} \\ f_M \end{array}\right] \tag{10.22}$$

A matrix of the type appearing in Eq. (10.22) has nonzero terms only on the main diagonal and the two diagonals on each side of the main diagonal. For this reason it is called a quidiagonal system, and the algorithm for its solution is called the ***quidiagonal algorithm***. A typical equation now takes the form

$$a_i u_{i-2} + b_i u_{i-1} + c_i u_i + d_i u_{i+1} + e_i u_{i+2} = f_i, \qquad 1 \leq i \leq M \tag{10.23}$$

with $a_1 = b_1 = a_2 = e_{M-1} = d_M = e_M = 0$.

The algorithm can now be described as follows. First compute

$$\delta_1 = \frac{d_1}{c_1}, \qquad \lambda_1 = \frac{e_1}{c_1}, \qquad \gamma_1 = \frac{f_1}{c_1} \tag{10.24a}$$

and

$$\mu_2 = e_2 - b_2\delta_1 \tag{10.24b}$$

$$\delta_2 = \frac{d_2 - b_2\lambda_1}{\mu_2}, \qquad \lambda_2 = \frac{e_2}{\mu_2}, \qquad \gamma_2 = \frac{f - b_2\gamma_1}{\mu_2} \tag{10.24c}$$

Then, for $3 \leq i \leq M - 2$, compute

$$\beta_i = b_i - a_i\delta_{i-2} \tag{10.25}$$

$$\mu_i = c_i - \beta_i\delta_{i-1} - a_i\lambda_{i-2} \tag{10.26a}$$

$$\delta_i = \frac{d_i - \beta_i\lambda_{i-1}}{\mu_i} \tag{10.26b}$$

$$\lambda_i = \frac{f_i - \beta_i\gamma_{i-1} - a_i\gamma_{i-2}}{\mu_i} \tag{10.26c}$$

Next compute

$$\beta_{M-1} = b_{M-1} - a_{M-1}\delta_{M-3} \tag{10.27a}$$

$$\mu_{M-1} = c_{M-1} - \beta_{M-1}\delta_{M-2} - a_{M-1}\lambda_{M-3} \tag{10.27b}$$

$$\delta_{M-1} = \frac{d_{M-1} - \beta_{M-1}\lambda_{M-2}}{\mu_{M-1}} \tag{10.27c}$$

$$\gamma_{M-1} = \frac{f_{M-1} - \beta_{M-1}\gamma_{M-2} - a_{M-2}\gamma_{M-3}}{\mu_{M-1}} \tag{10.27d}$$

and

$$\beta_M = b_M - a_M\delta_{M-2} \tag{10.28a}$$

$$\mu_M = c_M - \beta_M\delta_{M-1} - a_M\lambda_{M-2} \tag{10.28b}$$

$$\gamma_M = \frac{f_M - \beta_M\gamma_{M-1} - a_M\gamma_{M-2}}{\mu_M} \tag{10.28c}$$

Note that β_i and μ_i are used only to compute δ_i, λ_i, and γ_i and need not be stored after they are computed. The δ_i, λ_i, and γ_i must be stored as they are used in the backward substitution, which is described by

$$u_M = \gamma_M \tag{10.29a}$$

$$u_{M-1} = \gamma_{M-1} - \delta_{M-1}u_M \tag{10.29b}$$

and

$$u_i = \gamma_i - \delta_i u_{i+1} - \lambda_i u_{i+2} \tag{10.29c}$$

for $M - 2 \geq i \geq 1$.

GAUSS ELIMINATION ALGORITHM FOR BANDED MATRIXES

In many applications, the matrix **A** exhibits neither a strictly tridiagonal nor a strictly quidiagonal structure. Nevertheless, nonzero entries of **A** show a band structure. The Gauss elimination algorithm (or $\mathbf{LDL}^T$ decomposition) can be adapted to this case. Toward this adaptation, it is necessary to introduce some parameters characterizing the bandedness of **A**. Figure 10.1 shows a typical pattern of the entries of matrix **A** arising from a finite element analysis. Notice that

$$a_{ij} = 0, \qquad \text{for } j > i + \beta$$

where $2\beta + 1$ is called the ***bandwidth*** and β is called the ***half-bandwidth*** of the matrix **A**. For the matrix shown, $\beta = 3$. Another parameter useful in the design of storage allocation is the profile or ***skyline*** of **A**, which is characterized by the set of variables $s_j, j = 1, 2, \ldots, n$, where s_j is the row number of the first nonzero element in column j of **A**. Similarly, the variable $(j - s_j)$ can be used to define the ***column heights***. With the column heights defined, all entires below the skyline of **A** can now be stored as a one-dimensional array in ***S***; that is, the active columns of **A** are stored consecutively in S. In addition to S, it is necessary to define an array MAXS, which stores the addresses of the diagonal entries of **A** in S. That is, the address of the ith diagonal entry of **A**, a_{ii}, in S is MAXS(I). Referring to Figure 10.1, it can be seen that MAXS(I) is equal to the sum of the column heights up to the $(i - 1)$st column plus I. Hence the number of nonzero entries in the ith column of **A** is equal to MAXS(I + 1) − MAXS(I), and the entry addresses are MAXS(I), MAXS(I + 1), . . . , MAXS(I + 1) − 1. Thus the address array MAXS can be easily used to address each entry of **A** in S.

Now the $\mathbf{LDL}^T$ algorithm can be efficiently carried out on a computer by considering each column of **A** in turn, although Gaussian elimination is carried out by rows. The computational algorithm starts by setting $d_{11} = a_{11}$. Then the entries in the jth column of **D** and **L** (or, equivalently, of **U**) are calculated for $j = 2, 3, \ldots, n$ by following a sequence of calculations defined by

$$g_{s_j,j} = a_{s_j,j} \tag{10.30a}$$

$$g_{ij} = a_{ij} - \sum_{r=s_m}^{i-1} m_{ri} g_{rj}, \qquad i = s_j + 1, \ldots, j - 1 \tag{10.30b}$$

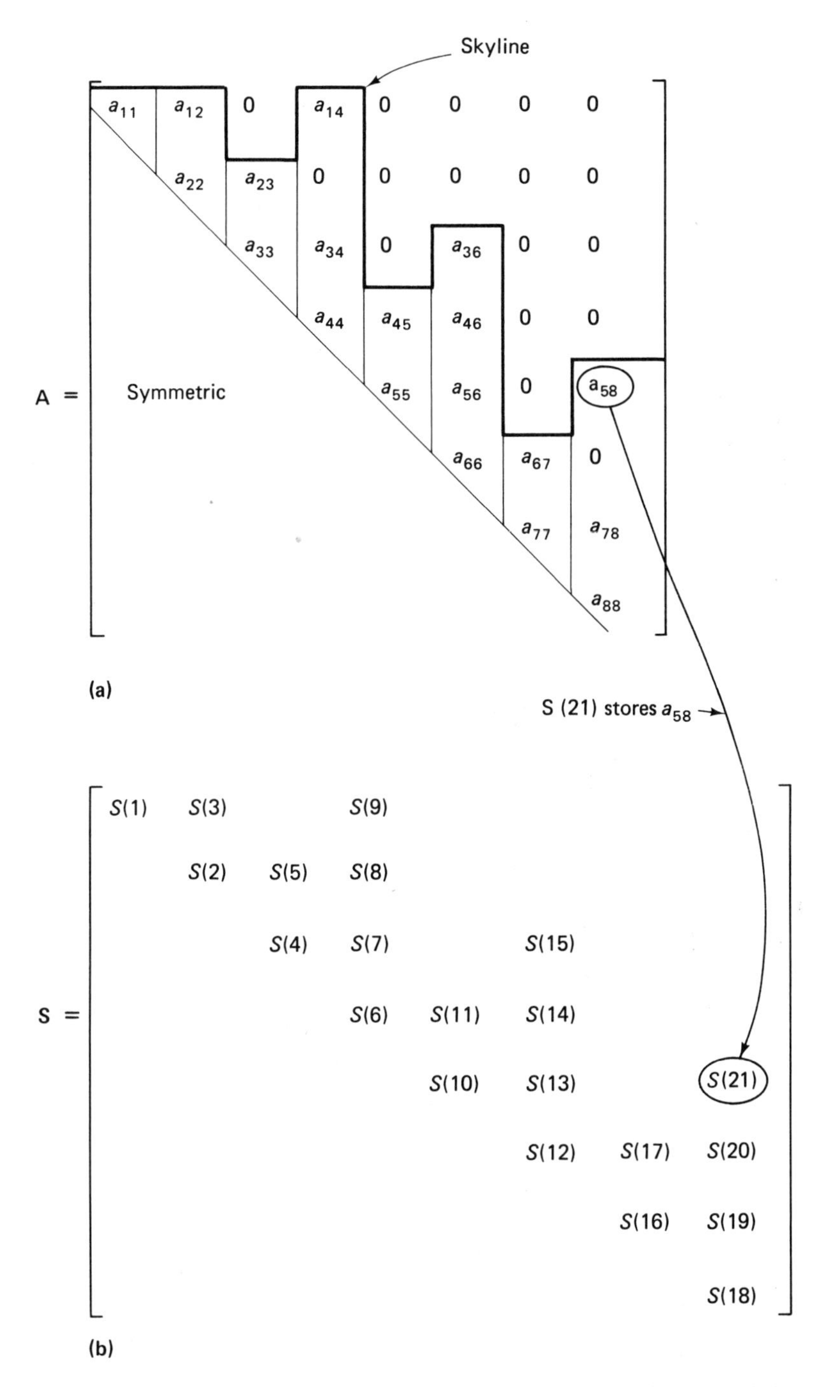

Figure 10.1 Storage scheme for Gaussian elimination with a banded matrix: (a) actual entries of **A**; (b) any **S** storing elements of **A**

and proceeding further with

$$m_{ij} = \frac{g_{ij}}{d_{ii}}, \qquad i = s_j, \dots, j-1 \tag{10.31}$$

$$d_{jj} = a_{jj} - \sum_{r=s_j}^{j-1} m_{rj} g_{rj} \tag{10.32}$$

where $s_m = \max\{s_i, s_j\}$, and the s_i's are the ***skyline parameters***, whose meaning will be explained presently. The computations on the right side of $\mathbf{Au} = \mathbf{b}$ are carried out via

$$y_1 = b_1 \tag{10.33a}$$

$$y_i = b_i - \sum_{r=s_i}^{i-1} m_{ri} b_r \tag{10.33b}$$

and the back substitution is carried out by evaluating successively $u_n, u_{n-1}, \dots, u_1$. This is achieved by first calculating $\bar{\mathbf{y}} = \mathbf{D}^{-1}\mathbf{y}$. Then setting $\bar{y}^{(n)} = \bar{y}$ and $u_n = \bar{y}_n^{(n)}$, the following calculations are carried out, for $i = n, \dots, 2$,

$$\bar{y}_r^{(i-1)} = \bar{y}_r^{(i)} - m_{ri} u_i, \qquad r = s_i, \dots, i-1 \tag{10.34a}$$

$$u_{i-1} = \bar{y}_{i-1}^{(i-1)} \tag{10.34b}$$

Notice that the entry m_{rj} when calculated for use in Eq. (10.32) immediately replaces g_{ij}, and d_{jj} replaces a_{jj}. Thus, at the end of the skyline reduction the entries d_{jj} are in storage locations previously occupied by a_{jj}, and m_{rj} are stored in locations occupied by $a_{rj}, j > r$. Similarly, $\bar{y}_k^{(j)}$, for all j, is stored in the location of y_k, the original storage location of b_k.

The preceding calculation procedure is now illustrated. Starting with $\mathbf{A}$, defined as

$$\mathbf{A} = \begin{bmatrix} 5 & -4 & 1 & 0 \\ & 6 & -4 & 1 \\ \text{symmetric} & & 6 & -4 \\ & & & 5 \end{bmatrix}$$

it is evident that $s_1 = 1$, $s_2 = 1$, $s_3 = 1$, and $s_4 = 2$. Using Eq. (10.30a) one gets, for $j = 1$,

$$d_{11} = a_{11} = 5$$

For $j = 2$, $s_j = s_2 = 1$. Therefore, $g_{s_j,j} = g_{12} = a_{12} = -4$. Now using Eqs. (10.31) and (10.32),

$$m_{12} = \frac{g_{12}}{d_{11}} = -\frac{4}{5}$$

$$d_{22} = a_{22} - m_{12} g_{12} = 6 - \left(-\frac{4}{5}\right)(-4) = \frac{14}{5}$$

Replacing a_{12} by m_{12} and a_{22} by d_{22}, the resulting matrix entries are

$$\left[\begin{array}{cc:cc} 5 & -\frac{4}{5} & 1 & 0 \\ & \frac{14}{5} & -4 & 1 \\ & & 6 & -4 \\ & & & 5 \end{array}\right]$$

where the dashed line is used to separate the reduced and unreduced columns of **A**.

For $j = 3$, analogous computations yield

$$g_{13} = a_{13} = 1$$

$$g_{23} = a_{23} - m_{12}g_{13} = -4 - \left(-\frac{4}{5}\right)(1) = -\frac{16}{5}$$

$$m_{13} = \frac{g_{13}}{d_{11}} = \frac{1}{5}$$

$$m_{23} = \frac{g_{23}}{d_{22}} = \frac{-(16/5)}{(14/5)} = -\frac{8}{7}$$

$$d_{33} = a_{33} - m_{13}g_{13} - m_{23}g_{23} = \frac{15}{7}$$

and the resulting matrix entries are

$$\left[\begin{array}{ccc:c} 5 & -\frac{4}{5} & \frac{1}{5} & 0 \\ & \frac{14}{5} & -\frac{8}{7} & 1 \\ & & \frac{15}{7} & -4 \\ & & & 5 \end{array}\right]$$

Repeating the procedure for $j = 4$, one gets

$$\begin{bmatrix} 5 & -\frac{4}{5} & \frac{1}{5} & 0 \\ & \frac{14}{5} & -\frac{8}{7} & \frac{5}{14} \\ & & \frac{15}{7} & -\frac{4}{3} \\ & & & \frac{5}{6} \end{bmatrix}$$

which can be seen to be identical to the matrix **U** calculated earlier.

The preceding discussion illustrates the importance of keeping the bandwidth and profile at a low value. For a simple illustration, consider a two-dimensional rectangular region. If there are fewer grid points in the horizontal than in the vertical direction, then the unknowns should be numbered consecutively along each row rather than each column. In general, finite element matrixes are less systematic in structure than finite-difference equations, and an optimal ordering, for reduced bandwidth, is far from evident.

STORAGE VERSUS SPEED

In designing computational strategies, one has to cope with two conflicting demands: speed and storage requirements. Speed is determined by the number of arithmetic operations to be performed. (An ***operation*** here is defined as one multiplication plus one addition. A division is counted as a multiplication and subtraction as an addition.) For a full matrix, the number of operations required during Gaussian elimination is of the order of $n^3/3$. If **A** is symmetric, then this number would be roughly $n\beta^2/2$, where β is the bandwidth. It should be emphasized that speed and storage requirements are intertwined. A banded matrix requires roughly $n\beta$ locations of storage if an optimal storage scheme is used. Therefore, if linear elements are used on a 50×50 square grid, $n = 2500$ and $\beta \simeq 50$. At about this point, one typically runs out of main storage and the problem becomes one of managing data in and out of main storage, which is in itself complex and time consuming.

It is at this stage that there is room for innovative ideas. For instance, one can develop algorithms for (1) reducing the bandwidth with a proper node labeling scheme, (2) reducing the profile, or (3) maintaining sparsity within the band during elimination, and so forth. One can develop new algorithms that will require a fewer number of operations. One promising direction would be to develop methods that are analogous to the fast Fourier transform in reducing the computational effort. Several ideas of this nature are being proposed, and the interested reader should refer to the literature on this subject.

10.3. ITERATIVE METHODS

An important class of approximate numerical methods for solving systems of equations $\mathbf{Au} = \mathbf{b}$ involves making an initial guess $u^{(0)}$ as to the solution u. These values $u^{(0)}$ are then substituted in $\mathbf{Au} = \mathbf{b}$, leading to a new, and hopefully better, estimate of the solution. If this procedure is repeated for a sufficiently large number of iteration cycles, a point will be reached at which the change in the solution effected in two successive cycles is smaller than a specified tolerance. The procedure is then said to have ***converged,*** and the solution thus obtained is accepted as the solution to the discrete problem $\mathbf{Au} = \mathbf{b}$. Thus an iterative method starts with an approximation to the true solution and attempts to develop a convergent sequence of approximations from a computational cycle that is repeated as often as necessary. More formally, to solve

$$\mathbf{Au} = \mathbf{b} \tag{10.35}$$

one attempts to write the matrix **A** as

$$\mathbf{A} = \mathbf{I} - \mathbf{P} \tag{10.36}$$

so that Eq. (10.35) becomes

$$\mathbf{u} = \mathbf{Pu} + \mathbf{b} \tag{10.37}$$

whence the iterative scheme

$$\mathbf{u}^{(i+1)} = \mathbf{Pu}^{(i)} + \mathbf{b} \tag{10.38}$$

is defined. Given $\mathbf{u}^{(0)}$, the initial guess, improved approximations $\mathbf{u}^{(1)}, \mathbf{u}^{(2)}, \ldots$ can be calculated using Eq. (10.38). Several questions can be raised in connection with the preceding iterative method. Does the limits $\lim_{i\to\infty} \mathbf{u}^{(i)}$ exist for each component of the vector **u**? If these limits exist, are they equal for every component of **u**? To answer these questions, one defines the error vector

$$\boldsymbol{\epsilon}^{(i)} = \mathbf{u}^{(i)} - \mathbf{u} \tag{10.39}$$

Subtracting Eq. (10.37) from (10.38)

$$\boldsymbol{\epsilon}^{(i+1)} = \mathbf{P}\boldsymbol{\epsilon}^{(i)} \tag{10.40}$$

from which it follows, by induction,

$$\boldsymbol{\epsilon}^{(i)} = \mathbf{P}^i\boldsymbol{\epsilon}^{(0)}, \qquad i \geq 0 \tag{10.41}$$

Therefore, for the required limits to exist, it is necessary to ensure that

$$\lim_{i\to\infty} \mathbf{P}^i\boldsymbol{\epsilon}^{(0)} = \mathbf{0} \tag{10.42}$$

for *all* $\epsilon^{(0)}$. This is equivalent to determining when

$$\lim_{i\to\infty} \mathbf{P}^i = \mathbf{0} \tag{10.43}$$

The iterative method defined in Eq. (10.38) is said to ***converge*** if the preceding condition is satisfied. It can be proved that this condition is equivalent to requiring that the ***spectral radius*** of **P** (i.e., the modulus of the largest eigenvalue of **P**) be less than unity. Several theoretical results concerning stability can be established if matrix **A** possesses some properties such as symmetry, diagonal dominance, irreducibility, consistent ordering, property A, and so forth. (See Exercise 9.)

The central idea behind iterative methods can be briefly presented by considering

$$\frac{\partial^2 u}{\partial x^2} + \frac{\partial^2 u}{\partial y^2} = k \tag{10.44}$$

over the square defined by $0 \leq x \leq 1, 0 \leq y \leq 1$. A general difference approximation to this can be written as

$$u_{l,m} = \alpha_1 u_{l+1,m} + \alpha_2 u_{l-1,m} + \alpha_3 u_{l,m+1} + \alpha_4 u_{l,m-1} + \alpha_0 k_{l,m} \tag{10.45}$$

For uniform fields and equal net spacings in the x and y directions, α_1, α_2, α_3, and α_4 will all equal $\frac{1}{4}$. For one-dimensional problems, some of these terms are equal to zero. For problems governed by Laplace's equation, rather than Poisson's equation, the last term on the right side of Eq. (10.45) equals zero. Equation (10.45) applies to each point in the finite difference grid interior to the field boundaries. For those points immediately adjacent to the boundaries, some of the potentials on the right side are the specified boundary conditions. The first step in the application of iterative techniques is to make an initial guess as to the potentials $u_{l,m}$ at the interior grid points. Evidently, the closer the guessed values are to the correct solution, the fewer the subsequent iterations that are required; but for greater convenience in programming, it is a frequent procedure to guess initially that all interior points have a potential of zero.

Jacobi Method. The simplest of the iterative methods is variously known as Jacobi's method or as the method of ***simultaneous displacements***. In this scheme the improved values for the potentials for any given iterative cycle are obtained merely by inserting the potentials obtained during the preceding iterative cycle into Eq. (10.45). Thus for the $(i + 1)$ cycle

$$u_{l,m}^{(i+1)} = \alpha_1 u_{l+1,m}^{(i)} + \alpha_2 u_{l-1,m}^{(i)} + \alpha_3 u_{l,m+1}^{(i)} + \alpha_4 u_{l,m-1}^{(i)} + \alpha_0 k_{l,m} \qquad (10.46a)$$

where the superscripts identify the iterative cycle. Thus, for example, in solving Laplace's equation and guessing zero potential for all interior net points, after the first iterative cycle all potentials $u_{l,m}^{(1)}$ will be equal to zero except those immediately adjacent to a field boundary. In subsequent cycles the potentials at net points lying progressively farther away from the field boundaries will approach their correct values.

Equation (10.46a) can be written compactly in matrix notation. Toward this end, if $\mathbf{A}$ is written as

$$\mathbf{A} = \mathbf{D} + \mathbf{L} + \mathbf{U}$$

where $\mathbf{D} = \text{diag}\,(a_{11}, a_{21}, \ldots, a_{nn})$, and $\mathbf{U}$ and $\mathbf{L}$ are, respectively, strictly upper and lower triangular parts of $\mathbf{A}$. Denoting $\tilde{\mathbf{A}} = \mathbf{D}^{-1}\mathbf{A}$, $\tilde{\mathbf{L}} = \mathbf{D}^{-1}\mathbf{L}$, and $\tilde{\mathbf{U}} = \mathbf{D}^{-1}\mathbf{U}$, and $\tilde{\mathbf{b}} = \mathbf{D}^{-1}\mathbf{b}$. Then Eq. (10.46a) can be rewritten as

$$\mathbf{u}^{(i+1)} = -(\tilde{\mathbf{L}} + \tilde{\mathbf{U}})\mathbf{u}^{(i)} + \tilde{\mathbf{b}} \qquad (10.46b)$$

$$= \mathbf{B}\mathbf{u}^{(i)} + \tilde{\mathbf{b}} \qquad (10.46c)$$

where $\mathbf{B} = -(\tilde{\mathbf{L}} + \tilde{\mathbf{D}})$ is called the Jacobi iteration matrix. Referring back to Eq. (10.41), it is evident that $\mathbf{B}$ stands for the iteration matrix $\mathbf{P}$ for the Jacobi method. For convergence, the modulus of the largest eigenvalue (i.e., the spectral radius) of $\mathbf{B}$ must be smaller than unity.

Gauss-Seidel Method. The Gauss-Seidel method, also known as the method of ***successive displacements***, represents a refinement of Jacobi's method. In the Gauss-Seidel method, improved values of the potentials at interior net points are used in the course of each iterative cycle as soon as these values become available. Thus, in calculating the potential $\mathbf{u}_{l,m}^{(1)}$ applying to the second point in the finite difference grid, the potential just calculated for the first point is used in Eq. (10.46) in place of $\mathbf{u}^{(0)}$, which constituted the initial guess. The formula used to calculate the solution at a particular net point therefore depends to some extent upon the manner in which all the points in the finite difference grid are scanned, since this determines which newly calculated values are available. If, for example, this scanning pattern is arranged so that the point (l, m) is treated after the treatment of points $(l - 1, m)$ and $(l, m - 1)$, but prior to the treatment of points involving the coordinates $l + 1$ and $m + 1$, the formula describing the calculation of the improved value of u at the $(i + 1)$st iteration at the grid point (l, m) is

$$u_{l,m}^{(i+1)} = \alpha_1 u_{l+1,m}^{(i)} + \alpha_2 u_{l-1,m}^{(i)+1} + \alpha_3 u_{l,m+1}^{(i)} + \alpha_4 u_{l,m-1}^{(i+1)} + \alpha_0 k_{l,m} \qquad (10.47a)$$

Or, in matrix notation,

$$\mathbf{u}^{(i+1)} = -(\mathbf{I} + \tilde{\mathbf{L}})^{-1}\tilde{\mathbf{U}}\mathbf{u}^{(i)} + (\mathbf{I} + \tilde{\mathbf{L}})^{-1}\tilde{\mathbf{b}} \qquad (10.47b)$$

where $\mathbf{G} = -(\mathbf{I} + \tilde{\mathbf{L}})^{-1}\tilde{\mathbf{U}}$ is called the Gauss-Seidel iteration matrix. It can be shown that the Gauss-Seidel method converges twice as fast as the Jacobi method so that the successive displacement of new values effects a considerable reduction in the number of iteration cycles. On the other hand, successive displacements imply a greater programming effort than does Jacobi's method. In general, however, the Gauss-Seidel method is preferred.

Successive Overrelaxation Method. The rate of convergence can be accelerated still further by multiplying each term on the right side of Eq. (10.46a) by a parameter ω, termed the relaxation factor. The improved formula then takes the form

$$u_{l,m}^{(i+1)} = \omega[\alpha_1 u_{l+1,m}^{(i)} + \alpha_2 u_{l-1,m}^{(i+1)} + \alpha_3 u_{l,m+1}^{(i)} + \alpha_4 u_{l,m-1}^{(i+1)} + \alpha_0 k_{l,m}] - (\omega - 1)u_{l,m}^{(i)} \tag{10.48a}$$

or in matrix notation

$$\mathbf{u}^{(i+1)} = (\mathbf{I} + \omega\tilde{\mathbf{L}})^{-1} \cdot \{(1-\omega)\mathbf{I} - \omega\tilde{\mathbf{U}}\}\mathbf{u}^{(i)} + (\mathbf{I} + \omega\tilde{\mathbf{L}})^{-1}\tilde{\mathbf{b}} \tag{10.48b}$$

where $\mathcal{L}_\omega = (\mathbf{I} + \omega\tilde{\mathbf{L}})^{-1} \cdot \{(1-\omega)\mathbf{I} - \omega\tilde{\mathbf{U}}\}$ is called the successive overrelaxation iteration matrix.

Here again the manner of scanning the finite difference grid determines which corrected $(i + 1)$ values are available for use in the right side of Eq. (10.48). Methods employing Eq. (10.48) are termed ***successive point overrelaxation methods*** (SPOR) or simply the successive overrelaxation (SOR) method. Evidently, if $\omega = 1$, this method reduces to the Gauss-Seidel method. For most problems, ω lies between 1 and 2.

A simple example can be used to illustrate the advantage of successive overrelaxation. Consider a two-dimensional uniform square field governed by Laplace's equation

$$\frac{\partial^2 u}{\partial x^2} + \frac{\partial^2 u}{\partial y^2} = 0 \tag{10.49}$$

where $\Delta x = \Delta y = M^{-1}$, and where $0 \leq x \leq 1$ and $0 \leq y \leq 1$. As a boundary condition let $u = 1000$ on the side of the square where $y = 1$, and let $u = 0$ at the other three sides of the square. If the x and y coordinates are divided into three equal line segments (i.e., $M = 3$), the interior net points will have the coordinates $(\frac{1}{3}, \frac{1}{3})$, $(\frac{2}{3}, \frac{1}{3})$, $(\frac{1}{3}, \frac{2}{3})$, and $(\frac{2}{3}, \frac{2}{3})$, and the coefficients of Eq. (10.46a) will take the values

$$\alpha_1 = \alpha_2 = \alpha_3 = \alpha_4 = \tfrac{1}{4} \tag{10.50}$$

With an initial guess of zero everywhere inside the field, the application of Gauss-Seidel method, that is, Eq. (10.47), results in convergence in seven iterations, while the application of SOR method, Eq. (10.48), with $\omega = 1.1$, results in convergence in only four iterations. (Do this test using a hand calculator.)

The improvement effected by successive overrelaxation is strongly dependent upon the choice of the relaxation factor ω. A general formula for the optimum value of ω, ω_b, is

$$\omega_b = \frac{2}{1 + \sqrt{1 - \mu_m^2}} \tag{10.51}$$

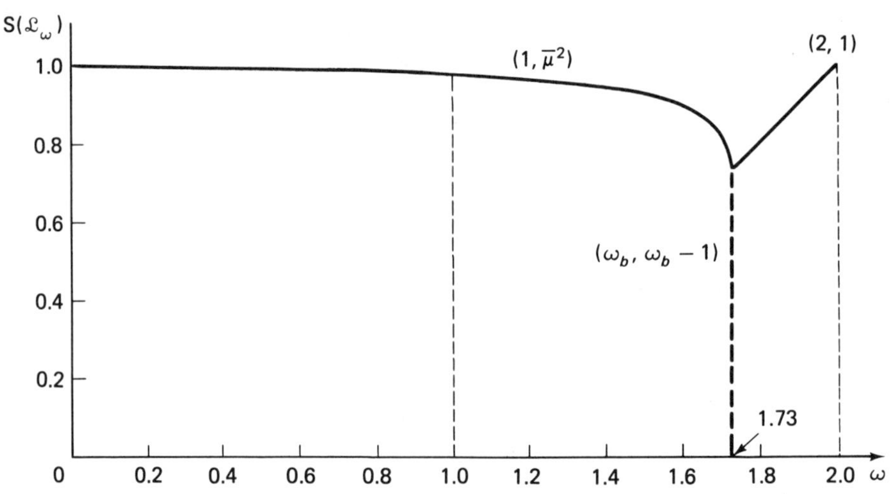

Figure 10.2 Variation of the spectral radius $S(\mathcal{L}_\omega)$, of $\mathcal{L}_\omega$, versus ω in the range $0 \leq \omega \leq 2$ for $\bar{\mu} = \cos \pi/20 = 0.987688$, where $\bar{\mu}$ stands for the spectral radius of the Jacobi matrix **B** (Reproduced with permission from *Iterative Solution of Large Linear Systems* by D. M. Young, © 1971, Academic Press, Inc., New York)

where μ_m is the eigenvalue of largest modulus of the Jacobi matrix **B**. An illustration of the importance of an optimum choice of ω is shown in Figure 10.2. This figure refers to the solution of a square field governed by Laplace's equation, Eq. (10.49), for $M = 20$. There are therefore $19^2 = 361$ interior finite difference net points, so that there result 361 simultaneous equations of the type of Eq. (10.48) with $\alpha_0 = 0$. The problem was solved using a number of different values for ω, and the number of iterations required to reduce the error to 2^{-31} was recorded. It should be pointed out that in engineering work a considerably larger error is satisfactory. Figure 10.2 demonstrates that for this particular problem ω_b equals approximately 1.73. The optimum successive overrelaxation method requires roughly $\frac{1}{10}$ the number of iterations required using the Gauss-Seidel method. Figure 10.2 also highlights an important difficulty in applying successive overrelaxation. Since ω_b, the optimum value of ω, is generally not known in advance and is extremely difficult to determine, particularly for nonuniform fields and nonuniform net spacing, it is highly unlikely that optimum use of successive overrelaxation will be made in solving any specific problem.

A number of variations of the preceding technique are possible. For instance, among equations derived from finite difference approximations, there is a degree of symmetry over alternate grid points. For example, in a two-dimensional Laplace's equation, each grid point is coupled to its four neighbors; that is, each variable at an odd-numbered grid point is coupled to variables at even-numbered grid points, and vice versa. One can take advantage of this symmetry by defining a double cyclic iteration, where all variables on the even points are improved first and are then used to improve the variable values at the odd points. To do this, the vector **u** is split into

$\mathbf{u}_{\text{even}}$ and $\mathbf{u}_{\text{odd}}$. Then a cyclic successive overrelaxation procedure can be defined as

$$\mathbf{u}_\nu^{(i+1)} = -\omega(\tilde{\mathbf{L}} + \tilde{\mathbf{U}})\mathbf{u}_{1-\nu}^{(i)} + (1-\omega)\mathbf{u}_\nu^{(i-1)} + \omega\tilde{\mathbf{b}}_\nu \tag{10.52}$$

where ν assumes the values 0 and 1 over successive iterative steps. The 0's stand for "even" and the 1's stand for "odd".

Cyclic Chebyshev Method. This method is the same as the one described in Eq. (10.52), except that the relaxation parameter is changed every step. That is,

$$\mathbf{u}_\nu^{(i+1)} = -\omega_i(\tilde{\mathbf{L}} + \tilde{\mathbf{U}})u_{1-\nu}^{(i)} + (1-\omega_i)\mathbf{u}_\nu^{(i-1)} + \omega_i\tilde{\mathbf{b}}_\nu \tag{10.53}$$

where $\nu = 0$ if i is even and $\nu = 1$ if i is odd. The relaxation factor ω_i is chosen such that

$$\begin{aligned} \omega_0 &= 1 \\ \omega_1 &= \frac{1}{1 - \frac{1}{2}\mu_m^2} \\ \omega_{i+1} &= \frac{1}{1 - \frac{1}{4}\mu_m^2\omega_i}, \qquad \text{for } i \geq 1 \end{aligned} \tag{10.54}$$

where μ_m is once again the eigenvalue of the largest modulus of the Jacobi matrix. Notice that the first step corresponds to the Gauss-Seidel method and, therefore, ω gradually increases.

Symmetric Successive Overrelaxation (SSOR) Method. The SSOR method is a modification of the SOR method. Here a complete iteration consists of two half-iterations. The first half is simply the SOR method itself, while the second half is the SOR method applied to the equations in reverse order.

Line Iterative Methods. A number of useful variations of the preceding techniques have been developed and applied, particularly to two-dimensional systems. One such family of computing schemes is known as the ***line iterative method*** and involves the improvement of the values of the approximate solution simultaneously on an entire line of points. The formula for ***simultaneous row iteration***, corresponding to Eq. (10.46), is

$$u_{l,m}^{(i+1)} = \alpha_1 u_{l+1,m}^{(i+1)} + \alpha_2 u_{l-1,m}^{(i+1)} + \alpha_3 u_{l,m+1}^{(i)} + \alpha_4 u_{l,m-1}^{(i)} + \alpha_0 k_{l,m} \tag{10.55}$$

in which improved values are not used until the end of a complete row iteration. The refinement in this technique embodied in the Gauss-Seidel method for point iteration becomes the iterative formula for ***successive row iteration***

$$u_{l,m}^{(i+1)} = \alpha_1 u_{l+1,m}^{(i+1)} + \alpha_2 u_{l-1,m}^{(i+1)} + \alpha_3 u_{l,m+1}^{(i)} + \alpha_4 u_{l,m-1}^{(i+1)} + \alpha_0 k_{l,m} \tag{10.56}$$

As indicated in Eq. (10.56), the rows are improved in order of increasing y.

Successive line overrelaxation (SLOR) can readily be effected by multiplying each term on the right side of Eq. (10.56) by the relaxation factor and adding the term $(1-\omega)u_{l,m}^{(i)}$ to the right side. It can be shown that Eq. (10.56) converges approximately twice as fast as Eq. (10.55), and that the improvement in convergence rate effected by the choice of an optimum of ω_b is of the same order as in successive point iteration.

Furthermore, it can be shown that the method of successive row (or line) overrelaxation converges approximately $\sqrt{2}$ as fast as successive point overrelaxation.

Alternating Direction Iterative Methods. Variations of the method of successive line overrelaxation have been introduced by D. W. Peaceman and H. H. Rachford as well as by J. Douglas. Their method involves essentially a row iteration followed by a column iteration. Each complete iterative cycle includes one row iteration and one column iteration. For example, for Laplace's equation in two dimensions, with a uniform mesh the iteration formulas are

$$u_{l,m}^{(i+1/2)} = u_{l,m}^{(i)} + r[u_{l+1,m}^{(i)} + u_{l-1,m}^{(i+1/2)} - 2u_{l,m}^{(i+1/2)}] + r[u_{l,m+1}^{(i)} + u_{l,m-1}^{(i)} - 2u_{l,m}^{(i)}] \quad (10.57a)$$

$$u_{l,m}^{(i+1)} = u_{lm}^{(i+1/2)} + r[u_{l+1,m}^{(i+1/2)} + u_{l-1,m}^{(i+1/2)} - 2u_{l,m}^{(i+1/2)}] + r[u_{l,m+1}^{(i+1)} + u_{l,m-1}^{(i+1)} - 2u_{l,m}^{(i+1)}] \quad (10.57b)$$

Equation (10.57a) defines a process that is similar to one step of simultaneous row iteration, while Eq. (10.57b) defines a process that is very similar to simultaneous column iterations. The tridiagonal algorithm can be used to effect the direct, non-iterative solution of the set of simultaneous equations applying to each row. Each iterative solution cycle then consists of the successive application of the algorithm to each row (or column) of the matrix. Consider, for example,

$$\frac{\partial u}{\partial t} = k\left(\frac{\partial^2 u}{\partial x^2} + \frac{\partial^2 u}{\partial y^2}\right) \quad (10.58)$$

over the rectangular region $0 \le x \le a$, $0 \le y \le b$, where u is known initially at all points within and on the boundary of the rectangle, and is known subsequently at all points of the boundary. An explicit finite difference scheme for Equation (10.58) is

$$\frac{u_{l,m,j+1} - u_{l,mj}}{(\Delta t)} = \frac{k}{(\Delta x)^2}(u_{l-1,m,j} - 2u_{l,m,j} + u_{l+1,m,j}) + \frac{k}{(\Delta y)^2}(u_{l,m-1,j} - 2u_{l,m,j} + u_{l,m+1,j}) \quad (10.59)$$

Even though this appears trivially simple, it is computationally laborious because the process is stable only if

$$k\left\{\frac{1}{(\Delta x)^2} + \frac{1}{(\Delta y)^2}\right\}(\Delta t) \le \frac{1}{2} \quad (10.60)$$

This condition forces one to take extremely small values of (Δt). Alternatively, one can use the Crank-Nicolson scheme to yield

$$\frac{u_{l,m,j+1} - u_{l,m,j}}{(\Delta t)} = \frac{k}{2}\left\{\left(\frac{\partial^2 u}{\partial x^2} + \frac{\partial^2 u}{\partial y^2}\right)_{l,m,j} + \left(\frac{\partial^2 u}{\partial x^2} + \frac{\partial^2 u}{\partial y^2}\right)_{l,m,j+1}\right\} \quad (10.61)$$

However, this requires the solution of $(L-1)(M-1)$ simultaneous algebraic equations for each step forward in time where $L(\Delta x) = a$ and $M(\Delta y) = b$. Unfortunately, the tridiagonal or quidiagonal algorithm cannot be applied to directly solve this system; one has to solve the equations iteratively.

To overcome this difficulty, special methods have been developed by Peaceman, Rachford, and Douglas. These techniques involve the application of the tridiagonal matrix algorithm successively to lines and columns of the array of grid points in the two-dimensional space domain. Their method can also be extended to three dimensional problems. In the Peaceman-Rachford method, which is also termed the ***alternating direction implicit method*** (or ADIM), the potentials at $t + \Delta t$ are obtained in two parts. First, the potentials at times midway between t and $t + \Delta t$ are obtained using the equation

$$\frac{u_{l,m,j+1/2} - u_{l,m,j}}{k(\Delta t)} = (\delta^2_{l,m,j+1/2} + \delta^2_{l,m,j})_x \tag{10.62}$$

where δ^2 symbolically stands for the second central difference, and the subscript x means that the differencing is done in the x-direction. Equation (10.62) has three unknowns and can therefore be solved by the tridiagonal matrix algorithm. A separate set of solutions using this algorithm is obtained for each line of points oriented in the x-direction. The terms with a subscript j of Eq. (10.62) all represent potentials at time t_j and are therefore known. In the second step of the calculation, a similar procedure is applied to all lines oriented in the y-direction, using the equation

$$\frac{u_{l,m,j+1} - u_{l,m,j+1/2}}{k(\Delta t)} = (\delta^2_{l,m,j} + \delta^2_{l,m,j+1/2})_y \tag{10.63}$$

Again there are three unknown terms, while the terms with the subscript $j + \frac{1}{2}$ have been determined in the previous calculation step described by Eq. (10.62). It is therefore now possible to apply the tridiagonal matrix algorithm separately to each line oriented in the y-direction. Hence, the combination of Eqs. (10.62) and (10.63) provides the required solution at time $t + \Delta t$ without iteration. Note that if the stability criterion developed is applied to either Eq. (10.62) or (10.63), these equations will be found to be only conditionally stable. However, the combination of Eqs. (10.62) and (10.63) is unconditionally stable for all ratios of $\Delta t/h^2$. This unconditional stability permits the control of the time increment Δt as determined by estimates of the truncation error due to the discretization of the time domain. To be absolutely sure of stability, it is advisable to change Δt only after the completion of the two half-steps represented by Eqs. (10.62) and (10.63). It can be shown, however, that this requirement can be relaxed somewhat, so that actually Δt can be changed within certain limits after each computation step. Actual implementation of this method is demonstrated in Section 13-3.

Even though extensive numerical analysis research has been done on iterative schemes, it is still quite difficult to come up with a firm recommendation of one of the preceding iterative schemes over another. Each seems to have certain advantages and disadvantages. One important criterion is certainly the rate of convergence. However, it is possible to calculate the convergence rates only for relatively simple cases. For example, in Table 10.1, some of the point iterative methods are compared in terms of the approximate relative rate of convergence for the solution of Laplace's equation with Dirichlet boundary conditions for the square with a square grid of size h and a

Table 10.1. Comparison of point iterative methods

Method	*Approximate Relative Rate of Convergence*
Jacobi	$\frac{1}{2}h^2$
Gauss-Seidel	h^2
Optimum SOR	$2h$
Semi-iterative	
General	h
Cyclic Chebyshev	$2h$
SSOR	$\frac{1}{2}h^2$

Table 10.2. Comparison of line iterative methods

Method	*Approximate Relative Rate of Convergence*
Single line (five-point formula)	
Jacobi	h^2
Gauss-Seidel	$2h^2$
SOR	$2\sqrt{2h}$
Single line (nine-point formulas for improved accuracy)	
Jacobi	h^2
Gauss-Seidel	$2h^2$
SOR	$2\sqrt{2h}$
Two-line SOR	$4h$

five-point difference formula. Similarly, in Table 10.2, some of the line iterative methods are compared.

The information presented in these two tables is generally not sufficient to make judgments about the efficiency of a particular method in a given application. However, it is useful to note that, in general, point iterative methods are easier to program and are more directly applicable to three-dimensional fields. For this reason, at the present time the successive point overrelaxation method is used most frequently. If the method of successive point overrelaxation is selected, there still remains the problem of determining a value for ω that is sufficiently close to the optimum value. Actually, convergence will take place over a very wide range of ω, but as illustrated in Figure 10.2, the number of iterations required for convergence differs enormously for relatively small differences of ω. G. E. Forsythe considers three practical approaches to estimating the optimum value for ω. The first method involves running the problem for a number of cycles using successive displacements with $\omega = 1$. From these runs he proposes a method of estimating the optimum ω. The second method involves the variation of ω over values near the optimum, that is, to carry out a series of experiments to pinpoint the optimum value. The third method involves an effort to approximate certain eigenvalues of a matrix describing a related problem with homogeneous boundary conditions ($u_B = 0$).

10.4. DIRECT VERSUS ITERATIVE METHODS

In the early days of digital computation, iterative methods became very popular primarily because they took much less storage than methods such as Gauss elimination. This fact made it practical to solve field problems on a stored program machine. Early iterative methods were primitive relaxation methods that converged very slowly unless an effective relaxation parameter could be found to accelerate convergence. There was no satisfactory method to resolve this problem until David Young developed a general theory of successive overrelaxation (SOR) that reduced the computation of the optimum relaxation parameter to a relatively straightforward algebraic evaluation involving eigenvalue bounds of a matrix called the "iteration matrix" associated with the iterative method under consideration.

The relevant eigenvalues are easily calculated only if matrix **A** exhibits what is called ***property A*** and if the grid is scanned in what is called a ***consistent order***. Property A can be heuristically explained as follows. Suppose the grid points are labeled black (B) and white (W) in a chessboard fashion. Now, if the difference approximation is such that the value of u at the black points can be evaluated exclusively in terms of the function values at the white points, or vice versa, then the matrix **A** is said to have property A. Thus, a five-point approximation to the two-dimensional Laplacian, for example, leads to a matrix with property A. However, a nine-point approximation to the two-dimensional biharmonic operator does not lead to a matrix with property A. It can be shown that if **A** has property A, then it is always possible to reorder the equations and unknowns such that the matrix of the reordered set has a tridiagonal or block-tridiagonal structure. If a matrix **A** has property A and if we are given a configuration of N grid points with the ordering indicated by the numbers $1, 2, \ldots, N$, then we can determine whether or not the matrix arising from a five-point difference equation is consistently ordered in the following way. Draw arrows between each pair of adjacent grid points in the direction of increasing labels. For each closed path consisting of mesh segments, if the number of arrows in the clockwise direction equals the number of arrows in the counterclockwise direction, then the corresponding matrix is consistently ordered. For a five-point difference approximation on a square mesh, the black-white ordering, the natural ordering, and ordering by diagonals all lead to consistently ordered matrixes.

The relevance of all this lies in the fact that, in property A matrixes that are consistently ordered, the eigenvalues λ of the SOR iteration matrix $(\mathbf{I} - \omega\mathbf{L})^{-1}\{(1 - \omega)I - \omega\tilde{\mathbf{U}}\}$ are related to the eigenvalues μ of the Jacobi iteration matrix $-(\tilde{\mathbf{L}} + \tilde{\mathbf{U}})$ by the equation

$$(\lambda + \omega - 1)^2 = \lambda\omega^2\mu^2, \qquad \text{if } |\mu| < 1 \tag{10.64}$$

Thus, if λ and μ are known, the relaxation parameter ω can be calculated. These ideas can be extended to the ADI methods also if the domain Ω is square and the matrix **A** can be split into the sum of commutative matrixes. It is useful to note that methods such as SOR have been successfully applied to equations not possessing property A, but few theoretical results concerning optimum ω are available.

The matrixes arising in finite element analysis in general do not possess property A, as they are obtained from an irregular subdivision of an irregular domain. Therefore, the theory of iterative methods developed for finite difference matrixes may not be applicable in general to finite element analysis. This fact probably played a key role in the widespread use of direct methods in finite element analysis. Of course, there are other reasons, also. For instance, storage is not as critical as for finite difference systems because, for a given accuracy, finite element systems tend to be lower in order. Finally, data management appears to be simpler for direct-elimination schemes than for iterative schemes.

10.5. VARIATIONAL METHODS

The iterative methods described in the preceding section all involve the assumption of an arbitrary solution distribution followed by the solution of each of the M equations, comprising the system of difference equations, for one of the unknowns. There results a new set of estimates for the solution at each of the grid points, which is then used to obtain even better estimates, and so on. An alternative mathematical approach begins likewise with the assumption of an arbitrary solution distribution. These values are inserted directly into $\mathbf{Au} = \mathbf{b}$, and a calculation is made as to the extent to which these equations are dissatisfied, that is, the extent to which the left side of these equations differs from the right side. An error term is thus calculated for each of the simultaneous equations. The assumed potentials are then modified so as to reduce or minimize this error term, and the procedure is repeated until convergence has been observed. Typically, the modification of an interior net point with coordinates i, j can be expressed as

$$u_{l,m}^{(i+1)} = u_{l,m}^{(i)} + \alpha_{l,m}^{(i)} p_{l,m}^{(i)} \tag{10.65}$$

where $u_{l,m}^{(i)}$ is the solution at the grid point after the ith iteration cycle, $u_{l,m}^{(i+1)}$ is the solution after the $(i + 1)$st cycle, $p_{l,m}$ is related to the error term after the ith iterative cycle, and $\alpha_{l,m}$ is a multiplier. Both $\alpha_{l,m}$ and $p_{l,m}$ depend upon the specific variational technique selected.

The best known and most general of the variational methods employing relations of the type of Eq. (10.65) is known as the ***method of steepest descent***. In this technique the correction term $\alpha_{l,m} p_{l,m}$ is given the value that will have the most pronounced minimizing effect upon the error. Methods of steepest descent and their application to simultaneous algebraic equations have been discussed in some detail by several authors. Their application to finite difference equations has been limited because of their relatively unfavorable convergence properties. The solution approaches the correct solution rather rapidly at first, but then tends to oscillate about the final value. On the other hand, steepest descent methods have the big advantage that they can be readily applied to nonlinear difference equations.

M. R. Hestenes and E. S. Stiefel have introduced a different variational approach, which has the advantage that the correct solution is approached monotonically in successive iteration steps; that is, the solution after each iteration represents a definite improvement over the solution of the preceding iterative step. Furthermore in this method, known as the ***method of conjugate gradients***, convergence is assured (in the absence of roundoff errors) in at most M iterations, where M represents the number of simultaneous equations. This method involves the use of improvement formulas of the type of Eq. (10.65) in which the terms $\alpha_{l,m}p_{l,m}$ in successive iterations are conjugates in the matrix sense.

The method of conjugate gradients is best discussed using matrix notation. Accordingly, the system of simultaneous algebraic equations to be solved is written as

$$\mathbf{Au} = \mathbf{b} \tag{10.66}$$

where $\mathbf{A}$ is a symmetric, positive-definite matrix. As a first step in the application of the conjugate gradient method, a guess is made as to the solution existing at all interior points; that is, $\mathbf{u}^{(0)}$ is arbitrarily assumed. The error in each equation, that is, the extent to which the left side and right side of the equation are unequal, is now calculated and expressed as a column matrix as

$$\mathbf{r}^{(0)} = \mathbf{b} - \mathbf{Au}^{(0)} \tag{10.67}$$

For this first step of the calculation *only*, the correction vector $\mathbf{p}$ in Eq. (10.65) is identical to the vector $\mathbf{r}$. That is,

$$\mathbf{p}^{(0)} = \mathbf{r}^{(0)} \tag{10.68}$$

The scalar multiplier α of Eq. (10.65) is calculated for the first step and every other step by the formula

$$\alpha^{(i)} = \frac{\mathbf{p}^{(i)T}\mathbf{r}^{(i)}}{\mathbf{p}^{(i)T}\mathbf{Ap}^{(i)}} \tag{10.69}$$

Here, the term $\mathbf{p}^{(i)T}$ signifies the transpose of the vector $\mathbf{p}^{(i)}$. Equation (10.65) can now be employed to predict the $(i + 1)$st set of values for the unknown $\mathbf{u}$. Accordingly,

$$\mathbf{u}^{(i+1)} = \mathbf{u}^{(i)} + \alpha^{(i)}\mathbf{p}^{(i)} \tag{10.70}$$

Equation (10.70) marks the end of the $(i + 1)$st iterative cycle. To embark on the succeeding cycle, it becomes necessary to derive new values for $\mathbf{r}$, α, and $\mathbf{p}$. The first step again is to calculate the error or residual vector $\mathbf{r}$:

$$\mathbf{r}^{(i+1)} = \mathbf{b} - \mathbf{Au}^{(i+1)} \tag{10.71}$$

After some algebraic manipulation, Eq. (10.71) can be written more conveniently as

$$\mathbf{r}^{(i+1)} = \mathbf{r}^{(i)} - \alpha^{(i)}\mathbf{Ap}^{(i)} \tag{10.72}$$

Now $\mathbf{r}^{(i+1)}$ is employed to derive a correction vector for $\mathbf{p}$;

$$\beta^{(i)} = -\frac{\mathbf{p}^{(i+1)T}\mathbf{Ap}^{(i)}}{\mathbf{p}^{(i)T}\mathbf{Ap}^{(i)}} \tag{10.73}$$

The new value for $\mathbf{p}$ can then be written as

$$\mathbf{p}^{(i+1)} = \mathbf{r}^{(i+1)} + \beta^{(i+1)}\mathbf{p}^{(i)} \tag{10.74}$$

$\alpha^{(i+1)}$ is then determined using an expression of the type of Eq. (10.69), in which all superscripts become $(i + 1)$. This is followed by a renewed application of Eq. (10.70).

Some of the salient features of this method are the following: (1) it requires at most M iteration steps; (2) the entire matrix $\mathbf{A}$ need not be stored as an array in memory; at each stage of the iteration it is necessary to compute only a product such as $\mathbf{Az}$ for a given vector $\mathbf{z}$. Unfortunately, the initial interest and excitement in the conjugate gradient method was dissipated, because in practice the numerical properties of the algorithm differed from the theoretical ones. For instance, even for small systems of equations ($M \leq 100$) the algorithm did not necessarily terminate in M iterations. In addition, for large systems of equations arising from a finite difference approximation of two-dimensional elliptic equations, competing methods such as the SOR required only $O(\sqrt{M})$ iterations to achieve a prescribed accuracy. For these reasons, this method was discarded in the 1960s as a useful method for solving linear equations except in conjunction with other methods. Interest in this method, however, continues to persist for solving nonlinear equations.

In recent times, there is a renewed interest in this method. It appears that the conjugate gradient method can be effectively used *as an iterative method* to solve large sparse systems. The primary attraction of this approach lies in two basic features. First, it does not require an estimation of parameters such as in SOR method. Second, this approach requires fewer restrictions on matrix $\mathbf{A}$ for optimal behavior than the SOR method. Current literature should be consulted for further details on these ideas.

SUGGESTIONS FOR FURTHER READING

1. *References to direct methods to solve linear algebraic equations are scattered throughout the literature. It is difficult to give a concise and complete list here. However, some general references are*

   J. H. Wilkinson, *Rounding Errors in Algebraic Processes*, H. M. Stationary Office, London, 1963.

   J. H. Wilkinson, *The Algebraic Eigenvalue Problem*, Oxford University Press, Inc., New York, 1965.

   G. E. Forsythe and C. B. Moler, *Computer Solution of Linear Algebraic Systems*, Prentice-Hall, Inc., Englewood Cliffs, N.J., 1967.

2. *Two standard references for iterative methods are*

   R. S. Varga, *Matrix Iterative Analysis*, Prentice-Hall, Inc., Englewood Cliffs, N.J., 1962.

   D. M. Young, *Iterative Solution of Large Linear Systems*, Academic Press, Inc., New York, 1971.

3. *For good source material on sparse matrix computations, consult*

   J. R. Bunch and D. J. Rose, *Sparse Matrix Computations*, Academic Press, Inc., New York, 1976.

4. *An authoritative description of the conjugate gradient method, developed by Hestenes, can be found in*

M. R. HESTENES, "The Conjugate Gradient Method for Solving Linear Systems," Proc. Symposium in Applied Mathematics, Vol. 6, pp. 83–102, *Numerical Analysis*, McGraw-Hill Book Co., New York, 1956.

5. *The field of numerical methods to solve partial differential equations is indeed very dynamic. For latest developments in algorithms as well as hardware architectures, consult*

R. VICHNEVETSKY (ed.), *Proceedings of the IMACS International Symposium in Computer Methods for Partial Differential Equations.* Copies of past proceedings can be obtained from the editor, Department of Computer Science, Rutgers University, New Brunswick, N.J., 08903, USA.

EXERCISES

1. (a) Write a FORTRAN program, in the form of a subroutine named ELIM, to solve a system $\mathbf{Ax} = \mathbf{b}$ by Gaussian elimination with partial pivoting. The program should allow multiple right sides. The subroutine receives from the main program four parameters called AB, N, NP, NDIM, where AB is the coefficient matrix augmented by the right side, N is the number of equations, NP is the total number of columns in the augmented matrix, and NDIM is the first dimension of the matrix AB in the calling program. The solution is to be returned in the space of the augmentation columns.

(b) What is the storage requirement for this problem for the case $n = 40$.

2. Some computer programs do not actually interchange the rows in partial pivoting. Alternatively, one keeps track of the order in which the rows are to be used in a vector whose elements represent row order. When a row interchange is required, only the elements in this ordering vector are changed. Discuss the advantages and disadvantages of this strategy by taking into your discussion issues such as computing time, program complexity, and the size of the matrix **A**.

3. (a) Write a program (in the form of a subroutine named TRIDG) to solve $\mathbf{Ax} = \mathbf{b}$ by Gaussian elimination, where **A** is now a tridiagonal matrix. Store the coefficients of **A** and the right side in an $n \times 4$ array, named AR. The first column of AR stores the lower diagonal, the second column of AR stores the main diagonal, the third column stores the upper diagonal, and the fourth column stores the entries of the right side. The parameters AR, N, and NDIM are passed from the main program, and the solution vector is returned to the fourth column of AR. The parameters N and NDIM are defined in Problem 1.

(b) What is the storage requirement for this problem for $n = 500$? Is this more or less than the one calculated in Problem 1.

4. Consider the problem of solving $\mathbf{Ax} = \mathbf{b}$, where **A** is in the banded format shown in (a) of the accompanying figure, where the X's stand for nonzero entries. Store this matrix in the format shown in (b). With this storage strategy, write a program to solve $\mathbf{Ax} = \mathbf{b}$ using Gaussian elimination.

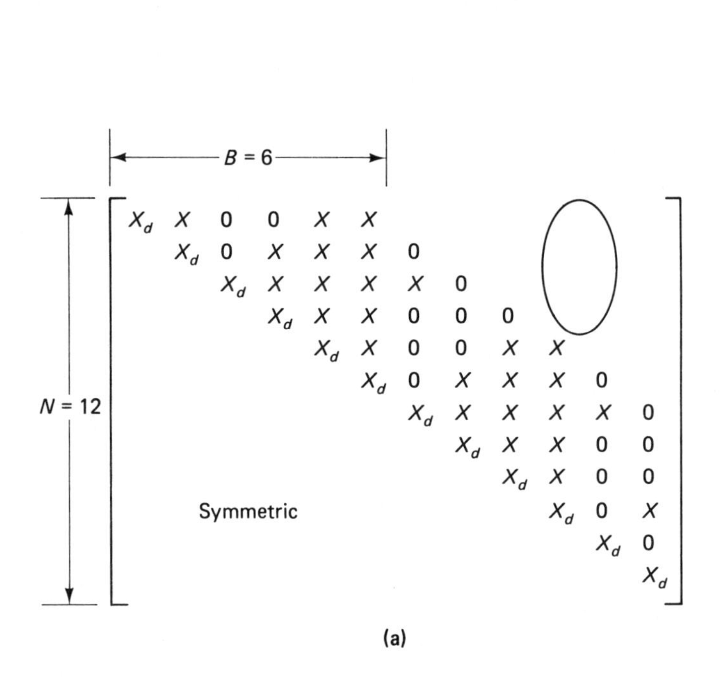

(a)

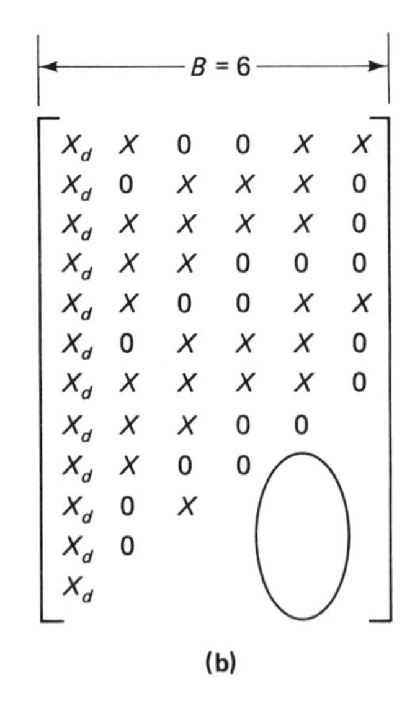

(b)

5. (a) Solve

$$\begin{bmatrix} 200 & -100 & 0 \\ -100 & 200 & -100 \\ 0 & -100 & 100 \end{bmatrix} \begin{bmatrix} x_1 \\ x_2 \\ x_3 \end{bmatrix} = \begin{bmatrix} -8 \\ -8 \\ -8 \end{bmatrix}$$

(*Answer:* $x_1 = -0.24$, $x_2 = -0.40$, $x_3 = -0.48$)

(b) Test the programs written in the previous exercises by using this as a test case.

6. (a) Write a subroutine, named GSIRN, to solve a system of n linear equations $\mathbf{Ax} = \mathbf{b}$ by Gauss-Seidel iteration. The parameters are A = coefficient matrix with rows rearranged such that the longest entries fall on the main diagonal (as far as possible), B = right side vector, X = first stores the initial guess and finally stores the answer, N = the number of equations, NDIM = first dimension of **A** in the calling program, NITER = a limit on the number of iterations, and TOL = a test value to stop the iterations.

(b) Test by first solving

$$\begin{aligned} 8x_1 + x_2 - x_3 &= 8 \\ 2x_1 + x_2 + 9x_3 &= 12 \\ x_1 - 7x_2 + 2x_3 &= -4 \end{aligned}$$

using paper and pencil.
(*Answer:* $x_1 = x_2 = x_3 = 1$)

7. B. Parlett in a paper (*SIAM Review*, Vol. 20, No. 3, pp. 443–456, 1978) presents the following quotation from an unknown source. (FE stands for finite-element.)

> Iterative techniques for processing large sparse linear systems were popular in the late 1950's and early 1960's (*and their decaying remains still pollute some computational circles*). When iterative methods finally departed from the finite element scene in the mid 1960's—having been replaced by direct sparse-matrix methods—the result was a quantum leap in the *reliability* of linear analysis packages, which contributed significantly to the rapid acceptance of FE analysis at the engineering group level. (This effect, it should be noted, had nothing to do with the relative Computational Efficiency, in fact iterative methods can run faster on many problems if the user happens to know the optimal acceleration parameters.) Presently, linear FE analyzers are routinely exercised as black box devices. . . .

Write a term paper touching upon the pros and cons of direct and iterative methods in the context of finite element and finite difference methods.

8. Iterative methods are often qualified by adjectivial prefixes such as linear, stationary, nonlinear, nonstationary, semi-, point, line, group, second-degree, and so on. Discuss briefly, giving examples, the meaning and scope of these methods.

9. By referring to the books by R. S. Varga and D. M. Young (see item 2, Suggestions for Further Reading), if necessary, explain the meaning of the following terms.

vector norm	irreducibility
matrix norm	diagonal dominance
spectral radius of a matrix	condition number
spectral norm of a matrix	average rate of convergence
consistent ordering	asymptotic rate of convergence

10. Test the relative rates of convergence of the Gauss-Seidel method and the SOR method by doing the problem described in Eq. (10.49). Use a hand calculator.

11. Figure 10.2 shows the variation of the spectral radius with ω. However, the discussion following Eq. (10.51) refers to the variation of the number of iterations, for convergence, with ω. What is the relation between the number of iterations for convergence and the spectral radius while solving $\mathbf{Ax} = \mathbf{b}$? The spectral radius of what matrix is of interest here?

12. Apply the conjugate gradient method, described in the text, to solve

$$3x_1 + x_2 = 5$$
$$x_1 + 2x_2 = 5$$

and demonstrate that the correct solution can be found in two iterative cycles.

13. Design a program that takes a matrix **A** and tests it for property A and consistent ordering. Test your program with suitably chosen test data.

14. Write a term paper on the use of linked-list data structure in representing and manipulating sparse matrixes. Write a few programs to implement your ideas. Discuss the relative merits of FORTRAN, PL/1 and PASCAL in this context.
(*Hint:* See E. Horowitz and S. Sahni, *Fundamentals of Data Structures*, Computer Science Press, Inc., Woodland Hills, Ca., 1976.)

11

Software Packages for Partial Differential Equations

11.1 INTRODUCTION

It was recognized during the early years of the development of electronic digital computers that the programming and coding of such computers would be inordinately difficult and error-prone without the availability of programming aids. As a result, there gradually emerged what has come to be known as a hierarchy of languages and software packages, all designed to facilitate the programming effort. At the lowest level of this hierarchy is the machine language, reflecting the command structure and detailed design of the computer. This is in fact the only "language" that a computer can recognize directly and requires that all instructions be provided as strings of "ones" and "zeros" and that all memory addresses likewise be provided in binary form. At the next level of the language hierarchy we find the ***assemblers***. Assembly languages permit a more compact and easily recognized abbreviation of each command and are available for virtually every commercial digital computer.

As far as most computer users are concerned, the most important programming aids are found at the third level of the hierarchy, the ***programming languages***. These include the scientific programming languages, which facilitate the preparation of a program in a compact manner using expressions and statements (syntax) that are relatively easy to learn and understand by scientifically oriented programmers. Languages falling into this category include FORTRAN, PL/1, ALGOL 60, ALGOL 68, and PASCAL. Compilers are available for each of these languages to translate the program prepared by the programmer into the assembly or machine language of the digital computer.

An overwhelming majority of all the programs prepared for the solution of partial differential equations are written in FORTRAN. Although by no means the most suitable language for this application, FORTRAN was the first language to reach maturity, so by the time that improved languages became available, a large investment in FORTRAN programs had already been made. Conversion to another language was deemed by many users to be excessively expensive. Furthermore, FORTRAN compilers are available for virtually all large digital computers, while other languages are not nearly as widely disseminated.

A variety of programming packages and programming systems has been prepared to facilitate the programming of partial differential equations in general-purpose scientific programming languages, such as FORTRAN, or to overcome some of the limitations of these languages. Some of these programming aids take the form of subroutines or libraries of subroutines that can be called by the FORTRAN programmer as required. These may include, for example, subroutines for successive overrelaxation, for the solution of tridiagonal linear algebraic systems, or for the Crank-Nicolson method. The availability of such subroutines can save substantial programming time and also avoids the introduction of programming errors. A second class of programming aids is located at the fourth level of the software hierarchy, the ***higher-level*** programming languages. A number of such higher-level languages has been developed for partial differential equations. These permit the user to formulate the problem in a more convenient and direct form than is possible in FORTRAN and provide a variety of algorithms for the solution of the equations, algorithms that may be selected by the user or by default. Frequently, such languages are accompanied by translator programs that produce a program in FORTRAN or another third-level language. A final category of programming aids for partial differential equations is comprised of programming systems prepared for users from a specific application area. Thus there are special software systems to facilitate the solution of the partial differential equations as they arise in structural engineering, fluid mechanics, meteorology, and so on.

Section 11.2 contains a brief description of three higher-level languages that have been developed to solve a wide variety of partial differential equations, together with some reasons that these languages were unable to acquire a wide acceptance on the part of the user community. In Section 11.3, a number of programming systems for the solution of elliptic partial differential equations or algebraic equations of the type that arise in the treatment of elliptic partial differential equations are discussed. Section 11.4 is devoted to the consideration of programming packages for parabolic and hyperbolic partial differential equations. Most of these packages use the "method of lines." Brief references to packages developed for specific application areas such as structural engineering and nuclear reactors are made in Section 11.5.

11.2. GENERAL-PURPOSE HIGHER-LEVEL LANGUAGES

Higher-level languages utilize syntax and semantics not found in third-level languages such as FORTRAN. They permit the user to make statements that are particularly convenient for the solution of partial differential equations in a very compact form. They therefore facilitate programming, particularly by inexperienced programmers, and thereby greatly reduce the time actually spent in programming and debugging. A single higher-level language that would be useful for all types of partial differential equations arising in science and engineering has been a tantalizing prospect for many years. In the 1960s a number of such languages were proposed and implemented, and three of these are described here in order to demonstrate their general approach. None has stood the test of time. While they were successful in solving relatively simple and straightforward elliptic, parabolic, and hyperbolic partial differential equations, and were therefore useful in an academic or teaching environment, they were too inefficient to compete with specialized program packages for the large "practical" problems. Invariably, in designing a general-purpose programming system, speed in the execution of the programs is sacrificed for programming convenience. The more general a package, the more inefficient is it likely to be for specific problems. For small problems, the loss of efficiency in executing programs is of minor importance, while the ability to permit an inexperienced programmer to obtain meaningful outputs is a great attraction. For the large problems that arise in fluid mechanics, nuclear reactor technology, structural engineering, and the like, on the other hand, it is of vital economic importance to make each solution run as brief as possible. Programming ease is therefore necessarily sacrificed, and only experienced programmers are employed.

PDEL

One of the first general-purpose languages to be developed in the late 1960s was designed by Cardenas and Karplus [1] at the University of California. It was designed as a fourth-level language and provided with a translator program to translate the source PDEL program into a third-level language. The third-level target language employed was PL/1, so that PDEL can be employed using any digital computer equipped with a PL/1 compiler. The utilization of PDEL takes the following steps:

1. The problem is formulated as a partial differential equation, together with all applicable boundary and initial conditions.
2. A finite difference grid is defined for the space as well as the time domains.

3. The mathematical model is expressed in PDEL as a sequence of formalized statements, and a separate punched card is prepared for each statement for digital computer input.

4. The PDEL translator is employed to convert this program into PL/1.

5. The regular PL/1 compiler is employed to translate the resulting program into assembly and eventually machine language.

6. The program now in machine code is executed, and the required solution is printed out.

PDEL statements are written in PL/1 preprocessor language; that is, they have the same syntax as PL/1 statements but are preceded by a % sign. There are ten functionally different types of statements: equation, dimension, parameter, geometry, boundary conditions, initial condition, control, output, declare, and comment. These statements are used to specify (1) whether the equation being solved is elliptic, parabolic, hyperbolic, or biharmonic; (2) whether the problem is formulated in one, two, or three space dimensions; (3) the parameters of the problem, which may be constants, time-varying, functions of the space variables, or functions of the dependent variable; (4) the geometry of the field, which may be regular or irregular; (5) the boundary conditions; (6) the initial conditions; (7) the finite difference grid spacings along the space and time coordinates; and (8) the nature of the desired printout. A number of different output formats are provided. These include numerical printouts of the potentials at all node points at the selected time levels, as well as a novel alphabetical printout. In the latter case, the dynamic range of the dependent variable is divided into 26 equally spaced levels, and a different letter of the alphabet is assigned to each of these levels.

As a simple example of the application of PDEL to a nonlinear partial differential equation, consider

$$\begin{aligned}
&\frac{\partial}{\partial x}\left(\sigma \frac{\partial \phi}{\partial x}\right) = k\frac{\partial \phi}{\partial t} \\
&\qquad \sigma = 1.0 \qquad k = 0.08\phi, \qquad 0 \le x \le 10 \\
&\qquad \phi(0, t) = \phi(10, t) = 0 \\
&\qquad \phi(x, 0) = 0.08\,\frac{5x - x^2}{2}
\end{aligned} \tag{11.1}$$

The x domain is to be divided into 20 equal intervals, and a time step of 0.125 second is to be used. It is desired to obtain a printout of the solution at every other point in the x domain for $t = 0$, 2, and 4, and a plot of ϕ versus t showing the value of the solution at every grid point for $t = 0, 0.125, \ldots, 4.00$. The following is the PDEL program that is prepared by the user.

PDEL Program	Statement #
`/ *PDEL PROGRAM FOR A NONLINEAR DIFFUSION EQUATION* /`	0
`% INCLUDE $PDEL(INITIAL);`	1
`% DIMENSION = '1';`	2
`% DECLARE (SIGMAX, KAPPA1) CHARACTER;`	3
`% EQUATION='SIGMAX*PX,PX,PHI=KAPPA1*PT,PHI';`	4
`% KAPPA1 = '0.08/PHI';`	5
`% SIGMAX = '1.0';`	6
`% SCOND='(*)=0.08*((5*X)-(X**2)/2)';`	7
`% GRIDPOINTSX='20';`	8
`% GRIDPOINTST = '32';`	9
`% BCOND = ' (0) = 0.0 ; (20) = 0.0';`	10
`% DELTAX = '0.5';`	11
`% DELTAT = '0.125';`	12
`% PRINTINTX = '2';`	13
`% PRINTINTT = '16';`	14
`% PLOTINTX = '2';`	15
`% PLOTINTT = '1';`	16
`% INCLUDE $PDEL(HEART);`	17

The statements above play the following role:

0	a comment
1	calls from the computer system the initialization part of the translator
2	defines spatial dimensionality
3	declares each parameter before being used
4	defines the equation
5	defines the parameter k
6	defines the parameter σ
7	indicates that the initial conditions at all grid points are defined by the function states
8	indicates the number of grid points in the x-direction
9	indicates the number of steps in the t-direction to be used
10	indicates that the boundary potentials are 0.0
11	indicates the spacing between grid points in the x-direction
12	indicates the spacing between grid points in the t-direction
13	indicates that the solution at every other grid point in the x-direction is to be printed out
14	indicates that the solution at every eight time levels (i.e., at $t = 0$, 2, and 4) is to be printed out

15, 16 indicate that the solution at every other point in the x-direction is to be plotted versus time at every time step (i.e., at $t = 0.125, 0.250, \ldots, 4.00$)

17 calls the part of the PDEL translator that performs the processing of the PDEL program

For a parabolic equation in two space dimensions such as

$$\frac{\partial}{\partial x}\left(\sigma_x \frac{\partial \phi}{\partial x}\right) + \frac{\partial}{\partial y}\left(\sigma_y \frac{\partial \phi}{\partial y}\right) = k\frac{\partial \phi}{\partial t} \tag{11.2}$$

$$k = f(\phi), \qquad 0 \le x \le X, \qquad 0 \le y \le Y$$

the applicable equation statement is

```
% EQUATION='SIGMAX*PX,PX,PHI+SIGMAY*PY,PY,PHI=K*PT,PHI';
```

By 1970, the equations shown in Table 11.1 had been implemented. Subsequently, the system was expanded to provide for equations formulated in cylindrical and spherical coordinates as well as biharmonic equations in one space dimension.

Table 11.1. Numerical algorithms used in PDEL to solve the basic equations

Type	*Partial Differential Equation*	*Numerical Algorithm*
Elliptic 1 dimension	$\frac{\partial}{\partial x}\left(\sigma \frac{\partial \phi}{\partial x}\right) = K$	Tridiagonal algorithm
Elliptic 2 dimensions	$\frac{\partial}{\partial x}\left(\sigma \frac{\partial \phi}{\partial x}\right) + \frac{\partial}{\partial y}\left(\sigma \frac{\partial \phi}{\partial y}\right) = K$	Successive point over-relaxation
Elliptic 3 dimensions	$\frac{\partial}{\partial x}\left(\sigma \frac{\partial \phi}{\partial x}\right) + \frac{\partial}{\partial y}\left(\sigma \frac{\partial \phi}{\partial y}\right) + \frac{\partial}{\partial z}\left(\sigma \frac{\partial \phi}{\partial z}\right) = K$	Successive point over-relaxation
Parabolic 1 dimension	$\frac{\partial}{\partial x}\left(\sigma \frac{\partial \phi}{\partial x}\right) = K\frac{\partial \phi}{\partial t} + F$	Crank-Nicolson, tri-diagonal algorithm
Parabolic 2 dimensions	$\frac{\partial}{\partial x}\left(\sigma \frac{\partial \phi}{\partial x}\right) + \frac{\partial}{\partial y}\left(\sigma \frac{\partial \phi}{\partial y}\right) = K\frac{\partial \phi}{\partial t} + P$	Alternating direction and tridiagonal algorithm
Parabolic 3 dimensions	$\frac{\partial \phi}{\partial x}\left(\sigma \frac{\partial \phi}{\partial x}\right) + \frac{\partial}{\partial y}\left(\sigma \frac{\partial \phi}{\partial y}\right) + \frac{\partial}{\partial z}\left(\sigma \frac{\partial \phi}{\partial z}\right) = K\frac{\partial \phi}{\partial t} + F$	Alternating direction and tridiagonal algorithm
Hyberbolic 1 dimension	$\frac{\partial}{\partial x}\left(\sigma \frac{\partial \phi}{\partial x}\right) = \frac{\partial}{\partial t}\left(K \frac{\partial \phi}{\partial t}\right) + K\frac{\partial \phi}{\partial t} + F$	Von Neumann and tri-diagonal algorithm
Hyperbolic 2 dimensions	$\frac{\partial}{\partial x}\left(\sigma \frac{\partial \phi}{\partial x}\right) + \frac{\partial}{\partial y}\left(\sigma \frac{\partial \phi}{\partial y}\right) = \frac{\partial}{\partial t}\left(K \frac{\partial \phi}{\partial t}\right) + K\frac{\partial \phi}{\partial t} + F$	Lees algorithm and tri-diagonal algorithm

J, K and F can be functions of x, y, z, t and ϕ Boundary Conditions: Dirichlet and Neumann

LEANS III

This language, developed by W. E. Schiesser [2] at Lehigh University, is based on FORTRAN, as are virtually all the other simulation packages discussed in this chapter. The system was designed to handle a broad class of partial differential equations, as well as mixed ordinary and partial differential equation models, which frequently arise in chemical engineering. For integration in the time domain, the method-of-lines algorithm was implemented.

To implement any specific mathematical model, the user of LEANS is required only to specify the parameters of the following partial differential equations:

$$A_2 \frac{\partial^2 u}{\partial t^2} + A_1 \frac{\partial u}{\partial t} = \frac{\partial (A_3 u)}{\partial x} + \left(\frac{1}{x^a}\right) \frac{\partial [x^a A_4 (\partial u/\partial x)]}{\partial x} + A_5 u + A_6 \qquad (11.3)$$

The coefficients $A_1, A_2, \ldots, A_6$ are programmable by the user and may be functions of the independent variables x and t and the dependent variable u. The coordinate factor a, programmable by the user, which may have the values 0, 1, or 2 for Cartesian, cylindrical, and spherical coordinates, respectively. A few special cases of Eq. (11.3) indicate the spectrum of equations that it encompasses:

Case I $A_2 = A_3 = A_5 = A_6 = 0,\ A_1 = A_4 = 1.$

Case Ia: $a = 0$ (parabolic heat conduction equation in Cartesian coordinates).

$$\frac{\partial u}{\partial t} = \frac{\partial^2 u}{\partial x^2}$$

Case Ib $a = 1$ (parabolic heat conduction equation in cylindrical coordinates).

$$\frac{\partial u}{\partial t} = \frac{\partial^2 u}{\partial x^2} + \frac{1}{x}\frac{\partial u}{\partial x}$$

Case Ic $a = 2$ (parabolic heat conduction equation in spherical coordinates).

$$\frac{\partial u}{\partial t} = \frac{\partial^2 u}{\partial x^2} + \frac{2}{x}\frac{\partial u}{\partial x}$$

Case II $A_1 = A_3 = A_5 = A_6 = 0,\ A_2 = A_4 = 1$; for $a = 0$ (hyperbolic wave equation).

$$\frac{\partial^2 u}{\partial t^2} = \frac{\partial^2 u}{\partial x^2}$$

Case III $A_2 = A_4 = A_6 = 0,\ A_1 = A_3 = A_5 = 1$ (first-order hyperbolic flow equation).

$$\frac{\partial u}{\partial t} = \frac{\partial u}{\partial x} + u$$

Equation (11.3) has an associated boundary condition:

$$B_1 \frac{\partial x}{\partial u} + B_2 u = B_3 \qquad (11.4)$$

where the coefficients B_1, B_2, and B_3 are programmable by the user and may be a function of the independent variable u. For example, $B_2 = u^3$, $B_3 = u_a^4$ corresponds to a nonlinear fourth-power radiation boundary condition.

Systems of simultaneous equations of the form of Eq. (11.3) are allowed along with ordinary differential equations of the initial-value type.

PDELAN

Gary and Helgerson [3] at the National Center for Atmospheric Research developed a higher-level language capable of handling more complicated numerical schemes with greater efficiency than is attained in PDEL or LEANS. A preprocessor is employed to convert the source program into a FORTRAN program. Unlike the preceding two languages, PDELAN is procedural in character, so all statements in the source program must be in the correct order. Moreover, it does not select any finite difference or integration scheme, but rather provides mesh and operator statements to simplify the coding of such schemes.

PDELAN contains three basic declarations, MESH, VARIABLE, and OPERAND. The first of these is employed to declare the dimensions of the finite difference grid. The second is employed to declare the dependent variables associated with each finite difference grid point. Since meshes are rectangular, variable statements can be translated directly into FORTRAN DIMENSION statements. The OPERATOR statement is employed to define the domain and the range of dependent variables or functions of such variables. A DOMESH statement is employed to operate upon previously defined meshes so that these can be translated into FORTRAN DO loops. This is illustrated in the following example taken from the PDELAN user's manual.

Consider the following simple differential equation with the given initial and boundary conditions

$$\frac{\partial u}{\partial t} + u\frac{\partial u}{\partial x} = 0 \tag{11.5}$$

$$u(x, 0) = f(x), \qquad 0 \leq x \leq 1$$

$$u(0, t) = g(t), \qquad 0 \leq T$$

We assume a mesh given by $x_i = (i - 1)\,\Delta x$, $t_n = n\,\Delta t$, where $1 \leq i \leq M$, $\Delta x = 1/(M - 1)$, $n \geq 0$. We make the definitions

$$u_{i,n} = u(x_i, t_n)$$

$$\delta(u)_i = \frac{u_i - u_{i-1}}{\Delta x}, \qquad 2 \leq i \leq M$$

Then we can write a simple finite difference scheme for this problem as follows:

$$\frac{u_{i,n+1} - u_{i,n}}{\Delta t} + u_{i,n}\delta(u_n)_i = 0$$

$$u_{i,0} = f(x_i), \qquad 1 \leq i \leq M$$

$$u_{1,n} = g(t_n), \qquad 1 \leq n$$

This can be rewritten as

$$u_{i,0+1} = u_{i,n} - \Delta t u_{i,n}\,\delta(u_n)_i$$
$$u_{i,0} = f(x_i)$$
$$u_{1,n} = g(t_n)$$

To write this in the PDELAN language, we must declare the mesh (x_i) and the operator δ. We represent the values $u_{i,n+1}$ by U2 (I) and $u_{i,n}$ by U1 (I). We assume $M = 100$. The program can then be written as follows:

```
C           SIMPLE PDE PROGRAM
      PROGRAM B1A
      MESH MA(100)
      VARIABLE (U1,U2) ON MA
      OPERATOR (MA) TO (MA(2),(100))       IS
     1 DX(W,I)=(W(I)-W(I-1))/DLX
C           INITIALIZE
      M=100 $ DLX=1./(M-1) $ NSTEP=100 $ DLT=.25*DLX
      DOMESH 10 MA(K)
      X=(K-1)*DLX
 10   U1=F(X)
      N=0
C           TIME STEP LOOP
 20   T=N*DLT
      U1(1)=G(T)
      DOMESH 30 MA(K=(2,100))
 30   U2=U1-DLT*U1*DX(U1)
      N=N+1
      DOMESH 40 MA(K)
 40   U1=U2
      IF(N .LE. NSTEP)GOTO 20
      END
```

The equivalent FORTRAN program is:

```
C           SIMPLE PDE PROGRAM -FORTRAN VERSION
      PROGRAM B1B
      DIMENSION U1(100),U2(100)
      M=100 $ DLX=1./(M-1) $ NSTEP=100 $ DLT=.25*DLX
C           INITIALIZE
      DO 10 K=1,M
      X=(K-1)*DLX
 10   U1(K)=F(X)
      N=0
C           TIME STEP LOOP
 20   T=N*DLT
      U1(1)=G(T)
      DO 30 K=2,M
```

```
30    U2(K)= U1(K) - DLT*U1(K)*(U1(K)-U1(K-1))/DLX
      N=N+1
      DO 40 K=1,M
40    U1(K)=U2(K)
      IF(N .LE. NSTEP)GOTO 20
      END
```

The MESH and OPERATOR declarations do not result in any output. The VARIABLE statement produces the following output:

```
DIMENSION U1(100),U2(100)
```

The DOMESH 10 statement generates the following DO statement:

```
DO 10 K = 1,100
```

Within the range of this DOMESH, any occurrence of an unsubscripted mesh variable will have the mesh subscript K added to it. Thus U1 becomes U1 (K).

The expansion of mesh operators (DX in this example)is central to the language. This is illustrated in the expansion of the range of the DOMESH 30 statement. This statement generates the following DO statement:

```
DO 30 K = 2,100
```

The statement labeled 30 is expanded by adding the subscript K to any occurrence of the mesh variables U1 and U2. The operator variable DX (U1) is expanded as follows. First DX (U1) is replaced by the defining expression for DX, which appears in the OPERATOR statement, that is,

```
(W(I)-W(I-1))/DLX
```

In this expression the ***marker*** W is replaced by the ***expression argument*** of the operator variable DX (U1), in this case U1. Thus, W is replaced by U1. The subscript for U1 is obtained by substitution of the mesh subscript K (found in the DOMESH statement) for the ***subscript argument*** I found in the defining expression. Thus, the preceding defining expression for DX expands to

```
((U1(K)-U1(K-1))/DLX)
```

Note that extra parentheses are added around the defining expression (superfluous in this case). Statement 30 expands to the following:

```
30    U2(K) = U1(K) - DLT*((U(K)-U(K-1))/DLX)
```

The expansion of the DOMESH range is the central part of the language.

11.3. SOFTWARE SYSTEMS FOR ELLIPTIC PARTIAL DIFFERENTIAL EQUATIONS

By the late 1970s most efforts to produce general-purpose partial differential equation packages had been abandoned or reduced in scope. Instead, numerous efforts were initiated to provide powerful software systems for specific classes of problems. With very few exceptions, these employ FORTRAN and are designed to permit the programmer to construct programs by linking suitable FORTRAN subroutines. Some of these subroutines are supplied with the programming system, and others must be specified by the user. This makes it necessary for the user to have considerable experience in FORTRAN programming and an understanding of the algorithms employed in the various subroutines. Machura and Sweet [4] provide an extensive survey of the state of the art of software systems for partial differential equations as of 1979.

Because elliptic partial differential equations occur in most physical and engineering disciplines, a large number of software systems have been designed for this general problem area. Many of these consist actually of subroutine packagcs for the solution of systems of algebraic equations. Others are designed to facilitate the formulation of these equations using finite difference or finite element approximations. The following discussion constitutes a sampling of programming systems developed in the late 1970s.

ELLPACK

Under the general coordination and leadership of J. R. Rice [5], the ELLPACK software project is directed to providing a framework for the development and evaluation of algorithms for the solution of partial differential equations. The emphasis in the late 1970s was on the solution of large, sparse linear systems of algebraic equations such as those arising in the solution of linear elliptic partial differential equations. The general approach is one of modularization. The overall programming task is broken up into five distinct stages, each including a number of programming tasks, and implemented as modules. The interface between these modules is carefully defined and fixed, thus permitting users to design and test individual modules for their requirements and to integrate them into the overall system. All existing modules are available to all users.

As an example of the ELLPACK approach, we briefly describe a program discribed by Houstis and Rice [6]. The program is designed to employ the method of collocation using bicubic piecewise Hermite polynomials for the solution of the linear two-dimensional elliptic equation

$$\alpha u_{xx} + 2\beta u_{xy} + \gamma u_{yy} + \delta u_x + \epsilon u_y + \zeta u = f \tag{11.6}$$

defined on a general two-dimensional domain and subject to mixed-type boundary conditions:

$$a u_x + b u_y + c u = g \qquad \text{on the boundary}$$

The following are the five major modules:

Input Module. The user must specify:

1. Region boundary definition: user supplied routine

```
SUBROUTINE BCOORD(P,X,Y)
```

to define the boundary as a parametric curve; the X, Y values are returned corresponding to the parametric value P. Boundaries with an arbitrary number of pieces are allowed.

2. A rectanguar mesh.
3. Problem definition: user supplied FORTRAN functions

```
FUNCTION COEF(X,Y,J),    J = 1, 2, . . . , 6 for α, β, . . . , ζ
FUNCTION BCOEF(X,Y,J),   J = 1, 2, 3 for a, b, c
```

which evaluate $\alpha, \beta, \ldots, \zeta$, a, b, c, and

```
FUNCTION F(X,Y,J),    J = 1, 2 for f, g,
```

which computes the functions f, g in the differential and boundary operators.

Region Module: Subroutine REGION. This module locates the region with respect to the grid and provides various information useful in further processing. The output from this module consists of grid specifications and boundary specifications in the form of arrays that include coordinates of boundary intersection points, their type, neighboring grid points, and so on.

Equation Construction Module. This module consists of three components:

1. SUBROUTINE CLDATA, which identifies the type of each grid element, calculates the number of boundary collocation points associated with each element, and calculates the boundary collocation points and the modified nodes of each element.
2. SUBROUTINE INDEXING, which numbers the nodes of the final mesh to be used later for the approximation of the differential operator, determines the element incidences, and estimates the bandwidth.
3. SUBROUTINE FORMEQ, which generates the system of the collocation equations.

Equation Solving Module: Subroutine BNDSOL. This module solves the system of collocation equations using Gaussian elimination.

Output Module. By appropriate commands the user can obtain

1. The output generated by the REGION module.

2. The input data to the subroutine FORMEQ.
3. The values of the approximate solution U and its derivatives U_x, U_y, U_{xy} evaluated at the nodes or at any set of points supplied by the user.
4. The execution time of each module.

Reference 6 includes examples of the application of this program to three relatively complicated field problems and includes the computer output obtained by the preceding procedure.

ITPACK

At the University of Texas, a group of researchers is engaged in the development of programs for the solution of linear, self-adjoint elliptic partial differential equations of the type

$$(au_x)_x + (bu_y)_y + cu = f \tag{11.7}$$

where a, b, c may be constants or functions of x and y. One objective of the ITPACK effort is to provide modules to be used within the ELLPACK system described in the preceding paragraph. To this end, the ITPACK codes are themselves organized into individual modules having the following functions:

Grid Definition

Generation of Nonzero Coefficients of the Linear System

Definition of the Ordering Vector for the Grid Points

Initialization of the Unknown Vector

Solution by an Iterative Algorithm

Output of Results

Grid definition is implemented by a separate subroutine that permits field boundaries that are horizontal, vertical, or 45° lines. Within this boundary the mesh spacing must be uniform, but holes within the region are permitted. For the solution of the algebraic equations, a variety of adaptive iterative algorithms is provided. These include adaptations of the Jacobi method, the conjugate gradient method, and the successive overrelaxation method. A report by Kincaid and Grimes [7] includes programming details and the results of the application of these algorithms to a variety of problems.

In a paper prepared by Eisenstat and others [8], the effectiveness of software packages for large, sparse linear systems prepared at four different academic institutions is compared.

LINPACK

Collaboration of researchers at the Argonne National Laboratory and numerous universities resulted in 1979 in a very powerful programming system. LINPACK is a collection of FORTRAN subroutines useful for the solution of a wide variety of classes of systems of simultaneous linear algebraic equations. The objective was to provide subroutines that are completely machine independent, fully portable, and run at near optimum efficiency. Different subroutines are intended to take advantage of different special properties of the linear systems, thereby conserving computer time and memory. As described by Dongarra and others [9], separate subroutines are provided for each of the following linear systems or problems:

General Matrixes

Band Matrixes

Positive Definite Matrixes

Positive Definite Band Matrixes

Symmetric Indefinite Matrixes

Triangular Matrixes

Tridiagonal Matrixes

Cholesky Decomposition

Q.R. Decomposition

Singular Value Decomposition

The scope of LINPACK therefore transcends in large measure the systems arising in the solution of elliptic partial differential equations. Moreover, the entire coefficient matrix is usually stored in the computer memory, although there are provisions for band matrixes and for processing large rectangular matrixes row by row. On most large sequential computers, LINPACK can therefore handle full matrixes only of order less than a few hundred and band matrixes of order less than several thousand. There are no subroutines for general sparse matrixes or for iterative methods for large problems. On the other hand, where the size of the field problem is such that it falls within these limits, the LINPACK subroutines prove highly useful. They embody a degree of software engineering not found in most other programming packages.

11.4. SOFTWARE PACKAGES FOR PARABOLIC AND HYPERBOLIC PARTIAL DIFFERENTIAL EQUATIONS

In the development of software packages for parabolic and hyperbolic partial differential equations, the designer is faced with the dual task of providing for a suitable discretization of the space domain as well as the time domain. Early software systems tended to favor finite difference methods for approximating the space domain. In recent years, there has been increasing emphasis on finite element methods, most frequently using the Galerkin method and splines. Although numerous methods for approximating the time derivatives of the partial differential equation are available, by the late 1970s virtually all widely disseminated packages made use of the method of lines.

In the method of lines, the partial differential equation is converted to a system of ordinary differential equations, one equation for each finite difference or finite element grid point in the space domain. Standard subroutines for the integration of ordinary differential equations are then invoked to perform the integration in the time domain. One reason for the overwhelming popularity of the method of lines is that very powerful algorithms for ordinary differential equations have been developed over the years. The basic algorithms include Euler's method, the trapezoidal rule, Runge-Kutta methods, forward integration (Adams-Bashforth), and predictor-corrector (Adams-Moulton) methods. Many of the algorithms based on these methods have provision for the automatic control of the step size in the time domain so as to keep truncation errors within specified bounds. Of particular utility in solving differential equations are the algorithms developed by C. W. Gear, known generally as Gear's methods. These algorithms are particularly useful in the treatment of stiff systems in which there is a wide spread in the location of the roots or characteristic values (i.e., the solution contains transient terms with widely ranging time constants). Under these conditions, other algorithms for integration fail to provide correct answers or make it necessary to take extremely short steps in time so as to avoid instability. Gear's methods provide an effective compromise of accuracy, stablity, and computer time. A number of useful variations of Gear's method have appeared in the literature and have been implemented in various software packages. Some of these are considered briefly next.

DSS/2

Developed at Lehigh University by Schiesser [10] and widely disseminated to universities and government installations, DSS/2 facilitates the solution of ordinary differential equations as well as partial differential equations via the method of lines. The approach is one of linked FORTRAN IV subroutines with emphasis on transportability and ease of use by students. It consists of a main program to which the user must supply three subroutines. For batch processing, the deck of cards submitted to the computer is organized as follows:

```
JOB OR ACCOUNTING CARD SPECIFYING MEMORY, CPU TIME, ETC
CONTROL CARD(S) TO CALL THE DSS BINARY
CONTROL CARD(S) TO COMPILE THE THREE USER SUBROUTINES
CONTROL CARD(S) TO LOAD AND LINK THE DSS/SUBROUTINE BINARIES
CONTROL CARD(S) TO EXECUTE THE COMPLETE BINARY
TERMINATOR FOR CONTROL CARD SECTION (IF REQUIRED)
   SUBROUTINE INITIAL
         .
         .
         .
   END
```

```
      SUBROUTINE DERV
            .
            .
            .
      END
      SUBROUTINE PRINT(NI,NO)
            .
            .
            .
      END
TERMINATOR FOR FORTRAN SECTION (IF REQUIRED)
THREE DATA CARDS
END OF RUNS
TERMINATOR FOR DECK (IF REQUIRED)
```

Four categories of information must be provided by the user:

1. The DERV subroutine to define the equations being solved.
2. The INITIAL subroutine to define initial conditions for all of the simultaneous ordinary differential equations.
3. The PRINT subroutine to define the format of the computer output.
4. Two data cards, which specify the numerical parameters of the problem being solved, and which provide control information for the execution of the program.

To illustrate the application of DSS/2, consider the following simple example of a one-dimensional nonlinear parabolic partial differential equation described in the DSS/2 manual.

$$\frac{\partial u}{\partial t} = \frac{\partial^2 u}{\partial x^2} \tag{11.8}$$

which is written as

$$\text{PU/PT} = \text{P2U/PX2} \tag{11.9}$$

with boundary conditions

$$\text{PU(0,T)/PX=0} \tag{11.10}$$

$$\text{PU(L,T)/PX} = \text{E*(UA**4} - \text{U(L,T)**4)} \tag{11.11}$$

$$\text{U(X,0)} = \text{U0} \tag{11.12}$$

where U = dependent variable
X,T = independent variables
PU/PT = first-order partial derivative of U with respect to T
P2U/PX2 = second-order partial derivative of U witth respect to X
L = maximum value of X
E,UA,U0 = constants

The spatial derivative, P2U/PX2, is approximated at a general point X=I*DX by a finite difference approximation, in this case a second-order central difference

$$\text{P2U/PX2} = \frac{\text{U(I+1,T)} - \text{2*U(I,T)} + \text{U(I}-\text{1,T)}}{\text{DX*DX}} \tag{11.13}$$

where DX = spatial integration interval in X
I = index denoting the spatial position of the terms in the finite difference approximation, Eq. (11.13)

Substituting in Eq. (11.9), a set of simultaneous ordinary differential equations results:

```
DU(I,T)/DT = (U(I+1,T) - 2*U(I,T) + U(I-1,T))/(DX*DX)
                         I = 0, 1, 2, 3, . . . , N          (11.14)
```

where DU(I,T)/DT is an ordinary or total derivative of U with respect to T at the point X=I*DX and, N is the total number of increments, each of length DX(i.e., L=N*DX). Since for I=0, the point U(−1,T) in Eq. (11.14) is outside the solution domain X = 0 to L, boundary condition (11.10) must be used to handle this point. A second-order, central difference approximation for Eq. (11.10) is

```
PU(0,T)/PX = (U(1,T)-U(-1,T))/(2*DX) = 0                    (11.15)
```

or

```
U(-1,T) = U(1,T)                                            (11.16)
```

Thus for I = 0, Eq. (11.14) becomes

```
DU(0,T)/DT = 2*(U(1,T) - U(0,T))/(DX*DX)                    (11.17)
```

Similarly, for I = N, U (N + 1, T) in Eq. (11.14) is outside the solution domain, and boundary condition (11.11) must be used to handle this point. A second-order, central difference approximation of Eq. (11.11) at X = N, with UA = 500, is

```
U(N+1,T) = U(N-1,T) + 2*DX*E*(500**4 - U(N,T**4)            (11.18)
```

Equation (11.18) can be used to eliminate the term U (I+1,T) in Eq. (11.14) when I = N. The initial condition, for U0 = 600, becomes

$$U(I,0) = 600$$

$$I = 0,1,2,3,\ldots,N \qquad (11.19)$$

Subroutine DERV to implement Eqs. (11.14) to (11.19) with L=1 is listed next along with the INITIAL and PRINT subroutines and the DATA cards.

```
      SUBROUTINE DERV
      COMMON/T/T,NFIN,NRUN/Y/U((11)/F/DUDT(11)/PARM/N,DX,DXS
C...
C...  EVALUATE THE TIME DERIVATIVES FOR 10 SECTIONS (11 GRID
C...  POINTS)
      DO 1 I=1,N
      IF(I.NE.1)GO TO 2
C...
C...  X=0,                 (NOTE - THE SPATIAL INCREMENT, DX,
C...  AND DX*DX = DXS, ARE SET IN SUBROUTINE INITIAL AND PASSED
```

```
C...     TO DERV THROUGH COMMON/PARM/TO ACHIEVE COMPUTATIONAL
C...     EFFICIENCY, I.E., THESE CALCULATIONS ARE DONE ONLY ONCE
C...     IN SUBROUTINE INITIAL WHICH IS ALL THAT IS REQUIRED SINCE
C...     DX AND DX*DX ARE CONSTANTS THROUGHOUT THE ENTIRE CAL-
C...     CULATION)
         DUDT(1)=2./DXS*(U(2)-U(1))
         GO TO 1
  2      IF(I.NE.N)GO TO 3
C...
C...     X=1 (NONLINEAR RADIATION BOUNDARY CONDITION).
C...
         DUDT(N)=(1./DXS)*(U(N-1)+2.*DX*1.73E-09*(500.**4-
        1 U(N)**4)-2.*U(N)+U(N-1))
         GO TO 1
C...
C...     INTERMEDIATE SPATIAL POSITIONS.
  3      DUDT(I)=(1./DXS)*(U(I+1)-2.*U(I)+U(I-1))
  1      CONTINUE
         RETURN
         END

         SUBROUTINE INITIAL
         COMMON/T/T,NFIN,NRUN/Y/U(11)/F/DUDT(11)/PARM/N,DX,DXS
C...
C...     SET THE NUMBER OF SPATIAL INTERVALS, COMPUTE THE SPATIAL
C...     INCREMENT
         N=10
         DX=1./FLCAT(N)
         DXS=DX*DX
C...
C...     EVALUATE THE INITIAL CONDITION VECTOR FOR 11 GRID
C..      POINTS, EQUATION (5.9)
         N=N+1
         DO 1 I=1,N
1        U(I)=600.
         RETURN
         END

         SUBROUTINE PRINT(NI,NO)
         COMMON/T/T,NFIN,NRUN/Y/U(11)/F/DUDT(11)/PARM/N,DX,DXS
         WRITE(NO,1)T
1        FORMAT(/,2X,7HTIME = ,E9.2,/,
        111H   X=0     ,11H   X=0.2   ,11H   X=0.4
        211H   X=0.6   ,11H   X=0.8   ,11H   X=1.0)
         WRITE(NO,2)(U(I),I=1,N,2)
2        FORMAT(6E11.3)
         RETURN
         END
```

The first data card provides the initial and final values of the independent variable, as well as the print interval. The second data card contains the number of first-order differential equations (11 in this case), the maximum ratio of the PRINT interval to the integration interval (in this case 64), the identification number of the integration algorithm to be used (1 in this case), a print option for error messages, relative or absolute error criterions to govern the automatic adjustment of the integration step (REL for relative in this case), and the magnitude of the maximum allowable error at each step in the time domain (0.001 in this case).

The integration algorithm for integrating in the time domain can be chosen from 14 Runge-Kutta methods of order 1 through 5 each with automatic step-size control. Extensions of the preceding approach to parabolic and hyperbolic differential equations in two and three dimensions in Cartesian, cylindrical, and spherical coordinates are straightforward and well documented.

FORSIM VI

This system, which was developed in evolutionary fashion at the Chalk River Nuclear Laboratories by the Atomic Energy of Canada and described by Carver [11], is also based on FORTRAN IV and subroutine oriented. As in DSS/2, the method of lines is used, and the user can select from a variety of integration algorithms, including Gear's method. A wider variety of partial differential equations can readily be handled, however, and there is provision for automatically providing discretization in space using various types of finite difference approximations as well as cubic spline approximations. For relatively simple problems, the user is required to prepare only one subroutine, UPDATE. The control card structure of FORSIM then takes the following form.

JOB,BXXX-name/proj.	Job account card
FTN	Compile UPDATE routine, object code to file LGO
ATTACH,FORSIM.	Access FORSIM control package
COPYL,FORSIM,LGO,EXEC.	Merge FORSIM and LGO on file EXEC.
RETURN,FORSIM,LGO	Release FORSIM and LGO
ATTACH,FORSIM,CY=1.	Access FORSIM library
LIBRARY(FORSIM)	Declare FORSIM library
EXEC.	Load and execute
7/8/9 (EOR)	
UPDATE Routine	
7/8/9	
FORSIM Data Deck	If required
7/8/9	
6/7/8/9 (EOF)	

The update routine specifies the form of the equation to be solved and is called at each integration step in the time domain to update the current values of all derivatives. Consider the following example taken from the FORSIM VI manual.

$$\frac{\partial u}{\partial t} = c\frac{\partial^2 u}{\partial x^2} \tag{11.20}$$

Spatial range	$0 < X < 1.0$, 11 points
Initial conditions	$U = 1.0$
Boundary conditions	$U(1) = U(N) = 0$
Diffusion coefficient	$C = 1$

The UPDATE subroutine for this problem is:

```
      SUBROUTINE UPDATE
C
C     SECTION 1 - STORAGE
C
      COMMON/INTEGT/U(11)
      COMMON/DERIVT/UT(11)
      COMMON/DERVX/UX(11)
      COMMON/DERVXX/UXX(11)
      COMMON/RESERV/T/CNTROL/INOUT
C
C     SECTION 2 - INITIAL CONDITIONS AND CONTROLS
C
      IF(T.NE.0)GO TO 100
      C=1.
      NPOINT=11
      NPDE=1
      DO 10 I=1, NPOINT
   10 U(I)=1.0
C
C     SECTION 3 - DYNAMIC SECTION
C
  100 CALL PARSET(NPDE,NPOINT,U,UT,UX,UXX)
C
C     SPECIFY EQUATION AT EACH SPATIAL POINT
C
      DO 110 I=1,NPOINT
  110 UT(I)=C*UXX(I)
C
      CALL PARFIN(NPDE,NPOINT,U,UT)
C
C     SECTION 4 - PRINTOUT
C
```

```
IF(INOUT.NE.1)RETURN
PRINT*,T
PRINT*,U
CALL FINISH(T,5.0,4HTIME)
END
```

The routine has four sections; the first contains common block storage, with additional blocks /DERVX/ and /DERVXX/ to ensure communication of the spatial derivatives $\partial u/\partial x$ and $\partial^2 u/\partial x^2$, respectively. The first derivative does not appear explicitly, but normally its associated block should be included. The second section sets up the initial values of U, C and two controls NPDE, the number of PDEs, and N, the number of points used. The dynamic section contains the components necessary to solve the PDEs and makes calls to two subroutines PARSET and PARFIN. PARSET establishes the boundary conditions and spatial derivatives. The equations are then stated in a DO loop and followed by a call to PARFIN for end processing. The fourth section prints the array U.

The preceding subroutine does not specify the integration method, the order of the spatial differentiation formulas, the spatial variable, or the boundary conditions; default values for all these have been used. The solution proceeds with Runge-Kutta-Fehlberg integration, three-point spatial formulas, and boundary conditions U (1) = U (N) = 0. Any of these may be changed by including further common blocks and common block elements.

OTHER METHOD OF LINES PACKAGES

Sincovec and Madsen [12] have developed a system called PDECOL. The spatial domain is discretized by the collocation method using B-splines as the basis functions. The user must specify the splines (breakpoints, degree of polynomials, order of continuity), but the collocation points are chosen automatically. Gear's method is employed to integrate the system of ordinary differential equations with respect to time.

Hyman [13] at New York University developed a subroutine package, which he calls MOLID. As the name implies, this program is employed to solve parabolic equations in one space dimension using the method of lines. A variety of programs is provided for automatically discretizing the space domain using difference equations of various orders as well as fast Fourier transformation approximations for periodic systems. Again, the Gear method is used to effect integration with respect to time.

From the Argonne National Laboratory comes a package called DISPEL described by Leaf and others [14]. The program is intended primarily for the kinetics-diffusion problems in one and two space dimensions that arise in studies of heat and mass transfer. Discretization of the space domain is achieved by using a Galerkin method in conjunction with B-splines of a specified order and smoothness. The resulting system of ordinary differential equations is integrated using a variation of the Gear method.

11.5. SPECIAL-PURPOSE PACKAGES FOR IMPLEMENTING THE FINITE ELEMENT METHOD

Most of the packages discussed in the preceding sections are intended to be general in purpose. That is, they are designed to facilitate the solution of field problems regardless of application area. Usually, a price in computing efficiency is paid for this feature. As a result, there have been intensive efforts in a number of fields of endeavor to develop software packages tailored specifically for users from a specific application discipline. This permits the inclusion of features and terminology that would have no place in a general-purpose package. The development of these packages has been particularly intensive in structural mechanics, nuclear reactor technology, and fluid mechanics—most often employing finite element approximations. Even a superficial survey of this activity would demand a separate chapter if not a separate textbook. This is beyond the scope of this text. However, as an example, a few of the larger and more important software packages aimed primarily at structural analysis problems are listed in Table 11.2. Many of these are available from government-sponsored program libraries, such as the COSMIC Library at the University of Georgia.

Table 11.2. Sample listing of finite element computer programs

Program	*Developer*
ANSYS	Swanson Analysis Systems, Inc.
ASKA Automatic System for Kinematic Analysis	J. H. Argyris and H. A. Kamel
ASTRA Advanced Structural Analyzer	The Boeing Co.
BEST 3 Bending Evaluation of Structures	Structural Dynamics Research Corp.
DAISY	H. A. Kamel for Lockheed Missile & Space Co.
ELAS and ELAS8[a] General-Purpose Computer Programs for Equilibrium Problems for Linear Structures	Jet Propulsion Lab.
FESS/FINESSE Systems of Programs	Civil Engineering Dept., Univ. of Wales, O. C. Zienkiewicz, Head
FORMAT Fortran Matrix Abstraction Technique	McDonnell Douglas Corp.
MAGIC Matrix Analysis Via Generative and Interpretive Computations	Bell Aerosystems
MARC Nonlinear Finite Element Analysis Program	Marc Analysis Research Corp.

Table 11.2. (continued)

Program	Developer
NASTRAN[a] NASA Structural Analysis	Computer Science Corp., The Martin Co., MacNeal-Schwendler Corp.
SAFE Structural Analysis by Finite Elements	Gulf General Atomic, Inc.
SAMIS[a] Structural Analysis and Matrix Interpretive System	Philco Div., Ford Motor Co.
SAP A General Structural Analysis Program	E. L. Wilson, Univ. of California, Berkeley
SBO-38	Martin Marietta Corp.
SLADE A Computer Program for the Static Analysis of Thin Shells	Sandia Lab.
STARDYNE	Mechanics Research Inc.
STARS 2 Shell Theory Automated for Rotational Structures	Grumman Aircraft Engineering Corp.
STRESS	M.I.T.
STRUDL (ICES)	M.I.T.
A Structural Design Language in the Integrated Civil Engineering System	Engineering Computer International

[a]Copies of these programs can be obtained from COSMIC Computer Center, Barrow Hall, University of Georgia, Athens. Ga. 30601.

Reproduced with permission from The Finite Element Method for Engineers by K. H. Huebner, © 1975, John Wiley & Sons, Inc, New York.

11.6. ROLE OF SYMBOLIC COMPUTATION IN THE DEVELOPMENT OF NUMERICAL SOFTWARE

Symbolic algebraic computation (or symbolic computation, for short) refers to the technique of manipulating, on a computer, symbolic expressions that may not necessarily have numerical values. Therefore, techniques of symbolic computation can be used, among other things, to perform algebraic manipulations of mathematical formulas. Crudely, one can think of symbolic computation as a computerized version of the traditional "paper and pencil" manipulations of algebraic expressions commonly arising in applied mathematics. Therefore, symbolic computation can significantly reduce the tedium of analytic calculations and increase their reliability. This capability permits one to carry on the analytic calculations before numerical computations start. The purpose of this section is to introduce this new tool and point out its potential in the numerical modeling and simulation of field problems.

A number of symbolic manipulation systems suitable for manipulating algebraic expressions have been developed over the past few years. A representative sample of these are listed as follows:

ALPAK	MACSYMA
ALTRAN	REDUCE
ASHMEDAI	SAC-1
CAMAL	SCHOONSCHIP
CONFORM	SCRATCHPAD
FORMAC	SYMBAL

Jenson and Niordson [15] performed a comparative study of some of these systems. At the present time, it appears that MACSYMA and REDUCE are two systems with relatively easy availability. Some rudimentary information about these two systems is now presented.

MACSYMA (Project MAC's symbolic manipulation system), probably one of the more versatile systems currently available, is a large interactive system written in MACLISP with an ALGOL-like syntax. MACSYMA (pronounced "maxima") is used for performing both symbolic as well as numerical mathematical manipulations. It has been under development by the Mathlab Group at MIT since 1969. The MACSYMA system is operated on DEC PDP-10 computers under ITS operating system and on Honeywell 6180 computers under the MULTICS operating system. It is accessible to a wide community of users through the ARPA network.

REDUCE (developed primarily by the Symbolic Computation Group at the University of Utah at Salt Lake City) is a program designed for general algebraic computations of interest to physicists, mathematicians, and engineers. Several versions of REDUCE are currently available. The IBM version is designed explicitly for use on IBM System/360 or 370 computers. Versions of REDUCE are also available for use on DEC systems 10 and 20 and for the UNIVAC 1100 series. REDUCE appears to be more readily available than MACSYMA.

What are some of the things a good symbolic manipulation language can do? A good system should be able to combine algebraic expressions through algebraic operations such as addition, multiplication, and exponentiation and to perform differentiation and integration in analytic form. It should be able to perform simplification of algebraic expressions and substitution of a specified "value" of x in $f(x)$. For example, MACSYMA has all the preceding capabilities. MACSYMA can also take a function $f(x)$ and expand it in Taylor series, and solve an ordinary differential equation by producing its general solution as well as a particular solution.

For example, if one wishes to work with the expression $(x + 1)^3$, after proper logging-in, one proceeds by typing a FORTRAN-like expression as follows:

```
(C1) (x+1) ** 3;
```

The ; terminates the input and prompts MACSYMA to display the expression. MACSYMA will come back with

```
(D1)                        (x+1)³
(C2)
```

The label (C1) is a name automatically assigned to the first command and the label (D1) to the result. The next input line is labeled (C2). To expand the expression in line (D1), one could enter on line (C2)

```
(C2) EXPAND (D1);
```

and MACSYMA responds with

```
(D2)                    x³+3x²+3x+1.
```

For a more revealing example, consider the problem of finding the first six terms of the Taylor series expansion of

$$f(x) = \frac{x^{5/2}}{\log(1 + x)\tan x^{3/2}} \tag{11.21}$$

This is truly a formidable task. Using MACSYMA, the transactions shown in Figure 11.1 could take place at a terminal. The meanings of these commands are self-explanatory. For example, after line (C16) is typed in by the user, MACSYMA responds with line (D16), which is simply a display of the input expression in human-readable form. To expand the given function in Taylor series, one enters the command shown in line (C18). Notice that lines (C17) and (D17) are not shown in the figure because they probably represent some transaction not relevant to our present interest. The meaning of line (C18) is: "Expand the entry in line (D16) in Taylor series about the point $x = 0$ and display the first six terms." The computer responds with the line (D18).

```
(C16) X↑3/(SQRT(X)◇TAN(X↑(3/2))◇LOG (1 + X));
```

(D16) $$\frac{X^{5/2}}{\text{LOG}(X+1)\ \text{TAN}(X^{3/2})}$$

```
(C18) TAYLOR(D16,X,0,6);
```

(D18)/T/ $$1 + \frac{X}{2} - \frac{X^2}{12} - \frac{7\,X^3}{24} - \frac{139\,X^4}{720} + \frac{67\,X^5}{1440} - \frac{3047\,X^6}{60480} + \ldots$$

Figure 11.1 Transactions at a terminal to evaluate $f(x)$ of Eq. (11.21) using MACSYMA

```
(C35)'DIFF(Y,X,2) -- 2♦ DIFF(Y,X) + 5♦Y = 0;
TIME = 6 MSEC
```

$$\text{(D35)} \qquad \frac{D^2Y}{DX^2} - 2\,\frac{DY}{DX} + 5\,Y = 0$$

```
(C36) ODE2(%,Y,X);
TIME = 77 MSEC.
```

(D36) $\qquad Y = \%E^X\ (\%K1\ SIN(2\ X) + \%K2\ COS(2\ X)\,)$

Figure 11.2 Transactions at a terminal to generate the general solution of Eq. (11.22) using MACSYMA. (Figures 11.1 and 11.2 are provided by the courtesy of Professor James Geer of the State University of New York at Binghamton)

As another example, Figure 11.2 shows the transactions involved in solving

$$\frac{d^2y}{dx^2} - 2\frac{dy}{dx} + 5y = 0 \tag{11.22}$$

The given equation is entered on line (C35). MACSYMA responds by displaying the equation in human-readable form. It takes 6 ms of CPU time to do this. Then, on line (C36) the user requests a solution to the equation displayed in the preceding step (% sign in line C36). The solution appears in 77 ms on line (D36). The equation in line (D36) is

$$y = e^x(k_1 \sin 2x + k_2 \cos 2x)$$

There are several reasons why symbolic computations are useful in the context of modeling and simulation of field problems. Some of the reasons cited by Brown and Hearn [16] are listed next.

1. Sometimes it is prohibitively expensive, or even impossible, to solve an essentially numerical problem by purely numerical means because it involves too many variables, requires too much accuracy, or is presented in an ill-conditioned or intractable form. However, a symbolic transformation may reduce the dimensionality, evade a large dose of roundoff, finesse the ill conditioning, and otherwise change the problem into one that can be solved by standard numerical methods.

2. The algebraic result obtained via symbolic computation can be subsequently evaluated over a wide range of parameter values.

3. Symbolic computation provides an opportunity for realizing the vital computational symbiosis between numerical experiments and symbolic theories. For example, to solve a system of partial differential equations by the finite element method, one first transforms the given continuum equations into an approximating set of algebraic equations, whose coefficients are integrals of certain "shape functions" and their derivatives. While these integrals are usually evaluated numerically, Noor and Andersen [17] have found that symbolic integration, performed by the MACSYMA system, is often considerably faster.

4. The idea of using symbolic computation to generate a needed numerical subroutine is a useful one. For example, in solving a system of nonlinear equations, each step requires a numerical approximation to the Jacobian matrix (i.e., the matrix of first partial derivatives). When numerical differentiation is used, it is generally considered too costly to evaluate the Jacobian afresh at each step, so one settles for rank-one updates, with perhaps an occasional restart in case of trouble. However, when the symbolic Jacobian matrix is available, it can often be evaluated together with the function vector at little extra cost, and thus one may achieve substantial improvements in both speed and accuracy.

5. Finally, in the realm of partial differential equations, Cloutman and Fullerton [18] have used symbolic multidimensional Taylor series expansions, computed by the ALTRAN system, to analyze the discretization and roundoff errors of various numerical methods and also, more importantly, to eliminate inaccurate or unstable methods prior to coding and testing, and to develop methods in which the lowest-order errors cancel each other out.

REFERENCES

1. A. F. Cardenas and W. J. Karplus, "PDEL—A Language for Partial Differential Equations," *Communications of ACM*, Volume 13, pp. 184–191, March 1970.

2. W. E. Schiesser, "A Digital Simulation System for Higher-Dimensional Partial Differential Equations," *Proceedings of the 1972 Summer Computer Simulation Conference*, San Diego, Calif., pp. 62–72, June 1972.

3. J. Gary and R. Helgerson, "An Extension of FORTRAN Containing Finite Difference Operators," *Software Practice and Experience*, Vol. 2, pp. 321–336, 1972.

4. M. Machura and R. A. Sweet, "A Survey of Software for Partial Differential Equations," *ACM Transactions on Mathematical Software*, Vol. 6, No. 4, pp. 461–488, December 1980.

5. J. Rice, "ELLPACK: A Research Tool for Elliptic Partial Differential Equation Software," in *Mathematical Software III*. New York: Academic Press, Inc., pp. 319–341, 1977.

6. E. N. Houstis and J. R. Rice, "Software for Linear Elliptic Problems on General Two-Dimensional Domains," in *Advances in Computer Methods for Partial Differential Equations II*, R. Vichnevetsky, ed., IMACS, pp. 7–12, 1977.

7. D. R. Kincaid and R. C. Grimes, "Numerical Studies of Several Adaptive Iterative Algorithms," Report CNA-126, Center for Numerical Analysis, University of Texas at Austin, August 1977.

8. S. Eisenstat, and others, "Some Comparisons of Software Packages for Large Sparse Linear Systems," in *Advances in Computer Methods for Partial Differential Equations III*, R. Vichnevetsky, ed., IMACS, pp. 98-106, 1979.

9. J. J. Dongarra and others, "LINPACK User's Guide," to be published by *Society for Industrial and Applied Mathematics*, 1979.

10. W. SCHIESSER, *DSS/2—An Introduction to the Numerical Solution of Lines Integration of Partial Differential Equations*, 2 volumes, Lehigh University, Bethlehem, Pa., 1976.

11. M. B. CARVER and others, "The FORSIM VI Simulation Package for the Automated Solution of Arbitrarily Defined and Partial and/or Ordinary Differential Equation Systems," Report No. AECL-5821, Atomic Energy of Canada Ltd., Chalk River, Ontario, 1978.

12. N. MADSEN and R. SINCOVEC, "General Software for Partial Differential Equations," in *Numerical Methods for Differential Systems: Recent Developments in Software Algorithms and Applications*, Academic Press, New York, pp. 229–242, 1976.

13. J. M. HYMAN, "The Method of Lines Solution of Partial Differential Equations," Report #COO-3077-139, ERDA Mathematics and Computing Laboratory, New York University, 1976.

14. G. K. LEAF, and others, "DISPEL: A Software Package for One and Two Spatially Dimensioned Kinetics-Diffusion Problems," Report No. ANL-77-12, Argonne National Laboratory, Argonne, Ill., 1977.

15. J. JENSON and F. NIORDSON, "Symbolic and Algebraic Manipulation Languages and Their Application in Mechanics," in *Structural Mechanics Software Series*, Vol. 1, pp. 541–576, University Press of Virginia, Charlottesville, Va., 1977.

16. W. S. BROWN and A. C. HEARN, "Application of Symbolic Algebraic Computation," *Proceedings, Computational Atomic and Molecular Physics Conference*, Nottingham, England, September 12–15, 1978.

17. A. K. NOOR and C. M. ANDERSEN, "Computerized Symbolic Manipulation in Structural Mechanics—Progress and Potential," *Computers and Structures*, Vol. 10, Nos. 1/2, pp. 95–118, April 1979.

18. L. D. CLOUTMAN and L. W. FULLERTON, "Some Applications of Automated Heuristic Stability Analysis," *Report* LA-6885-MS, Los Alamos Scientific Laboratory, Los Alamos, N.M., July 1977.

EXERCISES

1. Expand the discussion in Section 11.1 by describing in some detail various classes of members of the hierarchy of programming languages. Provide examples of at least five members of each level in the hierarchy, and discuss their advantages and limitations vis-à-vis languages belonging to other levels.

2. Prepare a table listing the relative advantages and disadvantages of the three higher-level language approaches represented by PDEL, LEANS III, and DSS/2 for solving field problems governed by parabolic partial differential equations.

3. By means of a survey of the current literature, determine what software packages for the solution of each of the following classes of partial differential equations has been imple-

mented since the preparation of the material of Chapter 11, and prepare brief summaries of their capabilities:
(a) Elliptic partial differential equations.
(b) Parabolic partial differential equations.
(c) Hyperbolic partial differential equations.

4. Write a term paper comparing any two of the software packages listed in Table 11.1.

5. Write a term paper discussing the role of symbolic computation in the context of modeling and simulation of field problems (a) using the finite difference method, and (b) using the finite element method.

6. Write a term paper comparing any two of the symbolic algebraic languages listed in Section 11.6.

12

Trends in Computer Hardware and Their Impact on Field Simulation

12.1. INTRODUCTION

Prior to World War II, numerical solutions of field problems, characterized by partial differential equations, were obtained with the aid of mechanical calculators or by a variety of analog techniques. Stimulated by military requirements during World War II, the first modern electronic digital computers began to make their appearance in the late 1940s and early 1950s. During that pioneering period a number of different approaches to digital computer organization and digital computing were investigated. Primarily as a result of the constraints imposed by the then-available electronics technology, the designers of digital computers soon focused upon the concept of computer system architecture championed by John von Neumann and first implemented in the computer constructed for the Institute of Advanced Studies at Princeton. Because of the pervasiveness of the Von Neumann architecture in digital computers during the 1950s and 1960s, most numerical analysts and other computer users concentrated their efforts on developing algorithms and software packages suitable for computers of that type. Virtually all the algorithms and numerical techniques discussed in this text are an outgrowth of that era of the computer field.

The large and powerful digital computers, which came into use in the late 1960s and 1970s, included numerous modifications and improvements over the computers of the earlier generation. However, most of these were relatively "transparent" to the programmer, affecting the speed and sometimes the precision of the computations without demanding any serious reevaluation or restructuring of the algorithms or programs. It was not until the mid 1970s that there appeared classes of digital computers that differed from the Von Neumann architecture so radically that the need for new algorithmic approaches became apparent. Of particular importance in this connection was the introduction of the techniques of multiprocessing and of pipelining in digital computer system design. It is the purpose of this chapter to provide a perspective on the potential impact of these major innovations in digital computer system design upon the digital computer treatment of partial differential equations. To this end, the basic concepts of the Von Neumann architecture are first briefly reviewed. This is followed by discussions of the meaning of the terms "multiprocessing" and "pipelining" in the present context. A number of major contemporary computer systems embodying these concepts are then described, together with comments regarding the types of algorithms that appear to be most suited for them.

To permit the rapid and automatic processing of numerical information, all digital computers must be capable of performing the following very general functions:

1. The acceptance of input data, including numerical information as well as coded instructions.
2. The storage of the input data, as well as data generated in the course of the calculations.
3. The channeling and directing, within the computer, of the data in such a manner that the proper calculations are performed in the sequence specified by the program.
4. The performance of all required arithmetic and logical operations.
5. The readout of the solution of the problem in a specified format.

These five functions are generally termed input, memory, control, arithmetic, and output, respectively. In the utilization of computers for the solution of numerical problems, the data stored within the memory and transferred between the other major functional units of the computer are of two types: numbers and instructions. These data are represented as sequences of binary digits. The devices employed for the storage of this information are termed ***registers***. A register may be visualized as an array of electronic elements each capable of making an unequivocal distinction between "zero" and "one." Each element thereby becomes the repository of one binary digit or ***bit***. Frequently, information in registers is organized in groups of eight bits, termed ***bytes***. One complete set of digits, corresponding to one number or one program instruction, is termed a ***word*** and may include from one to eight or more bytes, depending upon the design of the specific machine. A preliminary step in problem preparation is therefore the translation of all numerical data and instructions into binary form.

For the binary representation of numerical information, two principal alternatives exist: fixed point and floating point. In ***fixed-point*** operation a number may be represented by a binary digit indicative of the sign of the number, followed by the binary representation of the number itself. (There are also other ways of representing negative numbers.) In this type of representation, the radix point (the equivalent of the decimal point) is assumed to be located either at the extreme left or the extreme right of the number. If it is located at the extreme left, the machine is capable of representing numbers having magnitudes ranging from -1 to $+1$; if the radix point is at the extreme right, the machine is capable of representing only integers. In either event, the radix point is fixed, and it is up to the programmer and coder to assure that under no conditions will the need arise for the representation of a number outside of these limits. This involves a careful analysis and scaling of all steps of the computation so as to avoid register overflow.

In ***floating point operation*** a portion of the space of each register is reserved for the binary representation of an exponent. The number stored in the register actually has four parts: the sign bit, a bit representing the sign of the exponent, the exponent, and the most significant digit of the number itself, the mantissa. Figure 12.1 is a pictorial representation of a typical 32-bit register storing numbers in fixed-point and floating-point form. The register in Figure 12.1b has seven binary digits assigned to exponent representation.

12.2. THE VON NEUMANN ARCHITECTURE

Though differing in some details, the digital computers manifesting the by-now classic Von Neumann architecture are generally organized along the lines shown in Figure 12.2 and are well described in standard textbooks [1], [2].

The memory registers constitute the principal components of the memory unit and comprise the chief storage facility of the computer system. Modern general-purpose computers contain anywhere from 16,000 to several million memory registers, each capable of storing a multibyte word. Each of these registers is assigned an identification number, termed the ***address***, to permit it to be distinguished from the other registers. In digital computer operation there are only two basic tasks that the memory may be asked to perform: read and record. The record operation involves the taking of data from the outside, that is, from the input unit or the arithmetic unit, and registering this information in one of the memory registers. "Read" is the inverse operation and involves the furnishing of the information contained in a specific memory register to other units. The ***memory buffer*** serves as an intermediary in this process. This unit is a register that is employed to store, for a brief period of time, the word to be recorded in the memory or the word to be transmitted from the memory to other units. Its contents can be reduced to zero by a "clear" command from the control unit. The ***memory address register*** stores temporarily the address of the memory register to be involved in the next "read" or "record" operation. The contents of this register are modified by the control unit after each program step.

The arithmetic unit contains a special register termed the ***accumulator***. This register is in direct communication with the memory buffer in such a fashion that information may flow from the memory buffer into the accumulator ("operate") or from the accumulator to the memory buffer ("store"). The command for the perfor-

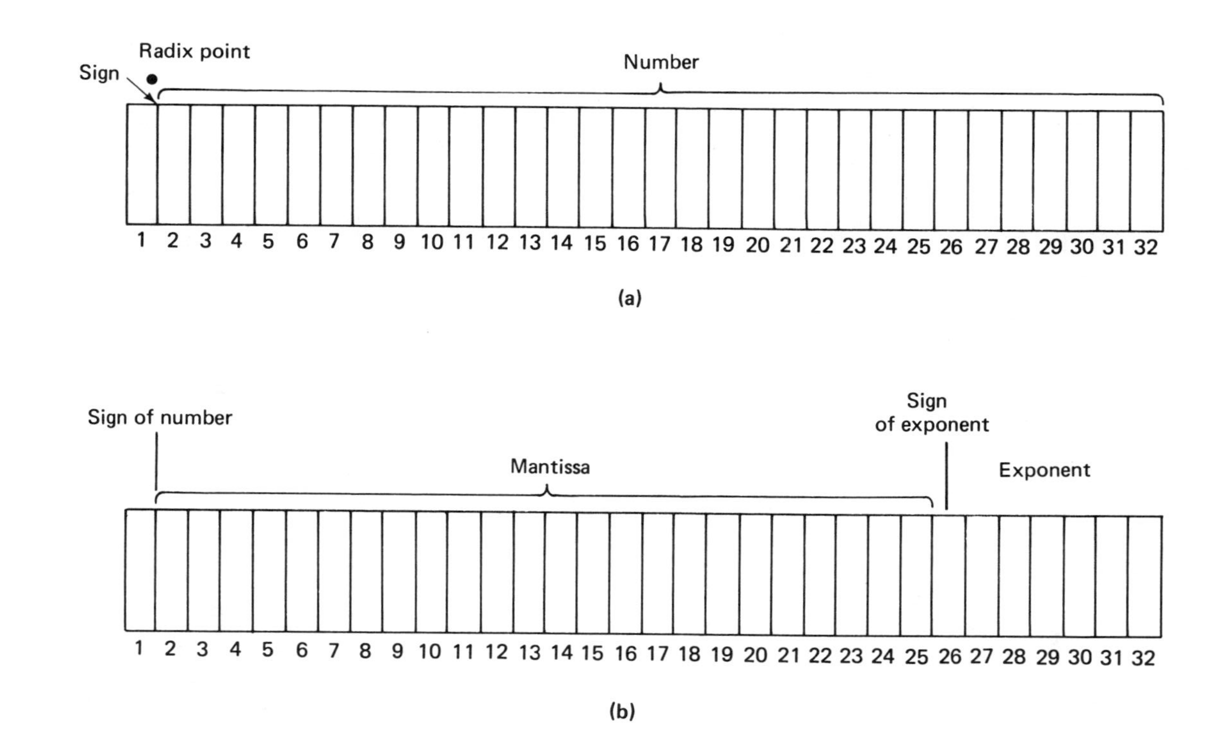

Figure 12.1 Numbers in binary form (a) Fixed-point format (b) Floating-point format

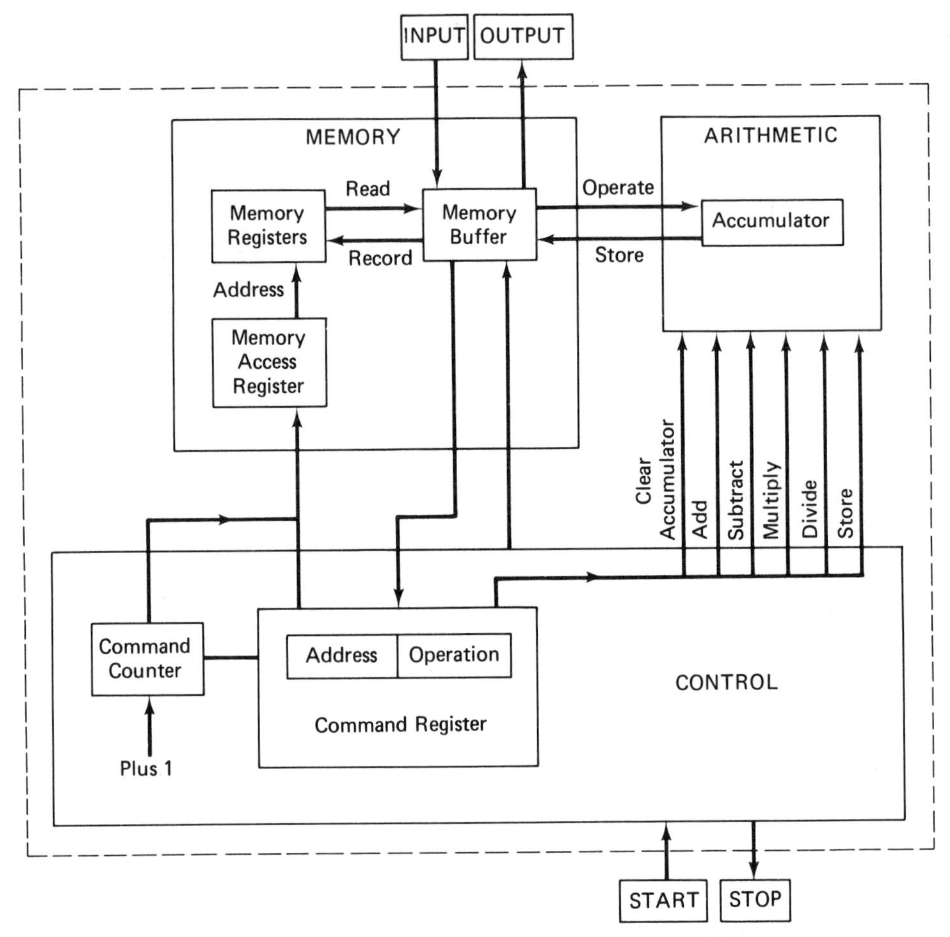

Figure 12.2 Basic Von Neumann digital computer architecture

mance of a specific arithmetic operation comes from the control unit. The desired operation, be it "add," "subtract," "multiply," "divide," or a wide variety of logical operations, is performed upon the numbers stored in the accumulator and the memory buffer. After the operation is finished, the result appears as the new contents of the accumulator, while the contents of the memory buffer remain unchanged. For example, if the memory buffer contains a number x and the accumulator a number y, the command "multiply" will cause the product xy to appear in the accumulator, while the number x remains in the memory buffer. The command "store" causes the contents of the accumulator to be transferred to the memory buffer; thence this number may either be recorded in a memory register for later reference, or it may be employed directly in the succeeding mathematical operation. The command "clear accumulator" causes the contents of the accumulator to become zero.

The control unit contains two principal registers: the command register and the command counter. The ***command register*** contains the address of the memory register to be involved in the next step of the computation sequence as well as the operation to be performed. Both of these items of information are coded in binary form. The contents of this register control the activity of the memory registers and the memory buffer as well as the arithmetic unit. The second control register, the ***command counter***, controls the order sequence. That is, it causes the contents of certain memory registers to be transferred into the command register via the memory buffer. For convenience, the command counter is frequently given a content of "zero" at the start of a calculation and increased by one integer at a time after each complete solution step, thereby activating the program commands located in corresponding memory registers 0000, 0001, 0002, 0003, 0004, ... Thus typical steps in the computing sequence could be as follows:

1. The contents of the command counter, namely, 0000 are transferred to the memory address register.
2. The data located in the memory register 0000 are transferred to the memory buffer.
3. The data are transferred from the memory buffer to the command register.
4. The command in the command register is executed.
5. The contents of the command counter are increased by "one."
6. The number 0001 is transferred from the command counter to the memory address register.

The execution of the operation, that is, step 4, may itself require a number of substeps.

A key feature of the Von Neumann organization is the "bottleneck" represented by the memory buffer. In effect, only one item of information at a time may be read out of memory, and only one arithmetic operation may be performed at any one time. All algorithms and programs must therefore specify a ***sequence*** of operations to be performed one at a time. For that reason, computers using a Von Neumann organization perform ***sequential computation***.

The overwhelming acceptance of the Von Neumann architecture also had a profound affect upon the design of programming languages and programming systems. Starting with the development of FORTRAN in the 1950s and continuing with ALGOL and PL/1 in the 1960s, all major programming languages intended to facilitate the solution of numerical problems were designed for procedural or sequential operation. That is, each of these languages is designed so as to carry out but a single arithmetic or logical operation at any one time, and it is up to the programmer to specify the appropriate sequence. The overwhelming majority of programs for the solution of partial differential equations are written in FORTRAN. This makes it relatively easy to transplant a program from one digital computer system to another, provided, however, that both computers manifest the Von Neumann architecture.

12.3. MULTIPLE PROCESSING ELEMENTS

The term "parallelism" in automatic computation refers in its most general sense to the simultaneous execution of two or more operations by a computer system, something not possible in the Von Neumann architecture. The present section is concerned with one approach to the attainment of parallelism—through the replication of one or more of the basic major functional units of the computer. Conventional digital processors contain a single control unit, a single arithmetic unit, and a single memory unit. Some recently introduced computers, on the other hand, have two or more of these basic architectural units, which are arranged so that they can operate simultaneously, thereby increasing the system's computational speed and power.

The term ***multiprocessor*** is usually reserved for computer systems in which there are two or more processing elements, each with individual memory, arithmetic, and control. Moreover, all these processing elements must have access to a large central memory so that they can all be working simultaneously on the same problem. The term ***array of processing elements*** is sometimes used to identify systems in which there is replication of memory and arithmetic but not of control. Flynn [3] provided a widely used technique for classifying such systems, while Enslow [4] presents a detailed survey of the many different implementations of this concept.

In a Von Neumann type computer, at any instant of time, there can be but a single command in the command register, and this command can effect an arithmetic or logical operation only upon the single word then resident in the accumulator. Such a machine organization is termed "single instruction stream-single data stream" (SISD).

An important class of digital computer systems is organized as shown in Figure 12.3. Here there are a number of processing elements each consisting of an arithmetic unit and a memory unit. The arithmetic and memory units are interconnected to form an array; hence a structure of this type is sometimes termed an array of processors or an array of processing elements. The control unit can activate any or all of the arithmetic units. Each active element of the array performs the same arithmetic or logic operation under command of the single control unit. Of course, each arithmetic element may be operating on different data in executing the instruction resident in the control unit. For this reason, this type of structure is termed "single instruction stream-multiple data stream" (SIMD).

SIMD computing systems were designed in the 1960s and early 1970s. The intent was to achieve high computing speed, particularly in the treatment of partial differential equations by finite difference methods, without the need to replicate the relatively expensive control unit. The ILLIAC IV computer described in Section 12.5 is an example of this approach. With the advent of low-cost minicomputers and microcomputers and the declining cost of digital hardware in general, replication of the control unit has become more and more feasible from the economic point of view. As a result, a wide variety of multiprocessor systems of the type shown in Figure 12.4 have been constructed or proposed. Each element of the array of processors now includes a control unit as well as an arithmetic and a memory unit. The elements of

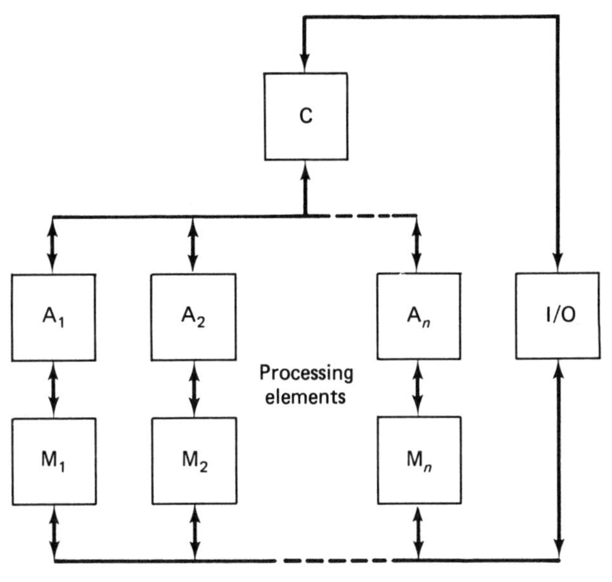

Figure 12.3 SIMD parallel structure

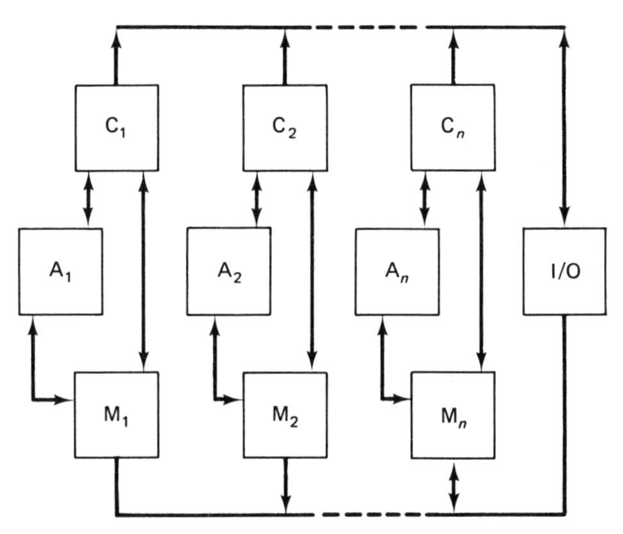

Figure 12.4 MIMD architecture

the array can therefore be full-fledged independent digital computers, and each processing element can be carrying out a different arithmetic or logic operation. For this reason, systems of this type are termed "multiple instruction stream-multiple data stream" (MIMD). Systems of this type are discussed in Section 12.8.

The replication of the memory units in SIMD and MIMD computers is an important step toward overcoming the "bottleneck" in Von Neumann architectures. In many respects, each processing element can proceed independently, thereby potentially effecting an appreciable increase in computing speed. Under these conditions, the major impediment to realizing the potential advantages of parallelism arises when it is necessary to exchange information between the various processing elements. The connectors, switching elements, and control circuits that are employed to transfer data from one processing element to another then act as effective constraints upon the computing speed. For this reason, multiprocessor systems are often classified in accordance with the manner in which the individual processing elements are interconnected. As discussed in detail by Enslow [4], three different approaches to interconnection can be recognized: common bus, cross-bar switch, and multiport memory.

In common bus multiprocessors, such as those shown in Figures 12.3 and 12.4, the processing elements time share common electrical connection paths. A number of these may be provided to facilitate higher throughput rates and to increase system reliability. Elaborate algorithms and hardware controllers must be provided to assure the achievement of the desired information flow.

In a cross-bar switching system, each arithmetic unit may be granted access to each memory unit through a switching matrix as shown in Figure 12.5. Numerous transfers can therefore occur simultaneously. This provides greater flexibility in system design and utilization, albeit at a considerably increased cost.

Multiport memory multiprocessors achieve an effect similar to that attained using a cross-bar switching matrix. In this case the memory units themselves are equipped with hardware units to control all switching functions, so that each memory module can communicate directly with each of the arithmetic units, as shown in Figure 12.6.

12.4. PIPELINING

The term ***pipeline*** is employed to designate parallelism of a different sort. A sequential computational procedure is broken down into stages, and separate hardware units are provided for carrying out the computations of each stage. When the pipeline is full, each hardware unit is engaged in processing different data, and as each completes its task, it passes the results of its computation to the next unit.

As an example of a typical pipelining application, consider the performance of floating-point addition. Assume that numbers are represented in floating-point form as shown in Figure 12.1b and that it is desired to add the numbers 987 and 65.4. The floating point representation of these numbers is (+.9870 + 03) and (+.6540 + 02). The sequential addition of these numbers involves four distinct operations:

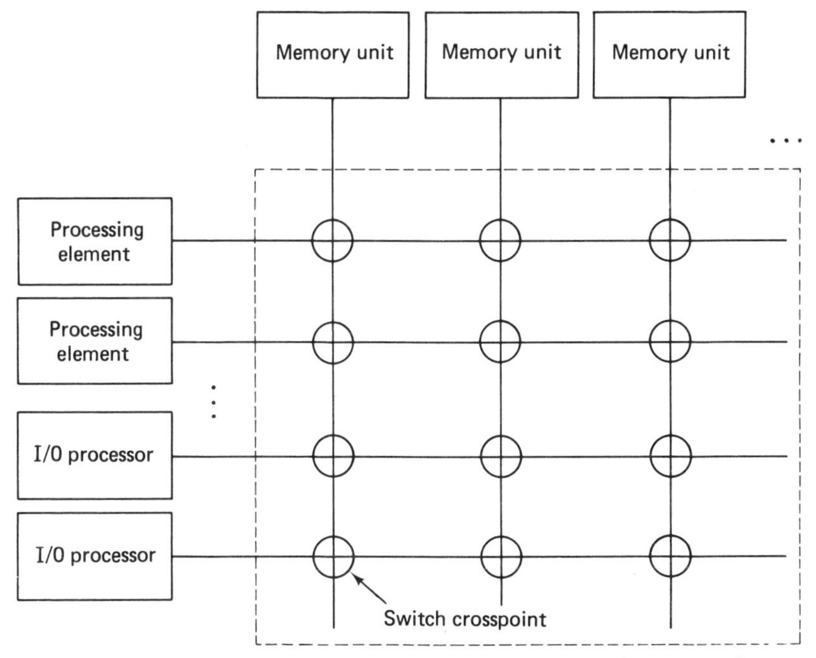

Figure 12.5 Crossbar switch method for interconnecting processing elements and memory units

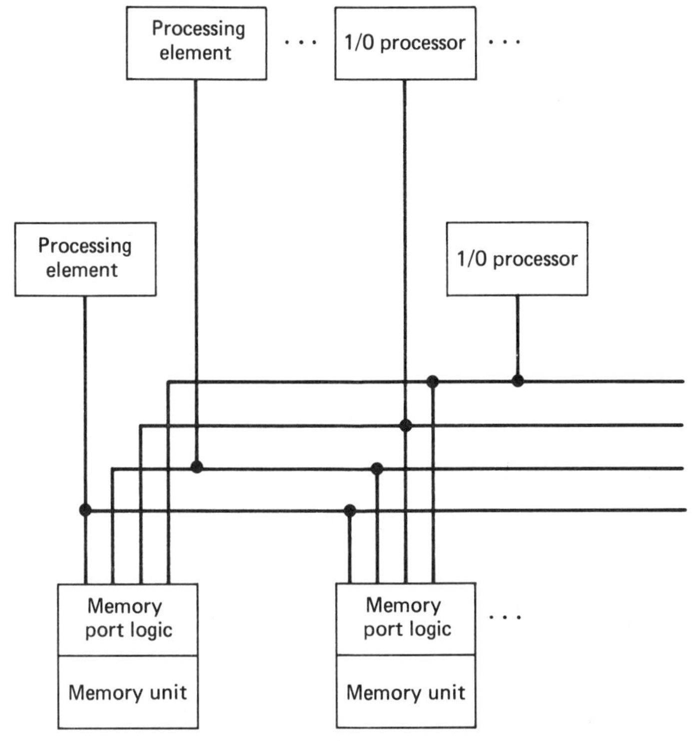

Figure 12.6 Ported memory structure

1. Modify one of the exponents so that the two numbers will have the same exponent: (+.6540 + 02) → (+.0654 + 03).
2. Add the mantissas: (+.9870) + (+.0654) → +1.0524.
3. Normalize: (+1.0524 + 03) → (+.10524 + 04).
4. Round off the result: (+.10524 + 04) → (+.1052 + 04).

Each of these operations can be regarded as a separate stage and so can be performed by a separate hardware unit. Assume that the characteristics of the electronic circuits are such that the slowest of these operations requires less than 50 nanoseconds (ns). It is then possible to pass data from one unit to the next and to commence work on a new pair of inputs every 50 ns. Once the pipeline is full, a new sum will appear at the output of the pipeline every 50 ns. This is illustrated in Figure 12.7, where summations are to be performed on two streams of numbers $a_1, a_2, a_3, \ldots$ and $b_1, b_2, b_3, \ldots$.

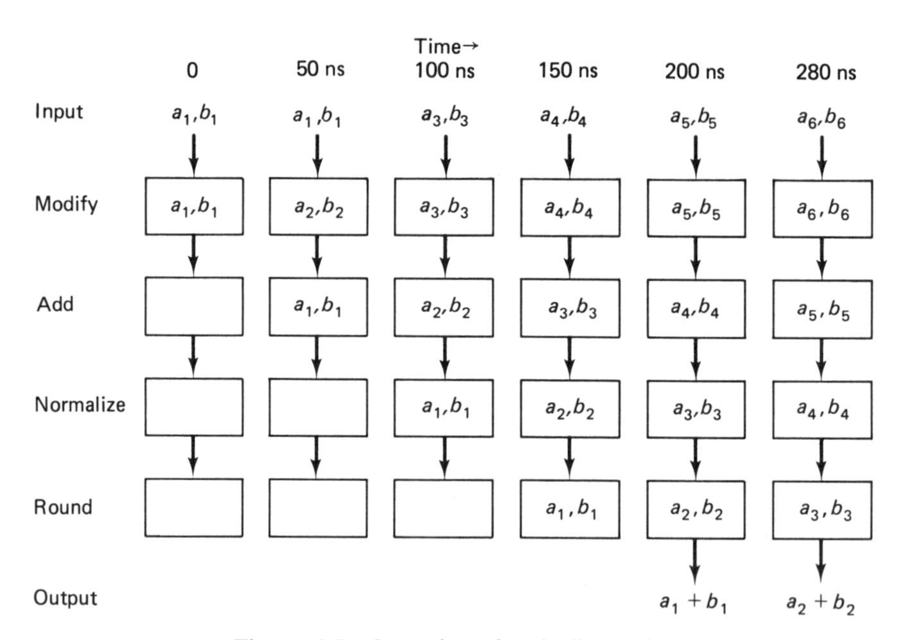

Figure 12.7 Operation of a pipeline adder

At time $t = 0$, the first pair of numbers to be added, a_1 and b_1, enter the pipeline, 200 ns later the sum $a_1 + b_1$ emerges from the pipeline adder. At that time each element in the pipeline is simultaneously operating on a different pair of numbers. Provided that the streams a and b are sufficiently long to keep the pipeline filled for a substantial length of time, the pipeline adder provides the same throughput as a single adder with a 50-ns addition time, or four times as rapidly as a conventional adder.

Pipelining techniques are particularly effective in enhancing execution speeds when processing loops. In FORTRAN, the DO command is employed to perform identical computations on arrays of numbers. For example, the two statements

```
    DO  100 I=1,75
100 A(I)=B(I)+5
```

are employed in FORTRAN to define an elementary loop operation. Two arrays A and B are involved. In this example, the index I ranges from 1 to 75 in steps of 1. The effect is to add the number 5 to each element of array B and to store the result in array A. In carrying out such an operation using a Von Neumann architecture, it is necessary to update the index after each computation and to expend an appreciable amount of computer time in loop control. By employing a pipelining approach, this "overhead" can be eliminated.

The pipelining concept has been extensively implemented, and a wide variety of computing systems utilizing pipelines are described in a detailed survey by Ramamoorthy and Li [5]. Several "supercomputer" systems employ several parallel pipeline processors. These include the STAR, the ASC, and the CRAY-1 computers discussed in Section 12.6. In matrix terminology, streams of numbers are termed vectors, and these pipeline computers are therefore referred to as vector processors.

12.5. THE ILLIAC IV: ARRAYS OF PROCESSING ELEMENTS

One of the most widely heralded of the supercomputers to be proposed in the 1960s was intended primarily to help in solving the massive field problems arising in meteorology, seismology, and fluid dynamics. Evolving from the earlier Solomon computers, ILLIAC IV was developed at the University of Illinois and is installed at the NASA-Ames Research Center in California. Since becoming fully operational in 1975, it has served as the chief testing ground for the SIMD architectural concept. At the same time it has provided computer engineers with an insight into the formidable difficulties inherent in achieving adequate performance, as described by Economidis [6] and by Thurber and Wald [7].

Initial plans for ILLIAC IV called for 256 parallel processing elements and four control units. The machine finally delivered by Burroughs Corporation actually contains only one control unit and 64 processing elements. Figure 12.8 is a simplified diagram of the ILLIAC IV organization. In line with the SIMD concept, each parallel processing element has an arithmetic unit, including the registers and logic necessary to execute a full set of instructions. In addition, each processing element has a separate fast semiconductor memory with 2048 64-bit registers.

The control unit is capable of executing some serial instructions, but most commands are transmitted to the processing elements. In the parallel mode, all processing elements are either executing a given instruction, or they are temporarily disabled by the control unit. The control unit also supervises the transfer of data between the various processing elements. The main memory of the system is a mag-

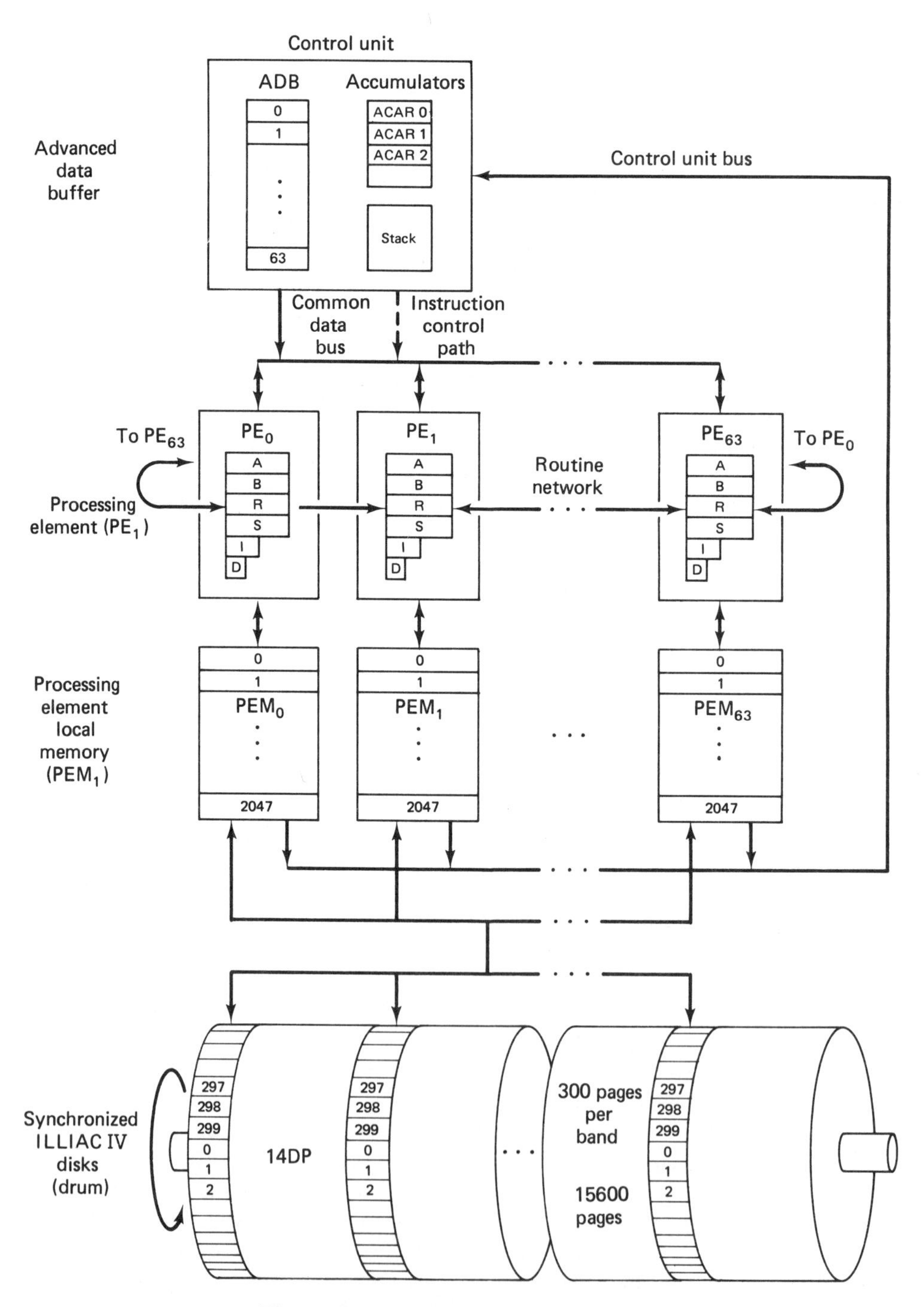

Figure 12.8 Organization of ILLIAC IV

netic drum with a capacity of 16 million words, which can be accessed by all the processing elements. A number of other memory devices, including magnetic discs and a laser memory, provide backup and establish a memory hierarchy. Distributed around the central memory is an array of mini- and midi-computers to facilitate communication with the external world and to perform other input and output operations. By means of the ARPA Network, the computer can be accessed via telephone lines by a large number of computer installations throughout the United States and the world.

Unlike serial or sequential computer operation, parallel processing makes it indispensable that the programmer comprehend the architectural features of the machine and that these be taken into account in designing algorithms and programs. In most current applications, the 64 processing elements can be considered to comprise an 8×8 array, as shown in Figure 12.9. Each processing element is connected only to its four nearest neighbors. Processing elements on the boundary of the array are connected to corresponding elements at the opposite boundary. Thus processing element 1 is connected to processing elements 57 and 64 as well as to elements 2 and 9. Data transfer between the processing elements and the magnetic drum memory is carried out in blocks of data, termed pages, each consisting of 1024 64-bit words.

Walkden and others [8] consider how some commonly used algorithms for solving

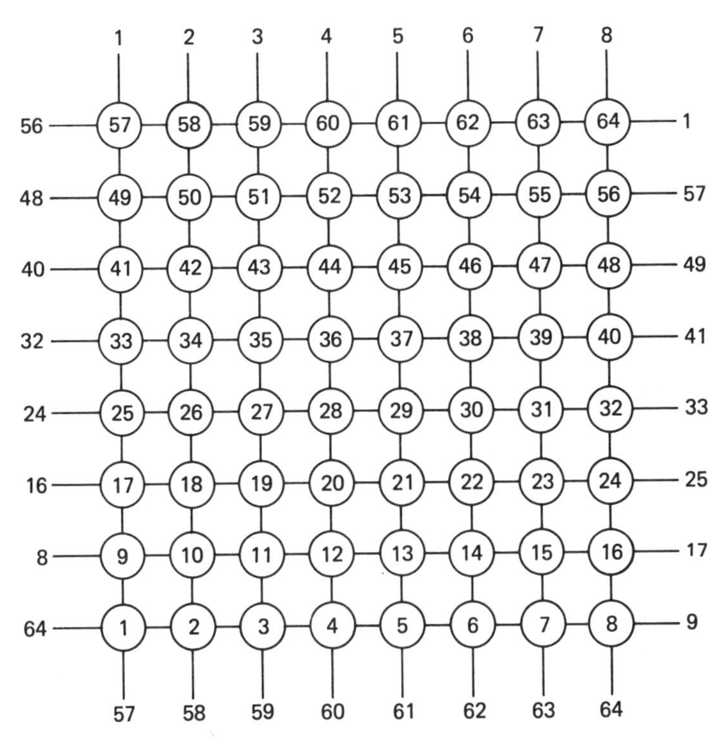

Figure 12.9 Organization of processing elements in ILLIAC IV

partial differential equations can advantageously be modified for parallel processing and provide a performance evaluation of ILLIAC IV, including comparisons with large serial computers. They conclude that, for problems of significant size, the speed advantage of ILLIAC IV compared to large serial computers such as the IBM 360/195 or the CDC 7600 is three to four at best. Reasons for this disappointingly small speed ratio (considering the size and cost of ILLIAC IV) include the following:

1. The time required to move blocks of data from the disk to the processing elements.
2. The need in many algorithms to pass data among processing elements that are not nearest neighbors (see Figure 12.9).
3. Difficulties in structuring data for storage and transfer, where the array of finite difference net points does not readily conform to the structure shown in Figure 12.9. This includes situations in which nets substantially larger than 8×8 are to be handled, so that each processing element is required to carry out the calculations for a number of net points and situations in which the finite difference net is not a square array of points.
4. Shortcomings in the processing elements in carrying out arithmetic operations.

Hopkins [9] describes the application of ILLIAC IV to a complicated seismology problem involving the solution of a hyperbolic partial differential equation in three dimensions. He reports attaining a speed-up of approximately 60 in comparing the ILLIAC IV solution to one obtained on a Univac 1108. The key to attaining this speed advantage lies in very careful data base design and in the adoption of a sophisticated data management scheme so as to minimize time lost in data transfers. Moreover, the finite difference grid selected was structured so as to conform as closely as possible to the organization of the ILLIAC IV processing elements.

Sameh and Kuck [10] present a comprehensive survey of the use of parallel computers to treat systems of linear algebraic equations, updating an earlier comprehensive survey of all published numerical algorithms for vector and parallel computers up to 1974 compiled by Poole and Voight [11]. Particularly intensive efforts have been devoted to solving tridiagonal systems of algebraic equations using parallel processors. This work has particular application in treating partial differential equations in one-space dimensions, as well as in two and three-dimensional systems using alternating direction algorithms.

12.6. STAR-100, ASC, AND CRAY-I: VECTOR PROCESSORS

The technique of pipelining, briefly described in Section 12.4, has been used to a limited extent in a variety of sequential computers, such as for example the IBM 360/91, in a way that does not substantially modify the basic Von Neumann architecture of these machines. In the mid 1970s, however, there appeared three "supercomputer" systems, which employ pipelines as major architectural elements and which therefore differ radically from classical architectural approaches. These are directed toward optimizing the processing of one-dimensional arrays of numbers, such as frequently arise in the solution of partial differential equations, and are therefore referred to as vector processors.

The general structure of the STAR-100 computer, manufactured by Control Data Corporation, is shown in Figure 12.10 and is described in considerable detail by Thurber [12]. This system is seen to be designed around two pipelined processors. Each contains a floating-point addition unit; one contains a multiply unit, while the other contains a divide unit. Buffering is the process of storing the results of a computation temporarily before passing them on to another structural element. Since the various elements in a pipeline may require different lengths of time to carry out their tasks, buffering is one of the principal problems encountered in the design of vector processors. In the STAR-100 system, the Write and Read buffers are employed to align the two vectors treated by the parallel pipeline processing units. There is also

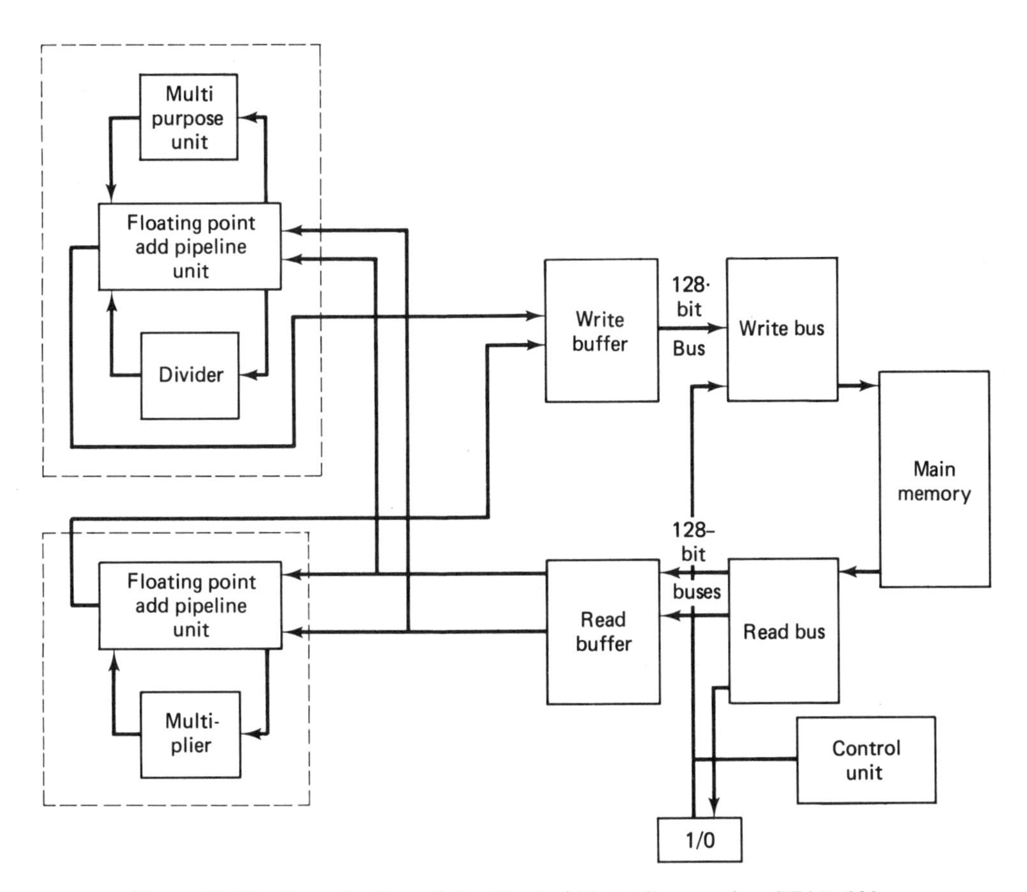

Figure 12-10 Organization of the Control Data Corporation STAR-100

an instruction buffer that is replenished from memory. The memory banks of the STAR-100 contain 32 interleaved units of 2048 512-bit words, with expansion possible to twice that size. Considerable effort has gone into the preparaton of software systems to facilitate the utilization of the STAR-100. These include a variety of FORTRAN extensions, making it possible to specify vector operations in a compact and efficient manner.

The design of the ASC (for advanced scientfic computer) manufactured by Texas Instruments, Inc., and also described by Thurber [12], is similar in many respects to the STAR-100. However, as shown in Figure 12.11, the ASC system contains four identical parallel pipeline units. Each pipeline unit contains pipeline hardware for floating-point and fixed-point addition and multiplication. When functioning in an optimum manner, these pipelines can produce a new result every 10 ns, corresponding

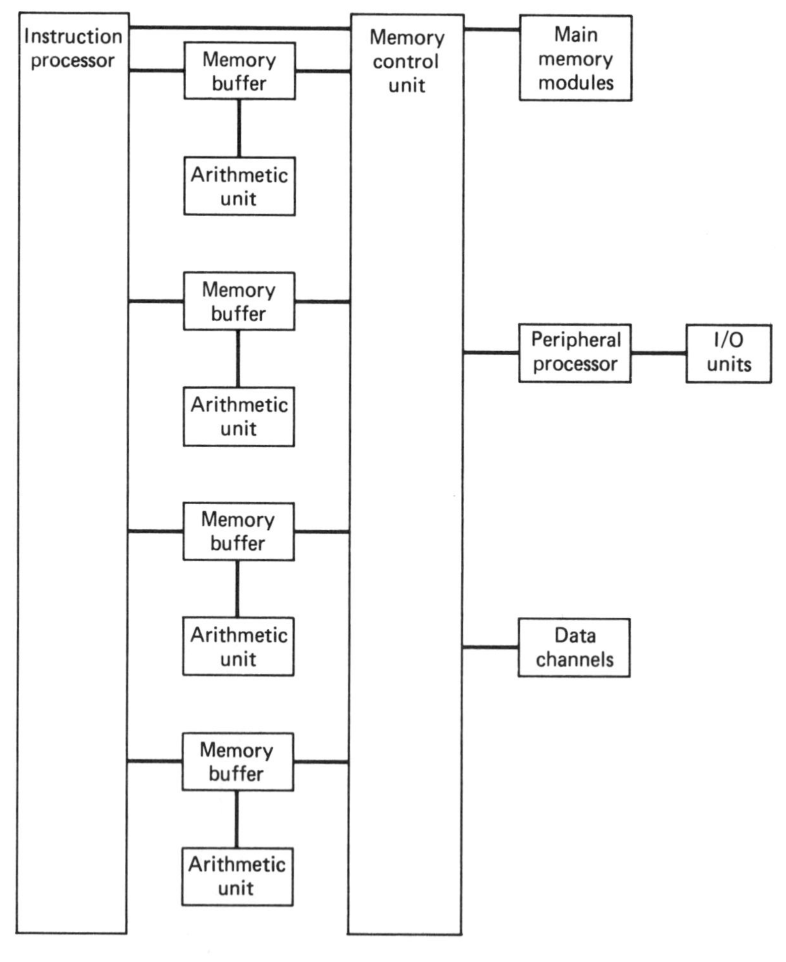

Figure 12.11 Organization of the Texas Instruments ASC

to 100 million results per second. As in the STAR-100, an extended FORTRAN package has been developed to facilitate programming.

Unlike the STAR-100 and the ASC, which were designed principally for vector operation and which rely upon peripheral processors for sequential computing, the CRAY-1 system manufactured by Cray Research, Inc., is designed for both scalar and vector operations. As described by Johnson [13] and shown in Figure 12.12, CRAY-1 contains both vector and scalar sections, the latter serving for sequential computations. As in the case of the other vector processors, CRAY-1 is highly effective in minimizing the "overhead" computations inherent in processing loops. In addition to its facility for nonvector processing, CRAY-1 also has improved channels for communicating with memory, thereby avoiding memory conflicts that arise in other vector processors. Vector processing is carried out with the aid of the eight vector registers, each containing 64 words. Fixed- or floating-point arithmetic operations are carried out using the pipeline ADD, MULTIPLY, and RECIPROCAL (divide) units. A new pair of operands can be accepted every 12.5 ns. The memory contains up to 1 million 64-bit words arranged in 16 banks.

Although vector processors have been in use since the mid 1970s, at the time of this writing no comprehensive theory for the design of algorithms for such computers had emerged. In fact, there were only very few reports of the solution of large and practical problems on such machines, and these usually employed adaptations of algorithms developed for serial processors. In constructing algorithms for vector

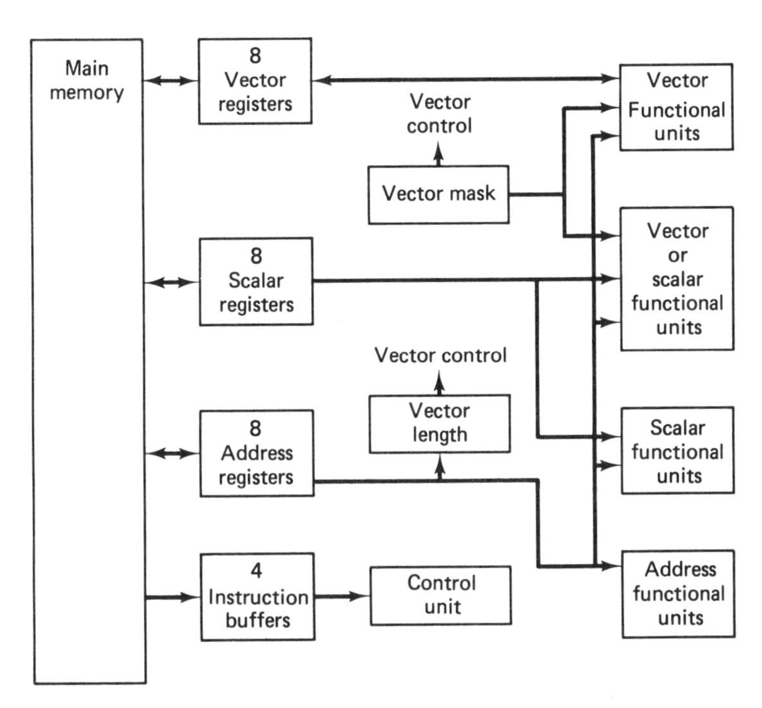

Figure 12.12 Organization of the Cray Research CRAY-1

processing, a basic aim is to perform the bulk of the computational task on one-dimensional arrays, that is, on vectors. When the length of these vectors is relatively short, the overhead involved in initiating vector instructions and in waiting for the pipeline to become full may well outweigh any advantages gained from vector processing, so vector processor computers may actually be slower than competing serial computers. As the pipeline becomes longer (say 100 to 1000 elements), vector processors become more and more effective. Ortega and Voight [14] present a detailed review of various approaches to the solution of partial differential equations on vector computers. The discussion in the rest of this section is based upon their paper.

In the solution of elliptic partial differential equations, on serial computers, there is a great advantage in dealing with tridiagonal matrixes, which have nonzero terms only on the main diagonal and on the elements immediately adjacent to the main diagonal. This permits the utilization of the tridiagonal algorithm to provide solutions for one-dimensional problems without iterations, and suggests the use of alternating direction methods for problems in two and three space dimensions. In the case of vector processors, on the other hand, no effective algorithms have as yet appeared to take full advantage of the special structure of tridiagonal matrixes. For one-dimensional elliptic equations, cyclic reduction techniques appear to be most effective. For problems in two and three dimensions, Jacobi iteration is often preferred to the Gauss-Seidel and the successive overrelaxation methods. By careful ordering of grid points and by clever programming techniques, alternating direction and successive overrelaxation methods can be successfully implemented on vector processors, albeit without achieving any spectacular increase in speed.

With regard to parabolic partial differential equations, comparisons of the relative effectiveness of various algorithmic approaches to simple problems exist. Consider the one-dimensional diffusion equation

$$\frac{\partial^2 u}{\partial x^2} = \frac{1}{\alpha}\frac{\partial u}{\partial t}, \qquad 0 < x < 1 \tag{12.1}$$

with initial and boundary conditions

$$u(x, 0) = g(x) \tag{12.2}$$
$$u(0, t) = \beta, \qquad u(1, t) = \gamma$$

where α, β, and γ are constants. The simplest explicit solution method characterized by

$$u_{l,j+1} = u_{l,j} + a(u_{l+1,j} - 2u_{l,j} + u_{l-1,j}), \qquad l = 1, \ldots, N \tag{12.3}$$

where

$$a = \alpha\frac{\Delta t}{\Delta x^2}$$

and $u_{l,j}$ and $u_{l,j+1}$ refer to the current and succeeding time levels, respectively, is most easily and directly handled on vector computers. The length of the vector in this case is equal to the number of interior grid points in the x domain, and five arithmetic operations are carried out on each element of the vector. By contrast, vector computers are not nearly as efficient for implicit schemes, such as the Crank-Nicolson method:

$$u_{l,j+1} = u_{l,j} + \frac{a}{2}(u_{l+1,j+1} - 2u_{l,j+1} + u_{l-1,j+1} + u_{l+1,j} - 2u_{l,j} + u_{l-1,j}) \qquad (12.4)$$

In this case a tridiagonal system of algebraic equations must be solved at each time level, which is not particularly easy using a vector processor. On the other hand, Eq. (12.4) is computationally stable regardless how large Δt is, whereas stringent limits on Δt are imposed by stability considerations in the case of Eq. (12.3). Depending upon the desired accuracy and the parameter α, it may be necessary to take an uneconomically large number of steps if the explicit method is used. The Dufort-Frankel method

$$u_{l,j+1} = u_{l,j-1} + 2a(u_{l+1,j} - u_{l,j+1} - u_{l,j-1} + u_{l-1,j}) \qquad (12.5)$$

is explicit, requiring only four vector operations per time step, and is also unconditionally stable. Unfortunately, Eq. (12.5) is not computationally consistent, so that the solution of the finite difference system does not necessarily approach the solution of Eq. (12.1) as Δx and Δt approach zero. Unless care is taken with the net spacings in the x-and t-directions, error terms proportional to $\partial^2 u/\partial t^2$ may appear in the solution. A comparison of the effectiveness of the CDC STAR-100 vector processor and the CYBER 175 serial computer in solving Eq. (12.1) using Eqs. (12.3) to (12.5) is presented in Table 12.1.

Table 12.1

	STAR-100 (μs)	*CYBER* 175 (μs)
Explicit, Eq. (12.3)		
$N = 50$	43	150
$N = 1000$	194	2500
Implicit, Eq. (12.4)		
$N = 50$	600	560
$N = 1000$	3900	11700
Dufort-Frankel, Eq. (12.5)		
$N = 50$	42	165
$N = 1000$	196	2800

From Ortega and Voight [14].

It is readily seen in all three cases that the relative effectiveness of the vector processor is far greater for 1000 interior net points than it is for 50 net points. In the case of 50 net points, almost 75% of the time for each vector operation is consumed by startup, while startup time drops to approximately 12% for 1000 interior points in the x domain. It can be seen that the times required for the two explicit methods characterized by Eqs. (12.3) and (12.5) are very nearly identical and far shorter than the times required by the implicit Eq. (12.4).

12.7. AP-120B AND AD-10: PERIPHERAL ARRAY PROCESSORS

In the late 1970s a number of manufacturers introduced a class of devices that they identified as ***array processors***. This designation is misleading, since the units are neither designed for the processing of multidimensional arrays of numbers, nor are they composed of arrays of processing elements, as is the case in the ILLIAC IV. As a class, they are intended to function as peripheral devices for conventional serial computers, and with the aid of internal parallelism greatly to accelerate the solution of certain problems. Array processors are generally structured as shown in Figure 12.13. They employ a multipath bus structure that permits the simultaneous perfor-

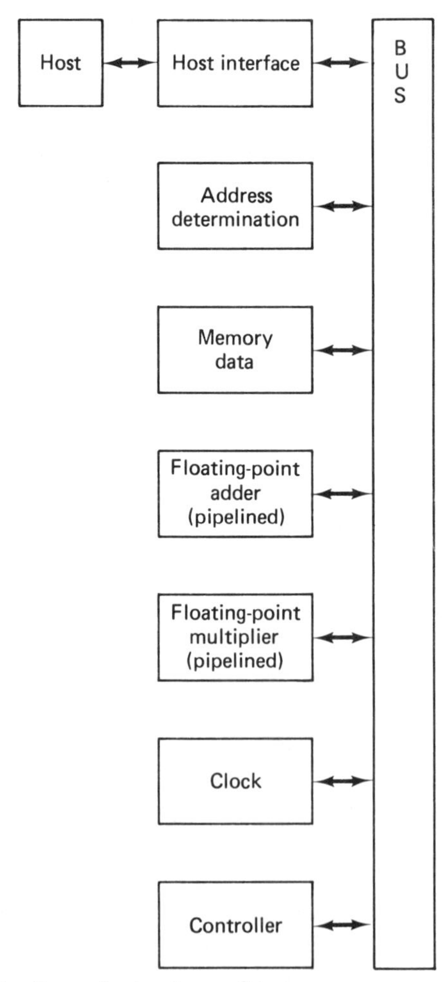

Figure 12.13 General structure of array processors functioning as peripheral elements

mance of fetches from a data memory, multiplication, and addition. Data to be processed are read into the peripheral processor by a host computer and subjected to a very rapid sequence of operations. Through the use of the latest solid-state circuitry, compact design, and extensive pipelining, these peripheral processors are able to exceed the speed of even the most powerful general purpose digital computers at a relatively modest cost. Manufacturers who market such array processors include IBM, Control Data Corporation, Data General Corporation, CSP, Inc., Applied Dynamics, Inc., and many others. Floating Point Systems, Inc., which produces units designated as AP-120B and AP-190, delivered well over 500 such units prior to 1980, and thereby assumed a leadership position in this specialized field.

Figure 12.14 shows a block diagram of Floating Point Systems AP-120B, described by Wittmayer [15]. The system elements are interconnected by multiple paths so that transfers can occur in parallel. All internal floating-point data are 38 bits in width (10-bit exponent and 28-bit mantissa). The interface unit is designed especially for the host computer, which is usually a minicomputer, and is organized so that either I/O or direct memory access channels can be utilized for data transfer. Instruction and data transfers take place at a 6-MHz rate, corresponding to a cycle time of 167 ns. The operation of the unit is controlled by the execution of 64-bit instruction words, which reside in the program memory. Additional control functions are pro-

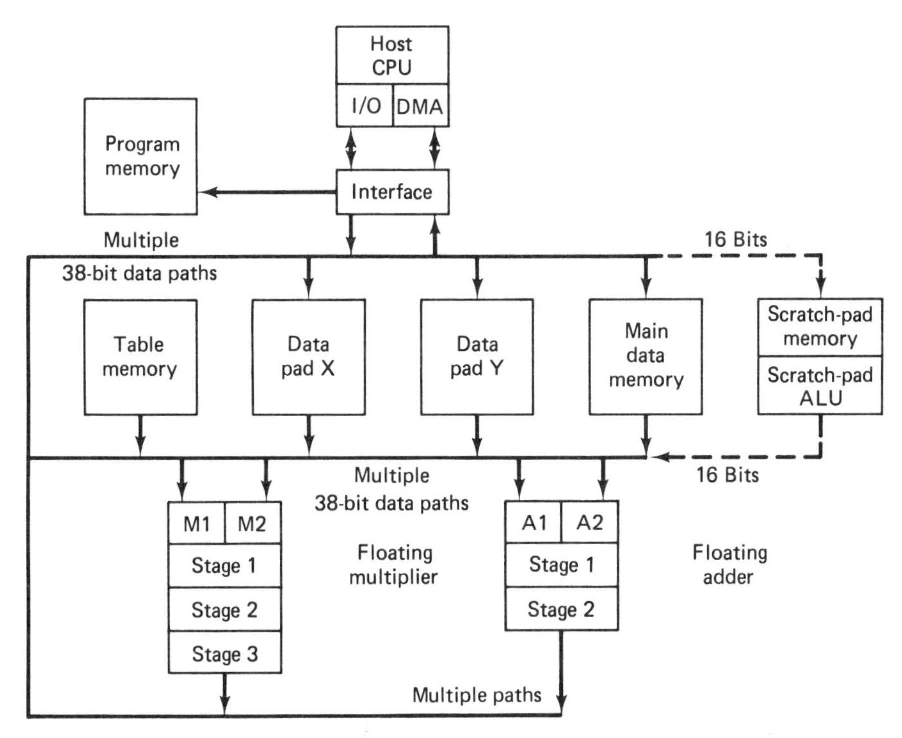

Figure 12.14 Organization of the floating point systems AP-120B

vided by the S-PAD unit, which performs integer address indexing, loop counting, and other tasks required by specific algorithms. Addition is performed in two pipelined stages, each of which takes 167 ns. Multiplication is performed in three pipelined stages, each requiring 167 ns. The data memory unit is the primary data store for the AP-120B and has a cycle time of 333 ns. so a new memory operation may be initiated every other machine cycle. To optimize the operation of the processor, it is necessary for the programmer to "look ahead" and initiate memory fetches prior to the actual time that arguments from the data memory are to be used in the calculations. The table memory unit employs more rapid and therefore more expensive circuitry and, as the name implies, is used to store data for table look-up. The DATA-PAD unit consists of two fast accumulator blocks, each with 32 floating-point locations. This unit serves primarily for the storage of intermediate results of computations.

During the 1970s the AP-120B and other peripheral array processors were used almost entirely for signal-processing applications, particularly those arising in medical tomography and seismic data handling. Algorithms for the solution of partial differential equations remained to be developed. One possible approach to this problem has been reported by Dawson, Huff, and Wu [16] in the use of the AP-120B for the simulation of physical phenomena in plasma systems.

The AP-120B was designed primarily for signal processing. By contrast, the AD-10 manufactured by Applied Dynamics, Inc., is a simulation-oriented peripheral processor intended primarily for the solution of systems of ordinary differential equations. As shown in Figure 12.15 and described by Gilbert and Howe [17], this unit has in many respects a structure similar to that of other array processors. The multibus is a parallel high-speed bus composed of 16 data lines, 18 address lines, and several control lines, and supports 20 data and address transfers per microsecond. All transfers, as well as all memory processor functions, are synchronously controlled by a 40-MHz master clock. In addition to the data memory and the pipelined arithmetic processors, the AD-10 also contains separate hardware units to permit memory mapping, breakpoint determinations in nonlinear function generation by table look-up, and a number of other functions of importance in solving nonlinear equations.

The AD-10 is unique among available peripheral processors in that it has a distributed program-control memory. Each functional unit has a separate instruction memory that controls its action during each clock cycle. Prior to a computer run, the host computer loads each of these program memories. The host interface controller couples the host computer to the AD-10, distributes the instructions to the appropriate processors, and loads data into the multiport data memory. The memory is organized in pages of 4096 16-bit words, and each page is ported separately so that it may be addressed independently. The arithmetic processor contains two adders and one multiplier, as well as a number of temporary storage registers. With the aid of pipelining, this unit can perform 20 additions and 10 multiplications per microsecond (μs). Unlike other array processors, the AD-10 is a fixed-point machine, making it necessary to scale all variables to assure that no register overflow or underflow occurs.

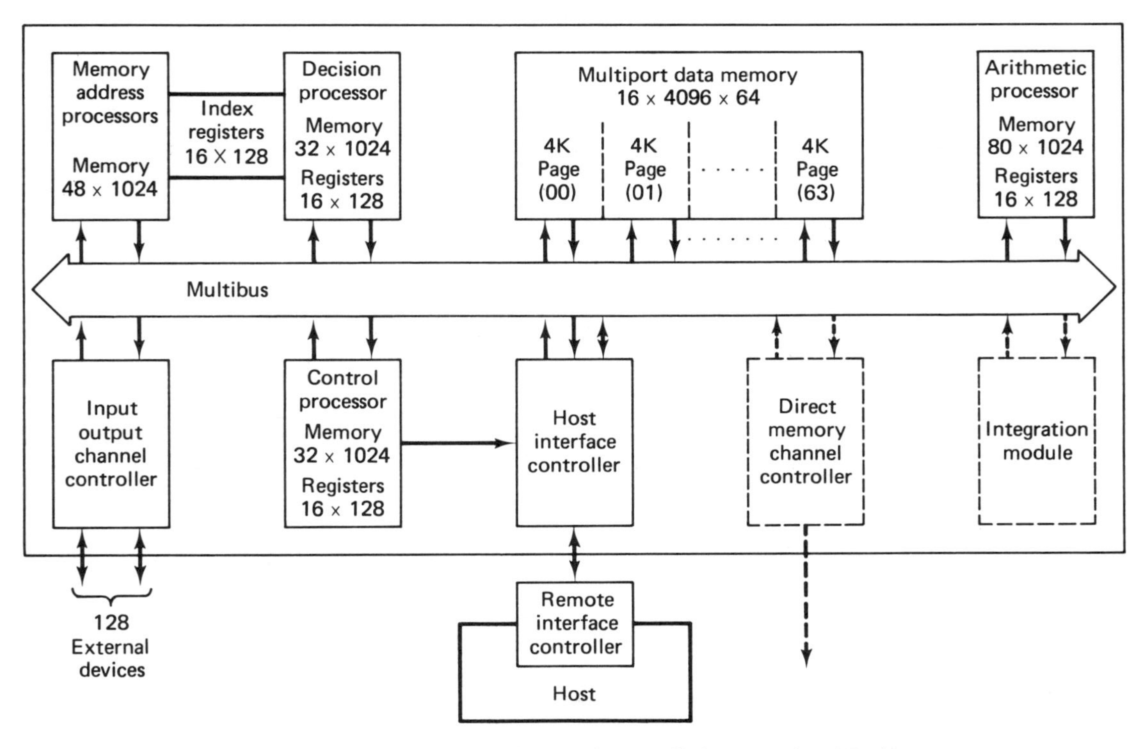

Figure 12.15 Organization of the Applied Dynamics AD-10

12.8. NETWORKS OF MICROPROCESSORS

The prospect of fashioning networks of general-purpose computers so as to attain greater speed through parallelism has been a tantalizing prospect for many years. A number of attempts to interconnect relatively inexpensive general-purpose minicomputers have been made over the years for various applications. With the increasing availability of very fast and very economical large-scale integrated (LSI) semiconductor units, it becomes feasible to contemplate the construction of networks of microprocessor units to form special-purpose simulators. Such parallel computer systems would be of the MIMD variety and would therefore be considerably more powerful than the earlier SIMD systems, such as ILLIAC IV, at greatly reduced cost. Each element of such an array of processors would contain individual arithmetic, memory, and control. Possibly, a favorable cost-speed tradeoff could be achieved by limiting the instruction repertoire of the control units to those commands that are required for the solution of partial differential equations.

Notwithstanding the many potentially attractive features of microprocessor networks, the development of actual prototype systems of this type has been frustratingly slow. Probably a major handicap in the development of this type of simulator system has been the fact that no manufacturer has found worthwhile a financial stake in the development of such systems. Indeed there are formidable hardware and software problems to be resolved. On the hardware side, it is necessary to select one of a number of ways of interconnecting the individual microprocessor units, and different simulation applications appear to demand different connection strategies. Also, substantially different algorithmic approaches appear indicated. For these and other reasons, large-scale simulation-oriented microprocessor networks remained in the proposal rather than the implementation stage throughout the 1970s.

Weissberger [18] presented a detailed analysis of the various considerations involved in the specification of a microprocessor network. He lists well over 50 commercially available microprocessor units that could be candidates for network applications. He then describes alternative ways of allocating tasks and of interconnecting and synchronizing the microprocessors comprising the network. The ability to identify and isolate component failures is an important attribute of multiprocessor systems. As the number of microprocessors in a network is increased, the probability that any one of them will be malfunctioning at any given time increases, and careful design techniques are required to maintain system reliability.

In 1977, a workshop sponsored by the Rand Corporation considered the feasibility of a special-purpose microprocessor network designed to solve the Navier-Stokes equations describing fluid flow. The report [19] produced by this workshop focused on a network simulator consisting of an array of 10,000 identical microprocessor units arranged in a 100×100 matrix. To minimize interconnection costs, each processor is limited to communicate only with its nearest neighbors. The proposed topology is therefore essentially similar to that shown in Figure 12.9, albeit much larger. For problems in three space dimensions, each processing element represents a point in a two-dimensional plane, data in the third dimension being carried along by the memory associated with each processing element. In this way, a three-dimensional finite difference grid containing approximately 1 million grid points can fit on this computer at one time, making it unnecessarry to communicate with a backup memory in the course of a computation. Each processing element would be capable of storing 256 64-bit words and of carrying out a 64-bit fixed-point multiplication in 5 μs. Interprocessor transfers are assumed to require 9 μs per word. The system would be much more powerful than ILLIAC IV and would far surpass the processing capability of any competitive system. Grosch [20] has made detailed studies of algorithms suitable for the proposed simulator. In particular, he considered the solution of Poisson's equation in three dimensions,

$$\frac{\partial^2 u}{\partial x^2} + \frac{\partial^2 u}{\partial y^2} + \frac{\partial^2 u}{\partial z^2} = u(x, y, z)$$

with various boundary conditions. A number of alternative algorithms were evaluated from the point of view of operations count and overhead, including the time expended for data transfers between processing elements.

Many other proposals for microprocessor networks have appeared in the literature. Most of these contain some novel architectural features. For example, the system proposed by Cyre and others [21] is composed of a three-dimensional array of microprocessor units arranged to permit rapid nearest-neighbor data transfer and featuring a special "pass-through" scheme to facilitate transfers to other than nearest-neighbor processing elements.

REFERENCES

1. A. S. Tanenbaum, *Structured Computer Organization*. Englewood Cliffs, N.J.: Prentice-Hall, Inc., 1976.
2. J. P. Hayes, *Computer Architecture and Organization*. New York: McGraw-Hill Book Co., 1978.
3. M. J. Flynn, "Some Computer Organizations and Their Effectiveness," *IEEE Transactions on Computers*, Vol. C-21, pp. 948–960, September 1972.
4. P. H. Enslow, "Multiprocessor Organization—A Survey," *ACM Computing Surveys*, Vol. 9, pp. 103–129, March 1977.
5. C. V. Ramamoorthy and H. F. Li, "Pipeline Architecture," *ACM Computing Surveys*, Vol. 9, pp. 61–103, March 1977.
6. T. Economidis, "Computing in the Range of 100 Megaflops," *Wescon 77 Convention Record*, San Francisco, September, 1977.
7. T. J. Thurber and L. D. Wald, "Associative and Parallel Processors," *ACM Computing Surveys*, Vol. 7, pp. 215–255, December 1975.
8. F. Walkden, H. A. J. McIntyre, and G. T. Laws, "A User's View of Parallel Processors," Institute for Advanced Computation, *Newsletter*, Vol. 1, No. 8, pp. 1–14, August 1977.

9. A. S. HOPKINS, "A Three-Dimensional Finite Difference Code for Seismic Analysis on the ILLIAC IV Parallel Processor," *Society of Automotive Engineers*, Aerospace Meeting, Paper No. 770956, Los Angeles, Calif., November 1977.

10. A. H. SAMEH and E. J. KUCK, "Parallel Direct Linear System Solvers—A Survey," *Transactions of IMACS*, Mathematics and Computers in Simulation, Vol. 19, pp. 272–277, December 1977.

11. W. POOLE and R. VOIGHT, "Numerical Algorithms for Parallel and Vector Computers: An Annotated Bibliography," *ACM Computing Reviews*, Vol. 15, pp. 379–388, 1974.

12. K. J. THURBER, "Parallel Processor Architectures—Part 1: General Purpose Systems," *Computer Design*, Vol. 18, pp. 89–97, January 1979.

13. P. M. JOHNSON, "An Introduction to Vector Processing," *Computer Design*, Vol. 17, pp. 89–97, February 1978.

14. J. M. ORTEGA and R. G. VOIGHT, "Solution of Partial Differential Equations on Vector Computers," Report Number 77–7, ICASE, March 30, 1977.

15. W. R. WITTMAYER, "Array Processor Provides High Throughput Rates," *Computer Design*, Vol. 17, pp. 93–100, March 1978.

16. J. M. DAWSON, R. W. HUFF, and C. C. WU, "Plasma Simulation on the UCLA CHI Computer System," *AFIPS, Proceedings National Computer Conference*, Vol. 47, pp. 395–407, 1978.

17. E. O. GILBERT and R. M. HOWE, "Design Considerations in a Multiprocessor Computer for Continuous System Simulation," *AFIPS, Proceedings National Computer Conference*, Vol. 47, pp. 385–393, 1978.

18. A. J. WEISSBERGER, "Analysis of Multiple-Microprocessor System Architectures," *Computer Design*, Vol. 16, pp. 151–163, June 1977.

19. E. C. GRITTON and others, "Feasibility of a Special-Purpose Computer to Solve the Navier-Stokes Equations," Rand Corporation, Report No. R-2183-RC, June 1977.

20. C. E. GROSCH, "Poisson Solvers on a Large Array Computer," Department of Mathematical and Computing Sciences, Old Dominion University, Norfolk, Va., TR78-4, October 1978.

21. W. R. CYRE and others, "WISPAC: A Parallel Array Computer for Large-Scale System Simulation," *Simulation*, Vol. 29, pp. 165–172, 1977.

EXERCISES

1. In a well-organized discussion, distinguish clearly between the terms "pipeline" and "parallelism" as they are used in the design of digital computer systems.

2. With the aid of a block diagram, describe how a pipelined floating-point multiplier could be designed.

3. It is stated in Section 12.6 that the Thomas (tridiagonal) algorithm for problems in one space dimension is not efficient when used in a pipeline "vector" processor. Explore this difficulty by describing in detail how the Thomas algorithm could be implemented in a vector processor.
4. Using Taylor series, prove that the Dufort-Frankel approximation, Eq. (12.5), to the one-dimensional diffusion equation, Eq. (12.1), is not consistent; that is, as Δx and Δt approach 0, the equation actually being solved may assume some of the properties of a hyperbolic partial differential equation. What measures must be taken to avoid this possibility? What is the effect of these measures upon computational efficiency?
5. Distinguish clearly between "arrays of processors," such as ILLIAC IV, and the "array processors" discussed in Section 12.7.
6. Survey the recent technical literature to determine which of the currently commercially available peripheral array processors is most effective for the solution of elliptic and parabolic partial differential equations.

13

Example Programs: Finite Difference Methods

13.1. INTRODUCTION

This chapter and the next are primarily aimed at those who have never programmed a digital computer to solve partial differential equations, but who do have experience in writing small segments of codes to solve relatively simple numerical analysis problems. The problems discussed here are complex enough to convey to the student the probable complexity of real-life problems; however, the programs supplied herewith are by no means sophisticated. That is, here emphasis is placed on the mechanics of translating an algorithm into a program that works at least for the specific cases discussed.

A word about the programming language chosen is in order. Although a number of programming languages are currently available, only some, such as APL, PL/1, ALGOL, PASCAL, and FORTRAN, are considered suitable for solving problems of the kind treated in this text. The choice of which language to use often depends upon the programmer's previous experience, the language in which most of the library programs are available, and the availability and efficiency of the respective compilers. Most of the library routines as well as a good number of special-purpose software packages (for a discussion of some of these, see Chapter 11) are in FORTRAN. Therefore, we decided to use FORTRAN in these example programs.

Traditionally, preparation of a program meant that one prepare a flow chart, and write the code while sprinkling the code with an occasional comment. As software development costs are mounting, writing and maintaining programs became a serious business. Recent research in the area of programming methodology led to the emergence of a new discipline called software engineering. Ideas such as top-down design, structured programming and design aids such as structure charts (or Nassi-Shneiderman charts), HIPO (hierarchy-input-process-output) charts, and pseudocodes are coming into widespread use, rendering the traditional tools, such as flow charts, obsolete. While traditional programming goals were correctness, efficiency, and creativity, modern software engineering discipline aims at correctness, clarity, comprehensibility, modifiability, maintainability, and the like. However, incorporating all these modern ideas in this text would be impractical. Serious programmers should consult standard textbooks on this subject.

For the purpose of this chapter, it is sufficient to narrow the scope of the discussion on program development to the following four topics: (1) design and documentation, (2) selection of an algorithm, (3) coding in some suitable source language, and (4) debugging and testing.

The design stage involves an identification of the functions to be performed (eg., generation of a grid or setting up of coefficient matrixes), possible implementation strategies, as well as documentation. In fact, documentation should go hand in hand with design and should precede the actual coding. Indeed, some believe that documentation is design! That is, documentation does not mean sprinkling the program liberally with comments. Documentation should include any and all information that renders the program readable and understandable. J. C. Reynolds says that there is a vast difference between a program being understandable and the program being correct. Information that is sufficient to determine the behavior of a computer will seldom be adequate to reveal the purpose of that behavior or to demonstrate that the behavior will satisfy that purpose. The program should bridge this gap by providing appropriate and adequate comments. Documentation should also include the nature of the functions to be performed, their scope and limitations, the inputs necessary to perform the function, the outputs generated, nature of the individual program modules, how these modules communicate with each other, and so forth. Imaginative selection of variable names, judicious use of comments, systematic numbering of the statements, indentation of the DO loops, and many other modern practices will of course improve the readability and therefore the understandability of a program.

The second stage of program development is the choice of an algorithm. Selection of an algorithm to meet the needs of a particular problem is not always a straightforward procedure. While selecting an algorithm, one is generally concerned about five important issues: stability, robustness, accuracy, adaptability, and speed. The final choice is essentially a compromise among these issues. Stability is essentially a measure of how sensitive an algorithm is to the accumulation of roundoff errors. Robustness is a measure of the ease with which a program can gracefully recover from arithmetic exceptions that may occur infrequently on unusual data. Accuracy is influenced by how errors (such as truncation and roundoff) are treated. A good algorithm should be designed not only to give an approximate solution to a given problem, but also to provide some measure of error in the computed value of the result. This is a difficult proposition while solving partial differential equations.

Without an adequate understanding of the errors, how do we know that our algorithm is producing accurate results? This question can be tackled by dividing it into two parts. If we assume that we have a correct program to implement the algorithm, then we can safely conclude that errors in the computed answers are due to truncation and roundoff. In the absence of rigorous methods to estimate these errors, how can we be sure that any and all errors in the computed result are only due to truncation and roundoff and not due to an incorrect program? Any attempt to answer this question will take us too far afield into the area of program validation and verification. For the purpose of meeting the requirements of this chapter, we can address the question of the accuracy of an algorithm by solving some problems whose exact solutions are known. This exact solution can be used not only to estimate the accuracy of an algorithm but also to verify that the program is indeed working. For this purpose, some selected equations and their exact solutions are summarized next.

1. $u_t = u_x, \qquad -\pi \leq x \leq \pi$
$u(x, 0) = \sin x$
$u(\pi, t) = u(-\pi, t)$
Exact solution: $u(x, t) = \sin (x + t)$
Reference [1]

2. $u_t = u_{xx}, \qquad -\pi \leq x \leq \pi$
$u(x, 0) = \sin x$
$u(\pi, t) = u(-\pi, t)$
Exact solution: $u(x, t) = e^{-t} \sin x$
Reference [1]

3. $u_t = u_{xx}, \qquad 0 \leq x \leq 1$
$u(x, 0) = 1$
$u(0, t) = u_x(1, t) = 0$
Exact solution:

$$u(x, t) = \sum_{n=1}^{\infty} c_n \exp\left[-\pi^2 \frac{(2n-1)^2}{4} t\right] \sin \frac{(2n-1)}{2} \pi x$$

where c_n are constants to fit the initial conditions.
Reference [2]

4. $u_t = \alpha^2 u_{xx}, \qquad 0 \leq x \leq L$

$$u(x, 0) = u_0 + \frac{u_1 - u_0}{L} x$$

where u_0 and u_1 are given constants

$$\frac{\partial u}{\partial x} = 0 \text{ at } x = 0 \text{ and } x = L \text{ for } t > 0$$

(or, equivalently, $u = u_0$ at $x = 0$ and $u = u_1$ at $x = L$)
Exact solution:

$$u(x, t) = u_0 + \frac{2(u_1 - u_0)}{\pi} \sum_{n=1}^{\infty} \frac{(-1)^n}{n} \sin\left(\frac{n\pi x}{L}\right) \exp\left(\frac{-\alpha^2 n^2 \pi^2 t}{L^2}\right)$$

Reference [3]

5. $u_{tt} = c^2 u_{xx}, \quad 0 \leq x \leq L$

$$u(x, 0) = \begin{cases} 2ax/L, & 0 \leq x \leq L/2 \\ 2a(L - x)/L, & L/2 \leq x \leq L \end{cases}$$

$u(0, t) = u(L, t) = 0, \quad$ for $t \geq 0$

Exact solution:

$$u(x, t) = \sum_{m=0}^{\infty} (-1)^m \frac{8a}{\pi^2(2m + 1)^2} \sin\left[\frac{(2m + 1)\pi x}{L}\right] \cos\left[\frac{(2m + 1)\pi ct}{L}\right]$$

Reference [3]

6. $u_{xx} + u_{yy} = -2, \quad 0 < x, y < 1$

$u(0, y) = u(1, y) = 0$

$\dfrac{\partial u}{\partial y}(x, 0) = \dfrac{\partial u}{\partial y}(x, 1) = 0$

Exact solution: $u(x, y) = x(1 - x), \quad 0 \leq x, y \leq 1$

Reference [4]

7. $u_{xx} + u_{yy} = 0$

$(x, y) \in$ circle centered at $(0.5, 0.5)$ and radius $= 0.5$

$\dfrac{\partial u}{\partial n} = g \quad$ with $u(0.5, 0) = 1$

Exact Solution: $u(x, y) = \tan^{-1}(y/x)$

Note: Here g denotes the function whose value is determined to make the problem have the specified solution.

Reference [5]

8. $u_{xx} + u_{yy} = 0, \quad 0 \leq x \leq a, \quad 0 \leq y \leq b$

$u(x, y) = 0 \quad$ on $x = 0$ and $x = a$ for $0 \leq y \leq b$

$u(x, y) = 0 \quad$ on $y = 0$ for $0 \leq x \leq a$

$u(x, y) = f(x) \quad$ on $y = b$ for $0 \leq x \leq a$

Case 1: If $f(x) = u_0 =$ a known constant

Exact solution:

$$u(x, y) = \frac{2u_0}{\pi} \sum_{n=1}^{\infty} \frac{[1 - (-1)^n]}{n \sinh(n\pi b/a)} \sinh\left(\frac{n\pi y}{a}\right) \sin\left(\frac{n\pi x}{a}\right)$$

Case 2: If $f(x) = \dfrac{4u_0 x(1 - x)}{a^2}, \quad u_0$ a given constant

Exact solution:

$$u(x, y) = \frac{16u_0}{\pi^3} \sum_{n=1}^{\infty} \frac{[1 - (-1)^n]}{n^3 \sinh(n\pi b/a)} \sinh\left(\frac{n\pi y}{a}\right) \sin\left(\frac{n\pi x}{a}\right)$$

Reference [3]

9. $u_{xx} + u_{yy} = 0, \quad 0 \leq x \leq a, \quad 0 \leq y \leq b$

$u(x, y) = 0 \quad$ on $x = 0$ and $x = a$ for $0 \leq y \leq b$

$\dfrac{\partial u}{\partial y} = 0 \quad$ on $y = 0$ for $0 \leq x \leq a$

$u(x, y) = f(x) \quad$ on $y = b$ for $0 \leq x \leq a$

Case 1: If $f(x) = u_0 =$ a known constant

Exact solution:

$$u(x, y) = \frac{2u_0}{\pi} \sum_{n=1}^{\infty} \frac{[1 - (-1)^n]}{n \cosh (n\pi b/a)} \cosh \left(\frac{n\pi y}{a}\right) \sin \left(\frac{n\pi x}{a}\right)$$

Case 2: If $f(x) = \dfrac{4u_0 x(a - x)}{a^2}$, u_0 a given constant

Exact solution:

$$u(x, y) = \frac{16u_0}{\pi^3} \sum_{n=1}^{\infty} \frac{[1 - (-1)^n]}{n^3 \cosh (n\pi b/a)} \cosh \left(\frac{n\pi y}{a}\right) \sin \left(\frac{n\pi x}{a}\right)$$

Reference [3]

10. $u_{xx} + u_{yy} = f$

$(x, y) \in$ ellipse centered at (0, 0) with major axis = 2 units and minor axis = 1 unit

$u(x, y) = g$ everywhere on the boundary

Exact solution: $u(x, y) = (e^x + e^y)/(1 + xy)$

Note: Here f and g denote functions whose values are determined to make the problem have the specified solution.

Reference [5]

13.2. EXAMPLE 1: A LINEAR HYPERBOLIC PDE

Consider the problem of solving

$$\frac{\partial^2 u}{\partial x^2} = \frac{\partial^2 u}{\partial t^2}, \qquad 0 \leq x \leq 1 \tag{13.1}$$

subject to the initial and boundary conditions

$$u(x, 0) = f(x) = \sin \pi x \tag{13.2a}$$

$$\frac{\partial u}{\partial x}(x, 0) = g(x) = 0 \tag{13.2b}$$

$$u(0, t) = u(1, t) = 0 \tag{13.2c}$$

This simple equation is chosen because its exact solution is known to be

$$u(x, t) = \sin \pi x \cos \pi t \tag{13.3}$$

Therefore, the accuracy of the computed solution can be determined by comparing it against the exact solution.

A first step of the numerical procedure, as usual, is discretization. A simple discrete model to Eq. (13.1) can be obtained by replacing u_{xx} and u_{tt} by central differences. This leads to

$$\frac{u_{I,J+1} - 2u_{I,J} + u_{I,J-1}}{(\Delta t)^2} = \frac{u_{I+1,J} - 2u_{I,J} + u_{I-1,J}}{(\Delta x)^2} \tag{13.4}$$

Solving for $u_{l,j+1}$,

$$u_{l,j+1} = \frac{(\Delta t)^2}{(\Delta x)^2}(u_{l+1,j} + u_{l-1,j}) - u_{l,j-1} + 2\left(1 - \frac{\alpha(\Delta t)^2}{(\Delta x)^2}\right)u_{i,j} \tag{13.5}$$

If the ratio $(\Delta t)^2/(\Delta x)^2 = 1$, the last term vanishes and we get a considerably simpler formula:

$$u_{l,j+1} = u_{l+1,j} + u_{l-1,j} - u_{l,j-1} \tag{13.6}$$

This formula can be used to solve Eq. (13.1) provided we can justify the selection of the preceding ratio.

There is yet one more problem before Eq. (13.6) can be implemented for a solution. The given initial condition in Eq. (13.2a) provides the value of u at $t = t_0 = 0$, that is, the value of $u_{l,0}$. To calculate $u_{l,1}$ from Eq. (13.6), we need not only $u_{l,0}$ but also $u_{l,-1}$, that is, the value of u at time $t = t_{-1}$. This can be obtained by using a fictitious time level at $t = t_{-1}$ and invoking the condition in Eq. (13.2b). Approximating u_t by a backward difference formula

$$\left.\frac{\partial u(x,0)}{\partial t}\right|_{l,0} = \frac{u_{l,0} - u_{l,-1}}{(\Delta t)} = 0 \tag{13.7}$$

we have

$$u_{l,-1} = u_{l,0} \tag{13.8}$$

However, Gerald [6] observes that this strategy is not good if $u_t \neq 0$ at the initial time. He suggests that a possible alternative is to generate the solution $u_{l,1}$ from $u_{l,0}$ using a simple integration scheme such as Simpson's one-third rule. However, because $u_t = 0$ in Eq. (13.2b), we propose to proceed with Eq. (13.8). Figure 13.1 shows a computer program as well as the correct and computed solutions.

```
C   PRCBLEM 1
C   THIS PROGRAM SOLVES THE ONE DIMENSIONAL HYPERBOLIC
C   ECUATICN SHCWN IN EQS (13.1) AND (13.2). THE INITIAL
C   DISPLACEMENT U IS DEFINED BY F(X).THE INITIAL VELOCITY
C   IS DEFINED BY G(X). FOR ANY GIVEN PROBLEM, F(X)
C   AND G(X) ARE SPECIFIED BY FUNCTION SUBPROGRAMS.
C
C   THE PARAMETERS ARE
C
C   X       DISTANCE FROM THE ORIGIN
C   DX      SPATIAL GRIC SIZE
C   L       LENGTH OF THE INTERVAL
C   N       NUMBER OF SUBDIVISIONS
C   T       TIME
C   TLAST   FINAL VALUE OF TIME FOR WHICH
C           A SOLUTION IS DESIRED.
C   F(X)    INITIAL DISPLACEMENT
C   G(X)    INITIAL VELOCITY
C   UE      DISPLACEMENT AT EVEN TIME STEPS
C   UC      DISPLACEMENT AT ODD TIME STEPS
C
```

Figure 13.1 (a) Program listing for example problem 1; (b) output of program in Figure 13.1a

```
 1           DIMENSION UE(100),UO(100),UEXCT(100)
       C
       C  DEFINE SOME DATA STATEMENTS FOR INITIAL CONDITION.
       C
 2           DATA X,N,T/0.0,20,0.0/
 3           DATA L,TLAST/1,1.0/
       C
       C  SET INITIAL CONDITIONS.
       C
 4           NP1=N+1
 5           DX=FLOAT(L)/FLOAT(N)
 6           UE(1)=0.0
 7           UE(NP1)=0.0
 8           UO(1)=0.0
 9           UO(NP1)=0.0
10           UEXCT(1)=0.0
11           UEXCT(NP1)=0.0
12           DO 10 I=2,N
13              X=X+DX
14              UE(I)=F(X)
15     10    CONTINUE
       C
       C  ECHO CHECK-PRINT INITIAL'U' VALUES
       C
16           WRITE(6,100)
17     100   FORMAT(1H1,'SOLUTION OF HYPERBOLIC PDE')
18           WRITE(6,105) (UE(I),I=1,NP1)
19     105   FORMAT(1H0,'ECHO CHECK: INITIAL U VALUES ARE',
          &         //(1H ,7F10.4))
       C
       C  NOW SET UP THE U-VALUES FOR THE FICTITIOUS
       C  TIME STEP AT T= -1 USING EQN(13.5)
       C
20           DO 20 I=2,N
21              UO(I)=UE(I)
22     20    CONTINUE
       C
       C  NOW CONSIDER THE SOLUTION AT T=-1 AND T=0
       C  AS THE STARTING VALUES AND PROCEED TO CALCULATE
       C  UO AND UE UNTIL 'TLAST' IS REACHED. FOR EACH
       C  TIME STEP CALCULATE THE EXACT SOLUTION
       C  BY EVALUATING EQN(13.6).
       C
23           DT=DX
24           T=DT
25     25    IF(T.GE.TLAST) STOP
       C
       C  CALCULATE SOLUTION AT ODD TIME STEPS
       C
26           DO 30 I=2,N
27              UO(I)=UE(I-1)+UE(I+1)-UO(I)
28     30    CONTINUE
29           WRITE(6,200) T
30     200   FORMAT('0','AT T = ',F8.3,'   THE CALCULATED ',
          &            ' SOLUTION IS')
31           WRITE(6,201) (UO(I),I=1,NP1)
32     201   FORMAT(1H0,7F10.4)
       C
       C  CALCULATION OF THE EXACT SOLUTION AT ODD STEPS.
       C
33           X=0.0
34           DO 35 I=2,N
35              X=X+DX
36              UEXCT(I)=E(X,T)
37     35    CONTINUE
38           WRITE(6,202) T
39     202   FORMAT('0','AT T = ',F8.3,'   THE EXACT ',
          &            ' SOLUTION IS')
```

Figure 13.1 *(continued)*

```
 40            WRITE(6,201) (UEXCT(I),I=1,NP1)
       C
       C  ADVANCE THE TIME STEP
 41            T=T+DT
       C
       C  CALCULATE SOLUTION AT EVEN TIME STEPS
       C
 42            DO 40 I=2,N
 43               UE(I)=UO(I-1)+UO(I+1)-UE(I)
 44    40      CONTINUE
 45            WRITE(6,200) T
 46            WRITE(6,201) (UE(I),I=1,NP1)
       C
       C  CALCULATION OF THE EXACT SOLUTION AT EVEN STEPS.
 47            X=0.0
 48            DO 45 I=2,N
 49               X=X+DX
 50               UEXCT(I)=E(X,T)
 51    45      CONTINUE
 52            WRITE(6,202) T
 53            WRITE(6,201) (UEXCT(I),I=1,NP1)
 54            T=T+DT
 55            GO TO 25
 56            END
       C
 57            FUNCTION F(X)
 58            F=SIN(3.1416*X)
 59            RETURN
 60            END
       C
 61            FUNCTION G(X)
 62            G=0.
 63            RETURN
 64            END
       C
 65            FUNCTION E(X,T)
 66            E=SIN(3.1416*X)*COS(3.1416*T)
 67            RETURN
 68            END
       $ENTRY
```

```
SOLUTION OF HYPERBOLIC PDE
ECHO CHECK: INITIAL U VALUES ARE

   0.0000     0.1564     0.3090     0.4540     0.5878     0.7071     0.8090
   0.8910     0.9511     0.9877     1.0000     0.9877     0.9511     0.8910
   0.8090     0.7071     0.5878     0.4540     0.3090     0.1564     0.0000
AT T =     0.050    THE CALCULATED  SOLUTION IS
   0.0000     0.1526     0.3014     0.4428     0.5733     0.6897     0.7891
   0.8691     0.9276     0.9634     0.9754     0.9634     0.9276     0.8691
   0.7891     0.6897     0.5733     0.4428     0.3014     0.1526     0.0000
AT T =     0.050    THE EXACT   SOLUTION IS
   0.0000     0.1545     0.3052     0.4484     0.5805     0.6984     0.7991
   0.8800     0.9393     0.9755     0.9877     0.9755     0.9393     0.8800
   0.7991     0.6984     0.5805     0.4484     0.3052     0.1545     0.0000
AT T =     0.100    THE CALCULATED  SOLUTION IS
   0.0000     0.1450     0.2864     0.4207     0.5447     0.6553     0.7497
   0.8257     0.8814     0.9153     0.9267     0.9153     0.8814     0.8257
   0.7497     0.6553     0.5447     0.4207     0.2864     0.1450     0.0000
AT T =     0.100    THE EXACT   SOLUTION IS
   0.0000     0.1488     0.2939     0.4318     0.5590     0.6725     0.7694
   0.8474     0.9045     0.9393     0.9511     0.9393     0.9045     0.8474
   0.7694     0.6725     0.5590     0.4318     0.2939     0.1488     0.0000
AT T =     0.150    THE CALCULATED  SOLUTION IS
   0.0000     0.1338     0.2643     0.3883     0.5027     0.6048     0.6919
   0.7621     0.8134     0.8447     0.8553     0.8447     0.8134     0.7621
   0.6919     0.6048     0.5027     0.3883     0.2643     0.1338     0.0000
```

Figure 13.1 *(continued)*

```
AT T =    0.150    THE EXACT   SOLUTION IS
    0.0000     0.1394     0.2753     0.4045     0.5237     0.6300     0.7208
    0.7939     0.8474     0.8800     0.8910     0.8800     0.8474     0.7939
    0.7208     0.6300     0.5237     0.4045     0.2753     0.1394     0.0000
AT T =    0.200    THE CALCULATED   SOLUTION IS
    0.0000     0.1193     0.2357     0.3463     0.4483     0.5393     0.6171
    0.6796     0.7254     0.7534     0.7628     0.7534     0.7254     0.6796
    0.6171     0.5393     0.4483     0.3463     0.2357     0.1193     0.0000
AT T =    0.200    THE EXACT   SOLUTION IS
    0.0000     0.1266     0.2500     0.3673     0.4755     0.5721     0.6545
    0.7208     0.7694     0.7991     0.8090     0.7991     0.7694     0.7208
    0.6545     0.5721     0.4755     0.3673     0.2500     0.1266     0.0000
AT T =    0.250    THE CALCULATED   SOLUTION IS
    0.0000     0.1019     0.2013     0.2958     0.3829     0.4606     0.5270
    0.5804     0.6196     0.6434     0.6514     0.6434     0.6196     0.5804
    0.5270     0.4606     0.3829     0.2958     0.2013     0.1019     0.0000
AT T =    0.250    THE EXACT   SOLUTION IS
    0.0000     0.1106     0.2185     0.3210     0.4156     0.5000     0.5721
    0.6300     0.6725     0.6984     0.7071     0.6984     0.6725     0.6300
    0.5721     0.5000     0.4156     0.3210     0.2185     0.1106     0.0000
```

Figure 13.1 *(continued)*

13.3. EXAMPLE 2: THE ADI METHOD FOR A TWO-DIMENSIONAL PARABOLIC PDE

Consider the problem of solving

$$\frac{\partial^2 u}{\partial x^2} + \frac{\partial^2 u}{\partial y^2} = \frac{\partial u}{\partial t} \tag{13.9}$$

over a square region $0 \leq x \leq 1$, $0 \leq y \leq 1$ with the initial condition

$$u(x, y, 0) = 0 \tag{13.10}$$

and the boundary conditions

$$\begin{aligned} u(x, y, t) &= 1, \qquad y = 1, \qquad t > 0 \\ u(x, y, t) &= 1, \qquad x = 1, \qquad t > 0 \\ \frac{\partial u}{\partial y} &= 0, \qquad y = 0, \qquad t > 0 \\ \frac{\partial u}{\partial x} &= 0, \qquad x = 0, \qquad t > 0 \end{aligned} \tag{13.11}$$

Step 1. To solve this problem, a finite difference grid of size $\Delta x = \Delta y = \frac{1}{10}$ is superimposed on the square, and the grid points are labeled as shown in Figure 13.2. This figure shows only the pattern of labeling, not the entire grid. For programming convenience, we set

$$u_{l,m,j} = u[(l-1)\,\Delta x, (m-1)\,\Delta y, (j-1)\,\Delta t] \tag{13.12}$$

Step 2. Application of ADIM requires two stages. In stage (a), one applies an implicit difference approximation for $\partial u/\partial x$ and an explicit difference approximation for $\partial u/\partial y$. This process leads to ten equations, one each at $l = 1$, 2, and 10. They are

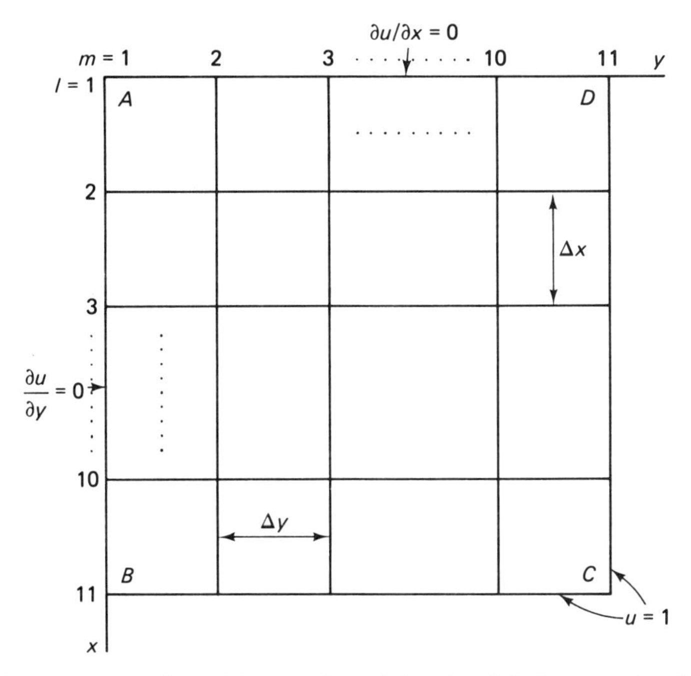

Figure 13.2 Region of interest for solving the diffusion equation by the ADIM method

At $l = 1$:

$(u_{2,m,j+1/2})$

$$\frac{u_{1,m,j+1/2} - u_{1,mj}}{(\Delta t/2)} = \frac{u_{0,m,j+1/2} - 2u_{1,m,j+1/2} + u_{2,m,j+1/2}}{(\Delta x)^2} + \frac{u_{1,m-1,j} - 2u_{1,m,j} + u_{1,m+1,j}}{(\Delta y)^2} \tag{13.13}$$

At $l = 2$:

$$\frac{u_{2,m,j+1/2} - u_{2,m,j}}{(\Delta t/2)} = \frac{u_{1,m,j+1/2} - 2u_{2,m,j+1/2} + u_{2,m,j+1/2}}{(\Delta x)^2} + \frac{u_{2,m-1,j} - 2u_{2,m,j} + u_{2,m+1,j}}{(\Delta y)^2} \tag{13.14}$$

At $l = 10$:

$$\frac{u_{10,m,j+1/2} - u_{10,m,j}}{(\Delta t/2)} = \frac{u_{9,m,j+1/2} - 2u_{10,m,j+1/2} + u_{11,m,j+1/2}}{(\Delta x)^2} + \frac{u_{10,m-1,j} - 2u_{10,m,j} + u_{10,m+1,j}}{(\Delta y)^2} \tag{13.15}$$

Notice that $u_{0,m,j+1/2}$ in Eq. (13.13) represents values of u at points outside the square region of interest. The points $(0, m, j + 1/2)$ are thus fictitious. The "values" of u at these fictitious points can be "calculated" using the boundary condition specified on the boundary AC, that is, $(\partial u/\partial x) = 0$. That is,

$$\frac{\partial u}{\partial x} = \frac{u_{2,m,j+1/2} - u_{0,m,j+1/2}}{2(\Delta x)} = 0 \tag{13.16}$$

or

$$u_{0,m,j+1/2} = u_{2,m,j+1/2}$$

Notice that an analogous situation did not arise at the boundary BD because the values of u along BD are specified. Now Eqs. (13.13) to (13.15) can be rearranged to give, with $r_1 = (\Delta t)/2(\Delta x)^2$ and $r_2 = (\Delta t)/2(\Delta y)^2$, the following equations:
At $l = 1$:

$$u_{1,m,j+1/2} - u_{1,m,j} = r_1(u_{2,m,j+1/2} - 2u_{1,m,j+1/2} + u_{2,m,j+1/2}) + r_2(u_{1,m-1,j} - 2u_{1,m,j} + u_{1,m+1,j})$$

That is,

$$0 + (2r_1 + 1)u_{1,m,j+1/2} + (-2r_1)u_{2,m,j+1/2} = u_{1,m,j} + r_2(u_{1,m-1,j} - 2u_{1,m,j} + u_{1,m+1,j})$$

That is,

$$0 + b_1 u_{1,m,j+1/2} + c_1 u_{2,m,j+1/2} = d_1 \tag{13.17}$$

At $l = 2$:

$$u_{2,m,j+1/2} - u_{2,m,j} = r_1(u_{1,m,j+1/2} - 2u_{2,m,j+1/2} + u_{3,m,j+1/2}) + r_2(u_{2,m-1,j} - 2u_{2,m,j} + u_{2,m+1,j})$$

That is,

$$(-r_1)u_{1,m,j+1/2} + (1 + 2r_1)u_{2,m,j+1/2} + (-r_1)u_{3,m,j+1/2} = u_{2,m,j} + r_2(u_{2,m-1,j} - 2u_{2,m,j} + u_{2,m+1,j})$$

That is,

$$a_2 u_{1,m,j+1/2} + b_2 u_{2,m,j+1/2} + c_2 u_{3,m,j+1/2} = d_2 \tag{13.18}$$

At $l = 10$:

$$u_{10,m,j+1/2} - u_{10,m,j} = r_1(u_{9,m,j+1/2} - 2u_{10,m,j+1/2} + u_{11,m,j+1/2}) + r_2(u_{10,m-1,j} - 2u_{10,m,j} + u_{12,m+1,j})$$

That is,

$$(-r_1)u_{9,m,j+1/2} + (1 + 2r_1)u_{10,m,j+1/2} = u_{10,m,j} + r_1(u_{11,m,j+1/2}) + r_2(u_{10,m-1,j} - 2u_{10,m,j} + u_{10,m+1,j})$$

That is,

$$a_3 u_{9,m,j+1/2} + b_3 u_{10,m,j+1/2} + 0 = d_3 \tag{13.19}$$

Equations (13.17) to (13.19) represent three equations in three unknowns and they can be solved using the tridiagonal algorithm.

Step 3. In stage (b), the ADIM requires that one use an explicit difference approximation for $\partial u/\partial x$ and an implicit approximation for $\partial u/\partial y$. This process leads to three equations, one each, at $m = 1$, 2, and 3. They are:
At $m = 1$:

$$\frac{u_{l,1,j+1} - u_{l,1,j+1/2}}{(\Delta t/2)} = \frac{u_{l-1,1,j+1/2} - 2u_{l,1,j+1/2} + u_{l+1,1,j+1/2}}{(\Delta x)^2} + \frac{u_{l,0,j+1} - 2u_{l,1,j+1} + u_{l,2,j+1}}{(\Delta y)^2} \tag{13.20}$$

At $m = 2$:

$$\frac{u_{l,2,j+1} - u_{l,2,j+1/2}}{(\Delta t/2)} = \frac{u_{l-1,2,j+1/2} - 2u_{l,2,j+1/2} + u_{l+1,2,j+1/2}}{(\Delta x)^2} + \frac{u_{l,1,j+1} - 2u_{l,2,j+1} + u_{l,3,j+1}}{(\Delta y)^2} \tag{13.21}$$

At $m = 10$:

$$\frac{u_{l,10,j+1} - u_{l,10,j+1/2}}{(\Delta t/2)} = \frac{u_{l-1,10,j+1/2} - 2u_{l,10,j+1/2} + u_{l+1,10,j+1/2}}{(\Delta x)^2} + \frac{u_{l,9,j+1} - 2u_{l,10,j+1} + u_{l,11,j+1}}{(\Delta y)^2} \tag{13.22}$$

As before, $u_{l,0,j+1}$ is replaced by $u_{l,2,j+1}$ because the condition along AB is $\partial u/\partial y = 0$. Now Eqs. (13.20) to (13.22) can be rearranged to give:
At $m = 1$:

$$u_{l,1,j+1} - u_{l,1,j+1/2} = r_2(u_{l,2,j+1} - 2u_{l,1,j+1} + u_{l,2,j+1}) + r_1(u_{l-1,1,j+1/2} - 2u_{l,1,j+1/2} + u_{l+1,1,j+1/2})$$

That is,

$$0 + (2r_2 + 1)u_{l,1,j+1} + (-2r_2)u_{l,2,j+1} = u_{l,1,j+1/2} + r_1(u_{l-1,1,j+1/2} - 2u_{l,1,j+1/2} + u_{l+1,1,j+1/2})$$

That is,

$$0 + b_1 u_{l,1,j+1} + c_1 u_{l,2,j+1} = d_1 \tag{13.23}$$

At $m = 2$:

$$u_{l,2,j+1} - u_{l,2,j+1/2} = r_2(u_{l,1,j+1} - 2u_{l,2,j+1} + u_{l,3,j+1}) + r_1(u_{l-1,2,j+1/2} - 2u_{l,2,j+1/2} + u_{l+1,2,j+1/2})$$

That is,

$$(-r_2)u_{l,1,j+1} + (1 + 2r_2)u_{l,2,j+1} + (-r_2)u_{l,3,j+1} = u_{l,2,j+1/2} + r_1(u_{l-1,2,j+1/2} + 2u_{l,2,j+1/2} + u_{l+1,2,j+1/2})$$

That is,

$$a_2 u_{l,1,j+1} + b_2 u_{l,2,j+1} + c_2 u_{l,3,j+1} = d_2 \tag{13.24}$$

At $m = 10$:

$$u_{l,10,j+1} - u_{l,10,j+1/2} = r_2(u_{l,9,j+1} - 2u_{l,10,j+1} + u_{l,11,j+1}) \\ + r_1(u_{l-1,10,j+1/2} - 2u_{l,10,j+1/2} + u_{l+1,10,j+1/2})$$

That is,

$$(-r_2)u_{l,9,j+1} + (1 + 2r_2)u_{l,10,j+1} = u_{l,10,j+1/2} \\ + r_2(u_{l,11,j+1}) + r_1(u_{l-1,10,j+1/2} - 2u_{l,10,j+1/2} + u_{l+1,10,j+1/2})$$

That is,

$$a_3u_{l,9,j+1} + b_3u_{l,10,j+1} + 0 = d_3 \tag{13.25}$$

Equations (13.23) to (13.25) are solved once again using a tridiagonal algorithm.

The program and the solution are shown in Figure 13.3. The program prints the values of temperature at each of the grid points for each time step Δt. Figure 13.3 (b) shows a sample solution.

```
      C PROBLEM NO. 2
      C  THIS PROGRAM SOLVES A 2-D PARABOLIC EQN. ON A SQUARE
      C  USING THE ADI METHOD.   SEE FIG. 13.2 FOR A DEFINITION
      C  OF THE REGION, THE GRID AND BOUNDARY CONDITIONS.
      C  SEE SECTION 13.2 FOR DISCUSSION AND NUMERICAL ANALYSIS.
      C  THE PARAMETERS ARE:
      C
      C  LMAX     NUMBER OF ROWS OF NODES IN THE GRID
      C  MMAX     NUMBER OF COLUMNS OF NODES IN THE GRID
      C  TMAX     NUMBER OF TIME STEPS
      C  DX       IS DELTA X, GRID SIZE IN X-DIRECTION
      C  DY       IS DELTA Y, GRID SIZE IN Y-DIRECTION
      C  DT       IS DELTA T, THE SIZE OF TIME STEP.
      C  R1       RELATED TO DX, DY AND DT
      C  R2       RELATED TO DX, DY AND DT
      C  U(L,M,J) IS THE VALUE OF THE SOLUTION
      C  AT GRID POINT(L,M) AT THE TIME LEVEL J.
      C  A( ), B( ), C( ) ARE THE COEFFICIENTS RESPECTIVELY
      C  ON THE LOWER, MAIN AND UPPER DIAGONALS.
      C  D( ) IS THE RIGHT SIDE OF AX = D.
      C
 1          5 FORMAT ('1'/'-',5X,'TEMPERATURES IN QUADRANT AT TIME =',
             &                F8.5,'--')
 2         10 FORMAT ('0',11F12.5)
 3            COMMON LAST,L,M,JPLUS1,U(11,11,50),A(50),B(50),C(50)
 4            COMMON D(50)
 5            INTEGER EVEN
      C
      C  ESTABLISH GRID SIZE
      C
 6            LMAX = 11
 7            MMAX = 11
 8            JMAX = 31
 9            DX = 1.0/(LMAX - 1)
10            DY = 1.0/(MMAX - 1)
11            DT = 0.05000
12            R1 = (DT/2.)/(DX*DX)
13            R2 = (DT/2.)/(DY*DY)
```

Figure 13.3 (a) Program listing for example problem 2; (b) output of program in Figure 13.3a

```
      C
      C   INITIALIZE U TO ZERO EVERY WHERE EXCEPT ALONG  BD AND CD
      C   BOUNDARIES WHERE IT IS SPECIFIED AS U=1.0
      C
14            DO 100 L = 1,LMAX
15            DO 99 M = 1,MMAX
16            U(L,M,1) = 0.0
17            BC = 1.0
18            IF (L .EQ. LMAX) U(L,M,1) = BC
19            IF (M .EQ. MMAX) U(L,M,1) = BC
20         99 CONTINUE
21        100 CONTINUE
      C
      C   PRINT INITIAL POTENTIAL DISTRIBUTION AS A CHECK
      C
22            WRITE(6,11)
23         11 FORMAT ('1',5X,'THIS IS AN ECHO CHECK FOR INITIAL ',
             &                    ' POTENTIAL DISTRIBUTION')
24            WRITE (6,10) ((U(L,M,1), M=1,MMAX), L=1,LMAX)
      C
      C THE COMPUTATION STARTS HERE
      C J=1 DEFINES AN ODD TRAVERSAL.
      C J=2 DEFINES AN EVEN TRAVERSAL OF GRID.
      C
25            DO 98 JPLUS1 = 2,JMAX
26            J = JPLUS1 - 1
27            EVEN = J/2
28            EVEN = EVEN*2
29            IF (J .EQ. EVEN) GO TO 95
      C
      C   THE FOLLOWING SEGMENT IS EXECUTED FOR ODD VALUES OF J
      C
30            STEP = 1.0
31            DO 97 M = 1,MMAX
32            U(LMAX,M,JPLUS1) = BC
33            LAST = LMAX - 1
34            DO 96 L = 1,LAST
35            IF (M .EQ. MMAX) U(L,M,JPLUS1) = BC
36            IF (M .EQ. MMAX) GO TO 96
      C
      C   FOR THE FICTITIOUS GRID POINTS ALONG A LINE
      C   BELOW AB (SEE FIG 13.1), WE INVOKE SYMMETRY.
      C       IF (M .EQ. 1) MLESS1 = M + 1 BECAUSE OF SYMMETRY.
      C
37            IF (M .EQ. 1) MLESS1 = M + 1
38            IF (M .GT. 1) MLESS1 = M - 1
39            MPLUS1 = M + 1
      C
      C   THE ENTRY A(1)=-R1 IS A DUMMY ENTRY SIMILARLY C(11)
      C   IS A  DUMMY CALCULATION
      C   THE FOLLOWING SIX INSTRUCTIONS IMPLEMENT
      C
40            A(L) = -R1
41            B(L) = +2.*R1 + 1.
42            IF (L .EQ. 1) C(L) = -2.*R1
43            IF (L .GT. 1) C(L) = -1.*R1
44            D(L) = (U(L,MLESS1,J) - 2.*U(L,M,J) +
             &          U(L,MPLUS1,J))*R2 + U(L,M,J)
45            IF (L .EQ. LAST) D(L) = D(L) + R1*U(LMAX,M,JPLUS1)
46         96 CONTINUE
47            IF (M .LT. MMAX) CALL TDA(STEP)
48         97 CONTINUE
49            GO TO 98
      C
      C
      C   THE FOLLOWING SEGMENT IS EXECUTED FOR EVEN VALUES OF J.
      C   THIS SEGMENT IS SIMILAR TO THE ODD SEGMENT ABOVE.
      C
```

Figure 13.3 *(continued)*

```
 50       95 STEP = 2.0
 51          DO 94 L = 1,LMAX
 52          U(L,MMAX,JPLUS1) = BC
 53          LAST = MMAX - 1
 54          DO 93 M = 1,LAST
 55          IF (L .EQ. LMAX) U(L,M,JPLUS1) = BC
 56          IF (L .EQ. LMAX) GO TO 93
     C
     C       IF (L .EQ. 1) LLESS1 = L + 1 BECAUSE OF SYMMETRY.
     C
 57          IF (L .EQ. 1) LLESS1 = L + 1
 58          IF (L .GT. 1) LLESS1 = L - 1
 59          LPLUS1 = L + 1
 60          A(M) = -R2
 61          B(M) = +2.*R2 + 1.
 62          IF (M .EQ. 1) C(M) = -2.*R2
 63          IF (M .GT. 1) C(M) = -1.*R2
 64          D(M) = (U(LLESS1,M,J) - 2.*U(L,M,J) +
            &         U(LPLUS1,M,J))*R1 + U(L,M,J)
 65          IF (M .EQ. LAST) D(M) = D(M) + R2*U(L,MMAX,JPLUS1)
 66       93 CONTINUE
 67          IF (L .LT. LMAX) CALL TDA(STEP)
 68       94 CONTINUE
 69          IF (J.NE.2.AND.J.NE.4.AND.J.NE.8.AND.
            &    J.NE.16.AND.J.NE.30) GO TO 98
 70          TIME = DT*(JPLUS1 - 1)/2.
 71          WRITE (6,5) TIME
 72          WRITE (6,10) ((U(L,M,JPLUS1), M=1,MMAX), L=1,LMAX)
 73       98 CONTINUE
 74          CALL EXIT
 75          STOP
 76          END
     C
 77          SUBROUTINE TDA(STEP)
     C
     C       TRIDIAGONAL ALGORITHM.
     C  THIS SUBROUTINE  IMPLEMENTS THE TRIDIAGONAL ALGORITHM.
     C  WHICH IS DESCRIBED IN CHAPTER 10.
     C  THE PARAMETER 'STEP' IS PASSED TO THIS SUBPROGRAM.
     C
 78          COMMON LAST,L,M,JPLUS1,U(11,11,50),A(50),B(50),C(50)
 79          COMMON D(50)
 80          DIMENSION BETA(100),GAMMA(100)
     C
     C THE FOLLOWING IMPLEMENTS EQNS (19) AND (20) OF CHAPTER 10.
     C
 81          BETA(1)  = B(1)
 82          GAMMA(1) = D(1)/B(1)
 83          DO 10 I = 2,LAST
 84          ILESS1 = I - 1
 85          BETA(I)  =  B(I) - A(I)*C(ILESS1)/BETA(ILESS1)
 86          GAMMA(I) = (D(I) - A(I)*GAMMA(ILESS1))/BETA(I)
 87       10 CONTINUE
     C
     C THE FOLLOWING IMPLEMENTS EQN (21) OF CHAPTER 10.
     C
 88          IF (STEP .LE. 1.5) U(LAST,M,JPLUS1) = GAMMA(LAST)
 89          IF (STEP .GT. 1.5) U(L,LAST,JPLUS1) = GAMMA(LAST)
 90          DO 20 K = 2,LAST
 91          I = LAST - K + 1
 92          IPLUS1 = I + 1
 93          IF (STEP .LE. 1.5) GO TO 1
 94          IF (STEP .GT. 1.5) GO TO 2
 95        1 U(I,M,JPLUS1) = GAMMA(I) - C(I)*U(IPLUS1,M,JPLUS1)/
            &                 BETA(I)
 96          GO TO 20
 97        2 U(L,I,JPLUS1) = GAMMA(I) - C(I)*U(L,IPLUS1,JPLUS1)/
            &                 BETA(I)
 98       20 CONTINUE
 99          RETURN
100          END
     $ENTRY
```

Figure 13.3 *(continued)*

```
SOLUTION OF 2-D PARABOLIC EQUATION
 THIS IS AN ECHO CHECK FOR INITIAL  POTENTIAL DISTRIBUTION     THIS IS AN ECHO CHECK FOR INITIAL  POTENTIAL DISTRIBUTION
0.00000   0.00000   0.00000   0.00000   0.00000   0.00000   0.00000   0.00000   0.00000   0.00000   1.00000
0.00000   0.00000   0.00000   0.00000   0.00000   0.00000   0.00000   0.00000   0.00000   0.00000   1.00000
0.00000   0.00000   0.00000   0.00000   0.00000   0.00000   0.00000   0.00000   0.00000   0.00000   1.00000
0.00000   0.00000   0.00000   0.00000   0.00000   0.00000   0.00000   0.00000   0.00000   0.00000   1.00000
0.00000   0.00000   0.00000   0.00000   0.00000   0.00000   0.00000   0.00000   0.00000   0.00000   1.00000
0.00000   0.00000   0.00000   0.00000   0.00000   0.00000   0.00000   0.00000   0.00000   0.00000   1.00000
0.00000   0.00000   0.00000   0.00000   0.00000   0.00000   0.00000   0.00000   0.00000   0.00000   1.00000
0.00000   0.00000   0.00000   0.00000   0.00000   0.00000   0.00000   0.00000   0.00000   0.00000   1.00000
0.00000   0.00000   0.00000   0.00000   0.00000   0.00000   0.00000   0.00000   0.00000   0.00000   1.00000
0.00000   0.00000   0.00000   0.00000   0.00000   0.00000   0.00000   0.00000   0.00000   0.00000   1.00000
1.00000   1.00000   1.00000   1.00000   1.00000   1.00000   1.00000   1.00000   1.00000   1.00000   1.00000

TEMPERATURES IN QUADRANT AT TIME = 0.05000--
0.01579   0.01737   0.02271   0.03398   0.05566   1.07278   0.17262   0.31467   0.57943   1.07278   1.00000
0.01737   0.01894   0.02428   0.03552   0.05717   1.07266   0.17394   0.31577   0.58010   1.07266   1.00000
0.02271   0.02428   0.02959   0.04077   0.06230   1.07227   0.17844   0.31949   0.58233   1.07227   1.00000
0.03398   0.03552   0.04077   0.05183   0.07311   1.07143   0.18790   0.32733   0.58720   1.07143   1.00000
0.05566   0.05717   0.06230   0.07311   0.09391   1.06983   0.20613   0.34243   0.59646   1.06983   1.00000
0.09644   0.09788   0.10279   0.11313   0.13304   1.06681   0.24041   0.37083   0.61389   1.06681   1.00000
0.17262   0.17394   0.17844   0.18790   0.20613   1.06118   0.30445   0.42387   0.64644   1.06118   1.00000
0.31467   0.31577   0.31949   0.32733   0.34243   1.05068   0.42337   0.52279   0.70714   1.05068   1.00000
0.57943   0.58010   0.58238   0.58720   0.59646   0.61389   0.64644   0.70714   0.82028   1.03110   1.00000
1.07278   1.07266   1.07226   1.07143   1.06983   1.06681   1.06118   1.05068   1.03110   0.99462   1.00000
1.00000   1.00000   1.00000   1.00000   1.00000   1.00000   1.00000   1.00000   1.00000   1.00000   1.00000
```

Figure 13.3 *(continued)*

```
TEMPERATUFES IN QUADRANT AT TIME = 0.10000--
0.09333   0.09942   0.11890   0.15546   0.21508   0.30495   0.42973   0.58082   0.70958   0.66423   1.00000
0.09942   0.10546   0.12481   0.16113   0.22035   0.30962   0.43356   0.58363   0.71153   0.66648   1.00000
0.11890   0.12481   0.14374   0.17927   0.23721   0.32455   0.44581   0.59264   0.71777   0.67370   1.00000
0.15546   0.16113   0.17927   0.21333   0.26886   0.35258   0.46891   0.60954   0.72948   0.68724   1.00000
0.21508   0.22035   0.23721   0.26886   0.32048   0.39828   0.50630   0.63710   0.74857   0.70932   1.00000
0.30495   0.30962   0.32455   0.35258   0.39828   0.46718   0.56283   0.67866   0.77736   0.74260   1.00000
0.42973   0.43356   0.44581   0.46880   0.50630   0.56283   0.64131   0.73634   0.81733   0.78881   1.00000
0.58082   0.58363   0.59264   0.60954   0.63710   0.67866   0.73634   0.80620   0.86573   0.84476   1.00000
0.70958   0.71153   0.71777   0.72948   0.74858   0.77736   0.81733   0.86573   0.90697   0.89245   1.00000
0.66423   0.66648   0.67370   0.68724   0.70932   0.74260   0.78881   0.84476   0.89245   0.87565   1.00000
1.00000   1.00000   1.00000   1.00000   1.00000   1.00000   1.00000   1.00000   1.00000   1.00000   1.00000
```

```
TEMPERATURES IN QUADRANT AT TIME = 0.20000--
0.40354   0.41076   0.43204   0.46633   0.51243   0.57037   0.64334   0.73701   0.84126   0.84398   1.00000
0.41076   0.41790   0.43892   0.47279   0.51833   0.57557   0.64766   0.74019   0.84319   0.84587   1.00000
0.43204   0.43892   0.45919   0.49183   0.53573   0.59090   0.66038   0.74958   0.84885   0.85144   1.00000
0.46633   0.47279   0.49183   0.52250   0.56376   0.61 60   0.68088   0.76469   0.85797   0.86041   1.00000
0.51243   0.51833   0.53573   0.56375   0.60144   0.64880   0.70845   0.78502   0.87024   0.87247   1.00000
0.57037   0.57557   0.59090   0.61560   0.64880   0.69054   0.74310   0.81057   0.88566   0.88762   1.00000
0.64334   0.64766   0.66038   0.68088   0.70845   0.74310   0.78673   0.84274   0.90508   0.90671   1.00000
0.73701   0.74019   0.74958   0.76469   0.78502   0.81057   0.84274   0.88404   0.93001   0.93121   1.00000
0.84126   0.84319   0.84885   0.85797   0.87024   0.88566   0.90508   0.93001   0.95775   0.95848   1.00000
0.84398   0.84587   0.85144   0.86041   0.87247   0.88762   0.90671   0.93121   0.95843   0.95919   1.00000
1.00000   1.00000   1.00000   1.00000   1.00000   1.00000   1.00000   1.00000   1.00000   1.00000   1.00000
```

Figure 13.3 (*continued*)

Solving a differential equation does not end with a printing of the results. It is very important that one spend some time looking at the output and make sure that the printed result is indeed a good approximation to the given problem.

13.4. EXAMPLE 3: FLOW AROUND A CYLINDER BY AN SOR-TYPE METHOD

Many interesting problems arise in the study of the flow of a fluid past a solid body [7]. The problem of calculating the alternating pressures generated by the flow of an ideal fluid through the blades of a turbomachine is a case in point. To calculate the pressures in such a problem, one first solves the Laplace's equation $\nabla^2\psi = 0$ for the stream function $\psi(x, y)$ from which velocities and the velocity potential $\phi(x, y)$ are derived. This velocity potential is then used in Bernoulli's equation in order to get the pressure P.

The use of ψ in the flow calculations limits the preceding procedure to two dimensions. There are clearly advantages in computing ϕ directly, as this allows, in theory at least, calculation of three-dimensional compressible flows. Furthermore, the need to integrate the velocities to calculate ϕ is avoided. However, use of $\phi(x, y)$ as a dependent variable makes it difficult to apply the boundary conditions at the blade surface. The purpose of this example is to address a simplified version of this problem of solving for $\phi(x, y)$ by considering a flow around an infinitely long circular cylinder whose axis is perpendicular to the direction of flow. A cross section of the cylinder and the flow around it are shown in Figure 13.4. Because of symmetry, it is sufficient to calculate the values of ϕ in one quarter of the region shown in Figure 13.5.

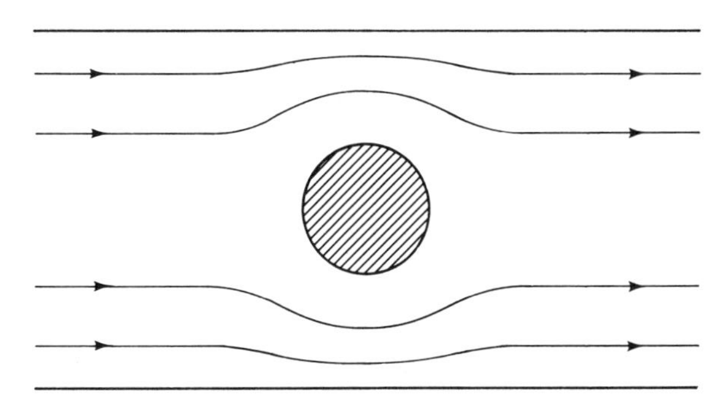

Figure 13.4 Cross-sectional view of potential flow past a right circular cylinder

Figure 13.5 shows one quarter of the surface of the cylinder and a finite region of flow in the vicinity of the cylinder. This region is denoted by EFGHCJ. Now the problem is to solve

$$\nabla^2\phi = 0 \tag{13.26a}$$

on the region *EFGHCJ*. What are the boundary conditions to this problem? If *EF* is sufficiently away from the cylinder, one can set

$$\frac{\partial \phi}{\partial x} = \text{velocity of flow past the line } EF \tag{13.26b}$$

If the flow is laminar, one can set

$$\frac{\partial \phi}{\partial y} = 0, \qquad \text{on } EJ \text{ and } FG \tag{13.26c}$$

assuming that *FG* is sufficiently far off from the cylinder. However, the problem of figuring the boundary conditions along *GH* and the curved surface *HCJ* is somewhat

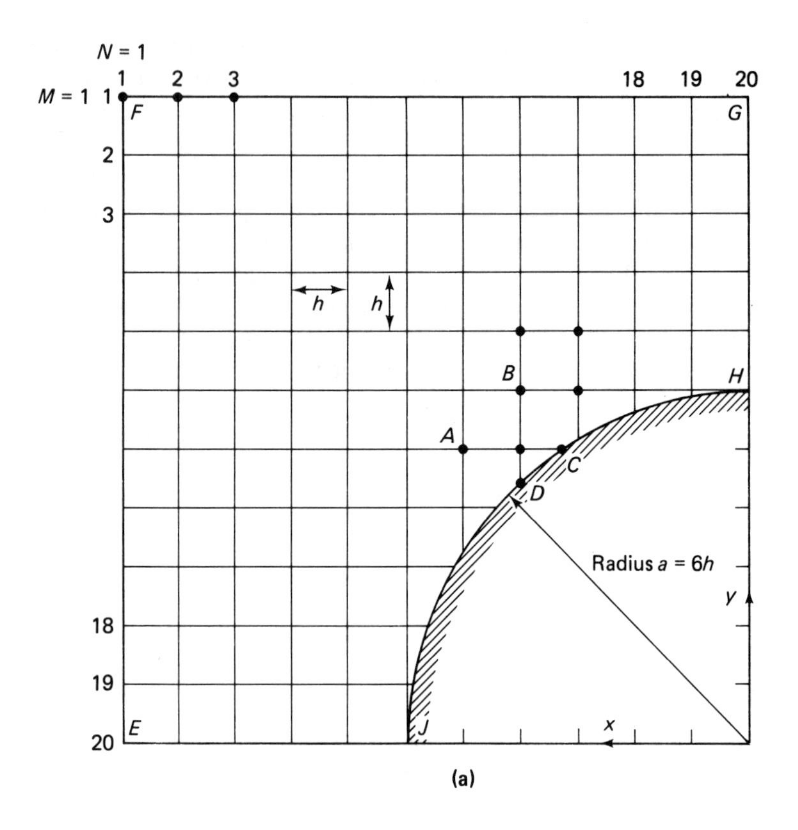

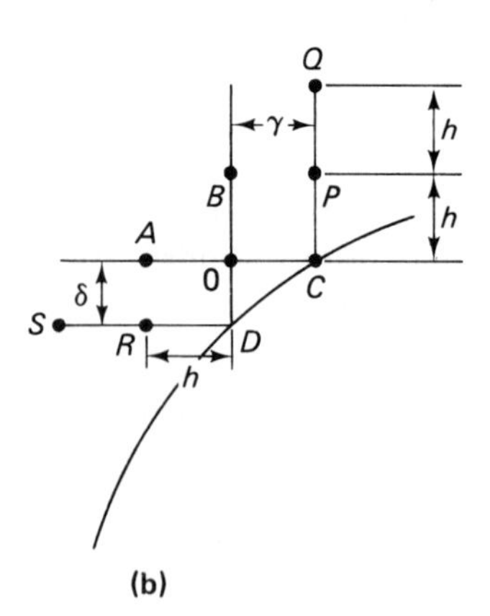

Figure 13.5 (a) Finite difference grid in a quadrant of the flow region; (b) details of the grid near the cylinder wall

difficult. The question of figuring out the boundary conditions along GH can be bypassed by considering one half, instead of one quarter, of the flow region. However, one has to still face the question along the curved surface HCJ. As the primary purpose of this example is to illustrate the solution of Laplace's equation with normal derivative boundary conditions, an artifice is used to bypass the question of figuring out the boundary conditions along the boundaries EF, FG, GH, and JE. The artifice is to use an expression for the exact solution of Eq. (13.26) in nondimensioned form, which is known to be

$$\phi = \frac{x}{a}\left(1 + \frac{a^2}{x^2 + y^2}\right) \tag{13.27}$$

where a is the radius of the cylinder. This expression is evaluated along EF, FG, and JE, and the values are assigned as the boundary conditions.

TREATMENT OF POINTS ON THE BOUNDARY OF THE CYLINDER

For convenience, the following assumptions are made. A square grid of size $\Delta x = \Delta y = h$ is superimposed on the region $EFGHCJ$. The radius a of the cylinder is assumed to be $6h$. The positive directions of x and y axes are taken away from the surface of the cylinder. Now consider a typical point O near the surface of the cylinder and its four neighboring points A, B, C, and D. As the value of ϕ at the point O depends upon the values of ϕ at the four neighboring points, it is necessary to know the values of ϕ at A, B, C, and D. Let the fluid velocity normal to the surface at the point C be q_n. Then

$$q_n = \left.\frac{\partial \phi}{\partial n}\right|_C = u_C \cos\theta + v_C \sin\theta = 0 \tag{13.28}$$

where u_C and v_C are the fluid velocities at the point C in the x and y directions, respectively, and θ is the angle between the positive direction of the x-axis and the outward normal drawn from the point C. If potential flow is assumed, then

$$u_C = \left.\frac{\partial \phi}{\partial x}\right|_C \quad \text{and} \quad v_C = \left.\frac{\partial \phi}{\partial y}\right|_C \tag{13.29}$$

Therefore,

$$q_n = \left.\frac{\partial \phi}{\partial x}\right|_C \cos\theta + \left.\frac{\partial \phi}{\partial y}\right|_C \sin\theta = 0 \tag{13.30}$$

or

$$\left.\frac{\partial \phi}{\partial x}\right|_C + \left.\frac{\partial \phi}{\partial y}\right|_C \tan\theta = 0 \tag{13.31}$$

Similarly,

$$\left.\frac{\partial \phi}{\partial x}\right|_D + \left.\frac{\partial \phi}{\partial y}\right|_D \tan\theta = 0 \tag{13.32}$$

The next step is to derive difference approximations for $\partial\phi/\partial x$ and $\partial\phi/\partial y$. In calculating $\partial\phi/\partial x$, for instance, our goal is to arrive at a second-order accurate expression in terms of ϕ values at the points A, O, and C. One of the many ways of doing this is to use a second-degree Lagrange interpolation formula for three unevenly spaced points:

$$\phi(x) = \frac{(x - x_1)(x - x_2)}{(x_0 - x_1)(x_0 - x_2)}\phi_0 + \frac{(x - x_0)(x - x_2)}{(x_1 - x_0)(x_1 - x_2)}\phi_1 + \frac{(x - x_0)(x - x_1)}{(x_2 - x_0)(x_2 - x_1)}\phi_2 \tag{13.33}$$

Differentiating Eq. (13.33) with respect to x,

$$\left.\frac{\partial\phi}{\partial x}\right|_x = \frac{(x - x_1) + (x - x_2)}{(x_0 - x_1)(x_0 - x_2)}\phi_0 + \frac{(x - x_0) + (x - x_2)}{(x_1 - x_0)(x_1 - x_2)}\phi_1 + \frac{(x - x_0) + (x - x_1)}{(x_2 - x_0)(x_2 - x_1)}\phi_2 \tag{13.34}$$

Evaluating this at the point C and identifying the points C, O, and A, respectively, as 0, 1, and 2, we have

$$\left.\frac{\partial\phi}{\partial x}\right|_C = \frac{-h + 2\gamma}{\gamma(h + \gamma)}\phi_C + \frac{h + \gamma}{h\gamma}\phi_0 - \frac{\gamma}{h(h + \gamma)}\phi_A \tag{13.35}$$

Similarly, one can show that

$$\left.\frac{\partial\phi}{\partial y}\right|_C = \frac{-3}{2h}\phi_C + \frac{2}{h}\phi_P - \frac{1}{2h}\phi_Q \tag{13.36}$$

The point D on the curved boundary can be treated in a similar manner to yield

$$\left.\frac{\partial\phi}{\partial x}\right|_D = -\frac{3}{2h}\phi_D + \frac{2}{h}\phi_R - \frac{1}{2h}\phi_S \tag{13.37}$$

$$\left.\frac{\partial\phi}{\partial y}\right|_D = \frac{-h + 2\delta}{\delta(h + \delta)}\phi_D + \frac{(h + \delta)}{h\delta}\phi_0 - \frac{\delta}{h(h + \delta)}\phi_B \tag{13.38}$$

Substituting $\partial\phi/\partial x|_C$ and $\partial\phi/\partial y|_C$ in Eq. (13.31), one can solve for ϕ_C. Similarly, substituting $\partial\phi/\partial x|_D$ and $\partial\phi/\partial y|_D$ in Eq. (13.32) gives the value of ϕ_D. Calculations of ϕ_D and ϕ_C are done, respectively, in subroutines 1 and 2.

THE COMPUTATION

A square grid of size $h = 1$ is superimposed on the field. The grid points are labeled along the x-axis as $N = 1, 2, \ldots, 20$ and along the y-axis as $M = 1, 2, \ldots, 20$. Boundary values computed from Eq. (13.27) were used at all grid points along the $M = 1$ and $N = 1$ lines. According to Eq. (13.27), $\phi(M, 20) = 0$ at all values of M.

Symmetry is required about the line $M = 20$. The values obtained by using Eqs. (13.31) and (13.32), as described in the preceding paragraph, are used along the curved boundary. Using a procedure analogous to the SOR method, the calculations are carried out as follows.

Step 1. Set up boundary values of ϕ along $N = 1$, $N = 20$, and $M = 1$ lines, and give all other grid points starting values corresponding to uniform stream flow.

Step 2. Perform the required computations for points near the cylinder in separate subroutines.

Step 3. Calculate the residual $R(M, N)$ at the interior grid point (M, N) using the formula

$$R(M, N) = \tfrac{1}{4}(\phi(M-1, N) + \phi(M+1, N) + \phi(M, N-1) + \phi(M, N+1)) \tag{13.39}$$

Calculate the residuals on the curved boundary HJ using the formula

$$R_0 = \frac{\gamma\delta}{\gamma + \delta}\left[\frac{\phi_A}{1+\gamma} + \frac{\phi_B}{1+\delta} + \frac{\phi_C}{\gamma(1+\gamma)} + \frac{\phi_D}{\delta(1+\delta)}\right] - \phi_0 \tag{13.40}$$

In using the preceding formula, one gets the values of ϕ_C and ϕ_D from the procedure indicated after Eq. (13.38). The sequence of calculation is as follows. Considering in turn each point outside the cylinder from $M = 2$ to 20 and $N = 2$ to 19 (N varying most rapidly) compute $R(M, N)$.

Step 4. Modify values of ϕ according to

$$\begin{aligned}
&\phi(M, N) \longleftarrow \phi(M, N) - 1.25R(M, N) \\
&\phi(M-1, N) \longleftarrow \phi(M+1, N) - \frac{1.25}{4}R(M, N) \qquad \text{if } M > 2 \\
&\phi(M, N-1) \longleftarrow \phi(M, N-1) - \frac{1.25}{4}R(M, N) \qquad \text{if } N > 2
\end{aligned} \tag{13.41}$$

This procedure is equivalent to a SOR procedure with $\omega = 1.25$.

A flow chart of the process described so far is shown in Figure 13.6. A FORTRAN program and results are shown in Figure 13.7.

13.5. EXAMPLE 4: A STEADY-STATE NAVIER-STOKES PROBLEM

We are indebted to D. Greenspan [8] for his permission to use this example here. The purpose of this example is to illustrate how one can approach a very troublesome problem in a heuristic manner. Consider a problem of steady-state, two-dimensional

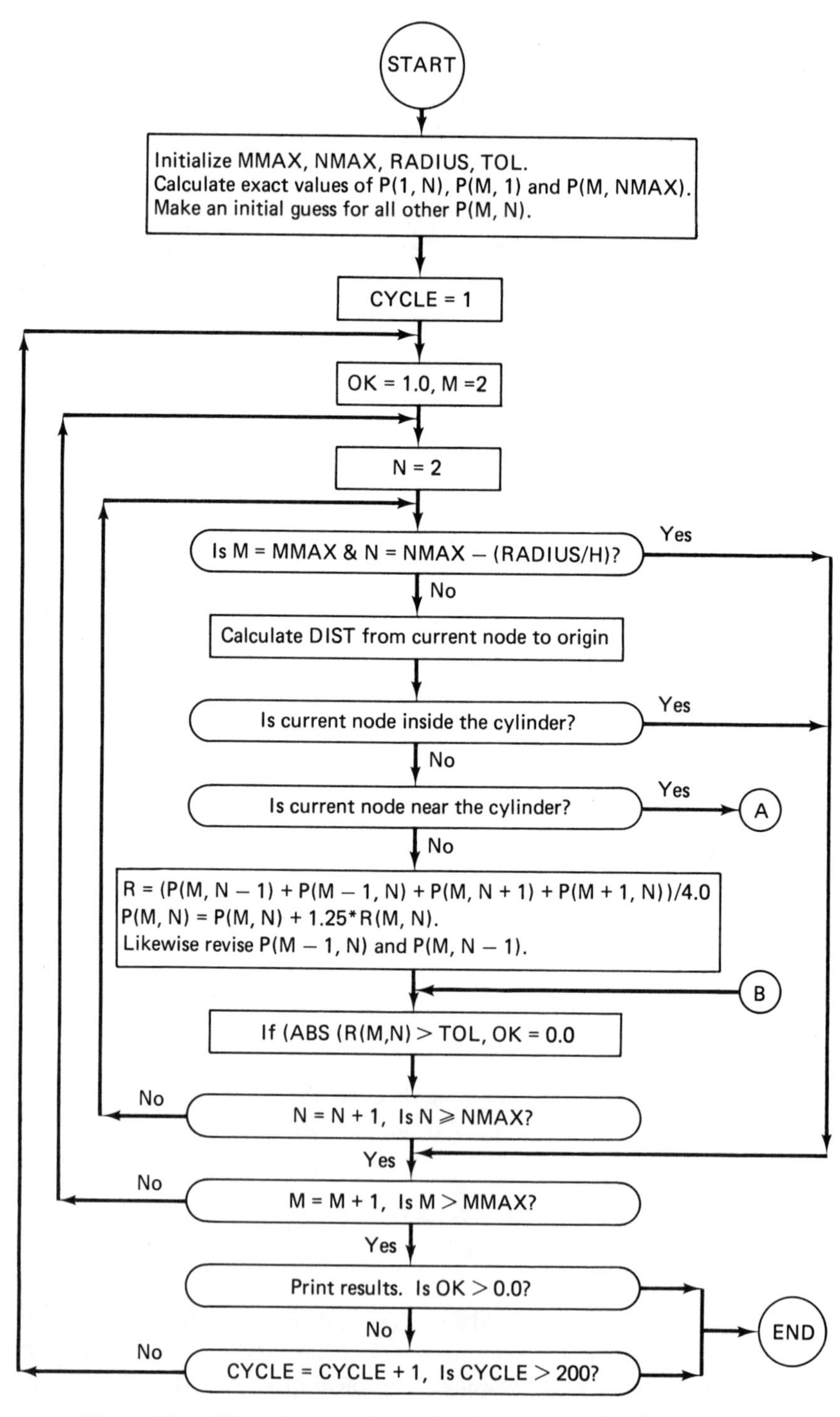

Figure 13.6 Flow chart used to write the program in Figure 13.7a

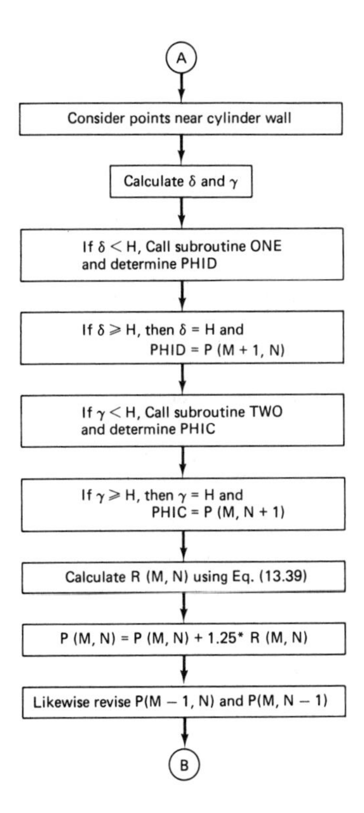

Figure 13.6 (*continued*)

fluid flow governed by the Navier-Stokes equations

$$\Delta\psi = -\omega \tag{13.42}$$

$$A\omega + \mathcal{R}\left(\frac{\partial\psi}{\partial x}\frac{\partial\omega}{\partial y} - \frac{\partial\omega}{\partial y}\frac{\partial\omega}{\partial x}\right) = 0 \tag{13.43}$$

on a square region Ω as shown in Figure 13.8. Here ψ is the stream function, ω is the vorticity, and $\mathcal{R}$ is the Reynold's number. In this problem, compressibility of the fluid is neglected and viscosity is assumed to play a major role. Such a situation often arises in the study of turbulent motion of liquids.

The boundary conditions to be satisfied on the boundary Γ are

$$\psi = 0, \qquad \frac{\partial\psi}{\partial x} = 0 \qquad \text{on } AD \tag{13.44}$$

$$\psi = 0, \qquad \frac{\partial\psi}{\partial y} = 0 \qquad \text{on } AB \tag{13.45}$$

$$\psi = 0, \qquad \frac{\partial\psi}{\partial x} = 0 \qquad \text{on } BC \tag{13.46}$$

$$\psi = 0, \qquad \frac{\partial\psi}{\partial y} = -1 \qquad \text{on } CD \tag{13.47}$$

```
C PROBLEM NO.3
C STREAM FLOW AROUND A CYLINDER BY FINITE DIFFERENCE METHOD
C THE SOLUTION IS CALCULATED ON A QUARTER OF THE REGION.
C THE PARAMETERS OF THE PROBLEM ARE:
C
C      P        FLUID PRESSURE
C      NMAX     MAXIMUM NUMBER OF GRID POINTS ALONG X-AXIS
C      MMAX       "       "    "   "     "     "   Y-AXIS
C      H        GRID SIZE ALONG X- AND Y- AXES.
C      RAD      RADIUS OF THE CYLINDER
C      TOL      ERROR TOLERANCE
C
   10 FORMAT ('1'/'-','CYCLE',I3,' -- VELOCITY POTENTIALS AT',
     &         'GRID POINTS')
   11 FORMAT ('0',10F6.3,2X,10F6.3)
   12 FORMAT ('1'/'-',10X,'VELOCITY POTENTIALS AROUND CYLINDER',
     &         '--' /'0',15X,'X',10X,'Y',10X,'PHI')
   13 FORMAT ('0',10X,3F10.3)
      INTEGER CYCLE,CYCLES
      DIMENSION R(20,20)
      COMMON INDEX,SAVE(100),P(20,20),MMAX,NMAX,H,RADIUS
C
C SET UP INITIAL INFORMATION
C
      MMAX = 20
      NMAX = 20
      H = 1.
      RADIUS = 6.0*H
      TOL = 0.0001
C
C CALCULATE COORDINATES FOR THE GRID POINTS.
C SET UP BOUNDARY CONDITIONS BY CALCULATING THE EXACT
C SOLUTION ALONG EF,FG,JE AND GH OF FIG.13.5 USING EQN.(13.27).
C
        WRITE(6,9)
9       FORMAT('1'/'-','STRAEM FLOW AROUND A CYLINDER')
        WRITE(6,8)
8       FORMAT('0','THE INITIAL POTENTIAL DISTRIBUTION IS',/)
      DO 100 M = 1,MMAX
      DO  99 N = 1,NMAX
      X = (NMAX - N)*H
      Y = (MMAX - M)*H
      IF (Y .EQ. 0.0) Y = 0.000001
C
C
C
C
C NOW CALCULATE EXACT SOLUTION
C     USE EXACT SOLUTION AT BOUNDARIES.
C
      P(M,N) = (1. + RADIUS**2/(X*X + Y*Y))*(X/RADIUS)
C
C MAKE INITIAL GUESS FOR POINTS ELSEWHERE BY CHOSING P=2.
C
      IF (M.NE.1.AND.N.NE.1.AND.N.NE.NMAX) P(M,N) = 2.
   99 CONTINUE
      WRITE (6,11) (P(M,N), N=1,NMAX)
  100 CONTINUE
C
```

Figure 13.7 (a) Program listing for example problem 3; (b) output of program in Figure 13.7a

```
C CALCULATION OF THE NUMERICAL SOLUTION STARTS HERE.
C THE CALCULATION IS ITERATED 200 TIMES.
C
      CYCLES = 200
      DO 400 CYCLE = 1,CYCLES
      OK = 1.0
      INDEX = 0
      DO 300 M = 2,MMAX
      MLESS1 = M - 1
      IF (M .LT. MMAX) MPLUS1 = M + 1
C
C     IF (M .GE. MMAX) MPLUS1 = M - 1 BECAUSE OF SYMMETRY.
C
      IF (M .GE. MMAX) MPLUS1 = M - 1
      DO 200 N = 2,NMAX
      IF (N .GE. NMAX) GO TO 300
      IF (M.EQ.MMAX .AND. N.EQ.(NMAX-(RADIUS/H))) P(M,N) = 2.0
      IF (M.EQ.MMAX .AND. N.EQ.(NMAX-(RADIUS/H))) GO TO 300
      NLESS1 = N - 1
      NPLUS1 = N + 1
      DIST = H*((MMAX-M)**2 + (NMAX-N)**2)**0.5
C
C IF CURRENT NODE IS INSIDE THE CYLINDER BODY, GO TO 300
C THIS FACT IS DETERMIND BY COMPARING DISTANCE OF A GRID
C POINT WITH RADIUS MINUS EPSILON. NOTE RADIUS=6*H
C
      IF (DIST .LT. (5.95*H)) GO TO 300
C
C IF CURRENT NODE IS OUTSIDE BUT CLOSE TO CYLINDER WALL
C GO TO 150 AND USE A SPECIAL NONLINEAR INTRPOLATION.
C
      IF (DIST .LT. (6.95*H)) GO TO 150
C
C IF CURRENT NODE IS FAR FROM THE CYLINDER WALL, SOLVE
C USING EQS.(13.39) AND (13.40).
C
      R(M,N) = P(M,NLESS1) + P(MLESS1,N) + P(M,NPLUS1) +
     &         P(MPLUS1,N)
      R(M,N) = R(M,N)/4. - P(M,N)
      P(M,N) = P(M,N) + 1.25*R(M,N)
      IF (M .GT. 2) P(MLESS1,N) = P(MLESS1,N) + R(M,N)*1.25/4.
      IF (N .GT. 2) P(M,NLESS1) = P(M,NLESS1) + R(M,N)*1.25/4.
      GO TO 200
  150 CONTINUE
C
C WE USE NONLINEAR INTERPOLATION FOR POINTS NEAR THE SURFACE
C
      DELTA = H*((MMAX-M) - ((RADIUS/H)**2 - (NMAX-N)**2)**0.5)
      GAMMA = H*((NMAX-N) - ((RADIUS/H)**2 - (MMAX-M)**2)**0.5)
      IF (DELTA .LT. H) CALL ONE (M,N,DELTA,PHID)
      IF (GAMMA .LT. H) CALL TWO (M,N,GAMMA,PHIC)
      IF (DELTA .GE. H) DELTA = H
      IF (DELTA .GE. H) PHID = P(MPLUS1,N)
      IF (GAMMA .GE. H) GAMMA = H
      IF (GAMMA .GE. H) PHIC = P(M,NPLUS1)
      R(M,N) = P(M,NLESS1)/(1.+GAMMA) + P(MLESS1,N)/(1.+DELTA)
      R(M,N) = R(M,N)+ (PHIC/GAMMA)/(1.+GAMMA) + (PHID/DELTA)/
     &         (1.+DELTA)
      R(M,N) = R(M,N)*GAMMA*DELTA/(GAMMA+DELTA) - P(M,N)
      P(M,N) = P(M,N) + 1.25*R(M,N)
      IF (M .GT. 2) P(MLESS1,N) = P(MLESS1,N) + R(M,N)*1.25/4.
```

Figure 13.7 *(continued)*

```
      IF (N .GT. 2) P(M,NLESS1) = P(M,NLESS1) + R(M,N)*1.25/4.
  200 IF (ABS(R(M,N)) .GT. TOL) OK = 0.0
  300 CONTINUE
C
      KEY = 20
      J = CYCLE / KEY
      J = J * KEY
      IF (CYCLE .LE. 3) J = CYCLE
      IF (OK   .GT. 0.5) J = CYCLE
      IF (J .EQ. CYCLE) WRITE (6,10) CYCLE
      IF (J .EQ. CYCLE) WRITE (6,11) ((P(M,N), N=1,NMAX), M=1,MMAX)
      IF (OK  .GT. 0.5) WRITE (6,12)
      IF (OK  .GT. 0.5) WRITE (6,13) (SAVE(I), I=1,INDEX)
      IF (OK  .GT. 0.5) GO TO 999
  400 CONTINUE
  999 CALL EXIT
      STOP
      END
C
      SUBROUTINE ONE (M,N,DELTA,PHID)
C
      COMMON  INDEX,SAVE(100),P(20,20),MMAX,NMAX,H,RADIUS
C
C DELTA IS LESS THAN H
C
      MLESS1 = M - 1
C
C     CALCULATE THETA.
C
      XD = H*(NMAX - N)
      YD = H*(MMAX-M) - DELTA
      ARG = YD/XD
      THETA = ATAN(ARG)
C
C     CALCULATE PHIPP AND PHIQQ USING THREE CLOSEST NODAL PHI VALUES.
C
      DO 100 K = 1,2
      J = N - K
      I2 = M + DELTA/H + 0.5
      I1 = I2 - 1
      I3 = I2 + 1
      DIST = H*((MMAX-I3)**2 + (NMAX-J)**2)**0.5
      IF (DIST .LE. RADIUS) I3 = I3 - 3
      Y1 = (MMAX - I1)*H
      Y2 = (MMAX - I2)*H
      Y3 = (MMAX - I3)*H
      C1 = (YD-Y3)*(YD-Y2)/((Y1-Y3)*(Y1-Y2))
      C2 = (YD-Y3)*(YD-Y1)/((Y2-Y3)*(Y2-Y1))
      C3 = (YD-Y1)*(YD-Y2)/((Y3-Y1)*(Y3-Y2))
      IF (K .EQ. 1) PHIPP = C1*P(I1,J) + C2*P(I2,J) + C3*P(I3,J)
      IF (K .EQ. 2) PHIQQ = C1*P(I1,J) + C2*P(I2,J) + C3*P(I3,J)
  100 CONTINUE
C
C     CALCULATE PHID.
C
      C = P(MLESS1,N)*(DELTA/H)/(DELTA+H) - P(M,N)*(DELTA+H)/(DELTA*H)
      C = C*TAN(THETA) - 2.*PHIPP/H + 0.5*PHIQQ/H
      PHID = C/(-1.5/H - TAN(THETA)*(H/DELTA+2.)/(H+DELTA))
C
```

Figure 13.7 *(continued)*

```
      INDEX = INDEX + 1
      SAVE(INDEX) = XD
      INDEX = INDEX + 1
      SAVE(INDEX) = YD
      INDEX = INDEX + 1
      SAVE(INDEX) = PHID
      RETURN
      END
C
      SUBROUTINE TWO (M,N,GAMMA,PHIC)
      COMMON INDEX,SAVE(100),P(20,20),MMAX,NMAX,H,RADIUS
C
C GAMMA IS LESS THAN H
C
      NLESS1 = N - 1
C
C     CALCULATE THETA.
C
      XC = H*(NMAX-N) - GAMMA
      YC = H*(MMAX-M)
      ARG = YC/XC
      THETA = ATAN(ARG)
C
C     CALCULATE PHIP AND PHIQ USING THREE CLOSEST NODAL PHI VALUES.
C
      DO 100 K = 1,2
      I = M - K
      J2 = N + GAMMA/H + 0.50
      J1 = J2 - 1
      J3 = J2 + 1
      DIST = H*((MMAX-I)**2 + (NMAX-J3)**2)**0.5
      IF (DIST .LE. RADIUS) J3 = J3 - 3
      X1 = (NMAX - J1)*H
      X2 = (NMAX - J2)*H
      X3 = (NMAX - J3)*H
      C1 = (XC-X3)*(XC-X2)/((X1-X3)*(X1-X2))
      C2 = (XC-X3)*(XC-X1)/((X2-X3)*(X2-X1))
      C3 = (XC-X1)*(XC-X2)/((X3-X1)*(X3-X2))
      IF (K .EQ. 1) PHIP = C1*P(I,J1) + C2*P(I,J2) + C3*P(I,J3)
      IF (K .EQ. 2) PHIQ = C1*P(I,J1) + C2*P(I,J2) + C3*P(I,J3)
  100 CONTINUE
C
C     CALCULATE PHIC.
C
      C = TAN(THETA) * (2.*PHIP-0.5*PHIQ)/H
      C = C + P(M,N)*(H+GAMMA)/(H*GAMMA)
      C = C - P(M,NLESS1)*(GAMMA/H)/(H+GAMMA)
      PHIC = C/(TAN(THETA)*1.5/H + (H/GAMMA+2)/(H+GAMMA))
C
      INDEX = INDEX + 1
      SAVE(INDEX) = XC
      INDEX = INDEX + 1
      SAVE(INDEX) = YC
      INDEX = INDEX + 1
      SAVE(INDEX) = PHIC
      RETURN
      END
//GO.SYSIN DD *
//
```

Figure 13.7 (*continued*)

```
STREAM FLOW AROUND A CYLINDER
THE INITIAL POTENTIAL DISTRIBUTION IS

 3.325  3.158  2.990  2.822  2.654  2.484  2.314  2.143  1.970  1.797
 3.333  2.000  2.000  2.000  2.000  2.000  2.000  2.000  2.000  2.000
 3.342  2.000  2.000  2.000  2.000  2.000  2.000  2.000  2.000  2.000
 3.351  2.000  2.000  2.000  2.000  2.000  2.000  2.000  2.000  2.000
 3.361  2.000  2.000  2.000  2.000  2.000  2.000  2.000  2.000  2.000
 3.371  2.000  2.000  2.000  2.000  2.000  2.000  2.000  2.000  2.000
 3.382  2.000  2.000  2.000  2.000  2.000  2.000  2.000  2.000  2.000
 3.392  2.000  2.000  2.000  2.000  2.000  2.000  2.000  2.000  2.000
 3.403  2.000  2.000  2.000  2.000  2.000  2.000  2.000  2.000  2.000
 3.414  2.000  2.000  2.000  2.000  2.000  2.000  2.000  2.000  2.000
 3.425  2.000  2.000  2.000  2.000  2.000  2.000  2.000  2.000  2.000
 3.435  2.000  2.000  2.000  2.000  2.000  2.000  2.000  2.000  2.000
 3.445  2.000  2.000  2.000  2.000  2.000  2.000  2.000  2.000  2.000
 3.454  2.000  2.000  2.000  2.000  2.000  2.000  2.000  2.000  2.000
 3.462  2.000  2.000  2.000  2.000  2.000  2.000  2.000  2.000  2.000
 3.469  2.000  2.000  2.000  2.000  2.000  2.000  2.000  2.000  2.000
 3.475  2.000  2.000  2.000  2.000  2.000  2.000  2.000  2.000  2.000
 3.479  2.000  2.000  2.000  2.000  2.000  2.000  2.000  2.000  2.000
 3.482  2.000  2.000  2.000  2.000  2.000  2.000  2.000  2.000  2.000
 3.482  2.000  2.000  2.000  2.000  2.000  2.000  2.000  2.000  2.000

CYCLE  1 -- VELOCITY POTENTIALS AT GRID POINTS
 3.325  3.158  2.990  2.822  2.654  2.484  2.314  2.143  1.970  1.797
 3.333  3.093  2.767  2.588  2.457  2.342  2.232  2.123  2.014  1.904
 3.342  2.982  2.592  2.405  2.296  2.218  2.150  2.086  2.024  1.961
 3.351  2.935  2.509  2.309  2.206  2.144  2.099  2.060  2.023  1.987
 3.361  2.916  2.469  2.259  2.157  2.102  2.067  2.041  2.019  1.998
 3.371  2.911  2.452  2.234  2.130  2.078  2.049  2.030  2.015  2.002
 3.382  2.913  2.445  2.222  2.116  2.064  2.038  2.022  2.012  2.004
 3.392  2.917  2.443  2.217  2.109  2.057  2.031  2.018  2.010  2.004
 3.403  2.923  2.444  2.215  2.106  2.053  2.028  2.015  2.008  2.004
 3.414  2.929  2.446  2.215  2.104  2.051  2.026  2.013  2.007  2.003
 3.425  2.936  2.448  2.215  2.104  2.051  2.025  2.013  2.006  2.003
 3.435  2.942  2.451  2.217  2.104  2.050  2.024  2.012  2.006  2.003
 3.445  2.949  2.455  2.218  2.105  2.050  2.024  2.012  2.006  2.003
 3.454  2.955  2.457  2.219  2.105  2.050  2.024  2.012  2.006  2.003
 3.462  2.961  2.460  2.221  2.106  2.051  2.024  2.012  2.006  2.003
 3.469  2.965  2.463  2.222  2.106  2.051  2.024  2.012  2.006  2.003
 3.475  2.970  2.465  2.223  2.107  2.051  2.025  2.012  2.006  2.003
 3.479  2.973  2.467  2.224  2.107  2.051  2.025  2.012  2.006  2.003
 3.482  3.037  2.517  2.254  2.124  2.060  2.029  2.014  2.007  2.003
 3.482  3.096  2.610  2.319  2.161  2.080  2.039  2.019  2.009  2.004

CYCLE  20 -- VELOCITY POTENTIALS AT GRID POINTS
 3.325  3.158  2.990  2.822  2.654  2.484  2.314  2.143  1.970  1.797
 3.333  3.167  3.000  2.834  2.667  2.499  2.331  2.163  1.993  1.821
 3.342  3.177  3.011  2.846  2.681  2.515  2.349  2.183  2.015  1.846
 3.351  3.187  3.023  2.859  2.695  2.531  2.368  2.204  2.038  1.871
 3.361  3.198  3.034  2.872  2.709  2.548  2.386  2.225  2.061  1.896
 3.371  3.209  3.046  2.885  2.724  2.565  2.405  2.246  2.085  1.922
 3.382  3.220  3.058  2.898  2.739  2.581  2.424  2.267  2.108  1.948
 3.392  3.231  3.071  2.911  2.754  2.598  2.443  2.288  2.132  1.974
 3.403  3.242  3.083  2.925  2.769  2.615  2.462  2.310  2.157  2.002
 3.414  3.254  3.095  2.938  2.784  2.632  2.481  2.332  2.183  2.031
 3.425  3.265  3.107  2.951  2.798  2.648  2.501  2.355  2.209  2.062
 3.435  3.276  3.118  2.964  2.813  2.665  2.520  2.377  2.236  2.094
 3.445  3.286  3.130  2.976  2.827  2.681  2.539  2.400  2.263  2.127
 3.454  3.296  3.140  2.988  2.840  2.697  2.558  2.423  2.291  2.160
 3.462  3.305  3.150  2.999  2.853  2.712  2.576  2.444  2.317  2.193
 3.469  3.312  3.158  3.009  2.864  2.725  2.592  2.464  2.341  2.224
 3.475  3.319  3.166  3.017  2.874  2.736  2.605  2.481  2.363  2.251
 3.479  3.323  3.171  3.024  2.882  2.746  2.616  2.494  2.379  2.273
 3.482  3.327  3.175  3.028  2.887  2.752  2.624  2.503  2.391  2.287
 3.482  3.328  3.177  3.031  2.890  2.755  2.627  2.507  2.395  2.292

                VELOCITY POTENTIALS AROUND CYLINDER--
                     X            Y           PHI
                   3.000        5.196        0.999
                   2.000        5.657        0.665
                   1.000        5.916        0.333
                   4.000        4.472        1.328
                   3.317        5.000        1.105
```

Figure 13.7 (*continued*)

```
STRAEM FLOW AROUND A CYLINDER
THE INITIAL POTENTIAL DISTRIBUTION IS

1.622  1.446  1.269  1.091  0.911  0.730  0.549  0.366  0.183  0.0
2.000  2.000  2.000  2.000  2.000  2.000  2.000  2.000  2.000  0.0
2.000  2.000  2.000  2.000  2.000  2.000  2.000  2.000  2.000  0.0
2.000  2.000  2.000  2.000  2.000  2.000  2.000  2.000  2.000  0.0
2.000  2.000  2.000  2.000  2.000  2.000  2.000  2.000  2.000  0.0
2.000  2.000  2.000  2.000  2.000  2.000  2.000  2.000  2.000  0.0
2.000  2.000  2.000  2.000  2.000  2.000  2.000  2.000  2.000  0.0
2.000  2.000  2.000  2.000  2.000  2.000  2.000  2.000  2.000  0.0
2.000  2.000  2.000  2.000  2.000  2.000  2.000  2.000  2.000  0.0
2.000  2.000  2.000  2.000  2.000  2.000  2.000  2.000  2.000  0.0

2.000  2.000  2.000  2.000  2.000  2.000  2.000  2.000  2.000  0.0
2.000  2.000  2.000  2.000  2.000  2.000  2.000  2.000  2.000  0.0
2.000  2.000  2.000  2.000  2.000  2.000  2.000  2.000  2.000  0.0
2.000  2.000  2.000  2.000  2.000  2.000  2.000  2.000  2.000  0.0
2.000  2.000  2.000  2.000  2.000  2.000  2.000  2.000  2.000  0.0
2.000  2.000  2.000  2.000  2.000  2.000  2.000  2.000  2.000  0.0
2.000  2.000  2.000  2.000  2.000  2.000  2.000  2.000  2.000  0.0
2.000  2.000  2.000  2.000  2.000  2.000  2.000  2.000  2.000  0.0
2.000  2.000  2.000  2.000  2.000  2.000  2.000  2.000  2.000  0.0
2.000  2.000  2.000  2.000  2.000  2.000  2.000  2.000  2.000  0.0

CYCLE    1  --  VELOCITY  POTENTIALS  ATGRID  POINTS
1.622  1.446  1.269  1.091  0.911  0.730  0.549  0.366  0.183  0.0
1.794  1.683  1.571  1.458  1.345  1.231  1.116  0.832  0.286  0.0
1.899  1.836  1.772  1.708  1.644  1.579  1.501  1.189  0.524  0.0
1.952  1.916  1.880  1.843  1.807  1.769  1.709  1.384  0.650  0.0
1.978  1.957  1.937  1.916  1.895  1.872  1.821  1.490  0.718  0.0
1.990  1.979  1.967  1.955  1.943  1.928  1.882  1.547  0.755  0.0
1.997  1.990  1.983  1.976  1.969  1.959  1.915  1.578  0.775  0.0
1.999  1.995  1.991  1.988  1.983  1.975  1.933  1.595  0.785  0.0
2.001  1.998  1.996  1.993  1.991  1.984  1.942  1.604  0.791  0.0
2.001  1.999  1.998  1.997  1.995  1.989  1.948  1.609  0.794  0.0
2.001  2.000  1.999  1.998  1.997  1.991  1.950  1.612  0.796  0.0
2.001  2.000  2.000  1.999  1.998  1.993  1.952  1.613  0.797  0.0
2.001  2.000  2.000  2.000  1.999  1.994  1.954  1.626  0.993  0.0
2.001  2.001  2.000  2.000  1.999  1.994  1.971  1.884  1.819  0.0
2.001  2.001  2.000  2.000  1.998  1.996  2.000  2.000  2.000  0.0
2.001  2.001  2.000  2.000  1.999  2.000  2.000  2.000  2.000  0.0
2.001  2.001  2.000  2.000  2.000  2.000  2.000  2.000  2.000  0.0
2.001  2.001  2.000  2.000  2.000  2.000  2.000  2.000  2.000  0.0
2.002  2.001  2.000  2.000  2.000  2.000  2.000  2.000  2.000  0.0
2.002  2.001  2.000  2.000  2.000  2.000  2.000  2.000  2.000  0.0

CYCLE   20  --  VELOCITY  POTENTIALS  ATGRID  POINTS
1.622  1.446  1.269  1.091  0.911  0.730  0.549  0.366  0.183  0.0
1.648  1.473  1.296  1.117  0.935  0.751  0.565  0.377  0.189  0.0
1.675  1.500  1.323  1.143  0.959  0.771  0.581  0.389  0.195  0.0
1.701  1.528  1.351  1.169  0.983  0.792  0.598  0.400  0.201  0.0
1.728  1.556  1.378  1.196  1.008  0.814  0.616  0.413  0.207  0.0
1.755  1.584  1.407  1.224  1.035  0.838  0.635  0.426  0.214  0.0
1.783  1.614  1.438  1.255  1.064  0.864  0.656  0.442  0.222  0.0
1.812  1.645  1.471  1.288  1.096  0.894  0.681  0.460  0.232  0.0
1.843  1.679  1.507  1.326  1.133  0.929  0.712  0.482  0.244  0.0
1.876  1.716  1.547  1.369  1.177  0.971  0.749  0.511  0.260  0.0
1.912  1.756  1.593  1.418  1.230  1.024  0.798  0.551  0.283  0.0
1.949  1.800  1.643  1.475  1.292  1.089  0.862  0.605  0.318  0.0
1.989  1.848  1.700  1.542  1.368  1.172  0.946  0.685  0.379  0.0
2.030  1.899  1.762  1.617  1.458  1.274  1.059  0.800  0.511  0.0
2.072  1.951  1.829  1.702  1.562  1.403  2.000  2.000  2.000  0.0
2.111  2.003  1.897  1.792  1.686  2.000  2.000  2.000  2.000  0.0
2.147  2.050  1.962  1.883  2.000  2.000  2.000  2.000  2.000  0.0
2.175  2.088  2.015  1.963  2.000  2.000  2.000  2.000  2.000  0.0
2.193  2.112  2.048  2.011  2.000  2.000  2.000  2.000  2.000  0.0
2.199  2.119  2.054  2.000  2.000  2.000  2.000  2.000  2.000  0.0
```

Figure 13.7 *(continued)*

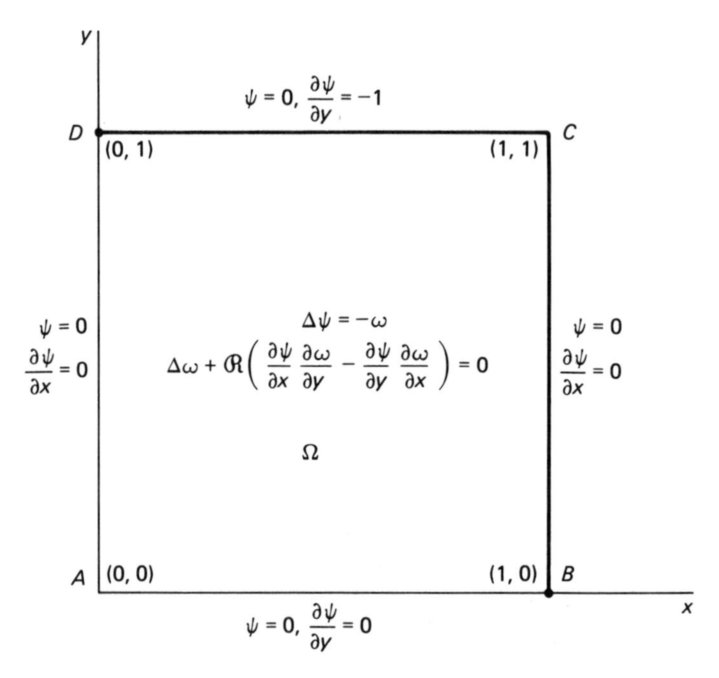

Figure 13.8 Region of interest and the boundary conditions for the Navier-Stokes problem

Several interesting features of this problem can be readily observed. Equations (13.42) and (13.43) represent a system of coupled partial differential equations in ψ and ω. However, if ω is known, then Eq. (13.42) is a linear elliptic equation in ψ, while if ψ is known, then Eq. (13.43) is a linear elliptic equation in ω. This fact can be used to our advantage in developing a computational algorithm. For instance, one can start with an initial guess $\omega^{(0)}$ and $\psi^{(0)}$. Then $\omega^{(0)}$ can be used in Eq. (13.42) to produce $\psi^{(1)}$, which in turn can be used to produce $\omega^{(1)}$. This process can be repeated iteratively, and the sequences $\{\omega^{(i)}\}$ and $\{\psi^{(i)}\}$ hopefully converge.

Difference Approximations. As usual, a square grid of size $h = \Delta x = \Delta y$ is superimposed on $\Omega + \Gamma$. For convenience, the set of internal grid points is denoted by Ω_h and boundary grid points by Γ_h. A typical internal grid point 0 and its four neighbors are labeled as shown in Figure 13.9. At a typical grid point 0, Eqs. (13.42) and (13.43) can be respectively approximated as

$$-4\psi_0 + \psi_1 + \psi_2 + \psi_3 + \psi_4 = -h^2\omega_0 \tag{13.48}$$

$$-4\omega_0 + \omega_1 + \omega_2 + \omega_3 + \omega_4 + h^2\mathcal{R}\left(\frac{\psi_1 - \psi_3}{2h}\frac{\omega_2 - \omega_4}{2h} - \frac{\psi_2 - \psi_4}{2h}\frac{\omega_1 - \omega_3}{2h}\right) = 0 \tag{13.49}$$

By writing the preceding type of equations at each internal grid point, and after one takes care of applying the normal derivative boundary conditions, one arrives at two

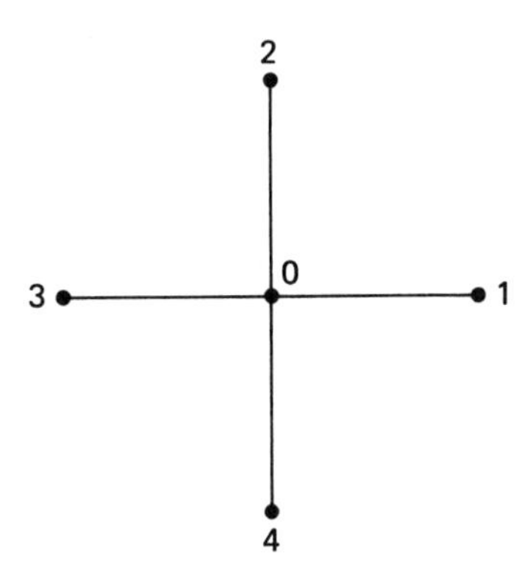

Figure 13.9 Typical internal grid point and its four neighbors

systems of equations. These systems can be solved using the iterative strategy suggested earlier. However, Greenspan reports this procedure diverged for $\mathcal{R} > 250$ and $h = \frac{1}{40}$. Actual observation of computer listings led him to the discovery that the divergence was due to a lack of diagonal dominance in the coefficient matrixes. Greenspan proceeded to correct this problem by approximating $\partial\omega/\partial x$ and $\partial\omega/\partial y$ in Eq. (13.43), not as indicated in Eq. (13.49), but as follows:

$$\begin{aligned}\left.\frac{\partial\omega}{\partial y}\right|_0 &= \frac{\omega_2 - \omega_0}{2h} \qquad \text{if } \psi_1 \geq \psi_3 \\ &= \frac{\omega_0 - \omega_4}{2h} \qquad \text{if } \psi_1 < \psi_3\end{aligned} \tag{13.50}$$

$$\begin{aligned}\left.\frac{\partial\omega}{\partial x}\right|_0 &= \frac{\omega_0 - \omega_3}{2h} \qquad \text{if } \psi_2 \geq \psi_4 \\ &= \frac{\omega_1 - \omega_0}{2h} \qquad \text{if } \psi_2 < \psi_4\end{aligned} \tag{13.51}$$

Therefore, depending on the relative magnitudes of the ψ_i's, Eq. (13.43) is approximated as

$$\left(-4 - \frac{\gamma\mathcal{R}}{2} - \frac{\delta\mathcal{R}}{2}\right)\omega_0 + \omega_1 + \left(1 + \frac{\gamma\mathcal{R}}{2}\right)\omega_2 + \left(1 + \frac{\delta\mathcal{R}}{2}\right)\omega_3 + \omega_4 = 0, \qquad \gamma \geq 0, \delta \geq 0 \tag{13.52}$$

$$\left(-4 - \frac{\gamma\mathcal{R}}{2} - \frac{\delta\mathcal{R}}{2}\right)\omega_0 + \left(1 - \frac{\delta\mathcal{R}}{2}\right)\omega_1 + \left(1 + \frac{\alpha\mathcal{R}}{2}\right)\omega_2 + \omega_3 + \omega_4 = 0, \qquad \gamma \geq 0, \delta < 0 \tag{13.53}$$

$$\left(-4 - \frac{\gamma\mathcal{R}}{2} - \frac{\delta\mathcal{R}}{2}\right)\omega_0 + \omega_1 + \omega_2 + \left(1 + \frac{\delta\mathcal{R}}{2}\right)\omega_3 + \left(1 - \frac{\gamma\mathcal{R}}{2}\right)\omega_4 = 0, \qquad \gamma < 0, \beta \geq 0 \tag{13.54}$$

$$\left(-4 + \frac{\gamma\mathcal{R}}{2} + \frac{\delta\mathcal{R}}{2}\right)\omega_0 + \left(1 - \frac{\delta\mathcal{R}}{2}\right)\omega_1 + \omega_2 + \omega_3 + \left(1 - \frac{\gamma\mathcal{R}}{2}\right)\omega_4 = 0, \qquad \gamma < 0, \delta < 0 \tag{13.55}$$

where $\gamma = \psi_1 - \psi_3$ and $\delta = \psi_2 - \psi_4$. This strategy resulted in convergence for $0 \leq \mathcal{R} \leq 10^5$ and $h = \frac{1}{10}$, but diverged for $h < \frac{1}{10}$. The divergence was predominant near the boundaries. Greenspan claims that this divergence is a result of his approximation of normal derivative boundary conditions being not good enough. So he suggested that the normal derivative boundary conditions be approximated with non-centered and one-sided difference quotients, as described next.

NORMAL DERIVATIVE BOUNDARY CONDITIONS

Consider the point 0 as lying on the boundary and two of its adjacent points, as shown in Figure 13.10. Referring to Figure 13.10a, one can write

$$\psi_1 = \psi_0 + h\psi_0' + \frac{h^2}{2}\psi_0'' + \cdots$$

$$\psi_2 = \psi_0 + (2h)\psi_0' + \frac{(2h)^2}{2}\psi_0'' + \cdots$$

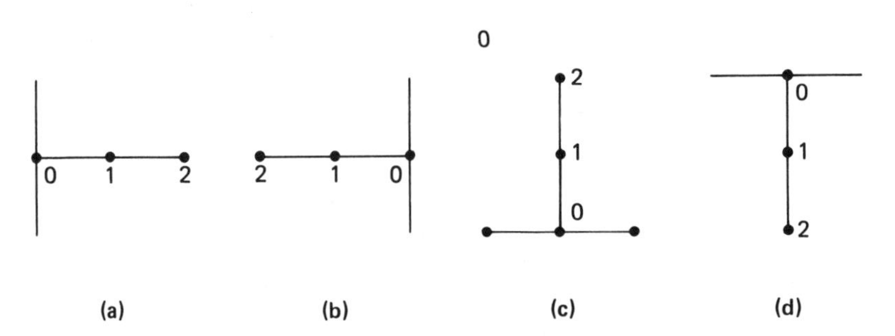

Figure 13.10 Treatment of normal derivative boundary conditions using grid points on only one side of the boundary

Multiplying the first equation by 4 and subtracting the second, and solving for ψ_0', one gets

$$\psi_0' = \frac{4\psi_1 - \psi_2 - 3\psi_0}{2h} + O(h^2)$$

Thus an approximation that is second-order accurate for $\partial\psi/\partial x$ at the point 0 in Figure 13.10a is

$$\left.\frac{\partial x}{\partial x}\right|_0 = \frac{1}{2h}(-3\psi_0 + 4\psi_1 - \psi_2) \tag{13.56}$$

Notice that the points 0, 1, and 2 are general and give relative orientation of internal grid points with respect to a grid point on the boundary. If the points 0, 1, and 2 are oriented as shown in Figure 13.10b, then $\partial\psi/\partial x$ at the point 0 would be

$$\left.\frac{\partial \psi}{\partial x}\right|_0 = \frac{1}{2h}(3\psi_0 - 4\psi_1 + \psi_2) \tag{13.57}$$

Similarly,

$$\left.\frac{\partial \psi}{\partial y}\right|_0 = \frac{1}{2h}(-3\psi_0 + 4\psi_1 - \psi_2) \qquad \text{in Figure 13.10c} \tag{13.58}$$

$$= \frac{1}{2h}(3\psi_0 - 4\psi_1 + \psi_2) \qquad \text{in Figure 13.10d} \tag{13.59}$$

Approximating the Laplacian on the Boundary. Greenspan's procedure requires, at a later stage, approximation of the Laplacian in terms of certain function values and normal derivatives. Toward this end, typical boundary configurations are shown in Figure 13.11. What we want is an approximation to $\psi_{xx} + \psi_{yy}$ at grid point 0 in terms of function values at grid points 0, 1, 2, 4 [see configuration (a)] and the derivative $\partial\psi/\partial x$ at 0. That is, we wish to determine parameters $\alpha_0, \alpha_1, \alpha_2, \alpha_4$, and α_5 [for configuration (a)], such that

$$\psi_{xx} + \psi_{yy}|_0 = \alpha_0\psi_0 + \alpha_1\psi_1 + \alpha_2\psi_2 + \alpha_4\psi_4 + \alpha_5\left.\left(\frac{\partial\psi}{\partial x}\right)\right|_0 \tag{13.60}$$

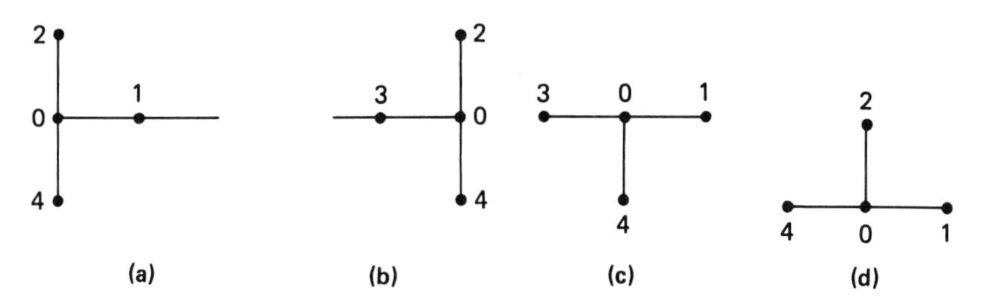

Figure 13.11 Approximating the Laplacian on the boundary

To determine the α's, we proceed by first expanding ψ_1, ψ_2, and ψ_4 in Taylor series as follows:

$$\psi_1 = \psi_0 + h\left.\frac{\partial\psi}{\partial x}\right|_0 + \left.\frac{h^2}{2!}\frac{\partial^2\psi}{\partial x^2}\right|_0 + \cdots$$

$$\psi_2 = \psi_0 + h\left.\frac{\partial\psi}{\partial y}\right|_0 + \left.\frac{h^2}{2!}\frac{\partial^2\psi}{\partial y^2}\right|_0 + \cdots$$

$$\psi_4 = \psi_0 - h\left.\frac{\partial\psi}{\partial y}\right|_0 + \left.\frac{h^2}{2!}\frac{\partial^2\psi}{\partial y^2}\right|_0 + \cdots$$

Substituting these in Eq. (13.60) and equating the coefficients of corresponding terms, one gets, after a little calculation,

$$\alpha_0 = -\frac{4}{h^2}, \qquad \alpha_1 = \frac{2}{h^2}, \qquad \alpha_2 = \alpha_4 = \frac{1}{h^2}, \qquad \alpha_5 = -\frac{2}{h} \tag{13.61}$$

Therefore, for the configuration in Figure 13.11a,

$$\psi_{xx} + \psi_{yy}|_0 = -\frac{4}{h^2}\psi_0 + \frac{2}{h^2}\psi_1 + \frac{1}{h^2}\psi_2 + \frac{1}{h^2}\psi_4 - \frac{2}{h}\left.\left(\frac{\partial\psi}{\partial x}\right)\right|_0 \tag{13.62a}$$

Analogous equations can be quickly derived for the other three configurations in Figure 13.11. The results are as follows. For the configuration shown in Figure 13.11b,

$$(\psi_{xx} + \psi_{yy})|_0 = -\frac{4}{h^2}\psi_0 + \frac{1}{h^2}\psi_2 + \frac{2}{h^2}\psi_3 + \frac{1}{h^2}\psi_4 + \frac{2}{h}\left(\frac{\partial \phi}{\partial x}\right)\bigg|_0 \qquad (13.62b)$$

For the configuration shown in Figure 13.11c,

$$(\psi_{xx} + \psi_{yy})|_0 = -\frac{4}{h^2}\psi_0 + \frac{1}{h^2}\psi_1 + \frac{1}{h^2}\psi_3 + \frac{2}{h^2}\psi_4 + \frac{2}{h}\left(\frac{\partial \psi}{\partial x}\right)\bigg|_0 \qquad (13.62c)$$

For the configuration shown in Figure 13.11d,

$$(\psi_{xx} + \psi_{yy})|_0 = -\frac{4}{h^2}\psi_0 + \frac{1}{h^2}\psi_1 + \frac{2}{h^2}\psi_2 + \frac{1}{h^2}\psi_3 - \frac{2}{h}\left(\frac{\partial \psi}{\partial y}\right)\bigg|_0 \qquad (13.62d)$$

The Numerical Algorithm. Now the following computational procedure is used to solve the given equation.

Step 1. Superimpose a square grid of size $h = 1/n$. Denote the set of interior grid points by Ω_h and the boundary grid points by Γ_h. Denote, further, by $\Omega_{h,1}$ those points of Ω_h whose distance from Γ is h, and denote by $\Omega_{h,2}$ those grid points whose distance from Γ is greater than h. (See Figure 13.12 for clarity.) These sets of grid points are referred to as outer and inner grid points in the program.

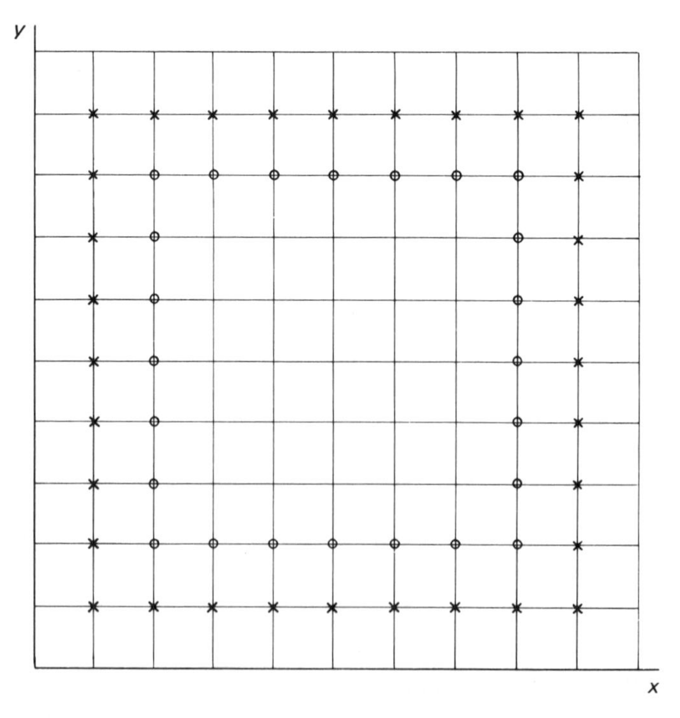

Figure 13.12 Explanation of outer and inner grid points. The set of points marked by x are $\Omega_{h,1}$ and the set of points marked by 0 are $\Omega_{h,2}$

Step 2. Set

$$\psi^{(0)} = c_1, \qquad \text{for all points in } \Omega_h \tag{13.63}$$

$$\omega^{(0)} = c_2, \qquad \text{for all points in } \Omega_h + \Gamma_h \tag{13.64}$$

where c_1 and c_2 are known constants. A modified overrelaxation procedure that does not require too much storage is now applied to get $\psi^{(1)}$ from $\psi^{(0)}$ and $\omega^{(0)}$. This procedure is described next.

Step 3. On Ω_h, set

$$\psi^{(1,0)} = \psi^{(0)} \tag{13.65}$$

and on $\Omega_{h,2}$ generate $\bar{\psi}^{(1,1)}$ by sweeping along each row of $\Omega_{h,2}$ from left to right, starting from the bottom row and proceeding to the top row, by the recursion formula

$$\bar{\psi}_0^{(1,i)} = (1 - r_\psi)\psi_0^{(1,i-1)} + \frac{r_\psi}{4}[\psi_1^{(1,i-1)} + \psi_2^{(1,i-1)} + \bar{\psi}_3^{(1,i)} + \bar{\psi}_4^{(1,i)} + h^2\omega_0] \tag{13.66}$$

where the overrelaxation parameter r_ψ is chosen to be $0 < r_\psi < 2$. After each sweep, $\psi^{(1,i)}$ is defined on $\Omega_{h,2}$ by the weighted average

$$\psi^{(1,i)} = \xi\psi^{(1,i-1)} + (1 - \xi)\bar{\psi}^{(1,i)}, \qquad 0 \le \xi \le 1 \tag{13.67}$$

This inner iteration process continues until

$$|\psi^{(1,k)} - \psi^{(1,k+1)}| < \epsilon \tag{13.68}$$

Once the inner iteration converges, we define

$$\psi^{(1)} = \psi^{(1,k)} \qquad \text{on } \Omega_{h,2} \tag{13.69}$$

Step 4. In order to define $\psi^{(1)}$ on $\Omega_{h,1}$, we apply Eqs. (13.56) to (13.59) and Eqs. (13.44) to (13.49) in the following manner.

At each point of $\Omega_{h,1}$ of the form (h, lh), $l = 2, 3, \ldots, n - 2$, set (see Figure 13.10a)

$$\psi_1^{(1)} = \frac{\psi_2^{(1)}}{4} \tag{13.70a}$$

while at each point of $\Omega_{h,1}$ of the form $(1 - h, lh)$, $l = 2, 3, \ldots, n - 2$, set (see Figure 13.10b)

$$\psi_1^{(1)} = \frac{\psi_2^{(1)}}{4} \tag{13.70b}$$

Similarly, at each point of $\Omega_{h,1}$ of the form (lh, h), $l = 1, 2, \ldots, n - 1$, set (see Figure 13.10c)

$$\psi_1^{(1)} = \frac{\psi_2^{(1)}}{4} \tag{13.70c}$$

Finally, at each point of $\Omega_{h,1}$ of the form $(lh, 1 - h)$, $l = 1, 2, \ldots, n - 1$, set (see Figure 13.10d)

$$\psi_1^{(1)} = \frac{h}{2} + \frac{\psi_2^{(1)}}{4} \tag{13.70d}$$

Thus Eqs. (13.69) and (13.70) define $\psi^{(1)}$ on all of Ω_h.

Step 5. Now construct the solution $\omega^{(1)}$ on $\Omega_h + \Gamma_h$ as follows. On Γ_h, solve Eqs. (13.42), (13.44) to (13.47), and (13.62) to yield at each point $(lh, 0)$, $l = 0, 1, 2, \ldots, n$ (see Figure 13.11b),

$$\bar{\omega}_0^{(1)} = -\frac{2\psi_2^{(1)}}{h^2} \tag{13.71a}$$

at the points $(0, lh)$, $l = 1, 2, \ldots, n-1$ (see Figure 13.11a),

$$\bar{\omega}_0^{(1)} = -\frac{2\psi_1^{(1)}}{h^2} \tag{13.71b}$$

at the points $(1, lh)$, $l = 1, 2, \ldots, n-1$ (see Figure 13.11c),

$$\bar{\omega}_0^{(1)} = -\frac{2\psi_3^{(1)}}{h^2} \tag{13.71c}$$

and at the points $(lh, 1)$, $l = 0, 1, 2, \ldots, n$ (see Figure 13.11d),

$$\bar{\omega}_0^{(1)} = \frac{2}{h} - \frac{2\psi_4^{(1)}}{h^2} \tag{13.71d}$$

Now define $\omega^{(1)}$ on Γ_h by the formula

$$\omega^{(1)} = \delta\omega^{(0)} + (1-\delta)\bar{\omega}^{(1)}, \qquad 0 \leq \delta \leq 1 \tag{13.72}$$

Step 6. Now proceed to determine $\omega^{(1)}$ on Ω_h by again using the modified relaxation procedure. At each point of Γ_h, set

$$\omega^{(1,0)} = \omega^{(1)}$$

while at each point of Ω_h set

$$\omega^{(1,0)} = \omega^{(0)}$$

Now, generate $\bar{\omega}^{(1,1)}$ by sweeping along each row of Ω_h, from left to right, starting from the bottom row to the top row, by the recursion formula

$$\bar{\omega}_0^{(1,i)} = (1 - r_\omega)\omega_0^{(1,i-1)} + \frac{r_\omega}{\sigma_0}[\sigma_1\omega_1^{(1,i-1)} + \sigma_2\omega_2^{(1,i-1)} + \sigma_3\bar{\omega}_3^{(1,i)} + \sigma_4\bar{\omega}_4^{(1,i)}]$$

where $0 < r_\omega < 2$ and

$$\sigma_0 = 4 + \frac{\mathcal{R}}{2}|\alpha| + \frac{\mathcal{R}}{2}|\beta|$$

$$\sigma_1 = \begin{cases} 1, & \beta \geq 0 \\ 1 + \frac{\mathcal{R}}{2}|\beta|, & \beta < 0 \end{cases}$$

$$\sigma_2 = \begin{cases} 1 + \frac{\mathcal{R}}{2}\alpha, & \alpha \geq 0 \\ 1, & \alpha < 0 \end{cases}$$

$$\sigma_3 = \begin{cases} 1 + \frac{\mathcal{R}}{2}\beta, & \beta \geq 0 \\ 1, & \alpha < 0 \end{cases}$$

$$\sigma_4 = \begin{cases} 1, & \alpha \geq 0 \\ 1 + \frac{\mathcal{R}}{2}|\alpha|, & \alpha < 0 \end{cases}$$

and where, as defined previously, $\alpha = \psi_1 - \psi_3$ and $\beta = \psi_2 - \psi_4$. After each such sweep, $\omega^{(1,i)}$ is defined on Ω_h by the weighted average.

$$\omega^{(1,i)} = \delta\omega^{(1,i-1)} + (1-\delta)\bar{\omega}^{(1,i)}, \qquad 0 \leq \delta \leq 1$$

where δ is the same weight as that used in Eq. (13.72). This inner iteration continues until, for the given tolerance ϵ, one has

$$|\omega^{(1,K)} - \bar{\omega}^{(1,K)}| < \epsilon$$

from which one defines on Ω_h

$$\omega^{(1)} = \omega^{(1,K)}$$

Step 7. The next step is to determine $\psi^{(2)}$ on Ω_h from $\omega^{(1)}$ and $\psi^{(1)}$ in the same fashion as $\psi^{(1)}$ was determined from $\omega^{(0)}$ and $\psi^{(0)}$. Then construct $\omega^{(2)}$ on $\Omega_h + \Gamma_h$ from $\omega^{(1)}$ and $\psi^{(2)}$ in the same fashion as $\omega^{(1)}$ was determined from $\omega^{(0)}$ and $\psi^{(1)}$. Proceed in this manner until

$$|\psi^{(m)} - \psi^{(m+1)}| < \epsilon \qquad \text{on } \Omega_h$$
$$|\omega^{(m)} - \omega^{(m+1)}| < \epsilon \qquad \text{on } \Omega_h + \Gamma_h$$

A flow chart, a computer program, and the results are shown in Figures. 13.13 and 13.14.

COMPUTER PROGRAM NOTATION

$\mathcal{R}$ = Reynolds number

H = interval length

MM,NN = number of intervals in each direction

MMM = $MM - 1$

NNN = $NN - 1$

TOL = tolerance for ψ and ω.

CI = initial guess for ψ

C2 = initial guess for ω

RP = $r\psi$ = overrelaxation factor for ψ

RW = $r\omega$ = overrelaxation factor for ω

XI = ξ = weight factor for ψ

DELTA = δ = weight factor for ω

OLDP(M,N) = $\psi(M, N)$ for previous cycle

OLDW(M,N) = $\omega(M, N)$ for previous cycle

BAR(M,N) = $\bar{\psi}(M, N)$ when calculating ψ
= $\bar{\omega}(M, N)$ when calculating δ

Q0, . . . , Q4 = $\sigma_0, \ldots, \sigma_4$

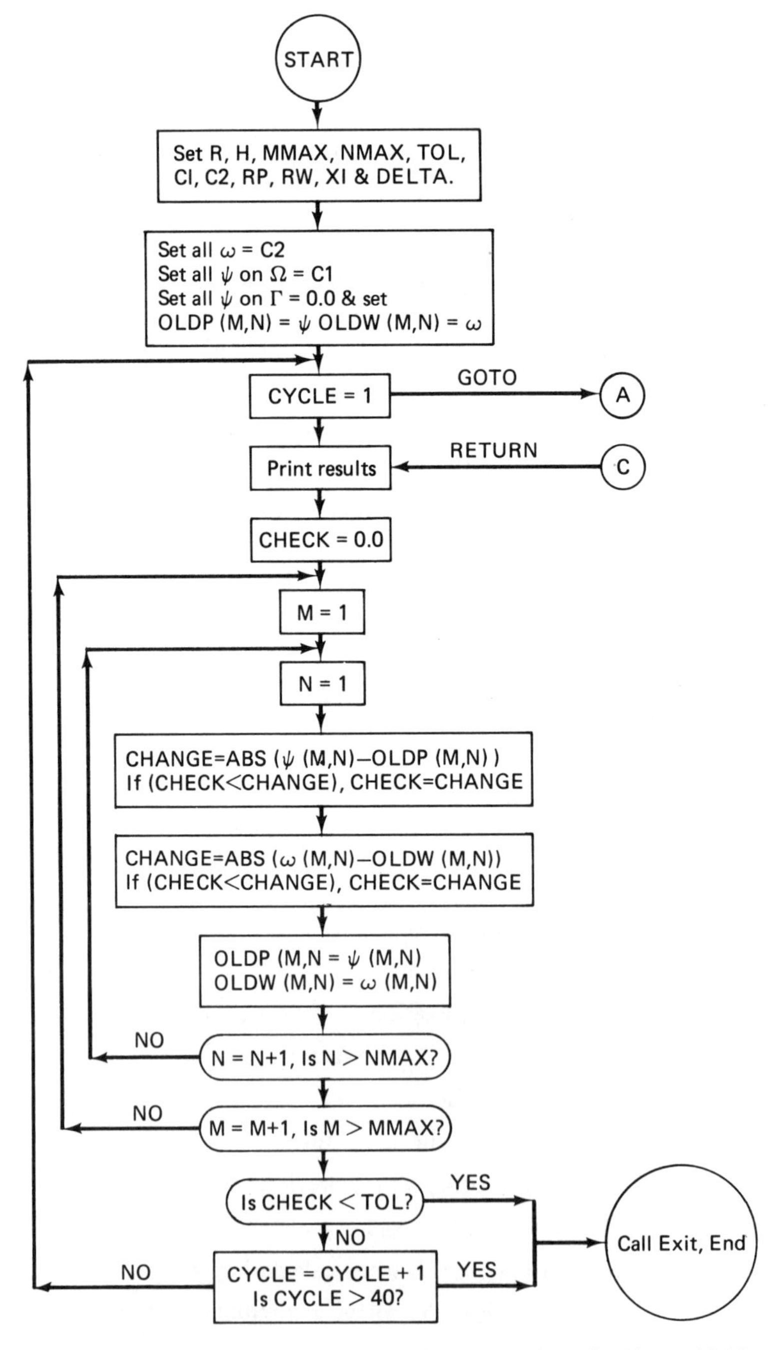

Figure 13.13 Flow chart used to write the program in Figure 13.14a

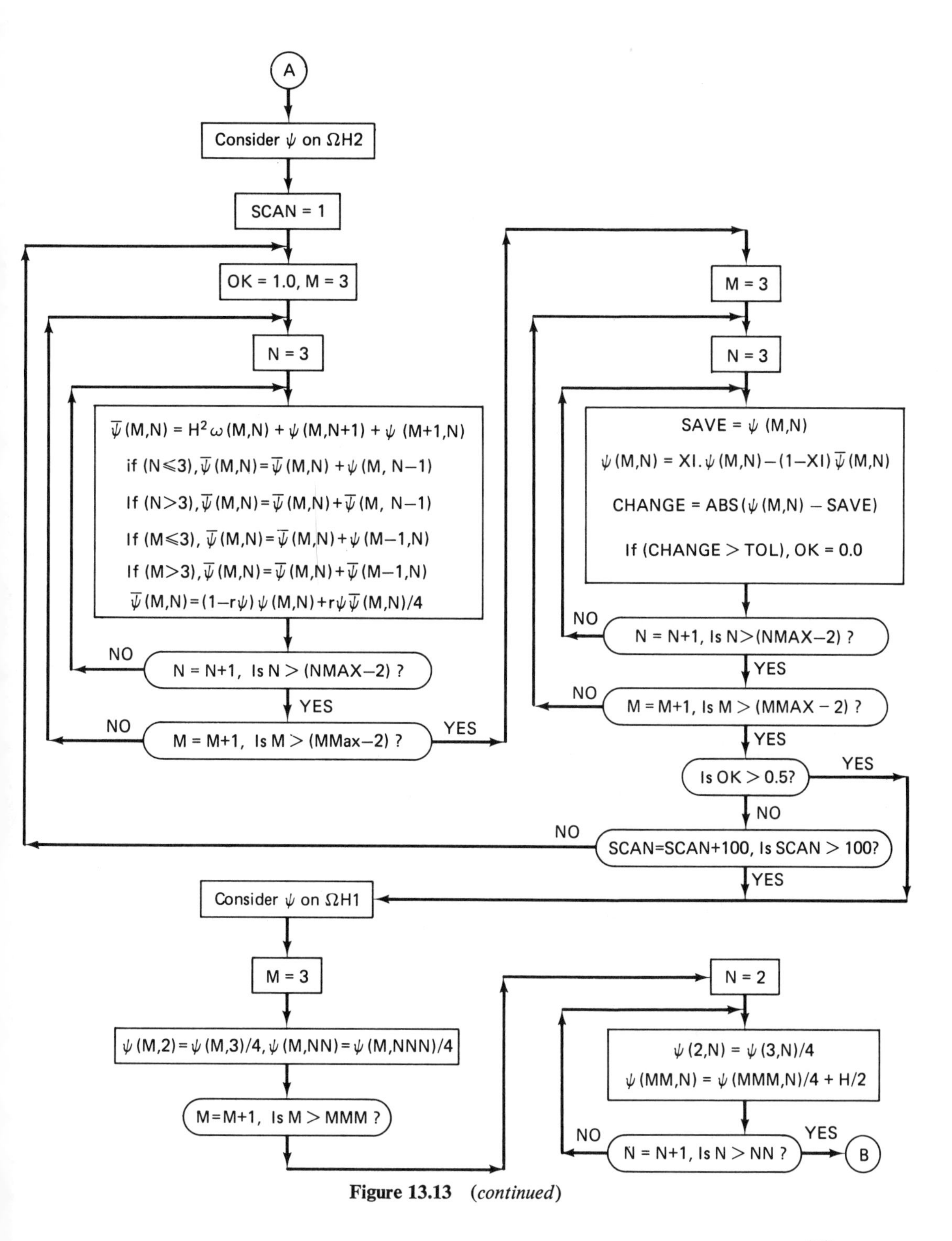

Figure 13.13 *(continued)*

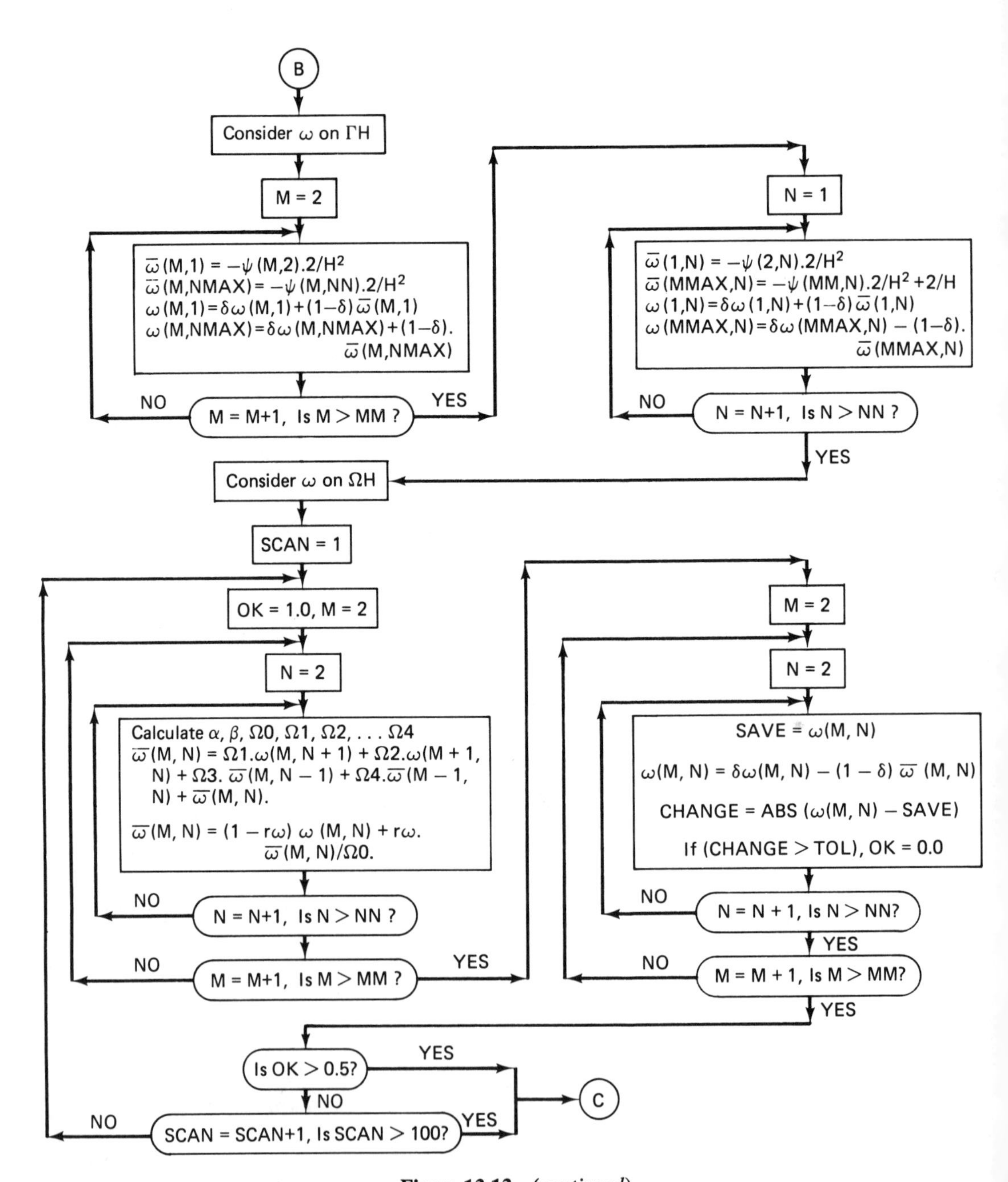

Figure 13.13 *(continued)*

```
$JOB
C PROBLEM NO.4
C STEADY-STATE NAVIER-STOKES PROBLEM ON A SQUARE.
C THE PARAMETERS OF THE PROBLEM ARE:
C
C      R          - REYNOLD'S NUMBER
C      H          - GRID SIZE
C      MMAX       -
C      NMAX       -
C      TOL        - ERROR TOLERANCE
C      C1,C2      - INITIAL GUESS PSI AND OMEGA
C      RP         - OVER RELAXATION FACTOR FOR PSI
C      RW         - OVER-RELAXATION FACTOR FOR OMEGA
C      XI         - WEIGHT FACTOR FOR PSI
C      DELTA      - WEIGHT FACTOR FOR OMEGA
C      OLDP(M,N)-PSI(M,N) FOR PREVIOUS CYCLE
C      OLDW(M,N)-OMEGA(M,N) FOR PREVIOUS CYCLE
C      B(M,N)     -PSI(M,N) WITH A BAR ON TOP OR
C                 -OMEGA(M,N) WITH A BAR ON TOP
C      Q0         - SIGMA WITH A SUBSCRIPT ZERO
C      Q1         -   "     "        "        ONE
C      Q2         -   "     "        "        TWO
C      ETC.
C
    9 FORMAT ('1'/' ',5X,'CYCLE',I3/'0',5X,'STREAM FUNCTION',
     &        'PSI --')
   10 FORMAT ('0',11F12.4)
   11 FORMAT ('0'/'0',5X,'VORTICITY, OMEGA --')
      DIMENSION PSI(50,50),OMEGA(50,50)
      DIMENSION OLDP(50,50),OLDW(50,50),BAR(50,50)
      INTEGER CYCLE,SCAN
C
C     SET UP OF INITIAL INFORMATION
C
      R = 10.
      H = 0.05
      MMAX = 21
      NMAX = 21
      MM = MMAX - 1
      NN = NMAX - 1
      MMM = MMAX - 2
      NNN = NMAX - 2
      TOL = 0.0001
      C1 = 0.0
      C2 = 0.0
      RP = 1.8
      RW = 1.0
      XI = 0.10
      DELTA = 0.70
C
C
C
C
C SETTING UP OF INITIAL GUESS BY
C FIRST IMPLEMENTING EQS.(13.63) - (13.65).
C
      DO 100 M = 1,MMAX
      DO 95 N = 1,NMAX
      OMEGA(M,N) = C2
      PSI(M,N) = C1
      IF (M.EQ.1 .OR. M.EQ.MMAX) PSI(M,N) = 0.0
      IF (N.EQ.1 .OR. N.EQ.NMAX) PSI(M,N) = 0.0
      OLDP(M,N) = PSI(M,N)
      OLDW(M,N) = OMEGA(M,N)
   95 CONTINUE
  100 CONTINUE
      DO 500 CYCLE = 1,40
C
C CONSIDER PSI AT POINTS ON THE INNER GRID
C THE INNER AND OUTER GRIDS ARE DEFINED IN STEP(1)
```

Figure 13.14 (a) Program listing for example problem 4; (b) output of program in Figure 13.14a; (c) streamlines for the Reynold's number $\mathcal{R} = 10$

```
C
      DO 400 SCAN = 1,100
      SAVE1 = SCAN
      OK = 1.0
C
C IMPLEMENT EQN.(13.66)
C
      DO 395 M = 3,MMM
      DO 390 N = 3,NNN
      MPLUS1 = M + 1
      MLESS1 = M - 1
      NPLUS1 = N + 1
      NLESS1 = N - 1
      BAR(M,N) = H*H*OMEGA(M,N) + PSI(M,NPLUS1) +
     &           PSI(MPLUS1,N)
      IF (N .LE. 3) BAR(M,N) = BAR(M,N) + PSI(M,NLESS1)
      IF (N .GT. 3) BAR(M,N) = BAR(M,N) + BAR(M,NLESS1)
      IF (M .LE. 3) BAR(M,N) = BAR(M,N) + PSI(MLESS1,N)
      IF (M .GT. 3) BAR(M,N) = BAR(M,N) + BAR(MLESS1,N)
      BAR(M,N) = (1. - RP)*PSI(M,N) + RP*BAR(M,N)/4.
  390 CONTINUE
  395 CONTINUE
C
C IMPLEMENT EQN.(13.67)
C
      DO 385 M = 3,MMM
      DO 380 N = 3,NNN
      SAVE = PSI(M,N)
      PSI(M,N) = XI*PSI(M,N) + (1. - XI)*BAR(M,N)
      CHANGE = ABS(PSI(M,N) - SAVE)
      IF (CHANGE .GT. TOL) OK = 0.0
  380 CONTINUE
  385 CONTINUE
      IF (OK .GT. 0.5) GO TO 401
  400 CONTINUE
  401 CONTINUE
C
C CONSIDER PSI AT POINTS ON THE OUTER GRID
C IMPLEMENT EQS.(13.70,A,B,C)
C
      DO 375 M = 3,MMM
      PSI(M,2)  = PSI(M,3)/4.
      PSI(M,NN) = PSI(M,NNN)/4.
  375 CONTINUE
C
C IMPLEMENT EQN. (13.70D)
C
      DO 370 N = 2,NN
      PSI(2,N)  = PSI(3,N)/4.
      PSI(MM,N) = PSI(MMM,N)/4. + H/2.
  370 CONTINUE
C
C IMPLEMENT EQS.(13.71 A,B,C)
C CONSIDER OMEGA AT POINTS ON SH
C
      DO 275 M = 2,MM
      BAR(M,1)    = -PSI(M,2) *2./(H*H)
      BAR(M,NMAX) = -PSI(M,NN)*2./(H*H)
      OMEGA(M,1)    = DELTA*OMEGA(M,1)    + (1.-DELTA)*
     &          BAR(M,1)
      OMEGA(M,NMAX) = DELTA*OMEGA(M,NMAX) + (1.-DELTA)*
     &          BAR(M,NMAX)
  275 CONTINUE
C
      DO 270 N = 1,NMAX
      BAR(1,N)    = -PSI(2,N) *2./(H*H)
      BAR(MMAX,N) = -PSI(MM,N)*2./(H*H) + 2./H
      OMEGA(1,N)    = DELTA*OMEGA(1,N)    + (1.-DELTA)*
     &                BAR(1,N)
      OMEGA(MMAX,N) = DELTA*OMEGA(MMAX,N) + (1.-DELTA)*
     &                BAR(MMAX,N)
```

Figure 13.14 *(continued)*

```
  270 CONTINUE
C
C     CONSIDER OMEGA ON RH.
C
      DO 300 SCAN = 1,100
      SAVE2 = SCAN
      OK = 1.0
C
      DO 295 M = 2,MM
C
      DO 290 N = 2,NN
      MPLUS1 = M + 1
      MLESS1 = M - 1
      NPLUS1 = N + 1
      NLESS1 = N - 1
      ALPHA = PSI(M,NPLUS1) - PSI(M,NLESS1)
      BETA  = PSI(MPLUS1,N) - PSI(MLESS1,N)
      Q0 = 4. + 0.5*R*ABS(ALPHA) + 0.5*R*ABS(BETA)
      IF (BETA  .GE. 0.0) Q1 = 1.0
      IF (BETA  .LT. 0.0) Q1 = 1.0 - 0.5*R*BETA
      IF (ALPHA .GE. 0.0) Q2 = 1.0 + 0.5*R*ALPHA
      IF (ALPHA .LT. 0.0) Q2 = 1.0
      IF (BETA  .GE. 0.0) Q3 = 1.0 + 0.5*R*BETA
      IF (BETA  .LT. 0.0) Q3 = 1.0
      IF (ALPHA .GE. 0.0) Q4 = 1.0
      IF (ALPHA .LT. 0.0) Q4 = 1.0 - 0.5*R*ALPHA
      BAR(M,N) = Q1*OMEGA(M,NPLUS1) + Q2*OMEGA(MPLUS1,N)
      BAR(M,N) = Q3*BAR(M,NLESS1)   + Q4*BAR(MLESS1,N)+
     &           BAR(M,N)
      BAR(M,N) = (1. - RW)*OMEGA(M,N) + RW*BAR(M,N)/Q0
  290 CONTINUE
  295 CONTINUE
      DO 285 M = 2,MM
      DO 280 N = 2,NN
      SAVE = OMEGA(M,N)
      OMEGA(M,N) = DELTA*OMEGA(M,N) + (1. - DELTA)*BAR(M,N)
      CHANGE = ABS(OMEGA(M,N) - SAVE)
      IF (CHANGE .GT. TOL) OK = 0.0
  280 CONTINUE
  285 CONTINUE
      IF (OK .GT. 0.5) GO TO 301
  300 CONTINUE
  301 CONTINUE
C
      WRITE (6,9) CYCLE
      WRITE (6,10) ((PSI(MMAX-L+1,N), N=1,NMAX,2),
     &              L=1,MMAX,2)
      WRITE (6,11)
      WRITE (6,10) ((OMEGA(MMAX-L+1,N), N=1,NMAX,2),
     &              L=1,MMAX,2)
      WRITE (6,10) SAVE1,SAVE2
      CHECK = 0.0
      DO 200 M = 1,MMAX
      DO 195 N = 1,NMAX
      CHANGE = ABS(PSI(M,N) - OLDP(M,N))
      IF (CHECK .LE. CHANGE) CHECK = CHANGE
      CHANGE = ABS(OMEGA(M,N) - OLDW(M,N))
      IF (CHECK .LE. CHANGE) CHECK = CHANGE
      OLDP(M,N) = PSI(M,N)
      OLDW(M,N) = OMEGA(M,N)
  195 CONTINUE
  200 CONTINUE
      IF (CHECK .LT. TOL) GO TO 1000
  500 CONTINUE
 1000 CALL EXIT
      STOP
      END
$ENTRY
//
/*
```

Figure 13.14 *(continued)*

```
CYCLE  1
STREAM FUNCTIONPSI --
 0.0          0.0          0.0          0.0          0.0
 0.0          0.0          0.0          0.0          0.0
 0.0          0.0          0.0          0.0          0.0
 0.0          0.0          0.0          0.0          0.0
 0.0          0.0          0.0          0.0          0.0
 0.0          0.0          0.0          0.0          0.0
 0.0          0.0          0.0          0.0          0.0
 0.0          0.0          0.0          0.0          0.0
 0.0          0.0          0.0          0.0          0.0
 0.0          0.0          0.0          0.0          0.0
 0.0          0.0          0.0          0.0          0.0

VORTICITY, OMEGA --
12.0000       6.0000       6.0000       6.0000       6.0000
 0.0          1.2199       3.1322       3.8267       4.0828
 0.0          0.7790       1.6727       2.2155       2.4795
 0.0          0.4174       0.8376       1.1455       1.3235
 0.0          0.1854       0.3697       0.5166       0.6113
 0.0          0.0686       0.1392       0.1989       0.2405
 0.0          0.0211       0.0439       0.0644       0.0796
 0.0          0.0053       0.0115       0.0173       0.0220
 0.0          0.0011       0.0025       0.0038       0.0050
 0.0          0.0002       0.0004       0.0007       0.0009
 0.0          0.0          0.0          0.0          0.0
 1.0000     100.0000

CYCLE  2
STREAM FUNCTIONPSI --
 0.0          0.0          0.0          0.0          0.0
 0.0          0.0180       0.0346       0.0424       0.0464
 0.0          0.0149       0.0386       0.0541       0.0628
 0.0          0.0124       0.0339       0.0497       0.0594
 0.0          0.0095       0.0266       0.0398       0.0483
 0.0          0.0068       0.0194       0.0294       0.0360
 0.0          0.0046       0.0133       0.0204       0.0251
 0.0          0.0029       0.0085       0.0132       0.0162
 0.0          0.0016       0.0047       0.0073       0.0090
 0.0          0.0005       0.0015       0.0023       0.0029
 0.0          0.0          0.0          0.0          0.0

VORTICITY, OMEGA --
20.4000       9.1183       8.1244       7.6544       7.4185
-1.0817       1.6620       4.4993       5.2880       5.4786
-0.8924       0.0806       2.2644       3.3157       3.7164
-0.7414      -0.5578       0.9472       1.8588       2.2950
-0.5679      -0.7515       0.2286       0.9041       1.2777
-0.4086      -0.7138      -0.1094       0.3453       0.6225
-0.2785      -0.5809      -0.2322       0.0467       0.2282
-0.1763      -0.4293      -0.2559      -0.1111      -0.0130
-0.0966      -0.2956      -0.2525      -0.2148      -0.1888
-0.0319      -0.1916      -0.2609      -0.3206      -0.3624
 0.0         -0.0319      -0.0889      -0.1387      -0.1734
35.0000     100.0000
```

Figure 13.14 *(continued)*

```
 CYCLE   1
 STREAM FUNCTIONPSI --
0.0           0.0           0.0           0.0           0.0           0.0
0.0           0.0           0.0           0.0           0.0           0.0
0.0           0.0           0.0           0.0           0.0           0.0
0.0           0.0           0.0           0.0           0.0           0.0
0.0           0.0           0.0           0.0           0.0           0.0
0.0           0.0           0.0           0.0           0.0           0.0
0.0           0.0           0.0           0.0           0.0           0.0
0.0           0.0           0.0           0.0           0.0           0.0
0.0           0.0           0.0           0.0           0.0           0.0
0.0           0.0           0.0           0.0           0.0           0.0
0.0           0.0           0.0           0.0           0.0           0.0

 VORTICITY, OMEGA --
6.0000        6.0000        6.0000        6.0000        6.0000        12.0000
4.1725        4.1699        4.0502        3.6470        2.4022        0.0
2.5843        2.5802        2.4403        2.0493        1.2253        0.0
1.4036        1.4050        1.3105        1.0664        0.6185        0.0
0.6587        0.6636        0.6173        0.4979        0.2873        0.0
0.2634        0.2680        0.2507        0.2032        0.1183        0.0
0.0887        0.0914        0.0865        0.0710        0.0421        0.0
0.0250        0.0261        0.0251        0.0210        0.0128        0.0
0.0058        0.0062        0.0061        0.0052        0.0032        0.0
0.0011        0.0012        0.0012        0.0010        0.0007        0.0
0.0           0.0           0.0           0.0           0.0           0.0


 CYCLE   2
 STREAM FUNCTIONPSI --
0.0           0.0           0.0           0.0           0.0           0.0
0.0478        0.0471        0.0440        0.0368        0.0199        0.0
0.0662        0.0645        0.0571        0.0423        0.0170        0.0
0.0632        0.0612        0.0528        0.0372        0.0140        0.0
0.0517        0.0497        0.0422        0.0290        0.0106        0.0
0.0386        0.0369        0.0310        0.0210        0.0075        0.0
0.0269        0.0257        0.0214        0.0143        0.0051        0.0
0.0174        0.0166        0.0137        0.0091        0.0032        0.0
0.0097        0.0092        0.0076        0.0050        0.0018        0.0
0.0031        0.0030        0.0024        0.0016        0.0006        0.0
0.0           0.0           0.0           0.0           0.0           0.0

 VORTICITY, OMEGA --
7.3320        7.3727        7.5624        7.9926        9.0057        20.4000
5.4773        5.3759        5.1301        4.5033        2.6709        -1.1943
3.7817        3.6254        3.2240        2.4358        1.0260        -1.0207
2.4105        2.2899        1.9397        1.3143        0.3707        -0.8390
1.4088        1.3463        1.1056        0.6824        0.0842        -0.6354
0.7392        0.7195        0.5766        0.3176        -0.0400       -0.4521
0.3145        0.3169        0.2442        0.1036        -0.0899       -0.3052
0.0385        0.0499        0.0260        -0.0292       -0.1080       -0.1930
-0.1707       -0.1552       -0.1400       -0.1263       -0.1160       -0.1060
-0.3755       -0.3544       -0.2992       -0.2167       -0.1219       -0.0337
-0.1867       -0.1773       -0.1461       -0.0964       -0.0337       0.0
```

Figure 13.14 (*continued*)

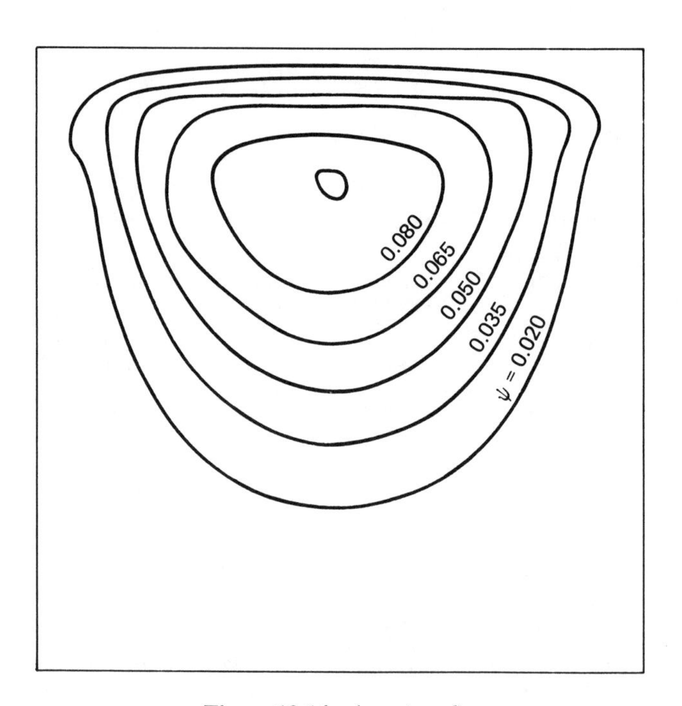

Figure 13.14 (*continued*)

REFERENCES

1. J. M. HYMAN, *The Method of Lines Solution of Partial Differential Equations*, Courant Institute of Mathematical Sciences, COO-3077-139, 1976.
2. W. F. AMES, *Numerical Methods for Partial Differential Equations*. 2nd ed., New York: Academic Press, Inc., 1977.
3. R. J. GRIBBEN, *Elementary Partial Differential Equations*. New York: Van Nostrand Reinhold Co., 1975.
4. A. RALSTON and H. S. WILF, eds., *Mathematical Methods for Digital Computers*, Vols. I and II. New York: John Wiley & Sons, Inc., 1967.
5. E. N. HOUSTIS, and others, "Evaluation of Numerical Methods for Elliptic Partial Differential Equations," *Journal of Computational Physics*, Vol. 27, pp. 323–350, 1978.
6. C. F. GERALD, *Applied Numerical Analysis*, 2nd ed. Reading, Mass.: Addison-Wesley Publishing Co., 1978.
7. R. PARKER and C. Y. MA, "Normal Gradient Boundary Conditions in Finite Difference Calculations," *International Journal of Numerical Methods in Engineering*, Vol. 7, pp. 395–411, 1973.
8. D. GREENSPAN, *Discrete Numerical Methods in Physics and Engineering*. New York: Academic Press, Inc., 1974.

SUGGESTIONS FOR FURTHER READING

1. *Students who are beginning to write large FORTRAN programs should cultivate good programming habits by implementing many of the good ideas discussed in*
   C. B. KREITZBERG and B. SHNEIDERMAN, *The Elements of FORTRAN Style*, Harcourt Brace Jovanovich, Inc., New York, 1972.
2. *An introduction to top-down design, modular programming, structured programming, and programming style can be found in*
   E. YOURDON, *Techniques of Program Structure and Design*, Prentice-Hall, Inc., Englewood Cliffs, N.J., 1976.
3. *For an excellent, easy to read, introduction to modern ideas of software development, consult*
   R. C. TAUSWORTHE, *Standardized Development of Computer Software*, Prentice-Hall, Inc., Englewood Cliffs, N.J., 1979.
4. *For a good source on a discussion of program portability, refer to*
   W. COWELL, ed., *Portability of Numerical Software*, Lecture Notes in Computer Science, Vol. 57, Springer-Verlag New York, 1977.
5. *Numerical software (i.e., the computer programs used in scientific and engineering computations) is rapidly increasing in quantity, scope, and complexity, and the evaluation of its*

performance or quality is a matter of growing economic importance. For a discussion of this subject, consult

L. D. FOSDICK, ed., *Performance Evaluation of Numerical Software*, North-Holland Publishing Co., New York, 1978.

EXERCISES

1. The accuracy of the results shown in Figure 13.2b is quite poor. Rewrite the appropriate segment of the program, to improve the accuracy, by calculating $u_{l,1}$ from $u_{l,0}$ using an integration scheme [rather than using Eqs. (13.6) to (13.8)] such as Simpson's one-third rule with the interval x_{l-1} to x_{l+1} subdivided into 20 parts. (For a description of Simpson's rules see reference 6.)

2. Solve, using a suitable finite difference method,

$$u_{xx} = u_{tt}, \qquad 0 \leq x \leq 1, \qquad 0 \leq t \leq 1$$

with the boundary and initial conditions on $u(x, t)$ specified as

$$u(0, t) = u(1, t) = 0, \qquad 0 \leq t \leq 1$$
$$u(x, 0) = \sin \pi x, \qquad \dot{u}(x, 0) = 0, \qquad 0 \leq x \leq 1$$

Compare the solution with the exact solution given by

$$u(x, t) = \sin \pi x \cos \pi t$$

3. Solve, using a suitable finite difference scheme,

$$u_{xx} = u_t, \qquad 0 \leq x \leq 1, \qquad t > 0$$

where $u(x, t)$ is specified as

$$u(x, 0) = \sin \pi x, \qquad 0 \leq x \leq 1$$
$$u(0, t) = 0, \qquad t > 0, \qquad x = 0 \text{ and } x = 1$$

Compare the calculated solution with the exact solution given by $u(x, t) = \exp(-\pi^2 t) \times \sin \pi x$. What is the accuracy of your solution at $t = 0.005$?

4. Derive the Crank-Nicolson equations for the problem in Exercise 3. With $\Delta x = 0.1$ and $\Delta t = 0.01$, solve from $t = 0$ to $t = 0.02$ using (a) Jacobi, (b) Gauss-Seidel, and (c) SOR iterative schemes. Compare the accuracies and the rates of convergence.

5. Reproduce the results of all the four example problems discussed in the text (a) by copying the programs given in the text and (b) by writing your own improved versions. Your improved versions must show improvement in one or more of the following areas: readability, modularity, structuredness, and accuracy. If a FORTRAN 77 compiler is available, write your revised programs in structured FORTRAN.

6. On a large graph sheet draw a quadrant similar to that shown in Figure 13.4a. Mark the grid points, but not the grid lines.
   (a) Plot the equipotential lines for $\phi = 0.5$, 1.0, 1.5, and 2.0 by using the exact solution shown in Eq. (13.27).

(b) On the same graph sheet plot the calculated values of ϕ that you obtained in Exercise 5.

7. Using the program in Figure 13.14 as a starting point, conduct an experiment by solving the Naviei-Stokes problem for Reynold's numbers $\mathcal{R} = 10, 100, 500, 1000, 3000, 100{,}000$ for the set of parameter values $h = 1/20$, $C_1 = C_2 = 0$, $r_\psi = 1.8$, $r_\omega = 1$, $\xi = 0.1$, and $\delta = 0.7$. Use a tolerance of 10^{-4} for both inner and outer iterations. Plot the equivorticity curves and streamlines in each case. How many iterations are required for convergence in each case? How much CPU time in each case? (*Hint:* Do not spend more than 10 seconds for each case.) Study the variation of convergence rate with variations in r_ψ and r_ω.

14

Example Programs: Finite Element Methods

14.1. INTRODUCTION

This chapter is primarily aimed at readers who have never programmed a digital computer to solve a field problem by the finite element method. If the reader is famil - iar with FORTRAN language, he should be able to follow all the coding.

The primary purpose of this chapter is to demonstrate the procedure for solving a couple of example problems. The scope of these programs is rather restricted; these programs have been constructed without the embellishments or sophisticated coding techniques found in advanced general-purpose finite element programs. The first problem is to solve a two-dimensional Laplace equation on a square region with Dirichlet boundary conditions. The second problem involves the determination of potential flow past a cylinder. Indeed, this second problem is the same as the one solved earlier via finite difference methods. These two examples are used not only to demonstrate programming but also as vehicles to bring forth several points of practical interest. Therefore, one has to pay attention to where discussion about points of general interest ends and specifics about the example problem begin.

It is important to note that almost the same program will be used to solve both the example problems. One difference is in the data supplied to each problem. A second difference is the application of normal derivative boundary conditions in the second problem. Finally, as the analytical solution is known for the first problem, the computed solution is compared with the "exact" solution. This is not always possible. The student is urged to step through the code carefully and note where these differences are.

Now attention will be focused on the first example problem, the solution of Laplace's equation

$$\frac{\partial^2 u}{\partial x^2} + \frac{\partial^2 u}{\partial y^2} = 0$$

on the square region Ω, $0 \leq x \leq 1$, $0 \leq y \leq 1$ with $u = 1$ on $(x = 0, 0 < y < 1)$ and $u = 0$ on the rest of the boundary.

14.2. DIVISION OF THE DOMAIN INTO FINITE ELEMENTS

The first step of the finite element method is the discretization of Ω. This discretization can be divided into two parts, the division of Ω into elements and the labeling of the elements and the nodes. The latter sounds simple but it has a profound influence on the computational efficiency.

The division of Ω into elements is accomplished by dividing Ω first into quadrilaterals and further subdividing these quadrilaterals into triangles. The grid lines should be drawn so that they separate regions where there is a change in geometry, applied load, or material properties.

Even though there exists more than one way to subdivide a quadrilateral into triangular subelements, we propose to subdivide a quadrilateral element into four triangular subelements as shown in Figure 14.1. This strategy permits the elimination of some of the variables, such as those associated with the nodes that are within the quadrilateral, at the element level. Thus the computations at the global level are somewhat simplified.

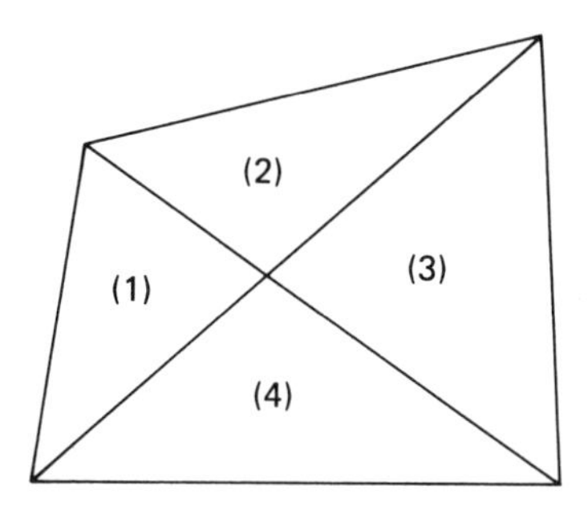

Figure 14.1 Typical quadrilateral element with four triangular subelements

If the region Ω has curved boundaries, as is the case with the second example problem, then the curvature along the boundary is approximated with straight line segments. This strategy is far more simple than using isoparametric elements, even though the latter choice would yield a better accuracy.

14.3. SELECTION OF THE APPROXIMATING POLYNOMIAL

The next important step is to decide on the nature of the interpolation polynomials to be used. This is important because the number of elements used and the order of the interpolating polynomial have an impact on the computational efficiency. In general, as the complexity of the elements is increased by adding more nodes and using higher-order polynomials, the number of elements needed to achieve a specified accuracy are less than would be required if simpler elements are used. But this does not mean that we should always favor higher-order elements over lower-order elements. Unfortuantely, no general guidelines for choosing the optimum element for a given problem can be presented because the type of element that yields good accuracy with low computing time is problem dependent.

In this present case, it has been decided to use quadratic triangular elements. That is, each side of the triangular element has three node points, one at each vertex and one at the midpoint of each side. A typical triangular element, with its nodes numbered, is shown in Figure 14.2a. If four such elements are assembled, we get a quadrilateral element shown in Figure 14.2b. Note that the direction of numbering of the nodes for these elements is arbitrary. Note also that for the numbering scheme shown, solution values at nodes 9, 10, 11, 12, and 13 can be eliminated at the element level because they do not play any role while assembling the quadrilaterals.

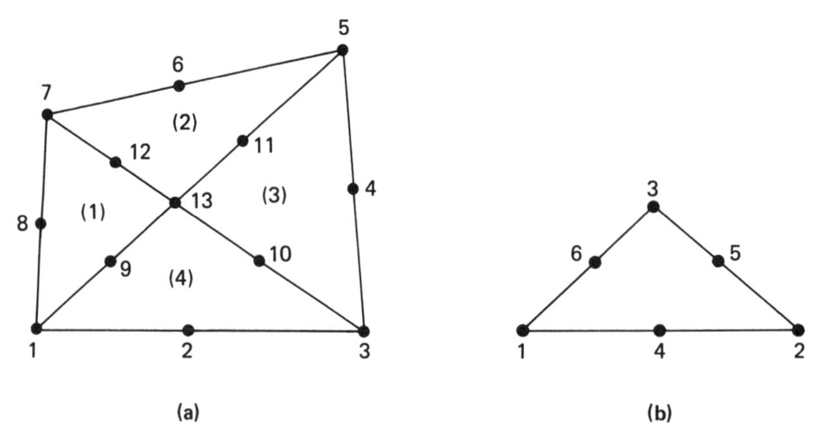

Figure 14.2 (a) Numbering the nodes of a typical quadrilateral element; (b) typical triangular subelement of the quadrilateral element shown in (a); note carefully the numbering patterns in both cases

14.4. LABELING THE NODES

Now that the decision to use quadrilateral elements has been made, the region Ω can be covered with quadrilaterals as shown in Figure 14.3 with all the corner nodes and midside nodes clearly marked. Now it is necessary to assign a number to each node.

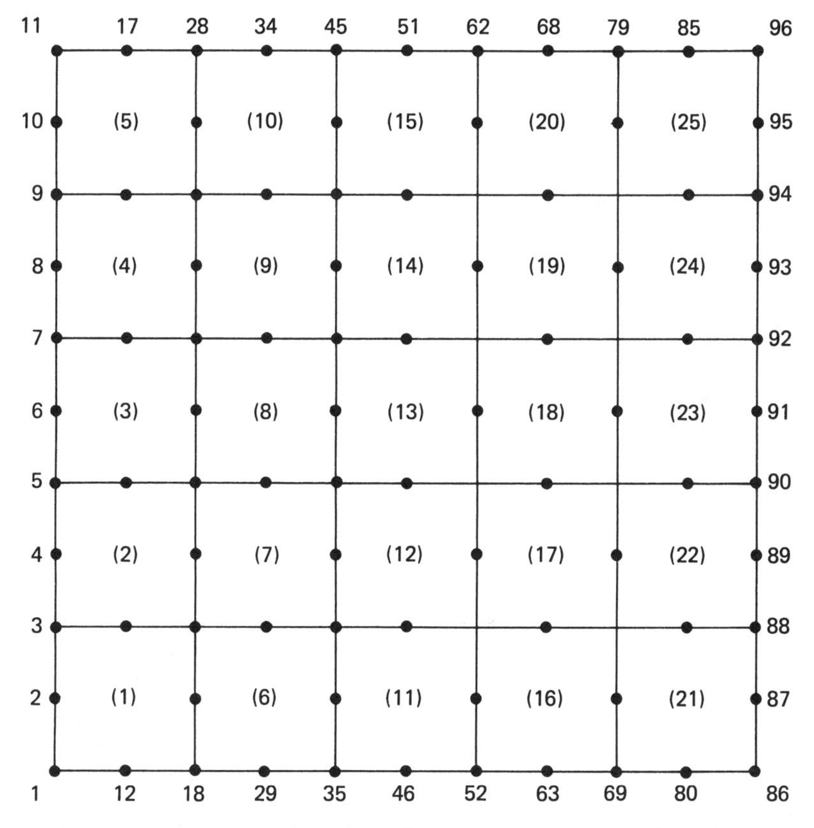

Figure 14.3 Square region of the first example problem with a finite element grid superimposed on it; element numbers are shown in parentheses

This process goes by the name "labeling." This operation would be trivial if the node numbers did not influence the computational efficiency. However, the set of linear algebraic equations $\mathbf{Au} = \mathbf{f}$ that arise in the solution process exhibits a band structure; that is, a majority of nonzero entries of $\mathbf{A}$ fall between two lines that can be drawn parallel on either side of the main diagonal. The distance between the main diagonal and one of these lines is called the ***bandwidth*** (some call this semibandwidth). This bandwidth B can be calculated using

$$B = (R + 1)N_2$$

where R is the largest difference between the node numbers in a single element and N_d is the number of ***degrees of freedom*** (i.e., the number of unknowns at each node). The minimization of B depends on the minimizing of R, which can be partially accomplished by labeling the nodes along the shortest dimension of the body. Labeling the elements, however, has no influence on the computational efficiency. As Ω for the first problem has a square shape with equal number of node points along x and y

directions, the order in which the nodes are numbered has no impact. However, note how the nodes are numbered along the y direction (the shorter dimension) in the second problem (see Figure 14.11).

14.5. PREPARATION OF INPUT DATA

Although the input of data to and output of final results from any finite element program is one of the technically less demanding aspects, it is very important to engineers. When the finite element method is employed in engineering analysis, the time spent in data preparation and interpretation of the results generally represents the major proportion of the total time. Therefore, any savings that can be made in these directions will be of direct benefit to the user.

For any finite element analysis, the input data required can be divided into three main classifications. First, obtain the data required to define geometry of the field and the nature of boundary conditions. Second, information regarding the material properties of the constituent materials must be prescribed. Finally, information about sources, sinks, and the like, must be provided.

The example problems presently under consideration are somewhat simpler. Nevertheless, to demonstrate the process of automating some routine jobs, it has been decided that it would be too laborious to enter geometric information about the coordinates of all the node points. To save the labor in the preparation of data cards, the program has been developed to provide two options: one for generating node point coordinates and one for generating node point numbers. These options will be referred to as ***generation options.*** In this way, only a limited number of data cards must be prepared.

Now we shall describe in detail how these data cards are prepared. A facsimile of the data cards is shown after the $ENTRY card at the end of the program listing shown in Figure 14.4. (This information should appear as shown in the source deck. However, no provision is made in the code to echo print this with the output.) As it is difficult to read the column numbers from this listing, the format of each card is given in parentheses right after the name of the card in the following description. From now onward, the material presented is best understood if it is read in conjunction with the program listing.

TITLE CARD (FORMAT: 18A4)

This is the first card of the data deck. This card appears immediately after the $ENTRY card at the end of the main program. The title card is used to identify a problem by an explanatory title. For exampe, the title chosen for the first example problem is 'LAPLACE EQ. DEFINED ON A UNIT SQUARE'. Note that this title gets printed on the top of the next sheet after the $ENTRY line.

Referring to the main program listing, we see that this title is read by instruction 7 and printed by instruction 10.

```
      C       FINITE ELEMENT SOLUTION OF LAPLACE EQUATION.
      C       BOUNDARY CONDITIONS CAN BE EITHER DIRICHLET
      C       OR NEUMAN TYPE. THE REGION CAN BE OF ANY SHAPE.
      C
 1            DATA A/'XXXX'/
 2            REAL*8 SUM
 3            DIMENSION TITLE(18),NPFSA(100),NPBOT(100),PBVAL(100),
             1NDIF1(100),NDIF2(100)
 4            COMMON N,NOD(150,8),X(502),Y(502),N1,N2,N3,AN,INDEX,
             1 T(6,6),I,SL3
 5            COMMON /Q/ CSL(13),CS(13,13),NPBO,NPBOA(100),BOV(100),
             1   ND(4,6)
 6            COMMON /SYMS/ SL(501),S(501,29),NNP,NCOL
      C
      C       READ AND PRINT CONTROL DATA
      C
 7        100 READ(5,110) (TITLE(I),I=1,18)
 8        110 FORMAT (18A4)
 9            IF (TITLE(1) .EQ. A) GO TO 4001
10            WRITE (6,120) (TITLE(I),I=1,18)
11        120 FORMAT (1H1,10X,18A4////)
12            READ(5,170) NNPC,NELEMC,NPBOC,NPFS
13        170 FORMAT(4I10)
      C
      C       READ AND/OR GENERATE CORNER POINT COORDINATES
      C
14            NCC=0
15        180 NCC=NCC+1
16            IF (NCC .GT. NNPC) GO TO 215
17            READ (5,190) N,X(N),Y(N),NPMIS
18        190 FORMAT (I10,2F10.0,I10)
19        200 FORMAT (1H ,13X,4HNODE,I4,5X,4HX = ,F8.4,5X,4HY = ,
             1F8.4)
20            WRITE (6,200) N,X(N),Y(N)
21            NI=N
22            IF (NPMIS .NE. 0) GO TO 210
23            GO TO 180
24        210 NCC=NCC+1
25            IF (NCC .GT. NNPC) GO TO 215
26            READ (5,190) N,X(N),Y(N),NPMIS
27            NE=N
28            NPG=(NE-NI)/2
29            DX=(X(NE)-X(NI))/FLOAT(NPG)
30            DY=(Y(NE)-Y(NI))/FLOAT(NPG)
31            DO 214 I=1,NPG
32            NG=NI+2*I
33            X(NG)=X(NI)+FLOAT(I)*DX
34            Y(NG)=Y(NI)+FLOAT(I)*DY
35        214 WRITE (6,200) NG,X(NG),Y(NG)
36            IF (NPMIS .EQ. 0) GO TO 180
37            NI=N
38             GO TO 210
39        215 NNP=N
40            IF (NNP .LE. 501) GO TO 219
41            WRITE (6,216)
42        216 FORMAT (1H1,10X,35HERROR...NO. OF NODES MORE THAN 501.)
43            GO TO 100
      C
      C       READ AND/OR GENERATE ELEMENT NODAL POINTS
      C
44        219  N=0
45            DO 222 M=1,NELEMC
46            N=N+1
47            READ (5,220) (NOD(N,I), I=1,8),NMIS
48        220 FORMAT (8I5,I10)
49            IF (NMIS .EQ. 0) GO TO 222
```

Figure 14.4 A Fortran program to solve either of the two example problems discussed in the text; the listing after the $ENTRY card at the end of the program shows the setup of the data cards

```
50          DO 221 K=1,NMIS
51          N=N+1
52          DO 221 J=1,8
53          NOD(N,J)=NOD(N-1,J)+2
54          IF (J .EQ. 2 .OR. J .EQ. 6) NOD(N,J)=NOD(N,J)-1
55      221 CONTINUE
56      222 CONTINUE
57          NELEM=N
58          IF (NELEM .LE. 150) GO TO 224
59          WRITE (6,223)
60      223 FORMAT(1H1,10X,27HERROR...NUMBER OF ELEMENTS ,
           117HGREATER THAN 150.)
61          GO TO 100
62      224 WRITE (6,225) NNP,NELEM
63      225 FORMAT(1H1,10X,25HNUMBER OF NODAL POINTS...,I4//11X,
           125HNUMBER OF ELEMENTS.......,I4//11X,8HELE. NO. ,5X,
           112HNODAL POINTS//)
64          DO 229 N=1,NELEM
65      229 WRITE (6,230) N,(NOD(N,I),I=1,8)
66      230 FORMAT (1H ,11X, 9I5)
      C
      CXXX  READ AND/OR GENERATE BOUNDARY VALUES FOR THE BOUNDARY
      C     WHERE NON-ZERO NORMAL COMPONENT IS SPECIFIED.
      C     THIS SECTION GENERATES VALUES AT EACH CORNER NODE.
      C
67          IF(NPBOC .EQ. 0) GO TO 22
68          WRITE (6,231)
69      231 FORMAT(1H1,29HNORMAL COMPONENTS ON BOUNDARY//1H0,
           15HNODES,5X,16HNORMAL COMPONENT//)
70          I=0
71          DO 235 J=1,NPBOC
72          READ (5,232) NSTART,BVAL,NBSAME,NINC
73      232 FORMAT (I10,F10.0,2I10)
74          DO 234 K=1,NBSAME
75          I=I+1
76          NPBOA(I)=NSTART+NINC*(K-1)
77          NB=NPBOA(I)
78          NE=NB+NINC
79          BOV(I)=BVAL
80          WRITE (6,233) NB,NE,BOV(I)
81      233 FORMAT (1H ,I3,1H-,I3,5X,F10.4)
82      234 CONTINUE
83          I=I+1
84          NPBOA(I)=NE
85          BOV(I)=0.
86      235 CONTINUE
87          NPBO=I
      C
      C        READ, BOUNDARY VALUES FOR PHI
      C
88     22   IF (NPFS .EQ. 0) GO TO 2000
89          DO 2 I=1,NPFS
90      2   READ(5,236) NPFSA(I),NPBOT(I),PBVAL(I),NDIF1(I),
           1NDIF2(I)
91     236  FORMAT(2I10,F10.0,2I10)
      C
      C     COMPUTE MATRIX BANDWIDTH AND CHECK TO SEE
      C     IF IT EXCEEDS 29
      C
92    2000  NCOL=1
93          DO 240 N=1,NELEM
94          DO 240 I=1,8
95          DO 240 J=1,8
96          NN=NOD(N,I)-NOD(N,J)+1
97     240  IF (NCOL-NN .LT. 0) NCOL=NN
```

Figure 14.4 *(continued)*

```
 98            IF (NCOL .GT. 29) WRITE (6,250)
 99        250 FORMAT (1H0,34HERROR...BANDWIDTH GREATER THAN 29.)
100            IF (NCOL .GT. 29) GO TO 100
      C
      C        ZERO OUT THE STIFFNESS AND LOAD MATRICES
      C
101            DO 290 I=1,NNP
102            SL(I)=0.0
103            DO 290 J=1,NCOL
104        290 S(I,J)=0.0
      C
      C           DEVELOP THE A AND B MATRICES FOR ELEMENT N AND
      C        ELIMINATE THE INTERIOR GRID POINTS
      C
105            N3=NNP+1
106            INDEX=0
107            IF(NPBOC .EQ. 0) INDEX=1
      C        CALCULATION OF RIGHT SIDE OF AU=F TYPE EQN.
108            DO 320 M=1,NELEM
109               N=M
110               CALL QUAD2D
111               DO 300 I=1,5
112                  IM=13-I
113                  IP=IM+1
114                  DO 300 J=1,IM
115                  CSL(J)=CSL(J)-CS(J,IP)*CSL(IP)/CS(IP,IP)
116                  DO 300 K=1,IM
117   300            CS(J,K)=CS(J,K)-CS(J,IP)*CS(IP,K)/CS(IP,IP)
      C
      C        ADD CONTRIBUTION OF ELEMENT N TO TOTAL A MATRIX
      C
118               DO 315 J=1,8
119                  NOROW=NOD(N,J)
120                  DO 310 K=1,8
121                  NOCOL=NOD(N,K)-NOROW+1
122   310      IF(NOCOL .GT. 0) S(NOROW,NOCOL)=S(NOROW,NOCOL)+CS(J,K)
123   315            SL(NOROW)=SL(NOROW)+CSL(J)
124   320      CONTINUE
      C
      C        SOLVE FOR THE SOLUTION AT THE NODAL POINTS
      C        WHILE VALUES FOR THE NODES FOR WHICH THE
      C        SOLUTION IS GIVEN ARE ELIMINATED
      C
125    325     IF(NPFS .EQ. 0) GO TO 329
126            DO 327 I=1,NPFS
127            NI=(NPFSA(I)-NPBOT(I))/(NDIF1(I)+NDIF2(I))+1
128            M=NPBOT(I)
129            DO 327 J=1,NI
130            CALL PBOUND(M,PBVAL(I))
131            M=M+NDIF1(I)
132            IF(J .NE. NI) CALL PBOUND(M,PBVAL(I))
133            M=M+NDIF2(I)
134    327     CONTINUE
135            GO TO 331
136    329     S(1,1)=1.0
137            SL(1)=0.0
138            DO 330 I=2,NCOL
139    330     S(1,I)=0.0
140    331     CALL SYMSOL
      C
      C           PRINT RESULTS
      C        NOTE THAT 'SUM' REPRESENTS THE EXACT SOLUTION
      C        AS THE EXACT SOLUTION TO THE SECOND EXAMPLE
      C        PROBLEM IS NOT KNOWN, THIS SUBROUTINE NEED
      C        NOT BE CALLED WHILE SOLVING PROBLEM 2
```

Figure 14.4 *(continued)*

```
      C       THEN, MODIFY INSTRUCTIONS 144 AND 146
      C
141           WRITE (6,11)
142       11  FORMAT(1H1)
143           DO 20 N=1,NNP
144           CALL ANSWER(X(N),Y(N),SUM)
145       20  WRITE(6,10)N,X(N),Y(N),SL(N),SUM
146       10  FORMAT (1H ,5X,8H NODE = ,I4,5X,8H X =     ,F8.4,
             15X,8H  Y  =   ,F8.4,5X,8H PHI =   ,F8.4,5X,9H EXACT = ,F  .4
147           CALL FINITE
148           GO TO 100
149     4001  STOP
150           END
      C
      C           SUBROUTINE TO DEVELOP ELEMENT MATRICES
      C
151           SUBROUTINE QUAD2D
152           COMMON N,NOD(150,8),X(502),Y(502),N1,N2,N3,AN,INDEX,
             1    T(6,6),I,SL3
153           COMMON /Q/ CSL(13),CS(13,13),NPBO,NPBOA(100),BOV(100),
             1    ND(4,6)
154           DO 100 J=1,13
155              CSL(J)=0.0
156              DO 100 K=1,13
157      100     CS(J,K)=0.0
158           CALL CENTER
159           DO 200 KT=1,4
160              I=KT
161              CALL TRIMAT
162              DO 200 J=1,6
163              NR=ND(KT,J)
164              IF(INDEX .EQ. 1) GO TO 190
165              IF(J .NE. 1) GO TO 190
166              DO 150 KR =1,NPBO
167                 K=NPBO+1-KR
168                 IF(N1 .NE. NPBOA(K)) GO TO 150
169                 DO 160 JR=1,NPBO
170                 JJ=NPBO+1-JR
171                 IF(N2 .NE. NPBOA(JJ)) GO TO 160
172                 DSL=SL3*BOV(K)/6.
173                 CSL(NR)=CSL(NR)+DSL
174                 NRF=ND(KT,4)
175                 CSL(NRF)=CSL(NRF)+4.*DSL
176                 NRFF=ND(KT,2)
177                 CSL(NRFF)=CSL(NRFF)+DSL
178                 GO TO 190
179      160        CONTINUE
180      150        CONTINUE
181      190     DO 200 K=1,6
182              NC=ND(KT,K)
183      200     CS(NR,NC)=CS(NR,NC)+T(J,K)
184           RETURN
185           END
      C
      C           SUBROUTINE TO DETERMINE THE CENTER OF ELEMENT N
      C
186           SUBROUTINE CENTER
187           COMMON N,NOD(150,8),X(502),Y(502),N1,N2,N3,AN,INDEX,
             1    T(6,6),I,SL3
188           M1=NOD(N,1)
189           M2=NOD(N,3)
190           M3=NOD(N,5)
191           M4=NOD(N,7)
192           X(N3)=0.25*(X(M1)+X(M2)+X(M3)+X(M4))
193           Y(N3)=0.25*(Y(M1)+Y(M2)+Y(M3)+Y(M4))
194           RETURN
195           END
```

Figure 14.4 *(continued)*

```
      C
      C       SUBROUTINE TO DETERMINE THE GEOMETRIC PROPERTIES
      C       OF A TRIANGULAR SUB-ELEMENT
      C
196           SUBROUTINE TRIMAT
197           COMMON N,NOD(150,8),X(502),Y(502),N1,N2,N3,AN,INDEX,
             1    T(6,6),I,SL3
198           N1=NOD(N,2*I-1)
199           N2=NOD(N,1)
200           IF(I .NE. 4) N2=NOD(N,2*I+1)
201           N4=NOD(N,2*I)
202           X(N4)=0.5*(X(N1)+X(N2))
203           Y(N4)=0.5*(Y(N1)+Y(N2))
204           A1=X(N3)-X(N2)
205           A2=X(N1)-X(N3)
206           A3=X(N2)-X(N1)
207           B1=Y(N2)-Y(N3)
208           B2=Y(N3)-Y(N1)
209           B3=Y(N1)-Y(N2)
210           AN=0.5*(A3*B2-A2*B3)
211           SL1=SQRT(A1**2+B1**2)
212           SL2=SQRT(A2*A2+B2*B2)
213           SL3=SQRT(A3**2+B3**2)
214           T(1,1)=3.*SL1**2
215           T(1,2)=-(A1*A2+B1*B2)
216           T(1,3)=-(A1*A3+B1*B3)
217           T(1,4)=4.*(A1*A2+B1*B2)
218           T(1,5)=0.
219           T(1,6)=4.*(A1*A3+B1*B3)
220           T(2,2)=3.*SL2**2
221           T(2,3)=-(A2*A3+B2*B3)
222           T(2,4)=T(1,4)
223           T(2,5)=4.*(A2*A3+B2*B3)
224           T(2,6)=0.
225           T(3,3)=3.*SL3**2
226           T(3,4)=0.
227           T(3,5)=T(2,5)
228           T(3,6)=T(1,6)
229           T(4,4)=8.*(SL3**2-A1*A2-B1*B2)
230           T(4,5)=8.*(A1*A3+B1*B3)
231           T(4,6)=8.*(A2*A3+B2*B3)
232           T(5,5)=8.*(SL1**2-A2*A3-B2*B3)
233           T(5,6)=8.*(A1*A2+B1*B2)
234           T(6,6)=8.*(SL2**2-A3*A1-B3*B1)
235           DO 100 J=1,5
236           JJ=J+1
237           DO 100 K=JJ,6
238   100     T(K,J)=T(J,K)
239           A12=1./(12.*AN)
240           DO 110 J=1,6
241           DO 110 K=1,6
242           T(J,K)=T(J,K)*A12
243   110     CONTINUE
244           RETURN
245           END
      C
      C       SUBROUTINE TO SOLVE A SET OF LINEAR ALGEBRIAC
      C       SIMULTANEOUS EQUATIONS WITH A COEFFICIENT MATRIX
      C       WHICH IS BOTH SYMMETRIC AND BANDED
      C
246           SUBROUTINE SYMSOL
247           COMMON /SYMS/ SL(501),S(501,29),NROW,NCOL
248           DIMENSION ST(29)
249           N=0
250       100 N=N+1
```

Figure 14.4 (*continued*)

```
      C
      C      REDUCE PIVOT EQUATION
      C
251          SL(N)=SL(N)/S(N,1)
252          IF (N-NROW) 150,300,150
253      150 DO 200 K=2,NCOL
254          ST(K)=S(N,K)
255      200 S(N,K)=S(N,K)/S(N,1)
      C
      C      REDUCE REMAINING EQUATION WITHIN SPAN
      C
256          DO 260 L=2,NCOL
257          I=N+L-1
258          IF (NROW-I) 260,240,240
259      240 J=0
260          DO 250 K=L,NCOL
261          J=J+1
262      250 S(I,J)=S(I,J)-ST(L)*S(N,K)
263          SL(I)=SL(I)-ST(L)*SL(N)
264      260 CONTINUE
265          GO TO 100
      C
      C      BACK SUBSTITUTION
      C
266      300 N=N-1
267          IF(N) 350,500,350
268      350 DO 400 K=2,NCOL
269          L=N+K-1
270          IF(NROW-L) 400,370,370
271      370 SL(N)=SL(N)-S(N,K)*SL(L)
272      400 CONTINUE
273          GO TO 300
274      500 RETURN
275          END
      C
      C         ADJUST  MATRIX  A  AND  B  FOR GIVEN VALUES OF PHI
      C
276          SUBROUTINE PBOUND(J,VAL)
277          COMMON /SYMS/ SL(501),S(501,29),NNP,NCOL
278          DO 100 I=2,NCOL
279          K=J-I+1
280          IF(K .LE. 0) GO TO 50
281          SL(K)=SL(K)-S(K,I)*VAL
282          S(K,I)=0.0
283     50   K=J+I-1
284          IF(K .GT. NNP) GO TO 100
285          SL(K)=SL(K)-S(J,I)*VAL
286          S(J,I)=0.0
287      100 CONTINUE
288          S(J,1)=1.0
289          SL(J)=VAL
290          RETURN
291          END
      C
      C      THIS SUBROUTINE CALCULATES AN APPROXIMATION
      C      TO THE EXACT SOLUTION BY ADDING 25 TERMS FROM
      C      AN INFINITE SERIES REPRESENTING THE SOLUTION
      C      DONOT CALL THIS SUBROUTINE WHILE SOLVING PROB.2
      C
292          SUBROUTINE ANSWER(X,Y,SUM)
293          REAL X,Y
294          REAL*8 PHI,SUM,SUMIND,TERM1,TERM2,DFLOAT,DSIN,DSINH
            &     ,XD,YD,DABS
295          INDMAX=25
296          PHI=3.14159265358979D0
297          SUM=0.0D0
```

Figure 14.4 *(continued)*

```
298          XD=X
299          YD=Y
300          DO 100 INDEX=1,INDMAX
301             SUMIND=DFLOAT(2*(INDEX-1)+1)
302             IF(DABS(DSINH(SUMIND*PHI)) .GT. 1.0D60) GO TO 100
303          TERM1=DSIN(SUMIND*PHI*YD)*DSINH(SUMIND*PHI*(1.0D0-XD))
304             TERM2=SUMIND*DSINH(SUMIND*PHI)
305             SUM=SUM+(TERM1/TERM2)
306    100      CONTINUE
307          SUM=(4.0D0/PHI)*SUM
308          RETURN
309          END
       C
       C     HERE WE SOLVE THE GIVEN PROBLEM BY A SIMPLE
       C     FINITE DIFFERENCE METHOD FOR COMPARISON.
       C
310          SUBROUTINE FINITE
311          REAL POTENT(17,17),DIFF
312          DO 100 I=1,17
313          DO 100 J=1,17
314    100      POTENT(I,J)=0.0
315          DO 150 I=2,10
316    150      POTENT(I,1)=1.0
317          DO 300 ITERAT=1,300
318             DIFF=0.0
319             DO 200 I=2,10
320             DO 200 J=2,10
321                OLD=POTENT(I,J)
322             POTENT(I,J)=0.25*(POTENT(I,J+1)+POTENT(I,J-1)
            1                 +POTENT(I+1,J)+POTENT(I-1,J))
323                DIFF=ABS(OLD-POTENT(I,J))+DIFF
324    200         CONTINUE
325             IF(DIFF .LT. 0.00001) GO TO 350
326    300      CONTINUE
       C
327    350   WRITE(6,500) ITERAT
328          DO 400 I=1,11
329             WRITE(6,450)(POTENT(I,J),J=1,11)
330    400   CONTINUE
331    450   FORMAT(11F8.5)
332    500   FORMAT(1H1,'THE NUMBER OF ITERATIONS IS ',I5)
333          RETURN
334          END
       C
       C     THIS DATA IS FOR ASSEMBLING TRIANGULAR SUBELEMENTS
       C     INTO A QUADRILATERAL ELEMENT.
       C
335          BLOCK DATA
336          COMMON/Q/ CSL(13),CS(13,13),NPBO,NPBOA(100),BOV(100),
            1   ND(4,6)
337          DATA ND /1,3,5,7,3,5,7,1,4*13,2,4,6,8,10,11,12,9,9,
            110,11,12/
338          END
       $ENTRY

             LAPACE EQ. DEFINED ON A UNIT SQUARE
                     12         5         0         4
                      1       0.0       0.0         1
                     11       0.0       1.0         0
                     18       0.2       0.0         1
                     28       0.2       1.0         0
                     35       0.4       0.0         1
                     45       0.4       1.0         0
                     52       0.6       0.0         1
                     62       0.6       1.0         0
                     69       0.8       0.0         1
```

Figure 14.4 (*continued*)

```
          79               0.8                 1.0                 0
          86               1.0                 0.0                 1
          96               1.0                 1.0                 0
     1    12      18      19      20      13       3       2                 4
    18    29      35      36      37      30      20      19                 4
    35    46      52      53      54      47      37      36                 4
    52    63      69      70      71      64      54      53                 4
    69    80      86      87      88      81      71      70                 4
          10                2              1.0                 1             1
          96               86              0.0                 1             1
          86                1              0.0                11             6
          96               11              0.0                 6            11
```

Figure 14.4 (*continued*)

CONTROL CARD (FORMAT: 4I10)

This is the second card of the data deck. This card appears immediately after the title card. From the format, it is evident that this card carries information about four integer variables. The fields in which these variables appear and their function are now described.

Columns 1–10. NNPC. This variable stands for the number of corner nodes at which coordinate values will be supplied.

Using this information, the coordinates of the remaining nodes can be generated by the program. For the first example problem, notice that this field contains the number 12 (see row 1, column 1 of the input data listing); that is, we have to supply the coordinate information for 12 nodes. The coordinates of the remaining 24 nodes at the corners of the quadrilateral will now be generated by the program.

Columns 11–20. NELEMC. This variable name stands for the number of elements for which nodal numbers will be provided.

Using this information, the program will generate the nodal numbers for the rest of the elements. For the first example problem, this field contains the number 5. That is, we have to supply the numbers of the corner and midside nodes of 5 of the elements. The node number for the rest of the elements will now be calculated by the program.

Columns 21–30. NPBOC. This variable stands for the number of boundary value cards on which nonzero values of the components of $\partial u/\partial n$ are specified.

For the first example problem under discussion, u is specified on all boundaries and $\partial u/\partial n$ is specified *nowhere*. Therefore, this entry is 0.

Columns 31–40. NPFS. This variable indicates whether u is specified on any part of the boundary. If NPFS $= 0$, then u is specified nowhere. If NPFS is nonzero, then u is specified somewhere.

For the first example problem, a 4 is entered in this field to indicate that u is specified.

Referring to the main program listing, we see that this control card is read by instruction 12.

CORNER NODE CARDS (FORMAT: I10, 2F10.0, I10)

This part of the data deck contains a group of cards containing information about the coordinates of corner nodes. The fields on these cards are defined as follows:

Columns 1–10. N. This variable defines the number of a corner node.

Columns 11–20. X. The x-coordinate of the node N.

Columns 21–30. Y. The y-coordinate of the node N.

Column 40. NPMIS. A 1 in this column means that we are using the generation option. That is, there is at least one corner node omitted between the value of N on the present card and the value of N on the next card. A 0 in this column means that the value of N on the next card logically follows the value of N on the present card.

Note that, if the generation option is not used at all (i.e., a 0 is punched in column 40 of every card in this category), then one card will be needed for each corner node. For the first example problem, this means a total of 36 cards. However, we propose to use the generation option. We had already indicated in columns 1 to 10 of the control card that coordinate information at 12 corner nodes will be provided. Therefore, we need 12 cards here. Using this information for the rest of the corner nodes, the program then prints the node numbers of each corner node and its coordinates. This output is displayed in Figure 14.5.

Referring to the main program listing, we see that the first of these 12 cards is read by instruction 17. If NPMIS $\neq 0$, the missing information is computed by instructions 24 to 38 located in a loop. This loop prints not only the information provided by us on the 12 data cards but also the intermediate information generated by the program as shown in Figure 14.5.

ELEMENT CARDS (FORMAT: 8I5, I10)

This group of data cards contains information about how the elements are situated relative to each other. The fields of these elements are defined as follows:

Columns 1–5. NOD(N, 1).

Columns 6–10. NOD(N, 2).

. .

. .

. .

Columns 36–40. NOD(N, 8).

} Numbers of the first, second, . . . , and eighth nodal points of the quadrilateral element N.

```
LAPACE EQ. DEFINED ON A UNIT SQUARE

NODE   1    X =   0.0000    Y =   0.0000
NODE   3    X =   0.0000    Y =   0.2000
NODE   5    X =   0.0000    Y =   0.4000
NODE   7    X =   0.0000    Y =   0.6000
NODE   9    X =   0.0000    Y =   0.8000
NODE  11    X =   0.0000    Y =   1.0000
NODE  18    X =   0.2000    Y =   0.0000
NODE  20    X =   0.2000    Y =   0.2000
NODE  22    X =   0.2000    Y =   0.4000
NODE  24    X =   0.2000    Y =   0.6000
NODE  26    X =   0.2000    Y =   0.8000
NODE  28    X =   0.2000    Y =   1.0000
NODE  35    X =   0.4000    Y =   0.0000
NODE  37    X =   0.4000    Y =   0.2000
NODE  39    X =   0.4000    Y =   0.4000
NODE  41    X =   0.4000    Y =   0.6000
NODE  43    X =   0.4000    Y =   0.8000
NODE  45    X =   0.4000    Y =   1.0000
NODE  52    X =   0.6000    Y =   0.0000
NODE  54    X =   0.6000    Y =   0.2000
NODE  56    X =   0.6000    Y =   0.4000
NODE  58    X =   0.6000    Y =   0.6000
NODE  60    X =   0.6000    Y =   0.8000
NODE  62    X =   0.6000    Y =   1.0000
NODE  69    X =   0.8000    Y =   0.0000
NODE  71    X =   0.8000    Y =   0.2000
NODE  73    X =   0.8000    Y =   0.4000
NODE  75    X =   0.8000    Y =   0.6000
NODE  77    X =   0.8000    Y =   0.8000
NODE  79    X =   0.8000    Y =   1.0000
NODE  86    X =   1.0000    Y =   0.0000
NODE  88    X =   1.0000    Y =   0.2000
NODE  90    X =   1.0000    Y =   0.4000
NODE  92    X =   1.0000    Y =   0.6000
NODE  94    X =   1.0000    Y =   0.8000
NODE  96    X =   1.0000    Y =   1.0000
```

Figure 14.5 Node numbers and coordinates of each of the corner nodes; this is part of the output generated by the program shown in Figure 14.4

Columns 41–50. NMIS. This variable is set equal to the number of succeeding elements whose numbers are not provided, and the generation option is to be used to obtain such information.

In the example problem, we have a total of 25 elements. However, only 5 element data cards are included in the data deck. Information about the remaining 20 elements will be generated by our program.

Referring to the program listing, we note that the 5 element data cards are read by instruction 47. Then instructions 49 to 66 perform the necessary calculations. The computed results are printed as shown in Figure 14.6.

As the program is now coded, to use the generation option the node point numbers of an element must be arranged so that the starting number is at the lower-left corner node and is followed by the other nodes in a counterclockwise direction.

NUMBER OF NODAL POINTS... 96

NUMBER OF ELEMENTS....... 25

ELE. NO.	NODAL POINTS							
1	1	12	18	19	20	13	3	2
2	3	13	20	21	22	14	5	4
3	5	14	22	23	24	15	7	6
4	7	15	24	25	26	16	9	8
5	9	16	26	27	28	17	11	10
6	18	29	35	36	37	30	20	19
7	20	30	37	38	39	31	22	21
8	22	31	39	40	41	32	24	23
9	24	32	41	42	43	33	26	25
10	26	33	43	44	45	34	28	27
11	35	46	52	53	54	47	37	36
12	37	47	54	55	56	48	39	38
13	39	48	56	57	58	49	41	40
14	41	49	58	59	60	50	43	42
15	43	50	60	61	62	51	45	44
16	52	63	69	70	71	64	54	53
17	54	64	71	72	73	65	56	55
18	56	65	73	74	75	66	58	57
19	58	66	75	76	77	67	60	59
20	60	67	77	78	79	68	62	61
21	69	80	86	87	88	81	71	70
22	71	81	88	89	90	82	73	72
23	73	82	90	91	92	83	75	74
24	75	83	92	93	94	84	77	76
25	77	84	94	95	96	85	79	78

Figure 14.6 Element numbers and the associated nodal points; this is part of the output generated by the program shown in Figure 14.4

For the first example problem, notice that the entries on the first card of this set are 1, 12, 18, 19, 20, 13, 3, and 2. These numbers correspond to the node numbers of element (1). Notice that the last entry on this card is 4. That is, element node numbers for the next four elements, 2, 3, 4, and 5, are to be generated via the generation option. Similarly, the second card of this set contains information about element 6, and so on.

NORMAL BOUNDARY VALUE CARDS (FORMAT: I10, F10·0, 2I10)

In this group, one data card is needed for each portion of the boundary on which a constant nonzero value of the components of $\partial u/\partial n$ exists. The fields of these cards are defined as follows:

Columns 1–10. NSTART. The variable specifies the node point number at which the specified boundary value is to begin.

Columns 11–20. BVAL. This is the value of $\partial u/\partial n$.

Columns 21–30. NBSAME. The number of the elements over which the same boundary value is to be specified.

Columns 31–40. NINC. This variable specifies the algebraic difference between two adjacent corner node numbers on this portion of the boundary.

In the first example problem, we have the values of u specified everywhere, and $\partial u/\partial n$ is not at all specified. Therefore, this group of cards does not appear in the data deck. This topic will be discussed further in the context of the second example problem.

CARDS FOR BOUNDARY WHERE u IS SPECIFIED (FORMAT: 2I10, F10·0, 2I10)

This group of data cards is used to input the values of u specified on each of the boundary segments. The format is as follows:

Columns 1–10. Node number of the top (right end) node.

Columns 11–20. Node number of the bottom (left end) node. By virtue of the manner in which the nodes are numbered, this number is always smaller than the number appearing in columns 1 to 10.

Columns 21–30. The specified value of u.

Columns 31–40. Number to be added to the number in columns 11 to 20 to get the next node number.

Columns 41–50. Number to be added to the number in columns 31 to 40 to get the next node number.

For the problem under consideration, there are four data cards in this group, one card for each side of the square. Specifically, the third data card of this group, for example, contains the numbers 86, 1, 0·0, 11, and 6. This card carries boundary data for the bottom boundary (or side AB). Note that 86 and 1 are the largest and smallest node numbers on this side. The entry 0·0 refers to the boundary condition on this side. The rest of the node numbers on this side can be obtained by adding 11 to 1 to get 12, then adding 6 to 12 to get 18, then adding 11 to 18 to get 29, and so forth.

Referring to the program listing, we note that these data are read via instructions 88 to 91. Instructions 92 to 100 perform a check on the matrix bandwidth.

As the program is coded now, only constant values of u along any given boundary segment can be included.

BLOCK DATA: SUBPROGRAM FOR SUBELEMENT ASSEMBLY

Finally, the BLOCK DATA subprogram appearing in instructions 335 to 337 specifies data necessary for assembling the triangular subelements into quadrilateral elements.

14.6. ASSEMBLY

This part of the program is concerned with (1) determining the geometric properties of the triangular subelements and (2) assembling these triangular subelements into quadrilateral elements. Our goal here is to generate the matrixes **A** and **B** for each quadrilateral element. The flow chart in Figure 14.7 shows the process involved at a

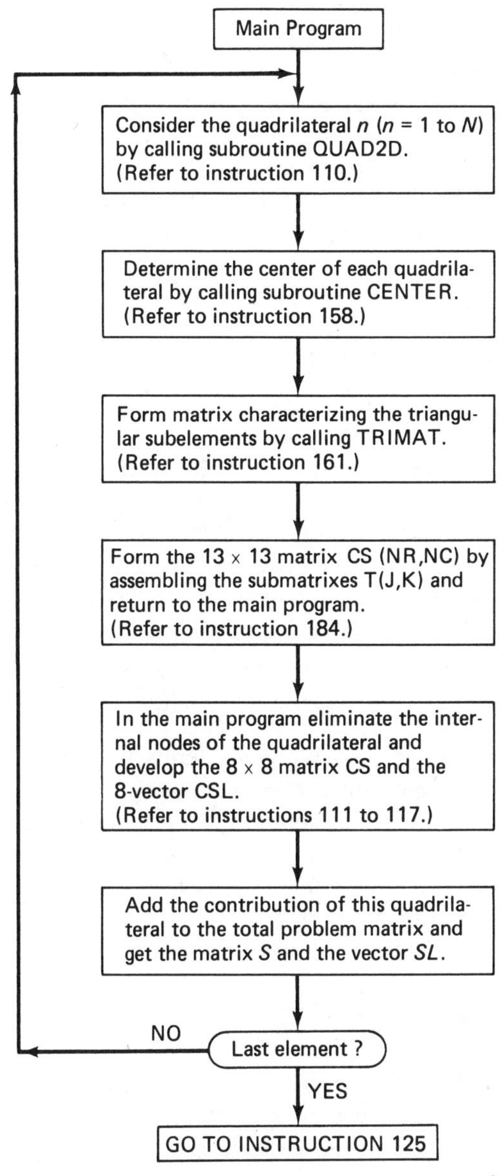

Figure 14.7 Flow chart showing the gross features of the element assembly process

gross level. The relevant program segments are in instructions 101 through 123 and in the subroutines QUAD2D, CENTER, and TRIMAT.

Subroutine CENTER determines the coordinates of the center (i.e., node 13 in Figure 14.2a) of each of the quadrilateral elements. Then subroutine TRIMAT calculates the entries of the 6×6 matrix $\mathbf{T}$ for each of the four triangular subelements. For a quadratic triangle, it can be shown that the entries t_{ij} of this matrix are (see Section 14.8)

$$t_{ij} = \frac{1}{12A^{(e)}}(a_i a_j + b_i b_j)$$

where $A^{(e)}$ is the area of the triangular subelement. When the control returns to the subroutine QUAD2D, we calculate the right side of $\mathbf{Au} = \mathbf{f}$ between instructions 159 to 183. Now control is transferred to instruction 111 in the main program. Now we eliminate the grid points inside the quadrilateral via instructions 111 to 117. Then, in instructions 118 to 124, the element matrixes are assembled to get the matrix S (NOROW, NOCOL).

14.7. SOLUTION OF EQUATIONS

The final step is to solve the system of algebraic equations (instructions 125 to 140) and print the results (instructions 141 to 150). Most of these instructions and the associated subroutines represent routine operations. So no attempt is made here to provide a commentary on them. Note, however, that the last subroutine (instructions 320 to 334) is a crude finite difference program. The purpose of this program is to serve as a rough check on the results obtained. In the first example problem, a better check can be used because an exact solution to it is available.

$$u(x, y) = \frac{4}{\pi} \sum_{\substack{n=1 \\ n;\ \text{odd}}}^{\infty} \frac{\sinh[(2n-1)\pi(1-x)] \sin[(2n-1)\pi y]}{(2n-1)\sinh[(2n-1)\pi]}$$

This solution is evaluated by adding the terms from $n = 1$ to $n = 25$. This calculation takes place in subroutine ANSWER, which starts at instruction 292.

Figure 14.8 shows the computed solution (PHI) and the exact solution at some of the nodes. Figure 14.9 shows the solution, at the same node points, obtained by a simple finite difference method. This later solution is supplied to facilitate a comparison between the two methods.

14.8. POTENTIAL FLOW PROBLEM

The second example problem is to determine the potential flow past a right circular cylinder with the direction of flow perpendicular to the axis of the cylinder. This problem is the same as the one treated in Section 13.4. As the problem was treated once before, a slight elaboration of the physics involved would be helpful.

```
NODE =   1   X =  0.0000   Y =  0.0000   PHI = -0.0000   EXACT =  0.0000
NODE =   2   X =  0.0000   Y =  0.1000   PHI =  1.0000   EXACT =  0.9826
NODE =   3   X =  0.0000   Y =  0.2000   PHI =  1.0000   EXACT =  1.0194
NODE =   4   X =  0.0000   Y =  0.3000   PHI =  1.0000   EXACT =  1.0146
NODE =   5   X =  0.0000   Y =  0.4000   PHI =  1.0000   EXACT =  0.9954
NODE =   6   X =  0.0000   Y =  0.5000   PHI =  1.0000   EXACT =  0.9855
NODE =   7   X =  0.0000   Y =  0.6000   PHI =  1.0000   EXACT =  0.9954
NODE =   8   X =  0.0000   Y =  0.7000   PHI =  1.0000   EXACT =  1.0146
NODE =   9   X =  0.0000   Y =  0.8000   PHI =  1.0000   EXACT =  1.0194
NODE =  10   X =  0.0000   Y =  0.9000   PHI =  1.0000   EXACT =  0.9827
NODE =  11   X =  0.0000   Y =  1.0000   PHI = -0.0000   EXACT =  0.0000
NODE =  12   X =  0.1000   Y =  0.0000   PHI = -0.0000   EXACT =  0.0000
NODE =  13   X =  0.1000   Y =  0.2000   PHI =  0.6864   EXACT =  0.6823
NODE =  14   X =  0.1000   Y =  0.4000   PHI =  0.7927   EXACT =  0.7923
NODE =  15   X =  0.1000   Y =  0.6000   PHI =  0.7927   EXACT =  0.7923
NODE =  16   X =  0.1000   Y =  0.8000   PHI =  0.6864   EXACT =  0.6823
NODE =  17   X =  0.1000   Y =  1.0000   PHI = -0.0000   EXACT =  0.0000
NODE =  18   X =  0.2000   Y =  0.0000   PHI = -0.0000   EXACT =  0.0000
NODE =  19   X =  0.2000   Y =  0.1000   PHI =  0.2698   EXACT =  0.2739
NODE =  20   X =  0.2000   Y =  0.2000   PHI =  0.4563   EXACT =  0.4563
NODE =  21   X =  0.2000   Y =  0.3000   PHI =  0.5576   EXACT =  0.5569
NODE =  22   X =  0.2000   Y =  0.4000   PHI =  0.6060   EXACT =  0.6060
NODE =  23   X =  0.2000   Y =  0.5000   PHI =  0.6211   EXACT =  0.6208
NODE =  24   X =  0.2000   Y =  0.6000   PHI =  0.6060   EXACT =  0.6060
NODE =  25   X =  0.2000   Y =  0.7000   PHI =  0.5576   EXACT =  0.5569
NODE =  26   X =  0.2000   Y =  0.8000   PHI =  0.4563   EXACT =  0.4563
NODE =  27   X =  0.2000   Y =  0.9000   PHI =  0.2698   EXACT =  0.2740
NODE =  28   X =  0.2000   Y =  1.0000   PHI = -0.0000   EXACT =  0.0000
NODE =  29   X =  0.3000   Y =  0.0000   PHI = -0.0000   EXACT =  0.0000
NODE =  30   X =  0.3000   Y =  0.2000   PHI =  0.3115   EXACT =  0.3122
NODE =  31   X =  0.3000   Y =  0.4000   PHI =  0.4525   EXACT =  0.4523
NODE =  32   X =  0.3000   Y =  0.6000   PHI =  0.4525   EXACT =  0.4523
NODE =  33   X =  0.3000   Y =  0.8000   PHI =  0.3115   EXACT =  0.3122
NODE =  34   X =  0.3000   Y =  1.0000   PHI = -0.0000   EXACT =  0.0000
NODE =  35   X =  0.4000   Y =  0.0000   PHI = -0.0000   EXACT =  0.0000
NODE =  36   X =  0.4000   Y =  0.1000   PHI =  0.1177   EXACT =  0.1180
NODE =  37   X =  0.4000   Y =  0.2000   PHI =  0.2178   EXACT =  0.2178
NODE =  38   X =  0.4000   Y =  0.3000   PHI =  0.2894   EXACT =  0.2895
NODE =  39   X =  0.4000   Y =  0.4000   PHI =  0.3316   EXACT =  0.3316
NODE =  40   X =  0.4000   Y =  0.5000   PHI =  0.3454   EXACT =  0.3453
NODE =  41   X =  0.4000   Y =  0.6000   PHI =  0.3316   EXACT =  0.3316
NODE =  42   X =  0.4000   Y =  0.7000   PHI =  0.2894   EXACT =  0.2895
NODE =  43   X =  0.4000   Y =  0.8000   PHI =  0.2178   EXACT =  0.2178
NODE =  44   X =  0.4000   Y =  0.9000   PHI =  0.1177   EXACT =  0.1180
NODE =  45   X =  0.4000   Y =  1.0000   PHI = -0.0000   EXACT =  0.0000
NODE =  46   X =  0.5000   Y =  0.0000   PHI = -0.0000   EXACT =  0.0000
NODE =  47   X =  0.5000   Y =  0.2000   PHI =  0.1526   EXACT =  0.1528
NODE =  48   X =  0.5000   Y =  0.4000   PHI =  0.2390   EXACT =  0.2391
NODE =  49   X =  0.5000   Y =  0.6000   PHI =  0.2390   EXACT =  0.2391
NODE =  50   X =  0.5000   Y =  0.8000   PHI =  0.1526   EXACT =  0.1528
NODE =  51   X =  0.5000   Y =  1.0000   PHI = -0.0000   EXACT =  0.0000
NODE =  52   X =  0.6000   Y =  0.0000   PHI = -0.0000   EXACT =  0.0000
NODE =  53   X =  0.6000   Y =  0.1000   PHI =  0.0562   EXACT =  0.0562
NODE =  54   X =  0.6000   Y =  0.2000   PHI =  0.1060   EXACT =  0.1060
NODE =  55   X =  0.6000   Y =  0.3000   PHI =  0.1444   EXACT =  0.1444
NODE =  56   X =  0.6000   Y =  0.4000   PHI =  0.1684   EXACT =  0.1684
NODE =  57   X =  0.6000   Y =  0.5000   PHI =  0.1765   EXACT =  0.1765
NODE =  58   X =  0.6000   Y =  0.6000   PHI =  0.1684   EXACT =  0.1684
```

Figure 14.8 Portion of the computed and exact solutions generated by the program in Figure 14.4 for the first example problem

```
THE NUMBER OF ITERATIONS IS    119
0.00000 0.00000 0.00000 0.00000 0.00000 0.00000 0.00000 0.00000 0.00000 0.00000 0.00000
1.00000 0.48892 0.28091 0.17884 0.12034 0.08289 0.05700 0.03803 0.02331 0.01107 0.00000
1.00000 0.67479 0.45587 0.31409 0.21965 0.15420 0.10708 0.07180 0.04412 0.02099 0.00000
1.00000 0.75436 0.55371 0.40201 0.28996 0.20719 0.14532 0.09798 0.06039 0.02877 0.00000
1.00000 0.78894 0.60258 0.45029 0.33098 0.23929 0.16901 0.11442 0.07068 0.03371 0.00000
1.00000 0.79882 0.61739 0.46559 0.34439 0.25000 0.17701 0.12001 0.07419 0.03541 0.00000
1.00000 0.78894 0.60258 0.45029 0.33098 0.23929 0.16901 0.11442 0.07068 0.03371 0.00000
1.00000 0.75436 0.55371 0.40201 0.28996 0.20719 0.14532 0.09798 0.06039 0.02877 0.00000
1.00000 0.67479 0.45587 0.31409 0.21965 0.15420 0.10708 0.07180 0.04412 0.02099 0.00000
1.00000 0.48892 0.28091 0.17884 0.12034 0.08289 0.05700 0.03803 0.02331 0.01107 0.00000
0.00000 0.00000 0.00000 0.00000 0.00000 0.00000 0.00000 0.00000 0.00000 0.00000 0.00000
```

Figure 14.9 Computed solution of the first example problem by a simple finite difference method

First, the fluid in question is assumed to be inviscid and incompressible. Inviscid fluids experience no shearing stress, and when they come into contact with a solid boundary, they slip tangentially along it without resistance. Dynamic aspects of fluid motion can be characterized by such adjectives as laminar or irrotational and turbulent or rotational. (Refer to a good book on fluid mechanics for a more precise definition.) Inviscid irrotational flow is called potential flow because the velocity field in the flow can be derived from a ***potential function***, traditionally denoted by the letter ϕ. For a two-dimensional, incompressible, irrotational flow, it can be shown, using conservation and continuity laws, that

$$\frac{\partial^2 \phi}{\partial x^2} + \frac{\partial^2 \phi}{\partial y^2} = 0$$

Two-dimensional flows can also be characterized by introducing a mathematical artifice known as the stream function, traditionally denoted by ψ. The stream function also satisfies Laplace's equation

$$\frac{\partial^2 \psi}{\partial x^2} + \frac{\partial^2 \psi}{\partial y^2} = 0$$

The potential function ϕ and the stream function ψ are related to the x- and y-components of velocity, denoted, respectively, by u and v, by

$$u = -\frac{\partial \psi}{\partial y}, \qquad v = \frac{\partial \psi}{\partial x}$$

$$u = -\frac{\partial \phi}{\partial x}, \qquad v = -\frac{\partial \phi}{\partial y}$$

Whether we use the potential function formulation or stream function formulation, mathematically the problem is the same as that of solving Laplace's equation, the difference arising only in the application of the boundary conditions.

Before applying the boundary conditions, let us observe that the actual solution domain is infinite in extent. For computational purposes, it is necessary to construct some finite domain. One procedure would be to enclose the cross section of the cylinder in a sufficiently large rectangular domain, as shown in Figure 14.10a. The nature of the boundary conditions for this rectangular domain is deduced as follows: On AF and BD, $\partial\phi/\partial n = \partial\phi/\partial x = -u_0$, the velocity of the fluid undisturbed by the solid body. Because the flow is laminar, there is no flow across the lines AB and FD. That is $\partial\phi/\partial n = \partial\phi/\partial y = 0$ along AB and FD. Because there can be no flow through the cylinder wall, $\partial\phi/\theta n = 0$ along the circumference of the circle. However, Laplace's equation subject to only Neumann boundary conditions has no unique solution.

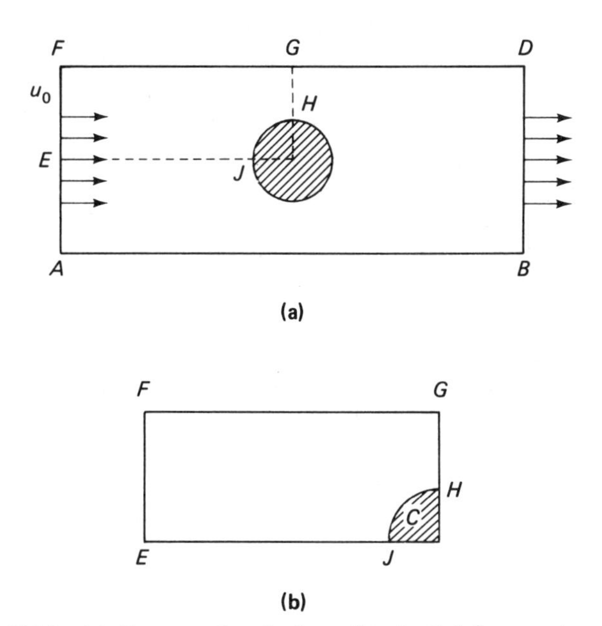

Figure 14.10 (a) Cross-sectional view of potential flow past a right circular cylinder; (b) quadrant of the region shown in (a)

There are several ways to overcome the preceding difficulty. One possibility is to take advantage of symmetry and consider only one quarter of the region, as shown in Figure 14.10b. The boundary conditions on *EF*, *FG*, *EJ*, and *JCH* are the same as those determined earlier. Now we have to determine the boundary conditions on the side *GH*. We use the following argument. Because *GH* is a line of symmetry, velocity on *GH* is totally in the x-direction. That is,

$$v = -\frac{\partial\phi}{\partial y} = 0$$

That is, ϕ is only a function of x on the line *GH*. But

$$\frac{\partial\phi}{\partial y} = -\frac{\partial\psi}{\partial x}$$

Therefore,

$$\frac{\partial \psi}{\partial x} = 0$$

That is, the stream function ψ is only a function of y on the line GH. Now using the fact that

$$u = -\frac{\partial \phi}{\partial x} = -\frac{\partial \psi}{\partial y} \neq 0 \qquad \text{on } GH$$

we observe that u can only be a constant on GH. That means the acceleration is zero which in turn means $\phi = 0$ on GH. Thus we have Neumann boundary conditions everywhere except on the line GH, where Dirichlet conditions apply.

Now we are ready for the discretization. The region is divided into quadrilateral elements. To avoid the need to use isoparametric elements, the curved boundary is approximated using straight line segments. Now the procedure is very similar to that used earlier. The region is divided into 45 elements as shown in Figure 14.11. The corner and midside nodes are numbered sequentially. The procedure for the preparation of input data is the same as the one used before, except that we now need nine different sets of data cards. A facsimile of the data deck for this problem is shown in Figure 14.12.

A new title card bearing the title 'POTENTIAL FLOW FOR A CYLINDER' is now prepared. Similarly, we have to provide information on the control card. Now let us decide to choose NNPC = 20, NELEMC = 9, and NPBOC = 1. This choice is not completely arbitrary. Ten corner nodes on the top and ten on the bottom justify the choice NNPC = 20. Nine elements along the x-direction justify the choice NELEMC = 9. The choice NPBOC = 1 allows us to specify that nonzero $\partial u/\partial n$ on EF.

Because NNPC = 20, after the control card we have 20 cards. Evidently, the generation option is being used. The relevant information for the rest of the corner nodes will be generated by the program. Then, as before, the program prints the node number of each of the corner nodes and their coordinates, as shown in Figure 14.13.

The next group of nine cards is the element cards. Information about the rest of the elements will be generated by the program and the results printed as shown in Figure 14.14.

The next data card is the normal boundary value card. Unlike the first example problem, a card is required here because $\partial\phi/\partial n \neq 0$ on the boundary EF. Because EF starts with node 1, a 1 is punched in the first field. Because the initial velocity u_0 has been chosen as unity, $\partial\phi/\partial n = -u_0 = -1$ on EF. This information is punched in the second field. The same boundary condition, that is, -1, is to be applied along the boundary of elements 1 through 5. Therefore, a 5 appears in the third field. Finally, the 2 in the last field of this card allows us to increment the corner node number from 1 to 3 . . . to 11. Now the program computes the boundary condition

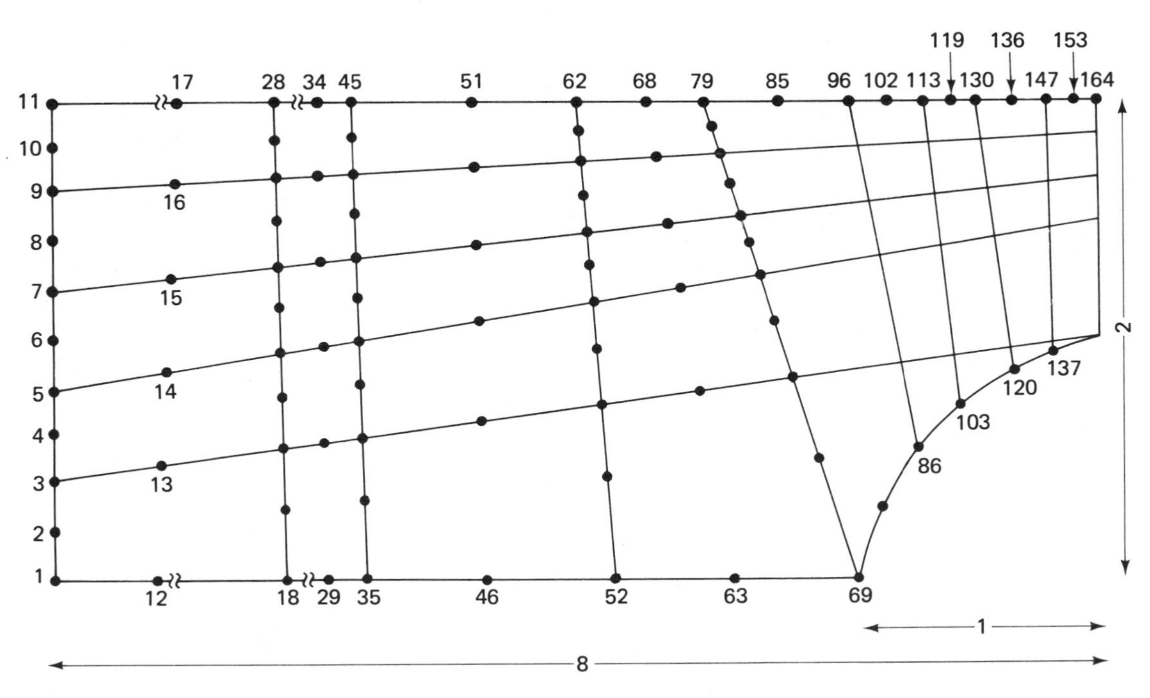

Figure 14.11 Finite element grid superimposed on the quadrant shown in Figure 14.10(b)

```
POTENTIAL FLOW FOR A CYLINDER
          20          9          1          1
           1        8.0        0.0          1
          11        8.0        2.0          0
          18        5.0        0.0          1
          28        5.0        2.0          0
          35        3.0        0.0          1
          45        3.0        2.0          0
          52        2.0        0.0          1
          62        2.1        2.0          0
          69        1.0        0.0          1
          79        1.6        2.0          0
          86        0.8        0.6          1
          96        1.0        2.0          0
         103        0.6        0.8          1
         113        0.7        2.0          0
         120        0.4     0.9165          1
         130        0.5        2.0          0
         137        0.2     0.9798          1
         147       0.21        2.0          0
         154        0.0        1.0          1
         164        0.0        2.0          0
    1   12   18   19   20   13    3    2          4
   18   29   35   36   37   30   20   19          4
   35   46   52   53   54   47   37   36          4
   52   63   69   70   71   64   54   53          4
   69   80   86   87   88   81   71   70          4
   86   97  103  104  105   98   88   87          4
  103  114  120  121  122  115  105  104          4
  120  131  137  138  139  132  122  121          4
  137  148  154  155  156  149  139  138          4
           1       -1.0          5          2
         164        154        0.0          1          1
```

Figure 14.12 Setup of data cards for the second example problem (the potential flow problem); while solving the second example program, replace the data cards (shown after the $ENTRY card) in Figure 14.4 with the deck prepared from the listing above

```
POTENTIAL FLOW FOR A CYLINDER


NODE    1      X =     8.0000      Y =    0.0000
NODE    3      X =     8.0000      Y =    0.4000
NODE    5      X =     8.0000      Y =    0.8000
NODE    7      X =     8.0000      Y =    1.2000
NODE    9      X =     8.0000      Y =    1.6000
NODE   11      X =     8.0000      Y =    2.0000
NODE   18      X =     5.0000      Y =    0.0000
NODE   20      X =     5.0000      Y =    0.4000
NODE   22      X =     5.0000      Y =    0.8000
NODE   24      X =     5.0000      Y =    1.2000
NODE   26      X =     5.0000      Y =    1.6000
NODE   28      X =     5.0000      Y =    2.0000
NODE   35      X =     3.0000      Y =    0.0000
NODE   37      X =     3.0000      Y =    0.4000
NODE   39      X =     3.0000      Y =    0.8000
```

Figure 14.13 Node numbers and coordinates for each of the corner nodes; this is part of the output generated by the second example problem, while Figure 14.5 shows the corresponding output generated by the first example problem

```
NODE  41    X =    3.0000    Y =    1.2000
NODE  43    X =    3.0000    Y =    1.6000
NODE  45    X =    3.0000    Y =    2.0000
NODE  52    X =    2.0000    Y =    0.0000
NODE  54    X =    2.0200    Y =    0.4000
NODE  56    X =    2.0400    Y =    0.8000
NODE  58    X =    2.0600    Y =    1.2000
NODE  60    X =    2.0800    Y =    1.6000
NODE  62    X =    2.1000    Y =    2.0000
NODE  69    X =    1.0000    Y =    0.0000
NODE  71    X =    1.1200    Y =    0.4000
NODE  73    X =    1.2400    Y =    0.8000
NODE  75    X =    1.3600    Y =    1.2000
NODE  77    X =    1.4800    Y =    1.6000
NODE  79    X =    1.6000    Y =    2.0000
NODE  86    X =    0.8000    Y =    0.6000
NODE  88    X =    0.8400    Y =    0.8800
NODE  90    X =    0.8800    Y =    1.1600
NODE  92    X =    0.9200    Y =    1.4400
NODE  94    X =    0.9600    Y =    1.7200
NODE  96    X =    1.0000    Y =    2.0000
NODE 103    X =    0.6000    Y =    0.8000
NODE 105    X =    0.6200    Y =    1.0400
NODE 107    X =    0.6400    Y =    1.2800
NODE 109    X =    0.6600    Y =    1.5200
NODE 111    X =    0.6800    Y =    1.7600
NODE 113    X =    0.7000    Y =    2.0000
NODE 120    X =    0.4000    Y =    0.9165
NODE 122    X =    0.4200    Y =    1.1332
NODE 124    X =    0.4400    Y =    1.3499
NODE 126    X =    0.4600    Y =    1.5666
NODE 128    X =    0.4800    Y =    1.7833
NODE 130    X =    0.5000    Y =    2.0000
NODE 137    X =    0.2000    Y =    0.9798
NODE 139    X =    0.2020    Y =    1.1838
NODE 141    X =    0.2040    Y =    1.3879
NODE 143    X =    0.2060    Y =    1.5919
NODE 145    X =    0.2080    Y =    1.7960
NODE 147    X =    0.2100    Y =    2.0000
NODE 154    X =    0.0000    Y =    1.0000
NODE 156    X =    0.0000    Y =    1.2000
NODE 158    X =    0.0000    Y =    1.4000
NODE 160    X =    0.0000    Y =    1.6000
NODE 162    X =    0.0000    Y =    1.8000
NODE 164    X =    0.0000    Y =    2.0000
```

Figure 14.13 (*continued*)

to be applied to each of the five elements and prints the results as shown in Figure 14.15.

The next data card allows for the specification of u as was done with the example problem 1.

A portion of the computed result is shown in Figure 14.16.

```
NUMBER OF NODAL POINTS... 164

NUMBER OF ELEMENTS....... 45

ELE. NO.       NODAL POINTS

  1    1   12   18   19   20   13    3    2
  2    3   13   20   21   22   14    5    4
  3    5   14   22   23   24   15    7    6
  4    7   15   24   25   26   16    9    8
  5    9   16   26   27   28   17   11   10
  6   18   29   35   36   37   30   20   19
  7   20   30   37   38   39   31   22   21
  8   22   31   39   40   41   32   24   23
  9   24   32   41   42   43   33   26   25
 10   26   33   43   44   45   34   28   27
 11   35   46   52   53   54   47   37   36
 12   37   47   54   55   56   48   39   38
 13   39   48   56   57   58   49   41   40
 14   41   49   58   59   60   50   43   42
 15   43   50   60   61   62   51   45   44
 16   52   63   69   70   71   64   54   53
 17   54   64   71   72   73   65   56   55
 18   56   65   73   74   75   66   58   57
 19   58   66   75   76   77   67   60   59
 20   60   67   77   78   79   68   62   61
 21   69   80   86   87   88   81   71   70
 22   71   81   88   89   90   82   73   72
 23   73   82   90   91   92   83   75   74
 24   75   83   92   93   94   84   77   76
 25   77   84   94   95   96   85   79   78
 26   86   97  103  104  105   98   88   87
 27   88   98  105  106  107   99   90   89
 28   90   99  107  108  109  100   92   91
 29   92  100  109  110  111  101   94   93
 30   94  101  111  112  113  102   96   95
 31  103  114  120  121  122  115  105  104
 32  105  115  122  123  124  116  107  106
 33  107  116  124  125  126  117  109  108
 34  109  117  126  127  128  118  111  110
 35  111  118  128  129  130  119  113  112
 36  120  131  137  138  139  132  122  121
 37  122  132  139  140  141  133  124  123
 38  124  133  141  142  143  134  126  125
 39  126  134  143  144  145  135  128  127
 40  128  135  145  146  147  136  130  129
 41  137  148  154  155  156  149  139  138
 42  139  149  156  157  158  150  141  140
 43  141  150  158  159  160  151  143  142
 44  143  151  160  161  162  152  145  144
 45  145  152  162  163  164  153  147  146
```

Figure 14.14 Element numbers and the associated nodal points, part of the output generated by the second example problem; Figure 14.6 shows the corresponding output generated by the first example problem

NORMAL COMPONENTS ON BOUNDARY

NODES	NORMAL COMPONENT
1- 3	-1.0000
3- 5	-1.0000
5- 7	-1.0000
7- 9	-1.0000
9- 11	-1.0000

Figure 14.15 Boundary conditions to be applied to each of the five elements along *EF* (see Figure 14.10b); the above results are computed by the program from the data given in the data cards

NODE =	1	X =	8.0000	Y =	0.0000	PHI =	7.3545
NODE =	2	X =	8.0000	Y =	0.2000	PHI =	7.3511
NODE =	3	X =	8.0000	Y =	0.4000	PHI =	7.3389
NODE =	4	X =	8.0000	Y =	0.6000	PHI =	7.3189
NODE =	5	X =	8.0000	Y =	0.8000	PHI =	7.2882
NODE =	6	X =	8.0000	Y =	1.0000	PHI =	7.2463
NODE =	7	X =	8.0000	Y =	1.2000	PHI =	7.1885
NODE =	8	X =	8.0000	Y =	1.4000	PHI =	7.1115
NODE =	9	X =	8.0000	Y =	1.6000	PHI =	7.0078
NODE =	10	X =	8.0000	Y =	1.8000	PHI =	6.9185
NODE =	11	X =	8.0000	Y =	2.0000	PHI =	6.8916
NODE =	12	X =	6.5000	Y =	0.0000	PHI =	6.0084
NODE =	13	X =	6.5000	Y =	0.4000	PHI =	6.0044
NODE =	14	X =	6.5000	Y =	0.8000	PHI =	5.9935
NODE =	15	X =	6.5000	Y =	1.2000	PHI =	5.9798
NODE =	16	X =	6.5000	Y =	1.6000	PHI =	5.9743
NODE =	17	X =	6.5000	Y =	2.0000	PHI =	5.9755
NODE =	18	X =	5.0000	Y =	0.0000	PHI =	4.7926
NODE =	19	X =	5.0000	Y =	0.2000	PHI =	4.7921
NODE =	20	X =	5.0000	Y =	0.4000	PHI =	4.7925
NODE =	21	X =	5.0000	Y =	0.6000	PHI =	4.7918
NODE =	22	X =	5.0000	Y =	0.8000	PHI =	4.7922
NODE =	23	X =	5.0000	Y =	1.0000	PHI =	4.7906
NODE =	24	X =	5.0000	Y =	1.2000	PHI =	4.7901
NODE =	25	X =	5.0000	Y =	1.4000	PHI =	4.7846
NODE =	26	X =	5.0000	Y =	1.6000	PHI =	4.7787
NODE =	27	X =	5.0000	Y =	1.8000	PHI =	4.7754
NODE =	28	X =	5.0000	Y =	2.0000	PHI =	4.7692
NODE =	29	X =	4.0000	Y =	0.0000	PHI =	3.9895
NODE =	30	X =	4.0000	Y =	0.4000	PHI =	3.9888
NODE =	31	X =	4.0000	Y =	0.8000	PHI =	3.9867
NODE =	32	X =	4.0000	Y =	1.2000	PHI =	3.9838
NODE =	33	X =	4.0000	Y =	1.6000	PHI =	3.9817
NODE =	34	X =	4.0000	Y =	2.0000	PHI =	3.9812
NODE =	35	X =	3.0000	Y =	0.0000	PHI =	3.1974
NODE =	36	X =	3.0000	Y =	0.2000	PHI =	3.1966
NODE =	37	X =	3.0000	Y =	0.4000	PHI =	3.1946
NODE =	38	X =	3.0000	Y =	0.6000	PHI =	3.1915
NODE =	39	X =	3.0000	Y =	0.8000	PHI =	3.1877
NODE =	40	X =	3.0000	Y =	1.0000	PHI =	3.1835
NODE =	41	X =	3.0000	Y =	1.2000	PHI =	3.1795
NODE =	42	X =	3.0000	Y =	1.4000	PHI =	3.1755
NODE =	43	X =	3.0000	Y =	1.6000	PHI =	3.1723
NODE =	44	X =	3.0000	Y =	1.8000	PHI =	3.1702
NODE =	45	X =	3.0000	Y =	2.0000	PHI =	3.1690
NODE =	46	X =	2.5000	Y =	0.0000	PHI =	2.8130
NODE =	47	X =	2.5100	Y =	0.4000	PHI =	2.8146
NODE =	48	X =	2.5200	Y =	0.8000	PHI =	2.8072
NODE =	49	X =	2.5300	Y =	1.2000	PHI =	2.7975

Figure 14.16 Portion of the computed solution of the second example problem

SUGGESTIONS FOR FURTHER READING

1. *The art of writing small segments of code and integrating them to solve a structural mechanics problem is nicely presented at an elementary level in*
   R. D. Cook, *Concepts and Applications of Finite Element Analysis*, John Wiley & Sons, Inc., New York, 1974.
2. *Students interested in refining their talents in writing fairly large finite element programs can benefit by reading*
   E. Hinton and D. R. J. Owen, *Finite Element Programming*, Academic Press, Inc., New York, 1977.
3. *Almost all the finite element books cited in Chapters 8 and 9 have chapters devoted to the programming task.*

EXERCISES

1. Using the program shown in Figure 14.4, verify the results presented in this chapter for the two example programs.

2. Do you think that the accuracy of the preceding solutions will improve by using more elements? Conduct a computational experiment to verify your prediction.

3. In Section 13.1, a number of PDE's and their exact solutions were given. Which of these problems can be solved using the program in Figure 14.4? Using appropriate data cards (to modify the program in Figure 14.4), solve as many problems as you can and compare your solution with the exact solution.

Author Index

Subject Index